TABLE

its symbol and the atomic number below its symbol.

may be found in the references listed at the end of Chap. 2.]

								4.003 He 2
			10.81 B 5	12.011 C 6	14.007 N 7	16.000 O 8	19.00 F 9	20.182 Ne 10
			26.98 Al 13	28.09 Si 14	30.97 P 15	32.066 S 16	35.453 Cl 17	39.946 A 18
58.71 Ni 28	63.54 Cu 29	65.38 Zn 30	69.72 Ga 31	72.59 Ge 32	74.91 As 33	78.96 Se 34	79.91 Br 35	83.80 Kr 36
106.4 Pd 46	107.87 Ag 47	112.40 Cd 48	114.82 In 49	118.69 Sn 50	121.74 Sb 51	127.61 Te 52	126.91 I 53	131.3 Xe 54
195.09 Pt 78	197.0 Au 79	200.6 Hg 80	204.37 Tl 81	207.2 Pb 82	209.0 Bi 83	210 Po 84	211 At 85	222 Rn 86

157.25 Gd 64	158.92 Tb 65	162.50 Dy 66	164.93 Ho 67	167.3 Er 68	169 Tm 69	173.04 Yb 70	174.98 Lu 71
247 Cm 96	249 Bk 97	251 Cf 98	254 E 99	255 Fm 100	256 Mv 101	255 (No) 102	257 Lw 103

hcp changing to fcc at higher temperatures

diamond structure

Physics of Solids

McGraw-Hill Series in Materials Science and Engineering

Avitzur *Metal Forming: Processes and Analysis*

Azároff *Introduction to Solids*

Barrett and Massalski *Structure of Metals*

Blatt *Physics of Electronic Conduction in Solids*

Brick, Gordon, and Phillips *Structure and Properties of Alloys*

Buerger *Contemporary Crystallography*

Buerger *Introduction to Crystal Geometry*

De Hoff and Rhines *Quantitative Microscopy*

Elliot *Constitution of Binary Alloys, First Supplement*

Gilman *Micromechanics of Flow in Solids*

Gordon *Principles of Phase Diagrams in Materials Systems*

Hirth and Lothe *Theory of Dislocations*

Murr *Electron Optical Applications in Materials Science*

Paul and Warschauer *Solids under Pressure*

Rosenfield, Hahn, Bement, and Jaffee *Dislocation Dynamics*

Rudman, Stringer, and Jaffee *Phase Stability in Metals and Alloys*

Shewmon *Diffusion in Solids*

Shewmon *Transformations in Metals*

Shunk *Constitution of Binary Alloys*

Wert and Thomson *Physics of Solids*

Physics of Solids

Second Edition

Charles A. Wert

Professor of Metallurgy
University of Illinois
Urbana, Illinois

Robb M. Thomson

Professor of Materials Science
State University of New York
Stony Brook, Long Island, New York

McGraw-Hill Book Company

New York St. Louis San Francisco Düsseldorf
Johannesburg Kuala Lumpur London Mexico
Montreal New Delhi Panama Rio de Janeiro
Singapore Sydney Toronto

This book was set in Modern, and was printed on permanent paper and bound by The Book Press, Inc. The designer was Richard Paul Kluga; the drawings were done by Harry Lazarus. The editors were Bradford Bayne and Andrea Stryker-Rodda. Adam Jacobs supervised production.

Physics of Solids

Library of Congress Catalogue Card Number: 77-98055

69435

1234567890 BPBP 79876543210

Preface

This book has developed out of a long period of teaching and research by the authors in the field of the physics of solids. It should express to the reader the pleasure we have enjoyed in being a part of the magnificent growth of the science and engineering of solids in the past twenty years. As was true of the first edition, we continue to owe a large debt to Frederick Seitz who largely established the intellectual climate in which we have worked for the major part of our professional lives.

The book is designed as a textbook at an intermediate level. The treatment can be understood by most undergraduates at the junior or senior level. Even so, the text is not a qualitative survey, we believe, and the writing is quantitative whenever possible. To be sure, the physical models and the analytical treatment are often not as sophisticated as the expert would like, though we believe the treatment is sound. We believe that little unlearning is required of the student as he goes on to graduate courses.

We have been encouraged, in writing the second edition, by the comments of many readers, who over the years have remarked about the physical clarity and style of the first edition. We have thanked them all individually whenever we could; to all, we repeat our thanks. We hope we have been able to maintain that tone in the second edition.

As before, we conceive of the book as consisting of three main parts. The first six chapters treat the crystal structure of solids and those properties which depend on crystal structure. The middle section describes the electronic structure of solids. The final chapters are applications of electronic structure to the electronic and magnetic properties of solids. The several sections are distinct, and considerable variation in usage is possible with proper selection of chapters.

We have not had unlimited time in writing the second edition. Therefore, we have put most of our effort in those areas which we feel have experienced most rapid development in the past seven years. These are (a) the properties of semiconductors and device development, (b) the band theory of solids and the electronic structure of alloys, (c) surface phenomena, and (d) thermodynamics of alloys.

To all who have helped in the writing, in reviewing sections of the manuscript and in preparation of the final draft, we offer our thanks.

Charles A. Wert
Robb M. Thomson

Contents

1 Introduction

1-1 Perspectives in Materials Science

The purpose of this book is to describe the physical nature of matter in the solid state. The physical view of a problem as contrasted to the chemical view or the engineering view typically emphasizes the particles and forces of the system. Thus the primary goal in the physical approach is usually the solution of the differential equations for the orbits of the particles of the system. In a system containing many particles, such as a solid, the number of equations is too large to permit an exact solution, hence the goal must be modified. Even with approximate solutions, however, a large number of useful theorems and general statements can be obtained. Of course, the very complexity of the large system of particles of which a solid is composed makes our study more interesting than the solution of a single differential equation. Since many of our readers do not intend to become practicing solid-state physicists, it is appropriate for us to inquire briefly in this introduction into the larger perspective in which the physical view of matter lies. We shall thus briefly trace the historical development of the physics of solids, then try to indicate something of the practical significance of this physical approach to solids.

The physics of solids is a part of the wider field of materials science, which includes, in addition to solid-state physics, parts of chemistry, electrical engineering, ceramics, and metallurgy. Although *materials science* is a relatively new term because of the modern focus on problems of materials design, the field of materials science is as old as civilization itself. In fact, the early stages of civilization are labeled in terms of the materials used in implements, e.g., Stone Age,

Bronze Age, etc., and the march of early civilization was tied stringently to man's facility in the production and utilization of his materials.

But it would be stretching a point to claim that the progress the ancients made in metal smelting was related to physics or was even scientific in character. It was, however, related to a practical kind of chemistry, and until very recent times (ca. 1930) the chemist made far greater contributions to the science of materials than did the physicist. The study of the chemical nature of solids, the discovery of purification processes, and the evolution of efficient production processes constituted most of the early progress in the field of materials.

The early contributions from physics were primarily from two directions. The field of elasticity was developed just before the turn of the century to a highly sophisticated mathematical apparatus, and the modern engineering mechanics of solids is an important application of that work. The second nonchemical approach to solids goes back even further and was the mathematics of symmetry as applied to crystals. At first, the main activity was in geology, where considerable effort was spent in cataloguing crystal types from the facets seen on crystal surfaces. However, crystal symmetry and crystallography in general began to have unexpected significance after the realization that the external symmetry of crystals is related to the far more important internal, or microscopic, symmetry of atom placement. This discovery was made with the help of X rays early in this century.

However, neither of these physical studies of solids involved the particle-and-force approach which has always characterized physics. Real success in this area had to await the development of a modern view of atoms contained in the science of quantum mechanics, which gave a satisfactory understanding of atoms about 1930. At that time, something approaching a breakthrough occurred, and a broad attack was made by physicists on the proper problems of the *physics* of a solid in terms of the orbits of electrons in the solid and the basic nature of the forces holding the atoms of the solid together. Finally by the end of the 1930s an outline was available which provided a remarkable synthesis of the general properties of solids from the physical view. A coherent structure had by then been constructed which, at least in principle, explained such things as metallic electrical conduction, the essential difference between a metal and an insulator, the specific heat of solids, and the optical properties of solids. This major synthesis involved the bulk, or average, properties of pure, or ideal, crystals.

A second phase of work began during the late 1930s and was taken up with great energy after 1945. It is the study of crystalline imperfections and their effects upon the physical properties of solids. For example, the ultimate mechanical strength of a solid depends primarily upon an imperfection in atom placement called a *dislocation*. Likewise the electrical properties of semiconductors are dependent upon minute quantities of certain impurities. Modern research in solids is now about equally divided between efforts directed toward a better understanding of perfect or ideal solids and efforts aimed at the determination of properties of solids containing imperfections. A modern example of the former is the study of superconductivity; of the latter, the study of atomic diffusion.

This description of how the physics of solids has developed, however, does not show its relation to those interested in more practical goals. Discussion of this point requires us to consider one of the crucial questions in modern technology: What is the nature of the relationship between applied and pure science? In concrete terms, how does basic knowledge concerning the physical character of a solid become directly useful, for example, to the person concerned with the design of an electrical circuit? This question has no simple answer, and it can be discussed on many levels. Here we can only indicate and suggest answers.

The solution of any specific problem has two general approaches. One can seek a solution by trial-and-error methods in the empirical approach. Many problems of an engineer are handled in this way. By this technique, a satisfactory process is slowly evolved or improved over a period of years. This is, for example, the method used by the ancients in the evolution of metal smelting. It was also the primary method used with great success by Edison and other inventors of the eighteenth and nineteenth centuries.

Until recently, the empirical method was used almost exclusively in the solution of practical engineering problems. Success in its use requires a highly developed intuition and broad practical experience. It also requires some knowledge of the basic processes involved in the problem, but a qualitative understanding is usually adequate. For the simpler everyday problems of an engineer, this approach is certainly the simplest and most efficient one.

Some problems, however, are much too difficult to be solved in a purely empirical manner. For them, a more systematic approach must be used. For example, suppose that a person is faced with the general problem of designing the most efficient possible transformer core. There are an infinite variety of different alloys which might be tried in a strictly empirical approach, and any attempt in this direction would surely be very wasteful. A more methodical approach might begin with an inquiry into the basic nature of magnetization in a solid and the cause for the energy loss in a hysteresis loop. With sufficient understanding of these basic phenomena, the labor needed to find the alloy with the correct properties will be greatly reduced. One additional advantage belongs to the fundamental approach. When enough basic knowledge is obtained to solve the immediate problem, the same knowledge will almost certainly be found useful in a wide range of other distantly related problems which the empirical approach would not have touched.

One can, however, easily misunderstand the application of the more fundamental approach to a technological problem. Very few engineering problems can be *completely* solved by the application of pure science. Pure science operates by the systematic simplification of natural processes, and in the development of any practical product, say, a heat-resistant alloy or a high-frequency transistor, specific design problems must be solved in the traditional manner of the engineer. The important point is that the fundamental approach is an essential part in the closely coupled chain in the development cycle from pure scientist through development engineer to design engineer. The fruits to be gained from a close coupling between pure and applied science are a relatively new discovery which has brought about the fast development cycles which are now characteristic of new discoveries. The time lag between the discovery of the nature of electromagnetic waves and the practical application to a radio was about thirty years, whereas the application of the transistor action to a device took only a few years.

This, then, is the nature of the world in which engineers increasingly find themselves. The basic approach to problems requires quick communication of problem areas and new ideas in both directions between pure and applied scientists. The engineer must have a thorough speaking acquaintance with the basic principles underlying the phenomena with which he deals, for two important reasons. First, he must be able to communicate with others over a wide range of the pure-applied spectrum and participate fully in the development process. The second is a corollary of the first: because of the vastly increased rate of technological development, an engineer trained narrowly in any given field becomes obsolescent, depending upon the field, in 10 to 15 years. The modern engineer can thus expect that within that period of time his whole field will have changed under his feet. He must have sufficient understanding of the basic science underlying any of the specific applied areas so that he can shift to the use of new processes involving a new set of scientific principles.

1-2 The Scope of This Book

As mentioned above, the physics of solids has two main branches, bulk (electronic) properties and imperfections. We have attempted, in writing this book, to produce a reasonable balance between them, but we have not tried to separate them rigidly. Nowhere is the treatment sufficiently complete to satisfy the specialist who is looking for a detailed understanding. For example, although the mechanical properties of solids are discussed in Chap. 6, this difficult and exceedingly complex subject could be given in detail only in a separate book similar in size to the present one. (As a matter of fact, the mechanical properties of a solid require a background of elasticity theory, and for this reason, the level of Chap. 6 is pitched somewhat lower than the majority of the rest of the book since we do not wish to include a long introduction to elasticity.) Likewise the electrical properties of a semiconductor are given in general outline, but the electrical engineer will have to study solid-state devices elsewhere.

The treatment is fixed at an intermediate level. The reader, as an upper-class undergraduate or an engineer or scientist not trained in the science of solids, should be able to reach two goals: (1) he should obtain a physical understanding of the properties of solids and (2) he should be adequately prepared for specialized study in specific areas of application. To implement the first of these objectives, we draw heavily upon physical models, some traditional, some of our devising. Generally the simplest model which illustrates the principle is selected. The second goal is reached by analytical discussion wherever it seems feasible. To be sure, this procedure often results in a rather unsophisticated analytical treatment, but the attempt is constantly made to set the level of the calculations close to the degree of sophistication of the model itself.

The concepts stated in the preceding paragraph are important for the successful use of the book. Certainly many of the models used here are improved upon in more advanced books, but we are satisfied that going through the physics of solids at an intermediate level gives a reader a fine base on which to build a more sophisticated understanding. The ability to generalize is an important attribute of any intellectual discipline; we hope that the models and examples used here will enable the reader to achieve this goal at an intermediate level.

1-3 Units

Mks units are used throughout the book. They are employed because engineers are now universally trained in their use. They are, unfortunately, not the most common or even the most appropriate units for atomic processes; therefore we cite in detail at appropriate points the relationship between the mks units and the cgs units.

Problems accompany each chapter. Most problems do not involve simply substituting numbers in formulas (although this exercise is useful at times). Many problems indeed serve to amplify the text and even to fill in parts of long analytical developments deleted from the text of the chapter.

1-4 The General Nature of Solids

The valence electrons, which are responsible for the electric, magnetic, and dielectric properties of matter, exist as a diffuse, nonuniform cloud of charge in solids, though the nuclei, heavier and slower-moving, occupy much more definite positions. Even though neither electrons nor nuclei can be seen individually by direct observation, a great deal can be learned about the regularity of charge distribution in solids by using observational techniques of varying resolving power.

The metal tungsten is a good choice to illustrate some important features of solids. To the unaided eye, a polished piece of tungsten (or any other metal) has a shiny appearance called the *metallic luster*. Usually no details of internal structure in metals can be seen by eye

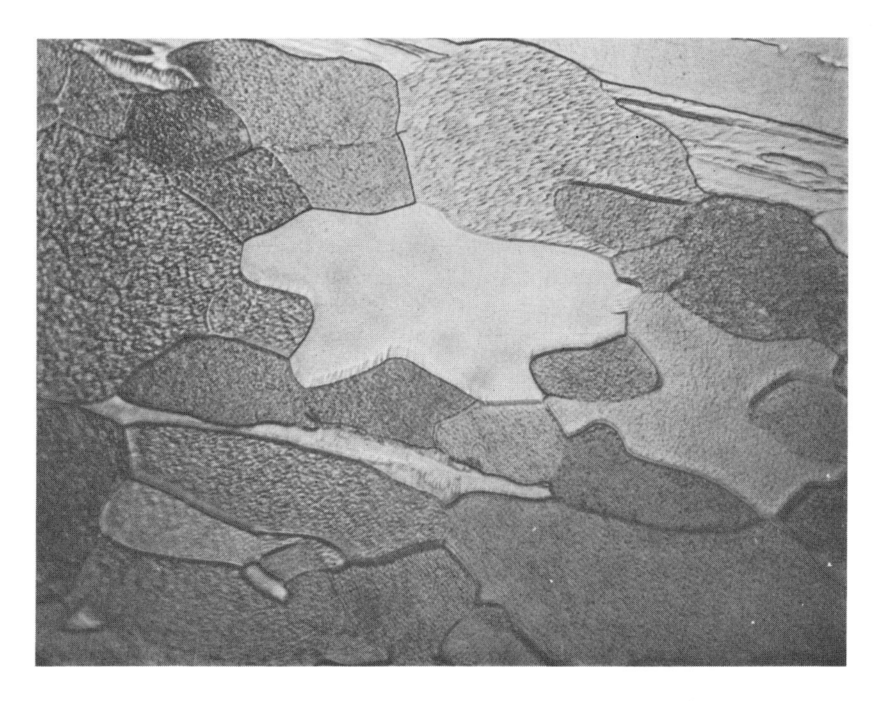

Figure 1-1 Photograph of grain structure of a polished tungsten surface at a magnification of about five hundred times. (*Photograph courtesy of R. Schnitzel.*)

alone. A piece of tungsten metal viewed in a particular way with an optical microscope (at a magnification of a few hundred times) is seen to consist of regions called *grains* (Fig. 1-1). Each grain is a volume of metal which has a repetitious pattern of atom placement with coordinates arranged in a certain way in space. The coordinate axes of the patterns of different grains are, in general, not the same, and the different grains thereby reflect light differently and appear to have various shades in the microscope. The electron microscope with its higher magnification, perhaps 50,000 times, sees even finer detail within the individual grains (Fig. 1-2). With its use, minute variations in the repetition pattern within individual grains can be seen.

The basic regularity of the positions of the atoms themselves becomes visible when the metal is viewed with the field-ion microscope at magnifications of about 10 million times. In Fig. 1-3, the image of a tiny tip of a fine tungsten needle is shown. The individual atoms are clearly visible, and the high degree of symmetry of atom positions within a grain of a metal is apparent. This metal grain is not perfect; the most prominent defects here are misplaced atoms, possibly defects called *interstitials*.

The instruments of highest spatial resolving power, the electron microscope and the field-ion microscope, are relatively new. Resolution

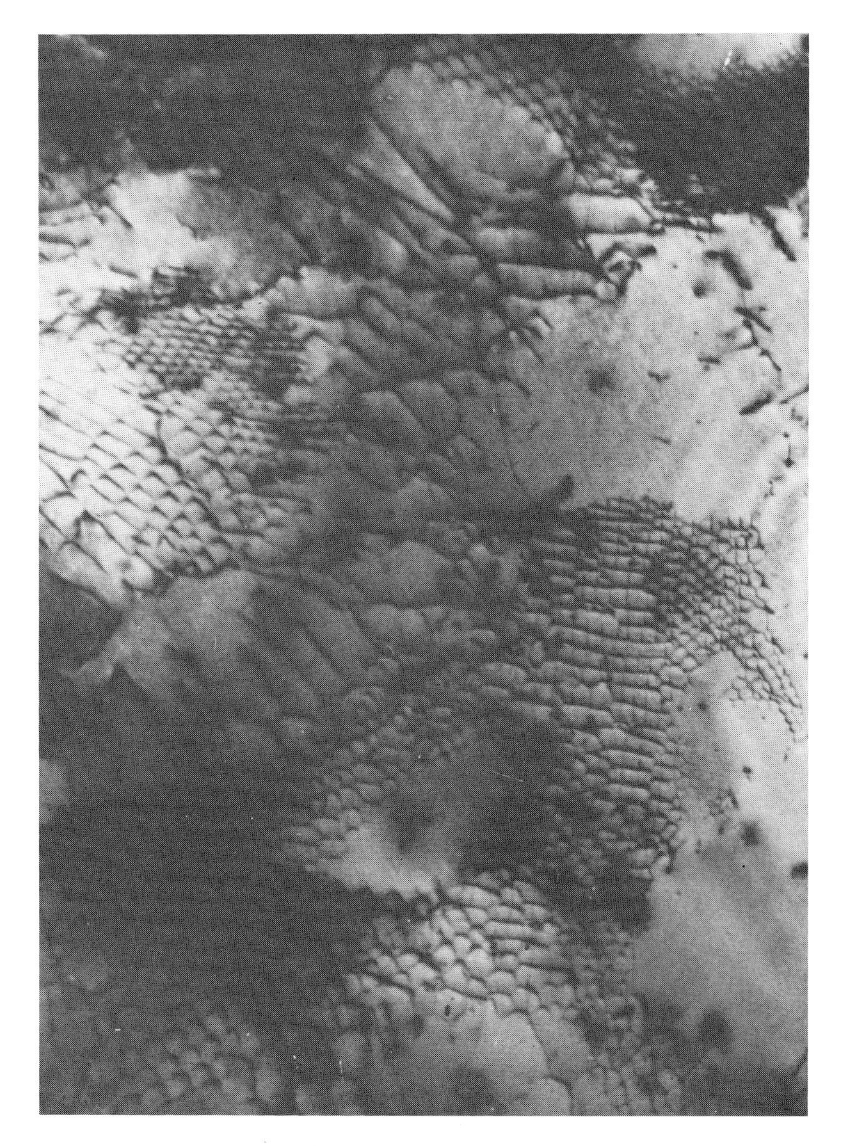

Figure 1-2 Electron micrograph of a single grain in a thin foil of tungsten at a magnification of 80,000 times. The dark lines are regions of local distortion of the repetitive arrangement of atom positions. (*Photograph courtesy of S. Weissmann.*)

such as that shown in Figs. 1-2 and 1-3 has provided direct justification for the correctness of models of atom placement in solids which had been deduced earlier by more indirect means. Undoubtedly, future techniques will provide even higher resolution of the details of the structure of solids.

Figure 1-3 A tungsten surface magnified about 4 million times with the field-ion microscope. The bright spots represent the positions of atoms on the surface of the solid. The high degree of regularity of atom placement is apparent. Mistakes in structure are occasionally found. The most obvious imperfections here are groups of misplaced atoms, the bright spots indicated by the arrows. (*Photograph courtesy of E. Müller.*)

With this introduction we turn to the detailed study of the properties of solids. The next five chapters are concerned with the crystalline nature of solids from the atomic point of view. Three chapters then describe the nature of electrons in metals. The remaining chapters discuss the finer details of electronic structure of metals, the electronic structure of insulators and semiconductors, and the maj or dielectric and magnetic properties which result.

PROBLEMS

1-1 Estimate an average grain diameter of the tungsten shown in Fig. 1-1. Assuming this cross section to be representative of the material in three dimensions, calculate the average volume of a grain. If the energy of a grain boundary is about 1 joule/m^2, calculate the total grain-boundary energy in a mole of tungsten with grains of this size.

1-2 Estimate from Fig. 1-2 the approximate size of each of the small subgrains in the structure. How many such subgrains would be found in an average grain of the tungsten of Fig. 1-1?

1-3 Estimate the distance between centers of tungsten atoms in Fig. 1-3. This distance is called the *atomic diameter.*

2

The geometry of perfect crystals

2-1 Symmetry of Crystals

All atoms are constructed of various elementary particles (electrons, protons, neutrons, etc.), and a complete description of a solid would simultaneously specify the condition of all these particles. However, such a description is unnecessarily complex for most purposes. To be sure, the atomic structure is important in dealing with electronic properties, and it is discussed in due time. The first part of this book, however, treats the geometrical arrangement of entire atoms in crystals. An approximation sufficiently accurate for this purpose is to suppose the atoms to be round, hard balls. These balls rest against each other in various geometrical arrangements, each solid having its own mode of atom placement.

The solids of primary interest for us have an arrangement of atoms (or molecules) in which the atoms are arranged in some regular repetitious pattern in three dimensions. Such solids are called *crystals*, and the arrangement of atoms is termed the *crystal structure*. The internal regularity of atom placement in solids often leads to a symmetry of their external shapes. Rock salt crystals, for example, are rectangular parallelepipeds with faces which are identical when looked at from several different directions; these crystals have a high degree of symmetry. Crystalline quartz crystals, though, have symmetry of a lower order. By examining external features only, crystallographers were able to build up a large body of knowledge about the symmetry properties of crystals long before modern methods were used to determine more directly the internal symmetry of atom arrangements.

Some elements and compounds display their crystal symmetry clearly when they are deposited under favorable conditions (e.g., rock salt, snow crystals, sulfur, calcite, and quartz). For many others, such as glasses and metals, the symmetry is usually not apparent. The glasses, indeed, are not crystalline at all; they are extremely viscous liquids. The metals are crystalline when solid, but shapes imposed by manufacturing operations usually render the crystalline nature of metallic objects invisible under casual observation. In reality, metals and alloys are usually composed of many tiny crystallites too small to be seen by the unaided eye, though metals deposited carefully (say, by vapor deposition) may assume the highly symmetrical shapes which we commonly associate with "crystals." The tiny crystallites of a metal (they may be as little as 0.001 in. in length) are indeed nearly perfect crystals, as X-ray evidence shows.

This chapter is concerned solely with the geometry of atom arrangements commonly observed. For simplicity of presentation, the solids used for examples are mostly those of the pure elements. The chapter is divided into several sections. First, a short, qualitative description is given of the nature of attractive and repulsive forces between atoms in a solid. The next sections deal with the geometrical arrangements of atoms in the solids of most interest in this book and detailed ways of describing the atom arrangements concisely and accurately. The final section contains a discussion of the principles of X-ray analysis, the method most widely used for determining crystal structure.

For some materials such as the common metals, the crystal structures are simple; for many compounds, they are complex. Most materials to be examined in detail have relatively simple structures. Fortunately, some of the mathematical relationships for crystal structures which are developed in this and later chapters have general enough validity to be useful in the understanding of more complex structures.

2-2 The Concept of the Bond

Since solids exist in equilibrium at temperatures lower than those of the corresponding liquids and gases, they must have the lowest (free) energy configuration at these low temperatures. Furthermore, since solids do possess symmetry, a symmetrical spatial arrangement of atoms must be of lower energy than a random spatial arrangement. Elementary reasoning shows why this should be true. Since the atoms of a solid do stick together and since they do not collapse to a very high density, a model of a solid must have at least two features. It must allow for attractive forces between atoms when they are far apart and for strong repulsive forces when they are too close together.

Such a model resembles in some respects the situation which exists among balls—say, marbles—covered with a sticky paste. The paste tends to attract the balls weakly when they approach each other and to hold them together. If the balls are forced too close together, strong elastic forces tend to repel them to some equilibrium distance (the distance between centers is about the diameter of the balls themselves). For just *two* balls, the lowest energy state occurs when they just touch. For *three* balls, the lowest energy configuration is a triangular array where each ball touches the other two. The lowest energy arrangement for *seven* balls in a plane is a hexagonal array in which six balls surround a central one (see Fig. 2-1). This array has 12 interacting pairs, the maximum number allowed for seven balls in a plane. Eight, nine, ten, . . . , N balls arrange themselves around the central seven, repeating endlessly the symmetry of the initial

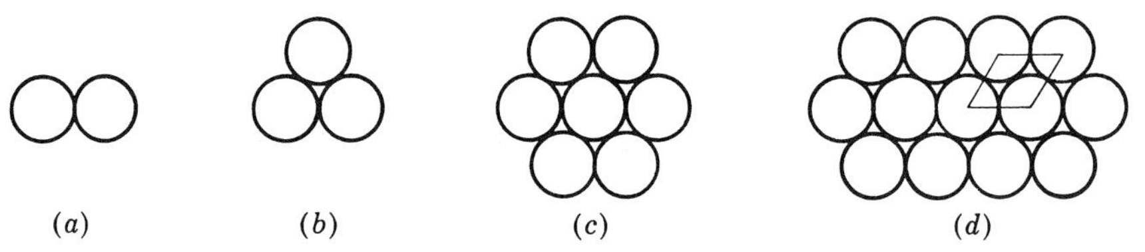

Figure 2-1 Close packing of round balls in a plane. (*a*) Two balls; (*b*) three balls; (*c*) seven balls; (*d*) many balls. The rhombus indicated in (*d*) is called a unit cell.

group. The number of interacting pairs is maximized (and the potential energy is minimized) by this perfect array. Though two-dimensional crystals of this type exist only in the imagination, three-dimensional crystals of low energy might be produced by piling such sheets of atoms on each other in a regular stacking sequence. More than one possible stacking sequence exists; the two most common are discussed in later sections.

The attractive force characteristic of this model has one natural attribute: it is an attractive force between adjacent (nearest) neighbors and not between distant neighbors (i.e., those more than one diameter distant). This nearest-neighbor attraction (and repulsion) is like the chemical bond of inorganic chemistry, and this bond concept is convenient in several later chapters. The common metals have a sublimation energy of several electron volts per atom so that the strength of an individual bond is about $\frac{1}{4}$ to $\frac{1}{2}$ ev. This discussion is not meant to imply that the origin of the bonds in metals is that of the electron transfer of inorganic compounds; the origin of the binding of like atoms to each other (say, for the pure metals or pure semiconductors) is not so simple. However, for convenience, the cohesive energy of a solid is sometimes expressed in terms of effective bonds between the individual atoms.

The concept of bonds helps to show that a perfect crystal is a minimum-energy configuration and that upsetting the perfect symmetry of a close-packed sheet (as portrayed in Fig. 2-1) absorbs energy. Suppose that the symmetry is reduced in the simplest way possible, by disturbing only one atom. By removing an atom from the interior of the sheet and putting it on the edge, the total number of bonds in the solid is reduced. Six bonds are broken when the atom is removed, and only two or three are re-formed when the atom is placed on the edge. The atoms in Fig. 2-2*a* must gain energy in going to the configuration of Fig. 2-2*b*, because the atoms around the hole cannot alter their local arrangement to re-form the missing bonds. Hence the high symmetry array is a configuration of minimum energy. (We have not shown that it is the lowest minimum, but the reader can easily verify this fact.) This argument carries over into three-dimensional structures.

Arrays of atoms which have long-range regularity of position permit the designation of elementary building blocks, called *unit cells*, with which the entire crystal can be constructed. Thus each atom

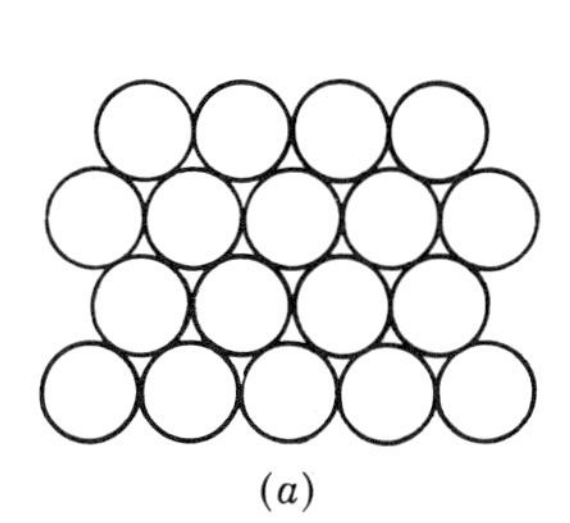

(*a*)

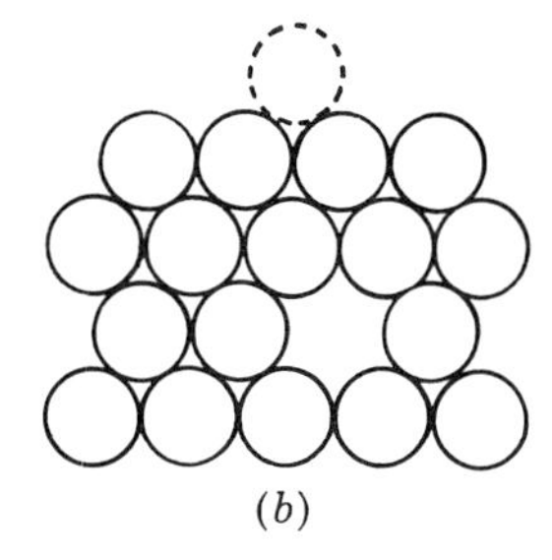

(*b*)

Figure 2-2 Formation of a vacant lattice site in a close-packed plane.

in the close-packed plane of Fig. 2-1*d* is surrounded by a regular hexagon of six other atoms. The position of every atom in this plane can be designated exactly by simply translating (i.e., moving without rotation) a rhombus along its bounding lines in the plane. The rhombus, outlined in Fig. 2-1*d*, is called a *unit cell;* the translations necessary to map out all atom positions are motions in appropriate directions of multiples of the interatomic spacing. Unit cells for three-dimensional crystals are a little more difficult to represent, for they must be parallelepipeds. A general arrangement (not for a structure of hexagonal symmetry) is sketched in Fig. 2-3. Here a unit cell is marked with heavy lines. The three edge lengths of this cell are fundamental translations, since the entire structure may be constructed by translating a single unit cell through unending repetition of these distances. The unit cell is, in general, defined as that volume of a solid from which the entire crystal can be constructed by translational repetition in three dimensions.

A given crystal type does not possess a unique unit cell; many unit cells exist for the same structure. For example, a cell of size $2a$, $2b$, $2c$ (Fig. 2-3) is as correct as one of size a, b, c. For many structures, cells of completely different geometry are possible. For some purposes a unit cell which contains the fewest atoms is desirable. For other purposes cells of somewhat larger volume are useful, for

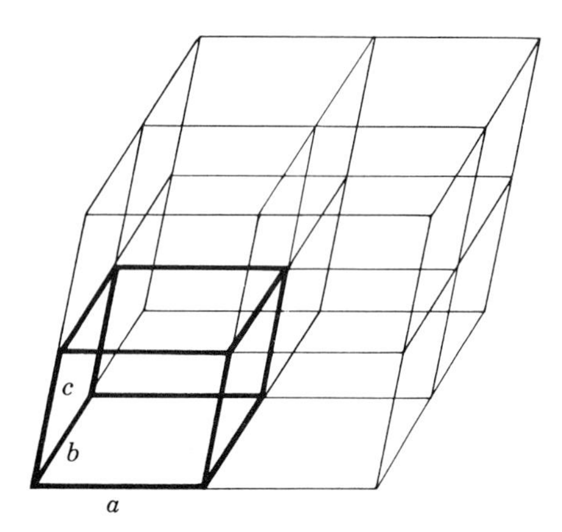

Figure 2-3 All of space can be filled by translating a unit cell along its edges.

they possess geometrical advantages (examples are the cubic structures). Most of the metals have smallest unit cells which contain only a few atoms. Some elements (such as manganese) and many compounds (such as Fe_3O_4) have many more atoms (dozens or even hundreds) in the smallest unit cell.

2-3 Observed Crystal Structures

The crystal structures of many materials of commercial importance are relatively simple. Some of them are even built up by the stacking of the two-dimensional close-packed hexagonal arrays of atoms, which we have already used extensively as examples. Examples of several types of structure are described briefly in the following paragraphs.

The hexagonal close-packed structure

The hexagonal close-packed structure, one of the common crystal types, is made by stacking close-packed planes in a simple sequence. This method of stacking has its basis in the manner in which two close-packed planes can rest against each other; each atom of one plane rests in a depression against three atoms of the other. The detailed geometry of two such planes is indicated in Fig. 2-4*a*. The light circles are the atoms in one close-packed plane A. The adjacent plane B is designated by the darkened circles. Observe that plane B has two possible positions, the one indicated and the other given by the positions C. For either choice, each atom of the second plane rests against three atoms of the first (and vice versa). The only important choice is the manner of stacking a third plane on the second. Clearly, there are two choices for the third plane also. Let the first two planes have the sequence AB. Then the third may be added in position A, so that the atoms of plane 3 are directly above those in plane 1. Alternatively, they may go in position C, so that the atoms of plane 3 are directly above interstices of plane 1. The

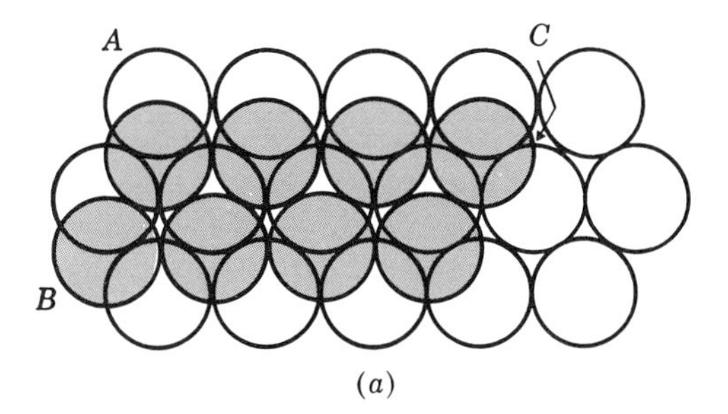

(*a*)

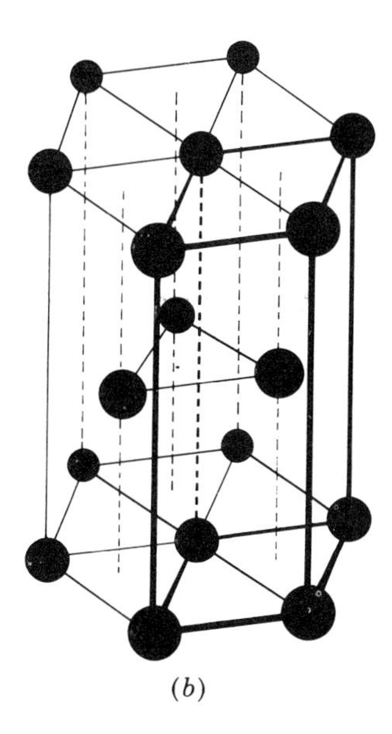
(*b*)

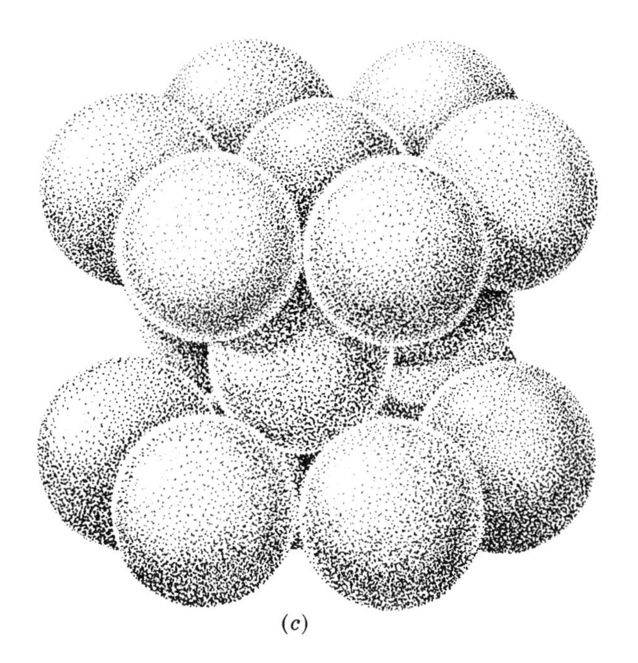
(*c*)

Figure 2-4 (*a*) Stacking of close-packed planes for the hcp crystal type. (*b*) The relative atom positions in the hcp crystal. The unit cell is outlined with heavy lines. (*c*) The hcp structure, with the atoms drawn to scale.

hexagonal close-packed structure (abbreviated hcp) has the *first* stacking sequence $ABAB \cdots$.

A small volume of the hcp structure (this volume contains the number of atoms in three unit cells) is sketched in Fig. 2-4*b*. Here the atom positions are drawn so that those in front will not obscure those behind. This structure is termed a close-packed structure because it possesses the maximum density of packing of hard spheres. Each atom has 12 nearest neighbors, 6 in its plane plus 3 in the plane above and 3 in the plane below. The distance a, the distance between the centers of adjacent atom sites, is called the *atomic diameter*. Figure 2-4*c* is a sketch of the structure with the atoms drawn to scale.

The face-centered cubic, or cubic close-packed structure

A related crystal structure is formed when close-packed planes are stacked in a slightly different sequence. In this case the first two planes go in the AB sequence; the third, however, is displaced relative to both A and B into the positions C. A fourth plane has the A position, and the repetition cycle $ABCABC \cdots$ is established. A unit cell of this crystal structure is shown in Fig. 2-5*a*. This cell is not the smallest possible cell, but it does have one convenient feature: the cell is a cube. An atom occupies each corner of the cube and the center of each face; hence the name *face-centered cubic* (fcc). This structure is symmetrical with respect to rotation by 90° about any one of the cube edges. The more conventional way to draw this cube (but one which does not bring out the stacking sequence nearly so well) is pictured in Fig. 2-5*b*. The face-centered cubic structure is also close-packed, each atom having 12 nearest neighbors. The atoms are in contact along the diagonals of the faces. A sketch of a unit cube of this structure showing the atoms to scale is seen in Fig. 2-5*c*.

The body-centered cubic structure

The fcc and hcp methods of stacking close-packed planes exhaust the possibilities for a simple stacking sequence. Mixed stacking sequences are possible—for example, $ABACA \cdots$—but none of the elements appear to crystallize with such complicated stacking patterns (except for particular metals in special temperature regions). Instead, many elements possess structures which are not formed of close-packed planes. One of these is the body-centered cubic, which is sketched in Fig. 2-6. This unit cell is cubic, with an atom at each corner and an atom in the center of the cube. This structure is not close-packed, since each atom has only eight neighbors. Note that atoms are in contact along body diagonals.

The simple cubic structure

Another possible structure is the *simple cubic* arrangement of atoms. The unit cube of this structure has atoms only at the corners of a

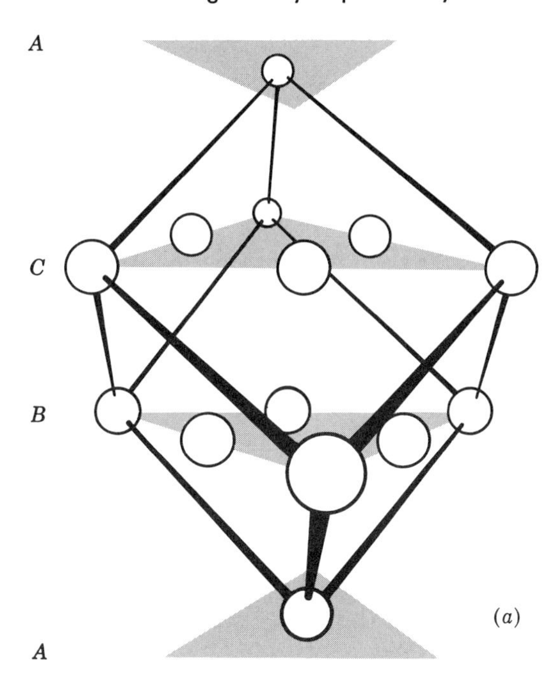

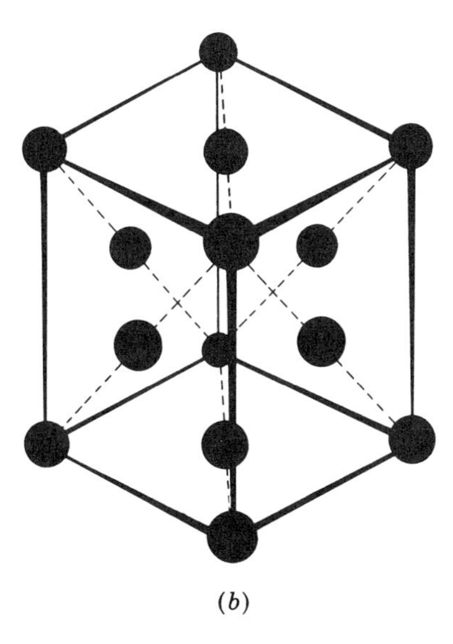

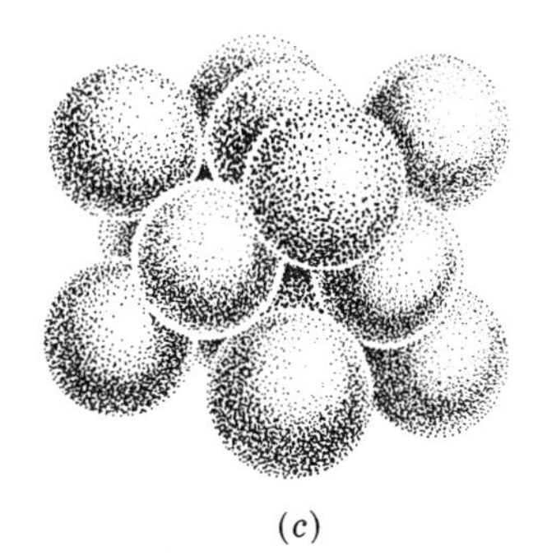

Figure 2-5 (*a*) The stacking of close-packed planes to form the fcc lattice. The unit cube is the volume enclosed by the heavy lines; the close-packed planes are designated by the shaded planes. (*b*) Unit cube of the fcc lattice. (*c*) Sketch of fcc lattice, showing atoms drawn to the scale of the cube edge.

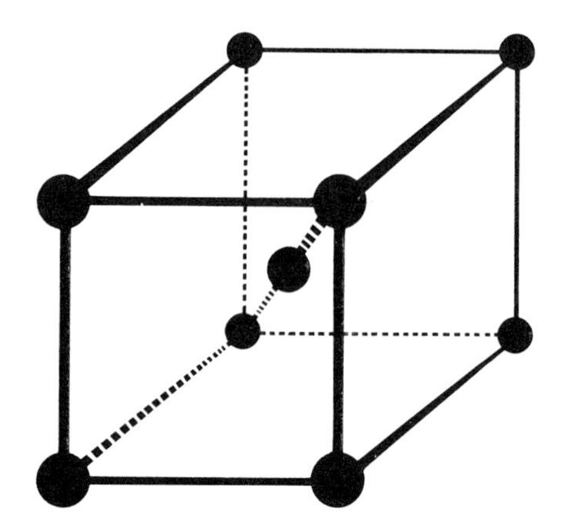

Figure 2-6 Position of atom centers on bcc lattice.

cube, and the atoms therefore touch along cube edges (Fig. 2-7). This structure is loosely packed; each atom has only six nearest neighbors. Only one element, Po (in a certain temperature region), exhibits this crystal structure. The simple cubic structure is introduced mainly because its simple geometry simplifies later semiquantitative calculations and discussion.

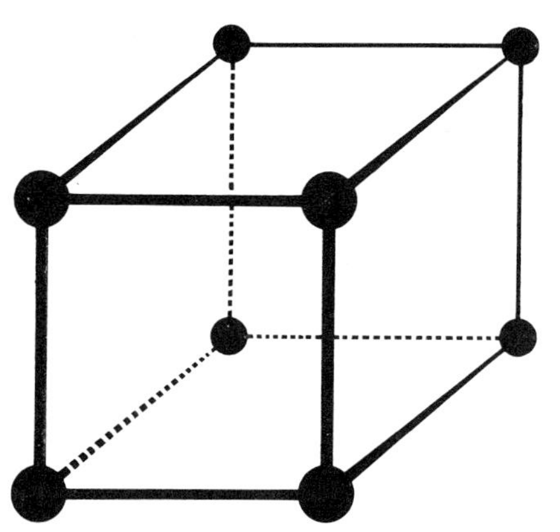

Figure 2-7 Position of atom centers on sc lattice.

Many other cubic unit cells bear a simple relationship to the simple cubic cell. Some can be reduced to sets of interpenetrating simple cubic sublattices. A body-centered cubic lattice can be considered to contain two identical simple cubic lattices, one consists of the corner atoms (Fig. 2-10), the other of the center atoms. Face-centered cubic lattices can be considered to contain four identical simple cubic lattices properly disposed with respect to one another. This point is taken up in a problem.

Other cubic structures

Many of the elements solidify in one of the structures just described, but several of the elements and all compounds have other structures. Some of these structures are derivatives or combinations of one of the basic cubic structures just described, in which one lattice is interpenetrated by another. A few of these structures are described in the following.

The diamond cubic structure is a combination of two interpenetrating face-centered cubic sublattices. The unit cell is sketched in Fig 2-8*a*. One sublattice has its origin at the point 0, 0, 0, the other

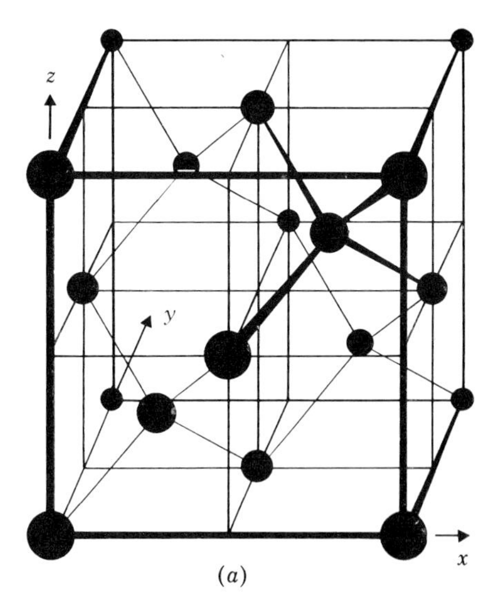

Figure 2-8 The diamond cubic structure. (*a*) The basic diamond lattice. The tetrahedral bond is heavily outlined.

at a point one-quarter of the way along the body diagonal (at the point $a/4$, $a/4$, $a/4$). The diamond cubic structure is loosely packed, since each atom has only four nearest neighbors. The bonding arrangement showing to scale the geometry of the nearest neighbors of an atom

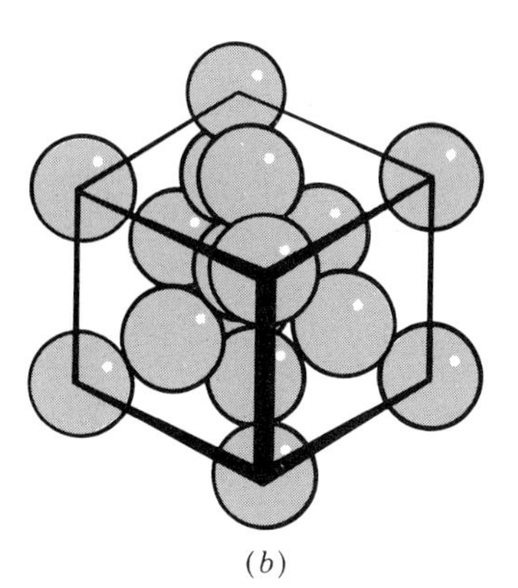

Figure 2-8(b) The diamond cubic structure to scale.

and the geometry of the unit cube is illustrated in Fig. 2-8*b*. If an atom is imagined to be at the center of a regular tetrahedron, then its neighbors are on the four corners. Ge, Si, and C (in the diamond phase) crystallize with the diamond cubic structure.

A new structure occurs if the two face-centered cubic sublattices are occupied by different elements. The resultant structure, termed the *zincblende* structure, is sketched in Fig. 2-8*c*. Important compounds which have this structure are the semiconductors InSb

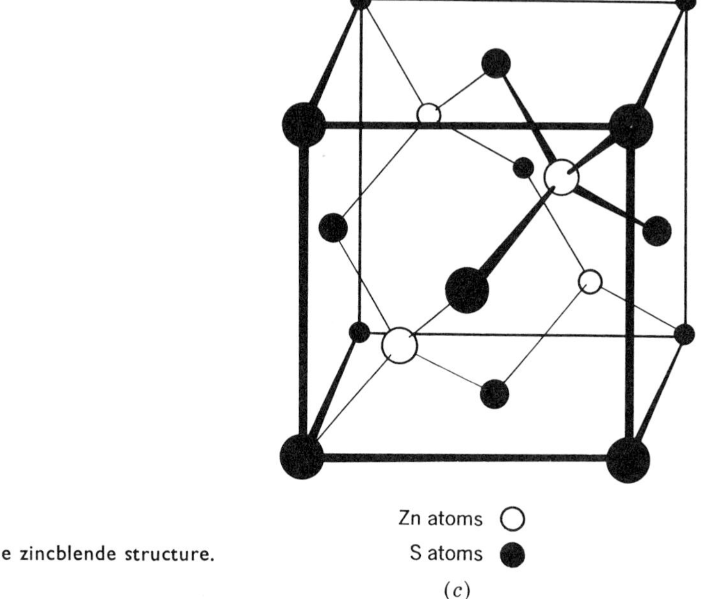

Figure 2-8(c) The zincblende structure.

and GaAs, also ZnS (for which it is named), CuCl, and many others.

Rock salt, NaCl, is another combination structure. This compound has the unit cell sketched in Fig. 2-9. It consists of two face-centered cubic sublattices, one of Na ions having its origin at the

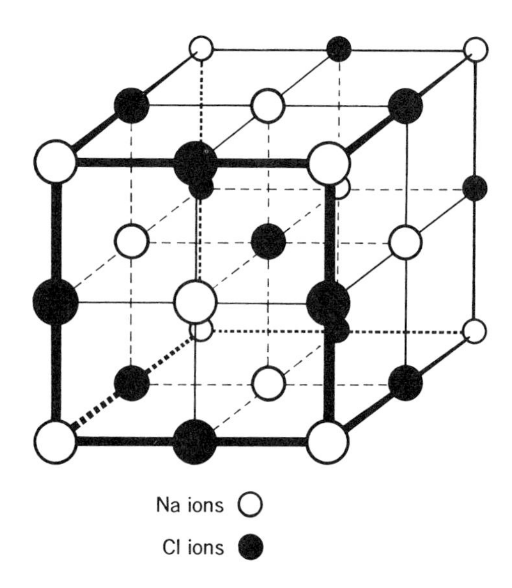

Figure 2-9 The NaCl crystal structure.

point 0, 0, 0, the other of Cl ions having its origin midway along a cube edge, say, at point $a/2$, 0, 0. This lattice is simple cubic if the difference between the Na and Cl ion positions is ignored.

Another combination cubic structure is that of CsCl. The lattice points of this compound are two interpenetrating simple cubic lattices; the corner of one sublattice is the body center of the other. One of the sublattices is occupied by Cs ions, the other by Cl ions. The resultant structure, termed the *cesium chloride* structure, is sketched in Fig. 2-10.

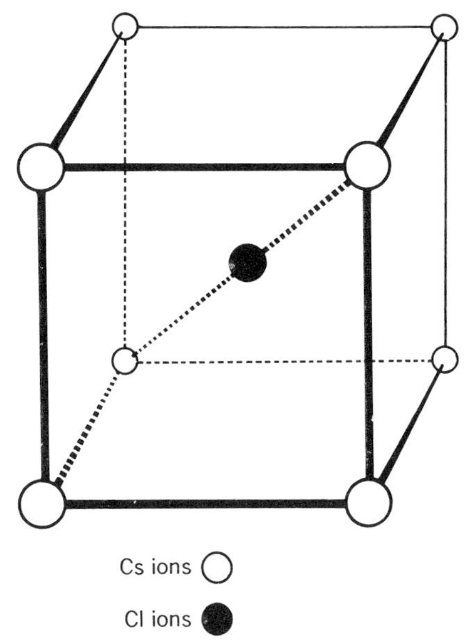

Figure 2-10 The CsCl crystal structure.

These examples are only a few of the types of structures observed. Many of the remainder are also simple, but some are extremely complicated. With a few exceptions, each solid considered in this book has one of the structures just described, so no further discussion will be given of other crystal types.

2-4 Summary of the Geometry of Simple Structures

The general features of the basic structures described in earlier paragraphs are such that numerous calculations concerning details of their geometry can be made. The calculations are especially easy for the cubic lattices. Properties of interest are the following: (1) the number of nearest neighbors (called the *coordination number*); (2) the *radius of the atom*, defined as half the distance between nearest neighbors in a crystal of a pure element, in terms of the cube edge a; (3) the relative *density* of packing (the fraction of volume occupied by spherical atoms compared with the total available volume of the structure). These quantities for the elementary cubic structures are listed in Table 2-1.

Table 2-1 Some details of elementary cubic structures

	Simple cubic	Bcc	Fcc	Diamond cubic
Coordination number	6	8	12	4
Atomic radius (a = cubic edge)	$a/2$	$a\sqrt{3}/4$	$a\sqrt{2}/4$	$a\sqrt{3}/8$
Atoms per unit cube	1	2	4	8
Density of packing	$\pi/6$	$\pi\sqrt{3}/8$	$\pi\sqrt{2}/6$	$\pi\sqrt{3}/16$

The reader is urged to calculate these quantities for himself. Cutaway views of the unit cells of the four structures are given in Fig. 2-11 to aid in the calculations.

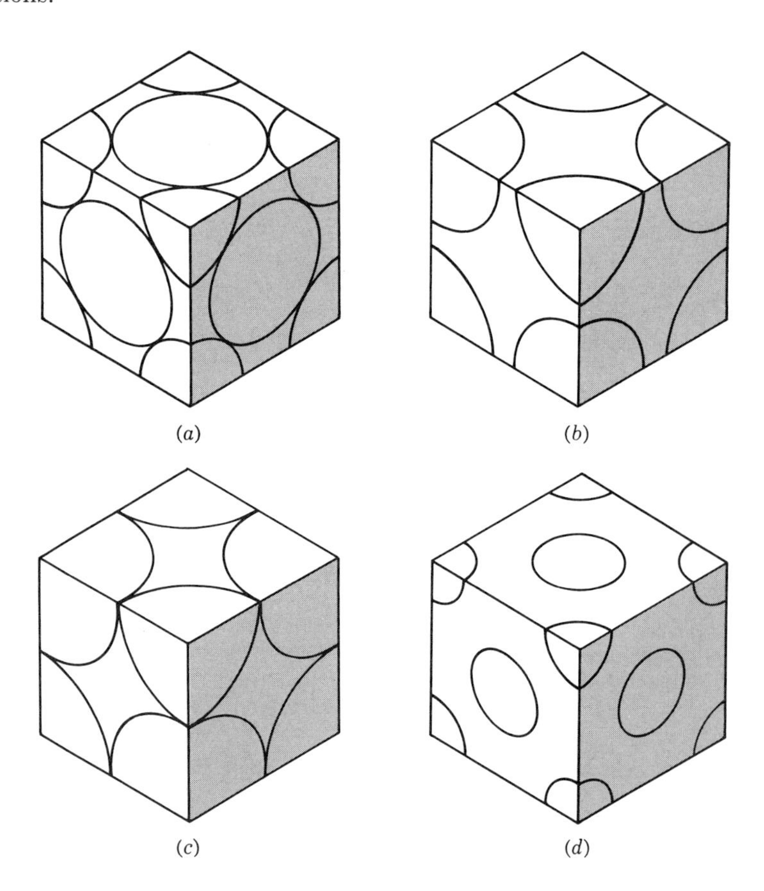

Figure 2-11 Cutaway views of four cubic structures. (*a*) fcc; (*b*) bcc; (*c*) simple cubic; (*d*) diamond cubic.

A compilation of some of the elements which solidify with the elementary crystal structures is given in the table on the inside back cover. Considerable regularity of structure is apparent with the elements arranged in the form of the periodic table. For example, the first IA subgroup are all bcc, etc. From these crystal structures and knowledge of the dimensions of the unit cells, atomic radii of the atoms may be calculated. Most of these radii are between 1 and 2A. The currently accepted values for many of the elements are presented in Table 2-2.

Table 2-2 Atomic radius of some of the elements

Element	Atomic radius, A	Element	Atomic radius, A
Al	1.43	Na	1.86
Ag	1.44	Ni	1.25
Au	1.44	Os	1.34
Be	1.11	Pb	1.75
C	0.72	Pd	1.38
Cb	1.43	Pt	1.38
Cd	1.49	Rh	1.34
Cr	1.25	Ru	1.32
Cs	2.63	Si	1.18
Cu	1.28	Sn (gray)	1.41
Fe	1.24	Ta	1.43
Ge	1.23	Ti	1.45
Hf	1.58	Th	1.80
Ir	1.36	Tl	1.70
K	2.31	V	1.32
Li	1.52	W	1.37
Mg	1.60	Zn	1.33
Mo	1.36	Zr	1.59

2-5 Crystallographic Notation

The symmetry characteristics of the many crystal types are described in a number of ways by crystallographers. The methods devised by them for this purpose have been aimed at systematizing crystallographic information about many elements and compounds. Most of these methods were devised prior to the use of modern atomistic techniques, but they are still applicable. A brief introduction into the formal language of crystallography aids later discussion.

One basic necessity of crystal analysis is the ability to describe directions in space. It is, for example, necessary in some instances to be able to define the direction of particular rows of atoms. A simple method of doing this is derived from vector notation. Since our concern is almost exclusively with cubic lattices, we define this method for the cube. Let the projections of a vector on the x, y, and z axes of the unit cube be some numbers x_0, y_0, and z_0. These projections are measured for convenience in units of the cube edge, a. There exists a number, say, r, such that the quotients x_0/r, y_0/r, and z_0/r are the lowest set of integers proportional to these projections. These quotients, which can be defined as u, v, and w, respectively, are called the *indices of the direction*. As an example, consider a vector whose projections on the three coordinate axes are $x_0 = 3a$, $y_0 = 4a$, $z_0 = 2.5a$. If these numbers are divided by $0.5a$, that is, $r = 0.5a$, three numbers 6, 8, 5 are found. They are in the same proportion as the original lengths; they are also the lowest set of integers with this property; therefore they are the indices of the original vector.

For convenience, indices of a direction are indicated with brackets $[uvw]$. Negative projections on any coordinate have negative indices and are indicated by a bar over the appropriate index number. Directions which are symmetrical with respect to permutation of the three axes are called a *set*. These directions are written collectively inside carets: $\langle uvw \rangle$.

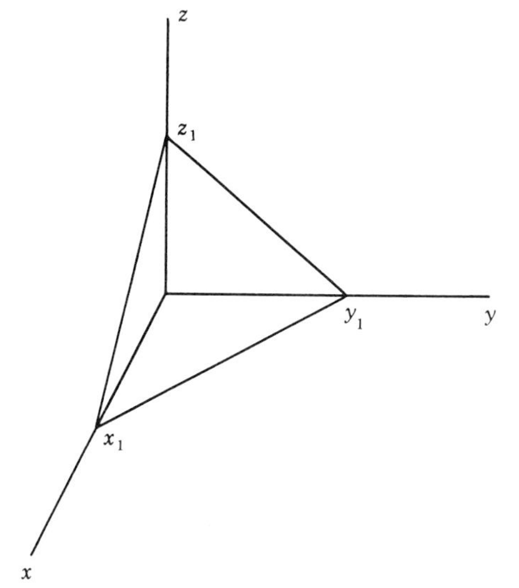

Figure 2-12 The intercepts of an arbitrary plane on rectangular coordinates.

Examples. The $+x$ axis has indices [100], the $-x$ axis, $[\bar{1}00]$. The $+y$ axis has indices [010]; the $-y$ axis, $[0\bar{1}0]$. The diagonal of the xy face has indices [110]; that of the xz face, [101]. The body diagonal of the cube in the positive quadrant has indices [111]; the opposite direction, $[\bar{1}\bar{1}\bar{1}]$. The eight body diagonals of the cube, [111], $[\bar{1}\bar{1}\bar{1}]$, $[\bar{1}11]$, $[1\bar{1}1]$, $[11\bar{1}]$, $[\bar{1}\bar{1}1]$, $[\bar{1}1\bar{1}]$, and $[1\bar{1}\bar{1}]$, are designated as a group $\langle 111 \rangle$; they are all permutations of the index numbers ± 1.

This notation has great simplicity. It has all the advantages of the vector notation except one: the magnitude of the vector is lost in the reduction of the projection to integers. The loss is not serious, since the distance in any direction can be specified when it is needed. The advantages of the index notation far outweigh this small disadvantage.

Sets of parallel planes of atoms can be specified by using three numbers called the Miller indices of a plane. These numbers are defined in terms of the intercepts of the plane on the coordinate axes. Let one plane of the set pass through the origin. Suppose that the adjacent plane has intercepts on the axes of x_1, y_1, and z_1, in units of the cube edge (Fig. 2-12). There exists a number, say, s, such that the product of s and the *reciprocals* of these intercepts are the lowest set of integers having the same ratios as $1/x_1$, $1/y_1$, and $1/z_1$. Define three numbers such that $h = s/x_1$, $k = s/y_1$, and $l = s/z_1$. The numbers h, k, and l are termed the *Miller indices of the plane* and are written in parentheses: (hkl). As an example, consider a plane which has intercepts $x_1 = 0.5a$, $y_1 = 1.25a$, and $z_1 = 1.5a$. Reduction of the fractions found by taking the reciprocals of these intercepts to integers can be done if s is selected as $7.5a$. Then $h = 7.5a/0.5a = 15$, $k = 7.5a/1.25a = 6$, $l = 7.5a/1.5a = 5$. This plane is (15,6,5), a high index plane relative to those of the cube faces.

The same general rules about index numbers apply to planes just as they do to directions. Negative index numbers are used to specify planes with negative intercepts; a plane with negative x intercept is written $(\bar{h}kl)$. Planes of a form—i.e., planes which are equivalent in the crystal—are enclosed in braces: $\{hkl\}$. For example, the planes $\{111\}$ embrace this group: (111), $(\bar{1}\bar{1}\bar{1})$, $(\bar{1}11)$, $(1\bar{1}\bar{1})$, $(1\bar{1}1)$, $(\bar{1}1\bar{1})$, $(11\bar{1})$, $(\bar{1}\bar{1}1)$.

Several features should be noted about this scheme in the cubic system. (1) Parallel planes equally spaced all have the same index numbers, (hkl), analogous to the index notation for directions, where all parallel rows of atoms have the same indices $[uvw]$. (2) A plane parallel to one of the coordinate axes has index number 0 for that direction. (3) A plane passing through the origin is defined in terms of a parallel plane having nonzero intercepts.

Examples. The cube face normal to the x axis is described by (100). Planes of this type are sketched in Fig. 2-13*a*. Planes normal to the y axis are designated (010) and are sketched in Fig. 2-13*b*. Planes having equal intercepts on the three axes are the (111) planes (Fig. 2-13*c*). Index numbers (110) specify planes with equal intercepts on x and y and parallel to the z axis (Fig. 2-13*d*). Some planes may have one or two negative intercepts. An example is the parallel family of $(\bar{1}10)$ planes (Fig. 2-13*e*). Finally a few of the planes (112) are sketched in Fig. 2-13*f*.

The Miller-index notation is especially useful for cubic systems. Among its desirable features are these: (1) The angle θ between two crystallographic directions $[u_1v_1w_1]$ and $[u_2v_2w_2]$ can be calculated easily. It is given by the expression

$$\cos\theta = \frac{u_1u_2 + v_1v_2 + w_1w_2}{(u_1^2 + v_1^2 + w_1^2)^{\frac{1}{2}}(u_2^2 + v_2^2 + w_2^2)^{\frac{1}{2}}} \tag{2-1}$$

(2) The normal to the plane with index numbers (hkl) is the direction $[hkl]$. This correspondence is an enormous aid in calculating angles between planes, since Eq. (2-1) can be used directly for this purpose. (3) The distance d between adjacent planes of index numbers (hkl) is given in terms of the cube edge a by the expression

$$d = \frac{a}{(h^2 + k^2 + l^2)^{\frac{1}{2}}} \tag{2-2}$$

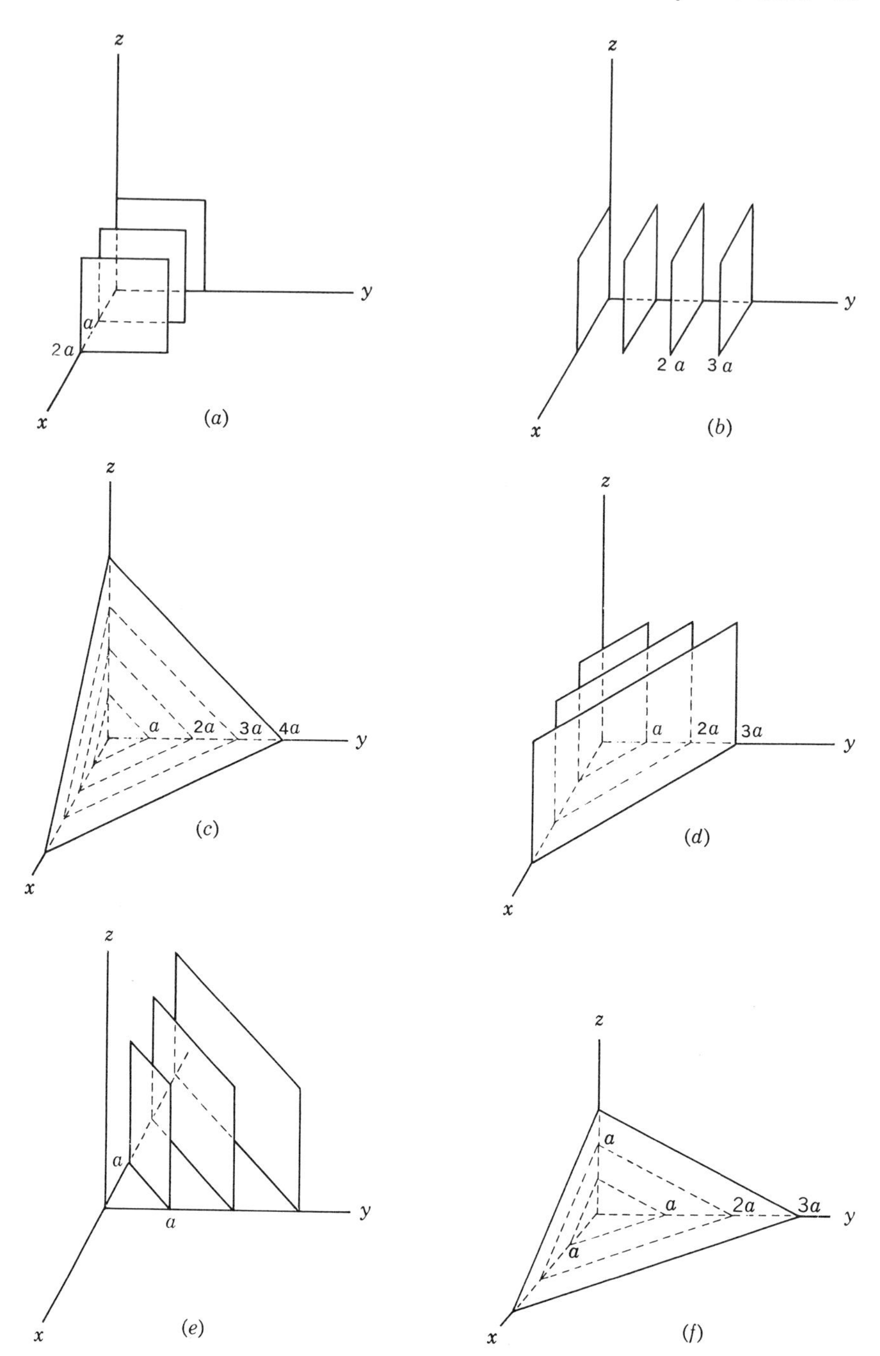

Figure 2-13 Examples of important cubic crystal planes. (*a*) (100) planes; (*b*) (010) planes; (*c*) (111) planes; (*d*) (110) planes; (*e*) ($\bar{1}$10) planes; (*f*) (112) planes.

Planes with low index numbers have wide interplanar spacing compared with those with high index numbers. The low-index planes also have a higher density of atoms per unit area than the high-index planes. Most planes which are important in determining the physical and chemical properties of solids are those with low index numbers. Rarely is a plane important if it has an index number larger than 3; only about 15 sets of planes are physically significant.

2-6 Determination of Crystal Structure

The general symmetry characteristics of crystals can be inferred from a combination of measurements of both the visible, external faces of crystals and bulk physical properties (such as electrical resistance and elastic constants). Symmetries determined in this way do not, however, fix the positions of atoms on the crystal lattice nor do they permit measurement of distances between atoms. The question is not one of experimental errors or the inability of crystallographers to interpret data accurately but is fundamental; a technique is required which permits either direct or indirect interpretation at the atomic scale of distance.

Field ion microscopy might have been used first to determine crystal structure. As an example, proper interpretation of such a picture as Fig. 1-3 describes much about the crystal structure of tungsten (see caption of that figure). Historically, however, events did not occur in that way because other techniques came into use before the ion microscope was invented. Shortly after the discovery of X rays, they were used to determine the simplest crystal structures, and by the 1930s X-ray diffraction techniques were well developed. At about that time electron diffraction methods were developed, and in the 1940s neutron diffraction methods became practical when large fluxes of neutrons from reactors became available. The common feature of all these techniques is the use of radiation or of particles with wavelengths comparable with the distance between atoms of the crystal, i.e., a few angstrom units.

The details of diffraction from crystals are complicated, but the elementary notions can be grasped using a concept systematized by W. H. Bragg. He reduced the problem to consideration of specular reflection off the infinite number of sets of lattice planes in the crystal, some of which are sketched for a cubic crystal in Fig. 2-13. Bragg showed that any diffracted ray can be regarded as though it were specularly reflected from one of these systems of planes. However, the wave can be reflected only if it satisfies the equation

$$n\lambda = 2d \sin \theta \qquad (2\text{-}3)$$

where λ = wavelength of the radiation

d = distance between adjacent planes

θ = grazing angle of incident (and outgoing) radiation

n = an integer

The important geometry of this arrangement is sketched in Fig. 2-14. The angle θ has a unique value for a given set of crystal planes, a given value of n and a given λ. Therefore an incident wave of a given wavelength must strike the crystal in some direction which lies in a cone which makes a definite angle with the given set of planes. Conversely, if a diffracted wave is observed, it may be concluded that the crystal possesses a set of planes with a normal in that direction which bisects the angle between the incident and diffracted waves. Furthermore, the spacing of these planes is related to λ and θ by Eq. (2-3).

Relation (2-3) shows why radiation of wavelengths a few angstrom units is most useful for crystal analysis. Interplanar spacing of important planes in crystals [d in Eq. (2-3)] is of order 2 to more than 25 A. Since sin θ can be no greater than 1, first order Bragg reflections off neighboring planes of a set require that λ be in this range.

The details of diffraction analysis of crystal structure are but extensions of the logic of the preceding paragraphs. Experimenters commonly use the "powder method," in which a collimated monochromatic beam of X rays or monoenergetic particles is incident on a finely pulverized specimen. Diffraction occurs off parallel planes of atoms in grains of the specimen which satisfy the Bragg condition. The diffracted beams lie on coaxial cones whose axes are the direction of the incident beam. The apex angle of a cone is 2θ; hence d for any diffracted beam can be determined by using Eq. (2-3) and the measured value of θ. A crystal structure is then assumed, and a consistent scheme of Miller indices is assigned to the diffracted beams. The size and shape of the unit cell are determined if this step is done consistently. If the chemical composition and density of the solid are measured, the number of atoms per unit cell can be determined. Finally, the detailed placement of the atoms in the unit cell is deduced from the relative intensities of the several diffracted lines.

The preceding operation is relatively simple for pure elements with only a few atoms per unit cell. For a metal such as Cu, the consistent assignment of Miller indices to an assumed crystal structure can be done quickly: the composition is, of course, known, and the density is easy to determine. On the other hand, determination of the crystal structure of compounds may be much more difficult:

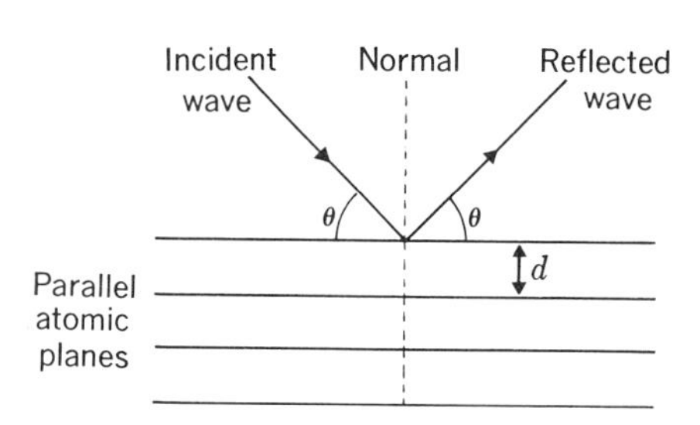

Figure 2-14 The geometry of Bragg's law.

finding the structure of a magnetic ferrite, say Fe_2NiO_4, requires that the positions of 56 atoms be determined.

X-ray crystallography is widely used for basic structure determination, but electron and neutron diffraction are indispensible for special purposes. Both the theory of diffraction and experimental methods are described in the references listed at the end of the chapter.

REFERENCES

1. C. S. Barrett and T. Massalski, *Structure of Metals*, 3d ed., McGraw-Hill Book Company, New York, 1966.
2. B. D. Cullity, *Elements of X-ray Diffraction*, Addison-Wesley Publishing Company, Inc., Reading, Mass., 1956.
3. L. V. Azároff, *Elements of X-ray Crystallography*, McGraw-Hill Book Company, New York, 1968.

2-1 From observation of the symmetry of the spots in Fig. 1-3, deduce both the crystal structure of tungsten and the approximate orientation of the crystal.

2-2 Aluminum has a density of about 2,699 kg/m^3. Its atomic weight is 26.97.

(*a*) About how many gram-moles of Al are contained in 1 m^3 of the solid?

(*b*) About how many atoms are contained in 1 m^3?

(*c*) Calculate the size of the unit cube for this fcc metal.

(*d*) Calculate the atomic radius of Al.

(*e*) What is the weight of a single atom of Al?

2-3 Iron is bcc below 910°C and is fcc above 910°C. The fractional change in density accompanying the transition from the bcc to the fcc phase is +1.0 per cent. Calculate the ratio of the atomic radii in the two phases.

2-4 The fcc structure has small voids (called interstices) between the atoms.

(*a*) Calculate the position of the largest of these voids.

(*b*) Consider the atoms in this structure to be spherical. Calculate the size of the largest spherical impurity which just fits into the largest void.

(*c*) How many neighboring atoms does the impurity atom touch?

2-5 Calculate the same quantities for the bcc and diamond structures.

2-6 The diamond cubic lattice may be considered to be a combination of two interpenetrating fcc sublattices. One sublattice has its origin at the point 0, 0, 0 and the other at a point one-quarter of the way along the body diagonal.

(*a*) Write the positions of all the atoms in this unit cell.

(*b*) How many atoms are contained in this unit cell?

2-7 Calculate the "bond angle" of solid Ge.

2-8 Prove that the direction [*hkl*] is the normal to the plane (*hkl*) for a cubic lattice.

2-9 An orthorhombic structure is one in which the unit-cell edges are mutually orthogonal but are unequal in length. Draw a (321) plane and a [321] direction. Find the angle between the [321] direction and the normal to the (321) plane for the ratio of unit-cell edge lengths of 1:1:2.

2-10 By writing the dot product of two vectors, show that the angle θ between two directions in a cubic lattice $[u_1v_1w_1]$ and $[u_2v_2w_2]$ is given by Eq. (2-1), i.e., that

$$\cos \theta = \frac{u_1u_2 + v_1v_2 + w_1w_2}{(u_1^2 + v_1^2 + w_1^2)^{\frac{1}{2}}(u_2^2 + v_2^2 + w_2^2)^{\frac{1}{2}}}$$

2-11 Show that the distance d between adjacent planes of types (hkl) in a cubic lattice of cube edge a is given by Eq. (2-2), i.e., that

$$d = \frac{a}{(h^2 + k^2 + l^2)^{\frac{1}{2}}}$$

2-12 How many planes of type $\{hkl\}$ are found in the cubic systems? Of type $\{hk0\}$?

2-13 Which planes in the fcc structure have the largest density of packing of atoms? Which planes in the bcc structure? In which directions in these planes is the linear density of atoms greatest?

2-14 The wavelength of a prominent X-ray line from a Cu target is known to be 1.537 A. This radiation incident on an Al crystal produces a diffracted beam off the (111) planes at a Bragg angle of 19.2°. Al is fcc, has density of 2,699 kg/m^3, and has atomic weight 26.98. From these experimental data, calculate Avogadro's number.

2-15 Calculate the number of nearest-neighbor bonds in one mole of a close-packed metal such as Cu. Calculate also the energy of a nearest-neighbor bond if the cohesive energy of Cu is 80,000 cal/mole.

2-16 The cube edge of bcc Fe at room temperature is 2.86 A.

(*a*) Calculate the spacing of the (110) planes in iron.

(*b*) Calculate the wavelength of X-radiation which would give a first-order diffracted beam at a Bragg angle of 55°.

(*c*) Calculate the kinetic energy of electrons and of neutrons for the same conditions of diffraction.

2-17 Zinc has the crystal structure hcp. At room temperature its unit cell parameters are $c = 4.947$ A and $a = 2.665$ A. The bulk density of zinc at room temperature is 7,134 kg/m^3.
Calculate:

(*a*) The atomic weight of zinc.

(*b*) The apparent ellipticity of the atoms.

3

Imperfections in crystals

3-1 General

The mathematically perfect crystal, which was described in the preceding chapter, is an exceedingly useful concept. With its use many properties of the lattice can be explained—such things as density, dielectric capacitivity, specific heat, and elastic properties. These are called *bulk*, or *structure-insensitive*, properties. However, all actual crystalline materials are not perfect. Mistakes in the structure, called *imperfections* (or *defects*), exist, generally in profusion. Although they affect such properties as those just mentioned only slightly, they have a strong influence upon many other properties of the crystal, properties such as strength, electrical conductivity, and hysteresis loss of ferromagnets. Those properties which are strongly sensitive to the state of perfection of the crystal are termed *structure-sensitive* properties.

The first departure from the ideal structure treated in this chapter is the thermal vibration of solids. This treatment is divided into two parts: a description of atomic vibrations using the ball-and-spring model and a more general description of the thermal energy of the solid. The latter is important because it permits the definition of local energy fluctuations which result in the production of the point defects.

The crystallographic point defects, vacancies and interstitials, are then considered as natural consequences of energy fluctuations. The more extensive crystallographic defects, line and surface defects, are also considered briefly.

Defects in crystal structure are not the only defects of importance; another type is a defect in electronic structure. The electronic defects are responsible for important electrical and magnetic properties.

The electronic defects are rather subtle in character; therefore they are taken up in detail through the latter part of the book after the ideal electronic structure is discussed.

To summarize, crystallographic defects are:

1. Thermal vibrations
2. Point defects
   a. Vacancies
   b. Interstitials
   c. Isolated impurities
3. Line defects: the dislocation
4. Surface defects
   a. External surfaces of solids
   b. Internal surfaces: grain boundaries and other internal boundaries

3-2 Thermal Vibrations

Atoms on a perfect lattice should occupy exactly the sites to which they are attached. Of course, no crystal is perfectly rigid, since it can be deformed by finite forces; hence atoms can be displaced from their ideal sites with a finite expenditure of energy. The forces holding atoms in position in the lattice are, in fact, weak enough so that the thermal energy of the atoms is sufficiently large to cause them to be displaced from equilibrium by sizable distances. The magnitude of this thermal vibration, some 5 to 10 per cent of the interatomic spacing at ordinary temperatures, is readily calculable (with suitable simplifying assumptions). Furthermore, several external manifestations of these vibrations, e.g., heat capacity, diffusion, and electrical resistivity, can be treated with sufficient accuracy to make elementary discussions meaningful.

A preliminary idea of the vibrations of atoms in a solid can be obtained by observing the nature of the interatomic forces. The equilibrium positions of the atoms are determined by a balance between the attractive and repulsive forces. At equilibrium the potential energy of the solid must be a minimum. The potential energy of an atom is sketched as a function of the lattice spacing of the solid in Fig. 3-1. For large lattice spacing (i.e., large total volume) the potential energy is arbitrarily set to zero, since the atoms then interact negligibly with each other. As the lattice spacing decreases, the potential energy decreases. That this must occur is clear, since solids exist even in the absence of external restraints. Finally, at some critical spacing the potential energy passes through a minimum and increases rapidly with further decrease in lattice spacing. That this increase must occur is also clear: solids have finite density.

The important region is that near the minimum, where no net force is exerted on an atom by its neighbors. If all the atoms of the solid except one are fixed in position and the atom in question moves

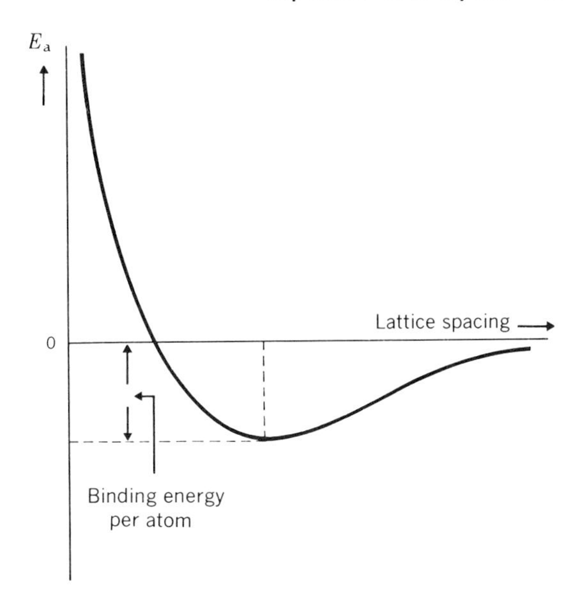

Figure 3-1 The potential energy of an atom, E_a, in a solid containing N atoms, as a function of the lattice spacing of the solid. The solid assumes the lattice spacing corresponding to the lowest energy.

in the x direction against its neighbors, as shown in Fig. 3-2, then unbalanced forces (represented by the extension and compression of springs) develop between the neighboring atoms along the x axis. Since the distance to neighbors in the y and z directions is assumed to be unchanged, the unbalanced force in these directions is zero. Figure 3-1 represents the energy change when the distances to *all* the neighbors of an atom vary; if only one varies, the energy $E_a(x)$ must be divided by z, where z is the number of nearest neighbors. (We assume that the potential energy for atoms farther away than the nearest neighbor can be neglected.) The change in energy, ΔE,

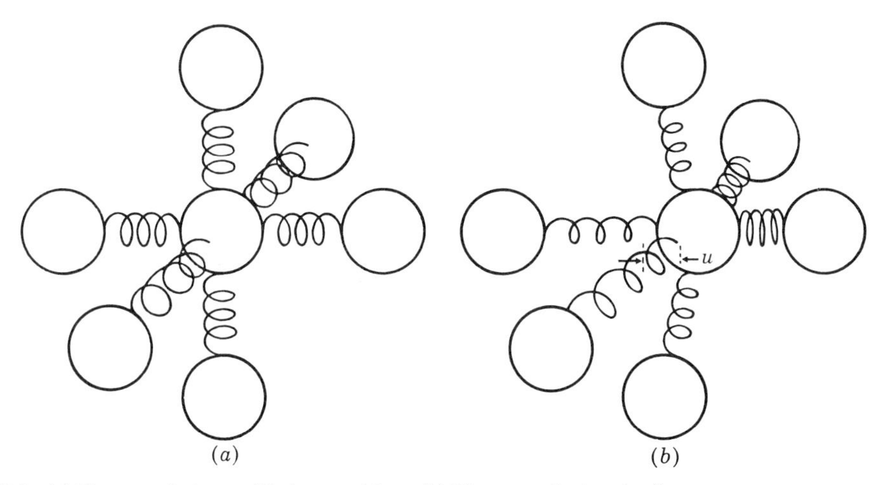

Figure 3-2 (*a*) The atom in its equilibrium position. (*b*) The atom displaced a distance u in the x direction. The springs in the y and z direction do not change their length to a first approximation.

which accompanies a change in atom position from x_0 to x is

$$\Delta E = \frac{2}{z}\{E_a(x_0 - u) + E_a(x_0 + u) - 2E_a(x_0)\} \tag{3-1}$$

where $u = x - x_0$. The first term is the energy of the bond with the right neighbor and the second is the energy of the bond with the left. The factor 2 appears because the movement of an atom against its neighbor causes both the energy of the displaced atom and that of its neighbor to change by an equal amount.

Since the function $E_a(x)$ can be expanded in a Taylor series about its minimum, the total energy change of the atom and its neighbors for small displacements is

$$\Delta E = \frac{4}{2z}\left(\frac{\partial^2 E_a}{\partial x^2}\right)_{x_0}(x - x_0)^2 = \frac{\alpha}{2}u^2 \tag{3-2}$$

where

$$u = x - x_0 \qquad \text{and} \qquad \alpha = \frac{4}{z}\left(\frac{\partial^2 E_a}{\partial x^2}\right)_{x_0}$$

An atom bound to a definite site by the potential-energy law of Eq. (3-2) is a simple harmonic oscillator. Hence, the vibrating atoms execute harmonic oscillations about their equilibrium positions. The force f on each atom as a function of its displacement is given by

$$f = \frac{-d(\Delta E)}{du} = -\alpha u \tag{3-3}$$

The differential equation describing the motion is then

$$m\frac{d^2u}{dt^2} = -\alpha u \tag{3-4}$$

where m is the mass of the atom. A solution is

$$u = A_0 \cos\sqrt{\frac{\alpha}{m}}\,t \tag{3-5}$$

The angular frequency ω of the atom is thus given by

$$\omega = \sqrt{\frac{\alpha}{m}}$$

and the frequency (called the *Einstein frequency*) by

$$\nu = \frac{1}{2\pi}\sqrt{\frac{\alpha}{m}} \tag{3-6}$$

Since the masses of atoms are known from experimental observations, the frequency of vibration can be calculated if the magnitude of the force constant, α, is known. Fortunately α may be estimated simply.

Suppose that Hooke's law can be applied to the unit cube of Fig. 3-3. The extension Δa which results when a force F is applied to the unit cube is given by the expression

$$Y\frac{\Delta a}{a_0} = \frac{F}{a_0^2} \tag{3-7}$$

In this expression a_0^2 is the area of the face of the cube of edge a_0. α is defined as that force necessary to stretch the cube unit distance. Since Young's modulus Y is of the order 10^{11} newtons/m^2, α is approximately 25 newtons/m.

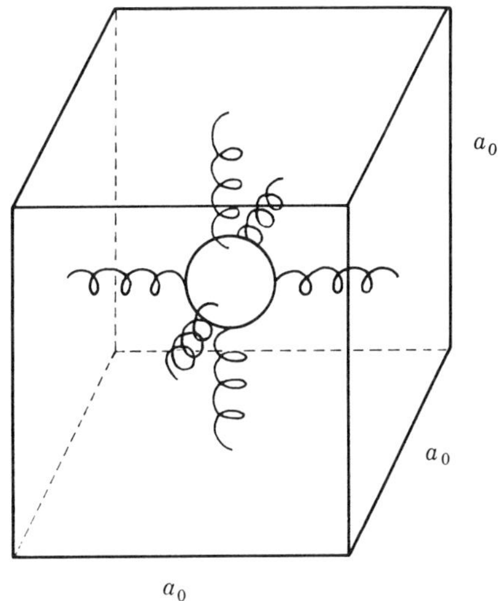

Figure 3-3 The unit cube around the atom.

The vibrational frequencies can now be calculated. For atoms in the middle of the periodic table, Cu, Fe, Ge, etc., the atomic masses are about 10^{-25} kg. The frequency of vibration, ν, of these atoms is therefore about 3×10^{12} cps, according to this model. More refined estimates of α place it a little higher than 25 newtons/m so that ν is nearer to 10^{13} cps. In the next section the amplitude of oscillation (at room temperature) is shown to be a few hundredths of an angstrom unit, i.e., a few per cent of the interatomic spacing.

The model of atom vibrations just described is adequate for many purposes. The assumption that the neighbors to a given atom are fixed is, of course, not true for real crystals: neighboring atoms move as well. When all possible motions of atoms and groups of atoms are considered, many other frequencies are found to exist, frequencies ranging from about the value of ν just calculated down to that of the fundamental acoustical mode of oscillation of the piece of solid. The vibrations at the lower frequencies are, in fact, just what we know as *sound waves*. Each separate mechanical oscillation at a given frequency is called a lattice vibration mode. In classical Newtonian mechanics, the amount of energy in a vibration mode is arbitrary, however in quantum mechanics this energy is quantized. A quantum of energy in a given vibration mode is called a *phonon*, in analogy to the photon which is the quantum of light energy at a given frequency. Thus the total displacement of any atom in the crystal is found by summing over all the phonons in the various modes. Although such summation can easily be carried out in principle, the calculation is complicated by the intricacy of the vibrational spectrum.

3-3 Thermal Energy of Solids

The most important ways in which a solid may absorb thermal energy are the following: (1) stimulation of atomic vibration, (2) stimulation of electronic motion or excitation, and (3) stimulation of molecular rotation. Mechanism (1) is common to all solids, since the atoms in all solids may be set into motion. It is also the most important of the three except in narrow temperature ranges where other effects may predominate.

The total energy of a solid consists of two parts, the thermal energy and any other energy which might exist at absolute zero; these two together are termed the *internal energy* E. This energy is a well-defined bulk property not much affected by small concentrations of crystalline defects. E is a function of temperature, of course, and it has been measured and tabulated for many solids. Historically, however, more attention has been paid to the temperature derivative of E, the heat capacity at constant volume,

$$C_v = \left(\frac{\partial E}{\partial T}\right)_v \tag{3-8}$$

The molar heat capacity C_v of solids ranges from zero to 6 or 8 cals/g-mole °K in the temperature regions of interest.† Furthermore, the heat capacity of solids is not constant with temperature, as is the case for an ideal monatomic gas (for which $C_v = 3$ cal/g-mole °K). A precise calculation of the heat capacity of a solid is extremely complicated, because such a calculation requires that the sum of both the kinetic and potential energies of every oscillating atom in the solid be known. We shall not carry out such a calculation. In fact it is not usually done in just this way: most calculations have been simplified by the use of extensive approximations about the behavior of the oscillating atoms. The most successful such theory is the Debye theory; the equation which results is plotted as the solid line in Fig. 3-4, along with data for C_v for Ag and Al. Clearly the fit of the theory to experimental observations is very good.

In Fig. 3-4 several temperature regions are of interest. At absolute zero the heat capacity is zero. At temperatures just above zero it climbs rapidly and is proportional to T^3 in this region. At high temperatures it reaches a nearly constant value of about

† Throughout this book we use the gram-mole (abbreviated g-mole).

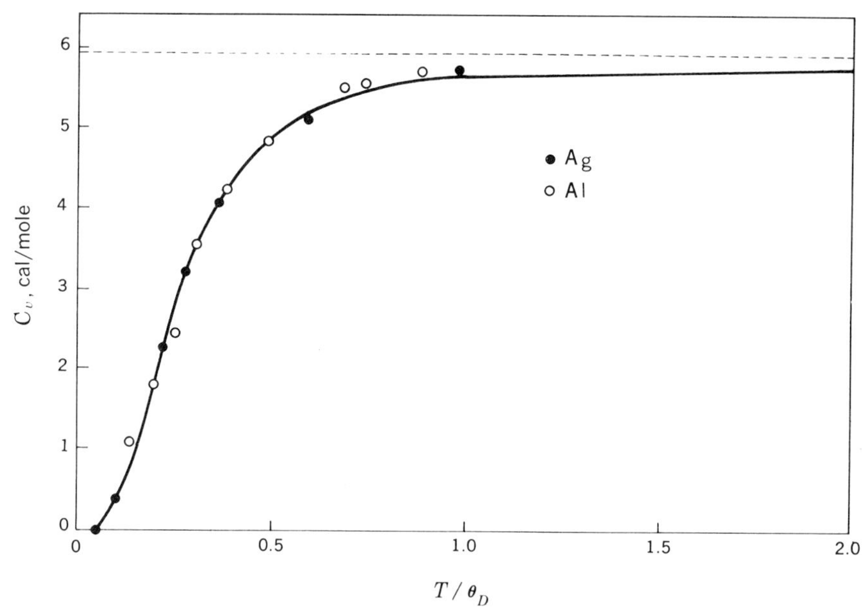

Figure 3-4 Comparison of the Debye heat-capacity curve and the observed heat capacities of Ag and Al. (*After Seitz.*)

6 cal/g-mole °K. Observe that this is a "reduced" temperature scale, that is, T divided by a constant. This constant θ_D, termed the *Debye temperature*, has a different value for each solid; at $T = \theta_D$ the heat capacity reaches about 96 per cent of its final value. A tabulation of Debye temperatures for a few materials is given in Table 3-1. By definition θ_D is related to the characteristic maximum vibration frequency of the atom ν_D, by the expression

$$\theta_D = \frac{h\nu_D}{k} \tag{3-9}$$

where h is Planck's constant and k is Boltzmann's constant. The quantity ν_D is in our approximation the quantity ν calculated using Eq. (3-6).

Table 3-1 Debye temperatures θ_D† of several materials

Metals:		Semiconductors:	
Hg	75	Sn (gray)	260
Pb	95	Ge	250–400
Na	160	Si	650
Au	170		
Sn	200		
Ag	230		
W	270	Insulators:	
Cu	340	H	100
Fe	360	AgBr	150
Al	375	NaCl	280
Be	1200	Diamond	1850

† Values of θ_D are given in degrees Kelvin.

In the early part of this century, heat-capacity data of the type portrayed in Fig. 3-4 were very puzzling. Calculations of classical mechanics gave the result that C_v should be a constant, independent of temperature, of magnitude 6 cal/g-mole °K. Thus, the calculations agreed with experiment at high temperatures but were in serious error at low temperatures. The trouble was resolved (by Einstein and Debye) with the application of quantum-mechanical ideas to the vibration of the atoms. The Debye temperature is an approximate dividing point between the high-temperature region, where the oscillators behave in the classical way, and the low-temperature region, where quantum effects are important.

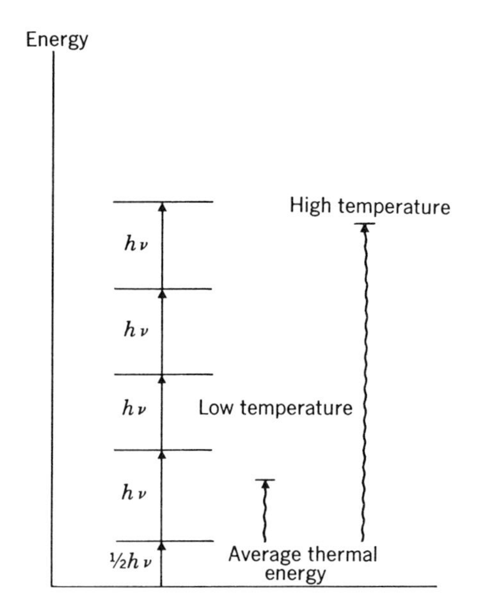

Figure 3-5 Thermal excitation of the energy levels of an oscillator. At low temperatures, the step between the lowest energy and the first excited energy is larger than the average thermal energy, kT. At high temperatures, kT is sufficient to populate many states, the discreteness of the levels becomes less important, and the system approaches the classical behavior.

While the details of the quantum effects cannot be shown in a simple way, their importance can be demonstrated physically. The heat capacity at low temperatures is small compared with its value at higher temperatures. Excitation of the oscillating atoms is relatively difficult at low temperatures, for the reason that the oscillating atoms may not have any arbitrary energy. Only discrete levels are allowed, and these are separated in energy by the factor $h\nu$ (Fig. 3-5). At low temperatures the raising of the energy of an oscillator with frequency ν from one level to the next is an infrequent occurrence, since the adjacent levels are far apart (about 0.05 ev for a solid like Cu) compared with the available thermal energy. Since these transitions are infrequent, few of the oscillators are excited, the heat capacity is low, and the internal energy is low. At high temperatures, i.e., temperatures above θ_D, the spacing between levels is relatively small compared with the thermal energy.

3-4 Heat Capacity of a Classical Solid

A detailed discussion of the heat capacity of a solid on a quantum-mechanical basis is not given here because interest throughout the book is generally in the region above the Debye temperature. However, the heat capacity can appropriately be considered in more detail from a classical view. Thereby more insight will be gained into the details of the oscillations of the atoms themselves.

The first step is to define the heat capacity of an oscillator. This is done by assuming that the heat capacity of the entire solid composed of N atoms may be divided equally among the $3N$ oscillators (each atom is considered to be three oscillators, since it may move in the three orthogonal directions). The problem then is to show why the heat capacity per oscillator should be $3R/3N$ ($R \approx 2$ cal/g–mole). To do this, we first discuss the heat capacity of the perfect gas, for the reason that the temperature scale is defined for a perfect gas. Thus, if the energy of the atoms of a perfect gas can be related to the gas temperature, those processes in the solid which absorb energy in an analogous way as the temperature of the solid is raised may be identified.

The equation of state of an ideal gas of volume V at pressure P and temperature T is

$$PV = RT \tag{3-10}$$

Its heat capacity is calculated by relating the pressure of the enclosed gas to its internal energy, then inserting this expression into Eq. (3-8). Not all steps are shown in full detail; some parts of the derivation are left as problems for the student.

Consider the pressure exerted by the gas on the walls of its container. If the area of one wall of the cubical container is 1 m^2, then the force on that wall is $F = P$. Suppose that N atoms of gas are enclosed; further let the motion be random so that $N/3$ of the atoms can be considered to be moving in the direction of each coordinate axis. If all atoms have the same speed v, they all have a common momentum p. At each collision of an atom with a wall, a transfer of momentum to the wall occurs of magnitude $2p$. Using Newton's law, force $= dp/dt$, one can write that for all N atoms

$$P = \frac{N}{3} mv^2 \tag{3-11}$$

where m is the mass of the atom. [Detailed calculation of Eq. (3-11) is left as a problem for the student.]

This expression can be converted to energy by observing that the kinetic energy E of each atom is $\frac{1}{2}mv^2$. Then Eq. (3-11) may be written

$$P = \tfrac{2}{3}NE \tag{3-12}$$

The value for P may then be substituted into Eq. (3-10) and the resulting expression rearranged to give

$$E_{\text{gas}} = \frac{3RT}{2N} \tag{3-13}$$

If N is Avogadro's number, the molar heat capacity C_v is just

$$C_v = \frac{3R}{2} \qquad \text{cal/g-mole °K}$$

or (3-14)

$$C_v = \frac{3k}{2} \qquad \text{cal/atom °K}$$

For an ideal gas, the heat capacity is independent of temperature. Its value of 3 cal/g-mole °K agrees well with measured values for the monatomic gases, and the thermal energy per atom for each coordinate of motion is $kT/2$.

The problem now is to derive for the solid an equation analogous to Eq. (3-14). Clearly such a derivation cannot proceed in the same way for the solid, since no atoms collide with the container walls and the pressure is zero. Although Eq. (3-14) might be thought to have no correspondence whatsoever with the solid, it has validity which is much more general than for the specific case for which it was derived. Every mode by which the system can absorb energy takes up $kT/2$ of thermal energy (as a high-temperature limit).

The energy modes of the harmonic oscillator are not difficult to determine. The oscillating atom has both kinetic and potential energy. Neither of these is constant; only the sum, the total energy E_T, is a constant. The kinetic energy during a vibration period varies from zero to E_T. Its average value indeed is just $E_T/2$, which also is the average value of the potential energy. Recall that, for the enclosed gas, the thermal energy per atom for each coordinate of motion is just $kT/2$. Recall also that for the gas all the thermal energy is kinetic energy; there is no potential energy. Suppose that, for an oscillator, the average kinetic energy $E_T/2$ has a value $kT/2$. Then the total thermal energy of each oscillator is kT, and the total thermal energy of the entire solid composed of N atoms is

$$E_{\text{solid}} = 3NkT \tag{3-15}$$

From this expression the molar heat capacity of the solid is

$$C_v(\text{solid}) = 3Nk \qquad \text{cal/g-mole °K} = 3R \qquad \text{cal/mole °K} \tag{3-16}$$

This expression gives the classical value of 6 cal/g-mole °K for temperatures higher than the Debye temperature. Observe that it is just twice the value of $3R/2$ for the classical gas, because the oscillator can also store heat as potential energy. An alternative way of stating the argument leading to Eq. (3-15) is that proposed earlier, that each method of energy absorption is capable of storing $kT/2$ of energy. Hence the thermal energy of a one-dimensional oscillator is $kT/2$ from kinetic energy and $kT/2$ from potential energy, and the thermal energy of the solid conceived as a set of $3N$ oscillators is again $3NkT$.

The reader should be cautioned that the remarks leading to Eq. (3-16) are qualitative in nature. The arguments are correct in principle, however.

The fact that the thermal energy of an oscillator has been shown to be about kT enables calculation to be made of the amplitude of the oscillation of an atom. At maximum displacement x_{max}, the energy is all potential. Since this energy is given by $\alpha x_{max}^2/2$ (see Sec. 3-2),

$$x_{max} = \left(\frac{2E_T}{\alpha}\right)^{\frac{1}{2}}$$

For an atom with a spring constant α of 25 newtons/m (as was estimated for α in Sec. 3-2), x_{max} is about 0.2 A at room temperature. This value agrees well with experimental measurements of atomic displacement by X-ray methods.

3-5 Fluctuations in Energy

Equations in the previous sections show that the thermal energy of a solid increases from zero at absolute zero to several hundred calories

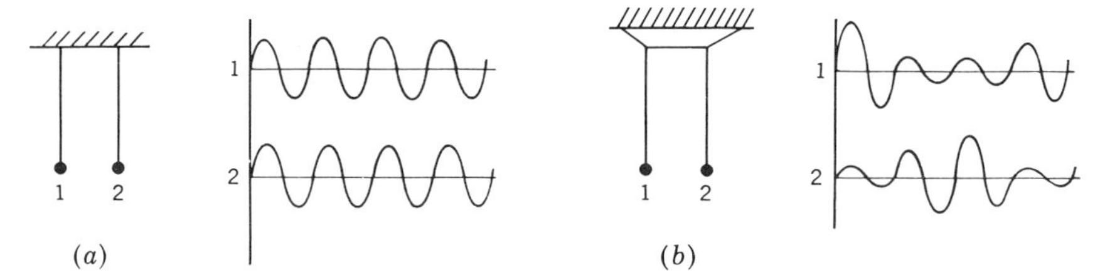

Figure 3-6 Oscillation of uncoupled and coupled vibrators.

per mole at room temperature. This is, on the average, about kT per atom per mode of oscillation (or about 0.025 ev per atom at room temperature). At first thought, one might suppose that the energy at elevated temperatures should be uniformly distributed, each atom always possessing the average amount, but this is not so. The vibrations of the atoms are characterized by fluctuations which occur in the energy of oscillation of a given atom.

The vibrations of the atoms of an actual crystal are complicated, but their qualitative behavior can be demonstrated simply by two ordinary pendulums. Let these pendulums be of the same length (Fig. 3-6a) and their motion be undamped. Then, if the two are set into oscillation with the same amplitude (hence the same energy), they swing indefinitely with the same energy. The total energy is fixed, and either of the two always possesses half (the average) of this total.

The situation is quite different, however, if the two pendulums are not independent but are coupled slightly. This coupling can conveniently be accomplished by connecting them to a wobbly support, e.g., through a string (Fig. 3-6*b*). If now the pendulums are started with some arbitrary phase difference, they do not continue to swing independently, for each senses the motion of the other through the supporting string, and the motion of each is affected by that of the other. The amplitude of each proceeds in time according to the sketches. Since the potential energy of either pendulum is proportional to the square of the amplitude at maximum, the energy of oscillation of either oscillates in time. Observe that (without damping) the total energy of the system is fixed and that the average energy per pendulum is just half this value. The instantaneous energy of either pendulum fluctuates periodically about this average (from zero to twice the average) and indeed has the average value only rarely.

(For electrical-engineering readers, a similar example is the oscillation of two coupled tank circuits. The current through each then has the appearance of the drawing in Fig. 3-6*b*.)

The vibrations of the atoms of an actual crystal are much more complicated than those in this simple example. Nevertheless, the physical principles are the same. As an atom vibrates, it exerts forces on its neighbors that affect the oscillations of the neighbors as well. The coupling between the two pendulums is represented in the crystal by these interaction forces between atoms. The motion of the atoms in the crystal ought then to behave qualitatively in the same way as the pendulums. Of course, the total energy of the crystal is fixed, but as time progresses, the vibrational energy of a particular atom fluctuates about the average energy in an irregular way. The only essential difference between the atoms of the crystal and the coupled pendulums is the vastly greater number of oscillating systems. Hence, instead of a regular periodic variation of the energy of an atom about its average value, the energy of an atom fluctuates in a random fashion because of the complexity of the system.

The energy fluctuations, if they are truly random, should be governed by the laws of chance. Thus, at any given instant of time, fluctuations of various amounts should occur in certain small regions of the crystal. It seems clear that large fluctuations from the average are less probable than small fluctuations, just as in tossing coins a long run of, say, heads is less likely than a short run of heads. The formula governing the probability of a fluctuation of a particular size, the Boltzmann equation, is

$$p(E) = A(T)e^{-E/kT} \tag{3-17}$$

Here $p(E)$ is the probability that a state with energy E is occupied by an atom, k is Boltzmann's constant, and T is the temperature. $A(T)$, a quantity characteristic of the entire system, is a function of both the temperature and other physical details in the system. It is included so that the integrated probability that the atom is in some state is unity. For any two states of the system having energies E_1 and E_2, the relative probability of occupancy of the states is

$$\frac{p(E_1)}{p(E_2)} = \frac{e^{-E_1/kT}}{e^{-E_2/kT}} \tag{3-18}$$

The range of fluctuations possible at a given temperature is important. A plot of Eq. (3-17) is given in Fig. 3-7; here $E = 0$ is

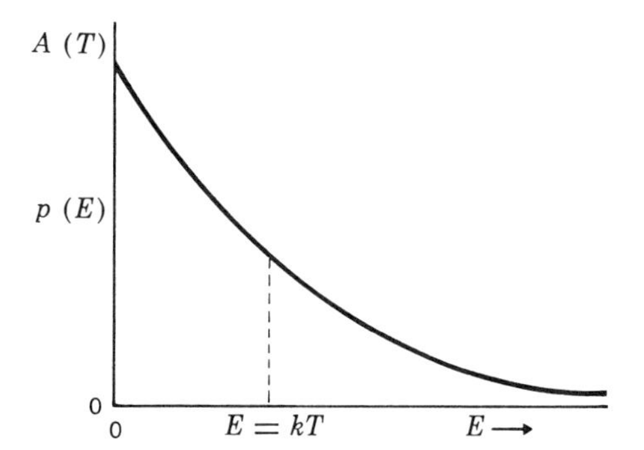

Figure 3-7 The Boltzmann probability function plotted as a function of the energy of the states.

taken as the energy of the lowest state. For $E = kT$ the value of $p(E)$ has fallen to some 37 per cent of its initial value, and for $E = 3kT$ it has fallen to about 5 per cent of its initial value. Hence fluctuations of more than a few kT are rare. Note that the chance of fluctuations of a given magnitude is higher as the temperature is higher. For this reason many physical and chemical processes have higher rates of reaction at high temperatures than they do at low temperatures. Important examples are diffusion in solids and conductivity in semiconductors.

Equation (3-17) was introduced to describe the energy fluctuation in time of a given atom, but it can also be interpreted in a slightly different, but equivalent, way. Since a solid is made up of atoms, each of whose vibrational energy fluctuates in just this way, Eq. (3-17) also describes the distribution of atoms in various energy states at any instant. The atoms can distribute themselves in the vibrational energy levels in a great variety of ways, of course, subject to the constraint that the total energy is constant. The term *entropy* is used as a measure of the number of ways in which this can be done. The entropy (denoted S) is a maximum when the number of ways of distributing the energy among the atomic vibrations is a maximum, i.e., when the randomness is maximum. A thermodynamic quantity called the *Gibbs free energy*, G, is defined in terms of the internal energy E, the pressure and volume of the system, and the entropy S.

$$G = E + PV - TS \tag{3-19a}$$

where T is the absolute temperature. The terms $E + PV$ are usually grouped together in a quantity called the *enthalpy*, H. For solids at pressures around one atmosphere, the term PV is usually small compared to E and TS. Therefore, one can neglect this term for solids near one atmosphere of pressure and use, if convenient, a thermodynamic quantity called the *Helmholtz free energy*

$$F = E - TS \tag{3-19b}$$

One is used to thinking of mechanical systems as being in their most stable state when the internal energy is lowest. For systems of low entropy, this is very nearly true. The exact condition for equilibrium for all systems is that the free energy should be a minimum.

3-6 Misplaced Atoms

The thermal vibration of atoms of a solid, important as it is, does not seriously disturb the perfection of the crystal. Each atom is, on the

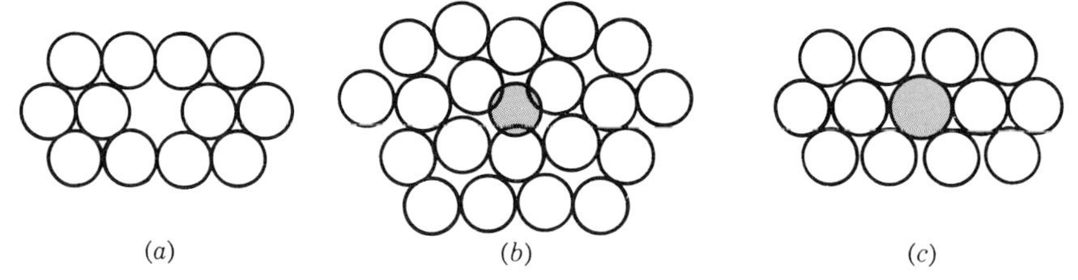

Figure 3-8 Three common point defects. (*a*) Vacancy; (*b*) interstitial; (*c*) impurity.

average, in its proper position. Each atom therefore has the requisite number of nearest neighbors at about the proper distance. Some of the other defects of the lattice fail to satisfy even these criteria. Atoms may have either the wrong number of nearest neighbors or the wrong nearest-neighbor spacing. Such defects may be classified into three groups on the basis of their geometry: point, line, and surface defects.

Point defects are lattice errors at isolated lattice points. For example, they may be *vacancies*, i.e., lattice sites at which atoms are missing. They may be *interstitials*, extra atoms inserted into the voids between the regularly occupied sites. Or they may be *impurities*, in which a lattice site is occupied by a foreign atom. These three defects are sketched in Fig. 3-8. Observe that they have dimensions of about an atomic diameter.

Dislocations are line defects having appreciable extension in only one dimension. Such defects are important in connection with mechanical properties of solids. A detailed description of dislocations is delayed until Chap. 6.

Surface defects are of two types, external and internal. The external type is just what its name implies, the imperfection represented by the discontinuity at the surface. Associated with the solid surface is a *surface energy*, which for metals is of order 1 joule/m^2. Internal surfaces exist where the crystal lattice changes from one orientation in space to another. Such boundaries are shown in cross section in Fig. 1-1 for the metal tungsten. These internal surfaces, too, have associated with them a surface energy of about 1 joule/m^2. Since the relative orientation of two adjacent crystal grains may have infinite variety, these grain boundaries may be infinitely varied. Most common solids are multigrain (i.e., polycrystalline) materials.

Since surfaces are abrupt changes in material density, in crystal structure, or in spatial orientation of grains, they exert a strong influence on many properties, mechanical, optical, and electrical. Their influence on mechanical properties is poorly understood and is not dealt with here. The chief features of the optical properties of surfaces are fairly well known; they are dealt with in later chapters. The effect of external surfaces on electrical properties is both pronounced and rather well known. The effect of internal surfaces is fairly weak and not well understood. Some of those features which are known are also discussed in later chapters.

3-7 Vacancies and Interstitials

Both vacancies and interstitials are defects of atomic dimension, so they cannot be seen with ordinary microscopes. The only instrument developed to date with which point defects can be observed directly is the field-ion microscope, in which vacancies on the emitter have been observed (see Fig. 3-9). Point defects play an essential role in some of the most important processes occurring in crystals. Consequently an extensive literature exists on indirect measurements of their properties.

Vacancies, which are considered first, are present in all crystals, no matter how carefully they may have been prepared. In fact, thermal fluctuations cause them to be produced and destroyed constantly in the crystal. Formally such a defect might be produced as in the two-dimensional crystal shown in Fig. 3-10. An interior atom might be plucked out of its regular lattice site and placed on the surface. Energy is required to perform this act. Since an accurate value of this energy E_v and the corresponding increase in enthalpy, H_v, are difficult to calculate, and since they can be measured accurately only with extreme care, the size of this energy expenditure is known only in a few cases. For most crystals it is of the order of 1 ev per vacancy (23,000 cal/mole).

Although accurate determination of the formation energy of a vacancy is indeed a difficult task, one can understand the physical origin of the main part. Suppose that, in the two-dimensional lattice of Fig. 3-10, each atom interacts only with its neighbors. To go back

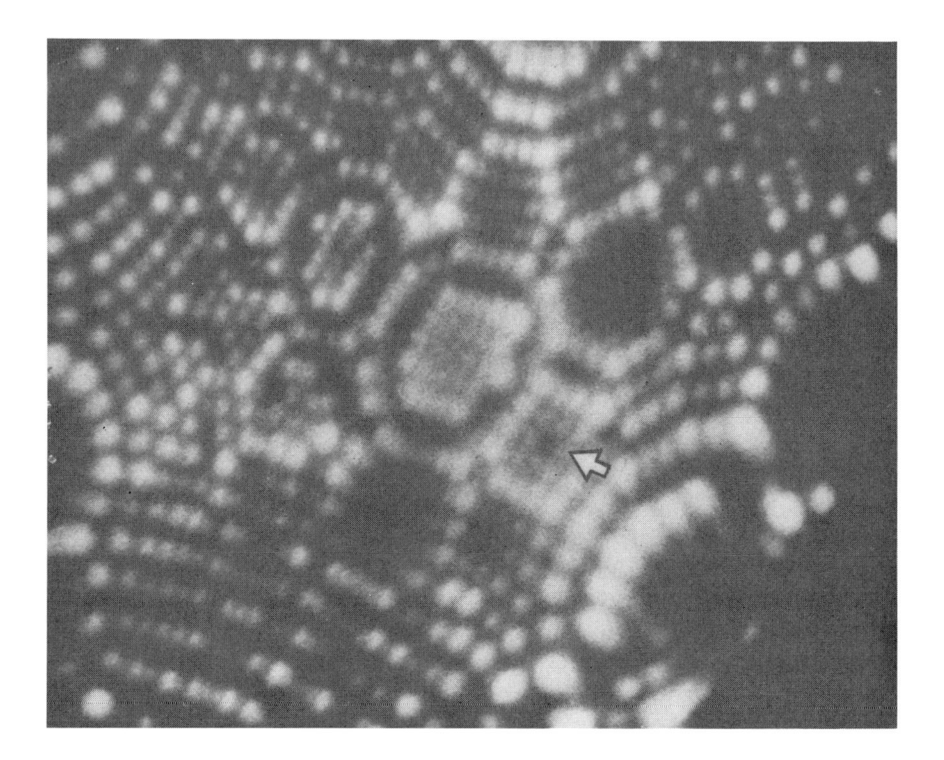

Figure 3-9 The arrow points to a vacancy in the (202) plane in the surface of a Pt crystal. A photograph made by E. Müller with the field-ion microscope. Magnification about 10^7 times.

to the discussion of Sec. 2-2, we call this interaction a *bond*. This is an extension of the concept of a true chemical bond, and it is not perfectly valid for a solid. Nevertheless its use enables us to carry

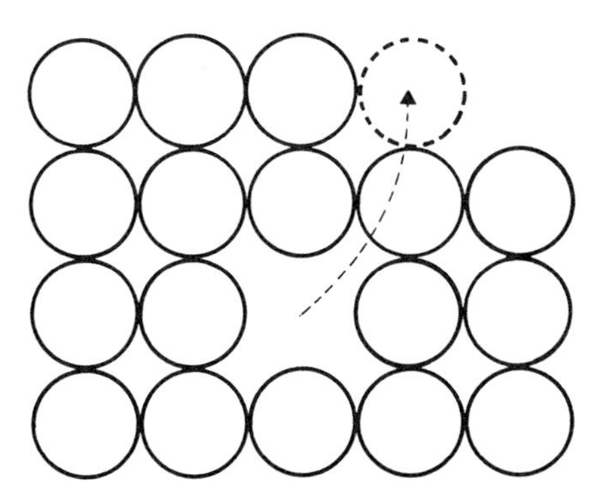

Figure 3-10 Geometry of formation of a vacancy in a cubic plane.

out a simple calculation of the vacancy energy. When an atom is plucked from the interior, four bonds are broken, and when it is replaced in the corner of the step on the surface, only two are re-formed. Hence the work necessary to make a vacancy is simply the energy of two bonds. In three-dimensional crystals, similar reasoning leads to a slightly higher number of unreplaced bonds, say, four to six. Since the energy of chemical bonds is of order $\frac{1}{2}$ ev per bond, this simple estimate gives energies of vacancy formation of order 2 to 3 ev per vacancy. This turns out to be an overestimate, because several factors have been disregarded. The chief neglected factor arises because atoms adjacent to the vacancy collapse slightly into it, reducing the energy to about 1 ev for ordinary metals. In the two-dimensional model, the vacancy formation energy with no relaxation is equal to the cohesive energy per atom. This result is a general principle. Where relaxations are important, however, the vacancy energy is reduced to perhaps half this value.

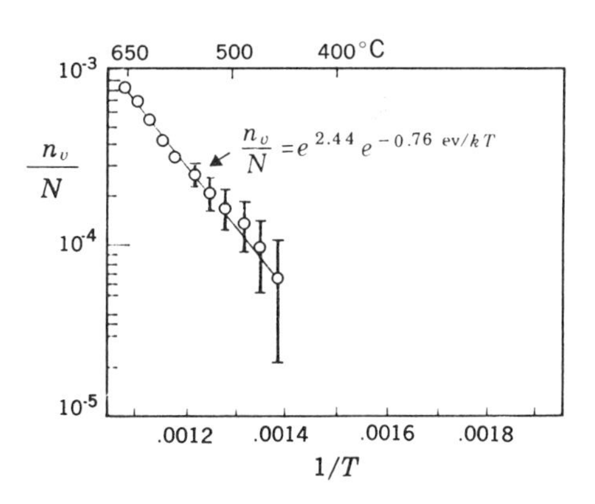

Figure 3-11 Plot of measured fraction of sites which are vacant in Al as a function of temperature. The slope of the line shows that the energy of formation of a vacancy in Al is about 3/4 ev. Note that the equation has an additional multiplying factor compared with Eq. (3-20). This result can be explained by a further refinement in the theory. (*Simmons and Balluffi.*)

The detailed way in which the crystal generates lattice vacancies is not that shown in Fig. 3-10. An atom in the interior of the crystal cannot leap to the surface, as has been proposed (unless it has the tremendous kinetic energy necessary to allow it to penetrate the intervening lattice). Nevertheless, if lattice sites in the crystal do have the requisite 1 ev of energy, vacancies will form. (This supposition is in the spirit of all thermodynamical calculations, which inquire only *what* equilibrium is, not *how* the system reaches equilibrium.)

Calculations of the thermal energy of atoms in a lattice show that the average vibrational energy of lattice atoms is much less than 1 ev at ordinary temperatures. Hence a lattice atom has the formation energy E_v only upon the occurrence of a large fluctuation of size E_v or greater. Since the relative probability of an atom having an energy E_v greater than the ground state is $e^{-E_v/kT}$, the chance that an atom site is vacant varies in the same way. In a crystal containing N atom sites, the number n_v of vacant sites is

$$n_v = Ne^{-E_v/kT} \tag{3-20}$$

Recent experiments show that this law holds remarkably well. Data for Al are presented in Fig. 3-11, where a quantity proportional to the logarithm of the number of vacancies is plotted as a function of $1/T$. The formation energy E_v is proportional to the slope of this line; the data of Fig. 3-11 yield a formation energy of 0.75 ev for a vacancy in Al. At the melting point of Al about 1 atom site in 10^3 is vacant; at room temperature only about 1 in 10^{12}. The fraction 1 in 10^3 is typical of solid metals at their melting points; the number at room temperature is even smaller for metals such as Ag and Cu, which have higher values of E_v and higher melting points. In spite of this relatively small concentration of vacancies, they play an essential role in several processes in crystals.

The astute reader will note that the experimental pre-exponential factor of the equation of the line drawn in Fig. 3-11 is not unity as is demanded by Eq. (3-20). The empirical number $e^{2.44}$ arises from the existence of an additional factor omitted in the derivation of Eq. (3-20), a factor called the *excess entropy of mixing*. From the point of view of statistical thermodynamics, Eq. (3-20) obtains its form from a balance between the internal energy and the entropy contribution to the free energy resulting from the addition of n_v vacancies in the crystal of N atoms. The increase in entropy which is used in development of Eq. (3-20) is only the mixing entropy, which can be approximated by the expression $\Delta S_{\text{mixing}} = -R(n_v/N)\ln(n_v/N)$. Other nonzero entropy contributions which might exist, $\Delta S_{\text{mixing}}^{\text{excess}}$ are simply added (usually in units of R by custom), and Eq. (3-20) must be modified to be

$$\frac{n_v}{N} = e^{\Delta S_{\text{mixing}}^{\text{excess}}/R}\, e^{E_v/RT} \tag{3-21}$$

In the present instance $\Delta S_{\text{mixing}}^{\text{excess}}$ is commonly called the vibrational entropy, ΔS_{vib}, since it is presumed to result from changes in vibrational energy levels of atoms in the solid resulting from changes in force constants between the atoms when vacancies are added to the perfect crystal. The size of this quantity cannot be calculated at the present time. In the instance shown in Fig. 3-11, the value of $\Delta S_{\text{vib}} = 2.44R$ (cal/mole °K) has been determined using the value of $n_{v/N}$ extrapolated to $1/T = 0$.

Interstitial atoms are extra atoms inserted into the lattice, but not on regular lattice sites (Fig. 3-8). They may be of two types: (1) atoms of the same type as the regular lattice atoms or (2) interstitial impurity atoms. Both types may exist for any lattice and indeed may coexist in the lattice.

Formation of the self-interstitial may be depicted as occurring by the process shown in Fig. 3-12*a*. An atom on the surface is transported into the crystal and placed in one of the interstices of the lattice. Again the energy of the crystal is increased when the interstitial is formed, because it is jammed into a region where the repulsive forces between the inserted atom and its neighbors are very large.

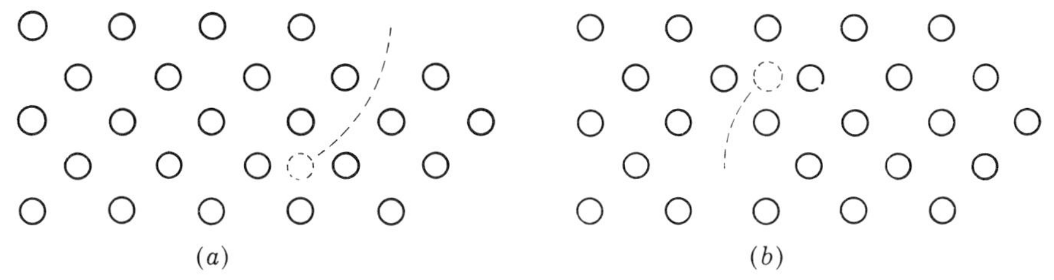

Figure 3-12 Formation of interstitials alone and in common with a vacancy.

(The energy of formation of interstitials is difficult to determine exactly.) In different lattices, say, metals as compared with semiconductors, the origin of the energy is different. In a tightly packed crystal like the fcc or hcp metal crystal, the energy of formation is large (perhaps 3 to 5 ev), whereas in an open structure like germanium the energy is smaller. The equilibrium number of interstitials, n_i, in a particular crystal (of N atom sites) is given by an exponential function,

$$n_i = aNe^{-E_i/kT} \tag{3-22}$$

where E_i is the formation energy of the interstitial and a is an integer (usually small) which gives the number of equivalent interstitial sites per lattice atom. In Prob. 3-6 the point is made that the larger formation energy of an interstitial compared with that of a vacancy makes the self-interstitial much less likely in most crystals than the vacancy.

An interstitial may be formed in a lattice by a lattice atom leaping directly into an interstitial site (Fig. 3-12*b*), leaving a vacancy behind. This defect, termed the *Frenkel* defect, has a formation energy equal to the sum of the vacancy energy and the interstitial energy.

The *impurity interstitial* exists when a foreign atom goes into a crystal in a lattice interstice. Since the formation energy of this interstitial is partly elastic-strain of the lattice, as was also true for the self-interstitial, small foreign atoms fit into the interstice easier than do big atoms. For the small impurity atoms, E_i should be low; the small atoms H, C, O, and N should therefore readily form interstitial defects in the metal lattices. In fact, they do so: metal crystals often absorb enormous quantities of these impurities. For example, Zr under proper conditions will absorb an atom of H for nearly every interstitial site.

Point defects often do not exist as isolated entities in crystals; often they are found in clusters. Consider, as an example, the presence in a crystal of clusters of divacancies, two nearest-neighbor vacancies. The thermodynamics describing the equilibrium number of divacancies is somewhat intricate, but if ΔS_{vib} for both monovacancies and divacancies are both zero, then equations describing the defect concentrations are relatively simple. They are (for defects in dilute concentration)

$$\frac{n_v}{N} = \exp\frac{-E_v}{RT}$$
$$\frac{n_{2v}}{N} = \frac{Z}{2}\exp\frac{-E_{2v}}{RT} \tag{3-23}$$

Here E_v is the formation energy of monovacancies (i.e., isolated vacancies), and E_{2v} is that of divacancies. The coordination number of the crystal, Z, is required to take account of the fact that a single lattice site may be a member of $Z/2$ crystallographically distinct divacancies. These two equations can easily be combined to give a single ratio of divacancies to monovacancies.

$$\frac{n_{2v}}{n_v} = n_v\frac{Z}{2}\exp\frac{E_{2v}+2E_v}{RT}$$
$$= n_v\frac{Z}{2}\exp\frac{E_2{}^b}{RT} \tag{3-24}$$

Here $E_2{}^b$ is called the binding energy of a divacancy, i.e., the energy necessary to dissociate a divacancy into two monovacancies in the crystal.

The concentration of the two defects depends on both E_v and E_{2v}, of course. If $E_2{}^b$ is zero, then $E_{2v} = 2E_v$ and the number of divacancies is based on random statistics of addition of vacancies

$$n_{2v} = n_v{}^2\frac{Z}{2}$$

However, $E_2{}^b$ is usually thought to be a positive quantity (for the metals at least) and n_{2v} is then greater than random. In Fig. 3-13, $\ln n_v$ and $\ln n_{2v}$ are plotted as a function of $1/T$ according to Eq. (3-23) in an fcc metal for which $E_v = 16{,}000$ cal/mole (0.7 ev per monovacancy) and $E_2{}^b = 4{,}500$ cal/mole (0.2 ev per divacancy).

3-8 Production of Point Defects

The preceding sections demonstrate that a crystal in thermal equilibrium possesses a finite number of vacancies and interstitials. In addition to point defects formed by thermal fluctuations, point defects may be created in the crystal by other means. One method of producing an excess number of point defects at a given temperature is by *quenching* from a higher temperature. If the temperature is lowered fast enough in the quench, some of the defects in equilibrium at the higher temperature may be retained at low temperature.

Another method of creating excess defects is by severe *deformation* of the crystal lattice, for example, by hammering or rolling. While the lattice still retains its general crystalline nature, numerous defects are introduced.

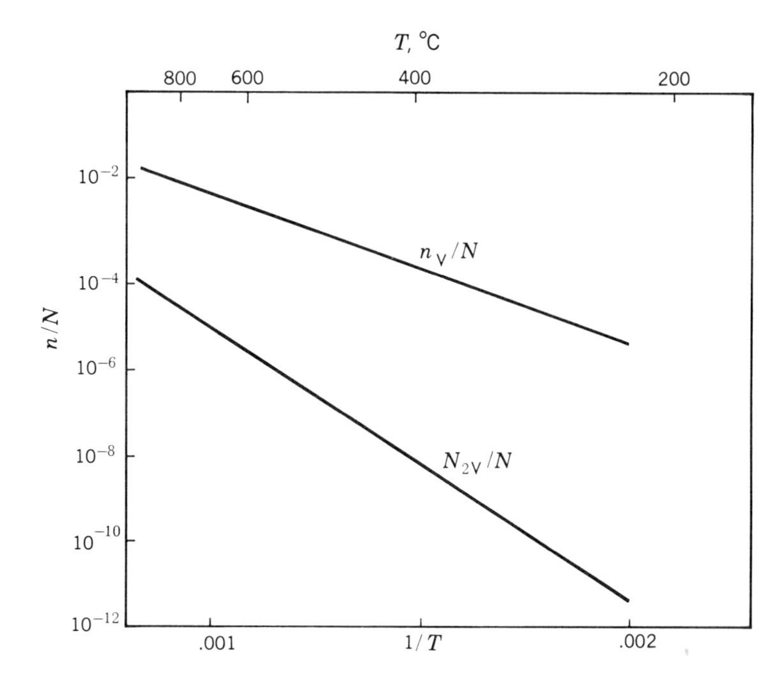

Figure 3-13 Fractional number of lattice sites occupied by single- and di-vacancies in an fcc metal for which E = 16,000 cal/mole and E = 4500 cal/mole.

A third method of creating excess point defects is by external *bombardment* by atoms or high-energy particles, for example, from the beam of a cyclotron or the neutrons in a nuclear reactor. The fast particles collide with the lattice atoms and displace them, thereby forming $v - i$ pairs, or Frenkel defects. The number produced in this manner is not dependent on temperature (except for annealing processes, which may destroy them) but depends only on the nature of the crystal and on the bombarding particles. An appreciable concentration of displaced atoms can be created by this bombardment, and significant changes in the properties of materials can be produced. So many defects can be produced in reactors, in fact, that concomitant changes must be taken into account in establishing safe operating conditions for some of the higher-energy power reactors and in selecting the materials out of which they are constructed.

3-9 Electronic Defects

Errors in charge distribution in solids are termed *electronic defects*. This does not imply that deviations from smooth uniformity of charge within the volume of a unit cell are defects. Far from it: the charge distribution is actually nonuniform. However, departures from the normal regularity of charge distribution or energy are electronic defects. Deviations of spatial distribution of charge commonly accompany the geometrical crystal defects just discussed. For example, an impurity atom may have a charge quite different from that of the host atoms and hence may produce a local electronic disturbance. Similarly a vacancy or interstitial is sure to produce electrical-charge deviations. In addition to these, the electrons in a geometrically perfect lattice may themselves move in such a way as to produce local fluctuations in charge. Furthermore, the electrons may absorb varying amounts of thermal energy so that their motion through the lattice is altered. Perhaps the most prominent example of this is the creation in semiconductors of positive- and negative-charge carriers; this effect is responsible for the operation of *p-n* junctions and transistors. These topics are treated in detail in later chapters.

REFERENCE

1. H. G. Van Bueren, *Imperfections in Crystals*, North Holland Publishing Company, Amsterdam, 1960.

PROBLEMS

3-1 The phonon of longest wavelength λ_{max} for a particular piece of a solid of length L has $\lambda_{max} = 2L$. The phonon of shortest wavelength λ_{min} has a wavelength about equal to c/ν, where ν has the value calculated by using Eq. (3-6). (c is the velocity of sound in the solid.) Calculate the approximate value of λ_{min} for a typical metal. How does this number compare with the interatomic spacing of the metal?

3-2 The classical theory of specific heats gives a heat capacity of 6 cal/g-mole °K. (*a*) Calculate the thermal energy of a mole of a metal at 300°K, using that value. (*b*) Au has a Debye temperature of about 170°K. From your knowledge of the shape of the Debye plot of heat capacity, estimate the true thermal energy of a mole of Au at 300°K. By about how much is the classical result in error?

3-3 Derive Eq. (3-11), using the assumptions outlined in the preceding paragraphs.

3-4 Equation (3-9) states that θ_D is proportional to ν_D (and hence about proportional to ν). Equation (3-6) states that ν is proportional to the square root of the force constant and inversely to the square root of the atomic mass. Hence θ_D should be high for light, stiff metals and low for heavy metals with low modulus. Is this generally true? Cite examples.

3-5 Show that the kinetic energy of an atom oscillating in simple harmonic motion has an average value (over one cycle) of half the total energy.

3-6 (*a*) Find the ratio PV to E for a perfect gas at room temperature and at one atmosphere of pressure.

(*b*) Find the same ratio for solid Cu, for which E has the value at room temperature of about 1200 cal/mole.

3-7 The specific heat at constant pressure is defined as

$$c_p = \left.\frac{\partial H}{\partial T}\right|_p$$

where H is the enthalpy of the material.

(*a*) Show that $c_p - c_v = R$ for a perfect gas.

(*b*) For a solid like Cu at room temperature, show that $c_p \approx c_v$.

3-8 (*a*) A vacancy in Al requires (according to Fig. 3-11) an expenditure of energy of about 0.75 ev. How many vacancies exist in thermal equilibrium at room temperature? At 550°C?

(*b*) An interstitial in Al requires an expenditure of energy of perhaps 3 ev. Calculate the ratio n_i/n_v at room temperature; at 550°C.

3-9 Suppose that a piece of Al is cooled rapidly from near its melting point so that all the vacancies characteristic of the high temperature are retained at room temperature. The number of vacancies at room temperature is in excess of the equilibrium number at room temperature. Suppose the vacancies "anneal out" in time so that the concentration comes into equilibrium at room temperature. If the annealing is adiabatic, does the solid heat up or cool down? By how much?

3-10 Formation or disappearance of vacancies in a solid causes its density to change. Excess vacancies in Al cooled from near its melting point anneal out in a few hours. Is the density change accompanying the annealing at room temperature an increase or a decrease? Is the change measurable by present experimental techniques?

3-11 One type of trivacancy in the fcc metals is a trio of nearest-neighbor vacancies lying in a {111} plane.

(*a*) How many crystallographically distinct types of vacancies of this type exist?

(*b*) Find an expression for the ratio of n_{3v}/n_v analogous to that of Eq. (3-24) for divacancies.

(*c*) Suppose that $E_v = 0.7$ ev and that $E_{3v} = 1.45$ ev. Make a plot of n_v/N and n_{3v}/N as a function of temperature similar to that of Fig. 3-13.

(*d*) Combining the results of part *c* with the data of n_{2v} from Fig. 3-13, plot the total number of vacant lattice points in the crystal as a function of T (or $1/T$).

3-12 A derivation of Eq. (3-20) can be made by using the fact that the free energy of a system must be a minimum at equilibrium.

Let E_v be the energy necessary to replace an atom in a crystal by a vacancy. Then n_vE_v is the increase in internal energy, dE, when n_v vacancies form.

The entropy increase dS when n_v vacancies form is given by the following expression

$$dS = k \ln w$$

where w is the number of distinct ways in which n_v indistinguishable vacancies can be distributed among N lattice sites. Therefore, at constant temperature, the change in free energy, dF, which accompanies the addition of n_v vacancies to the lattice is given by

$$dF = dE - T\,ds = n_vE_v - Tk \ln w$$

At equilibrium $d(dF)/dn_v = 0$.

(*a*) Show that $w = N!/n_v!\,(N - n_v)!$.

(*b*) Using Stirling's approximation, show that

$$k \ln w = k[N \ln N - n_v \ln n_v - (N - n_v) \ln (N - n_v)]$$

if $n_v \gg 1$, $N \gg 1$, $N - n_v \gg 1$, so that Stirling's approximation is valid.

(*c*) In terms of dE and dS write both dF and $d(dF)/dn_v$.

(*d*) By setting $d(dF)/dn_v = 0$, show that Eq. (3-20) results,

$$\frac{n_v}{N} = e^{-E_v/kT}$$

if $N \gg n_v$, so that $(N - n_v) \approx N$.

4

Diffusion

4-1 General

The tacit view taken in previous chapters was that any given atom has a particular lattice site assigned to it. Aside from thermal vibration about its mean position on the crystal lattice, the atom was assumed not to move. In actual fact the atoms do have more freedom, and they wander from one lattice site to another. This motion is called *diffusion*.

The fact that diffusion in solids occurs is strange, at first sight, because it is not a part of our everyday experience. A layer of silver atoms plated on steel knives and forks remains stable, apparently forever. Chromium plated on automobile bumpers diffuses into the underlying steel to no observable extent. In both these instances the random mixing of the surface and interior atoms is too slow at ordinary temperatures to produce easily observed effects. At high temperatures (1,000°C, for example), however, these platings are largely absorbed into the steel in a few weeks. Another example of a layered structure is the transistor. In such a device, impurity atoms must be arranged in carefully controlled geometry in the bulk germanium. These impurity layers can be formed by diffusion in a few minutes near the melting point of germanium. At room temperature, however, the geometry of this device is stable virtually forever.

This chapter has the purpose of pointing out the chief principles of the diffusion process. Emphasis is placed on two aspects: (1) the crystallographic and geometrical features and (2) the effect of temperature. Treatment of these topics also leads to consideration of the detailed atomic motions which occur during diffusion.

The detailed discussion begins with a short description of possible atomic motions which can result in diffusion. They are not equally probable, however, hence only the most important ones are considered in detail. Consideration of the energies of diffusion in substitutional structures by vacancy motion leads to an exponential temperature dependence which is verified by experiment. Next, the statistical averaging of random, successive atom jumps shows how long-range atom movements are related to the individual atom jumps. This statistical treatment results in formulas which can be directly correlated to massive transport of matter. A final section on diffusion in liquids points out that atom movements need not take place only on a crystal lattice: the same kind of statistical treatment is possible as long as the individual jumps of atoms are random both in length and in direction.

4-2 Crystallographic Features

Of course, one cannot look into a crystal and see the individual atomic motions which result in long-range atom movement. Nevertheless, several general crystallographic features are evident from a consideration of crystal geometry. First of all, diffusion almost certainly takes place in elementary steps of length approximately one atomic diameter, i.e., several angstrom units. The atoms move in discrete jumps from one lattice position to an adjacent one. These elementary jumps, when added together, permit the atoms to travel large distances.

Once these postulates are accepted, the question remains of the detailed mechanism by which the individual atomic jumps occur. Several possibilities exist: vacancy motion, interstitial motion, or some sort of atom-interchange mechanism. Since all three of these mechanisms are geometrically possible, they must all occur to some extent. The real question, then, concerns their relative importance in a given system at a given temperature.

Sketches of the motion of atoms necessary to allow diffusion by the methods cited above are presented in Fig. 4-1. Diffusion by *vacancy motion* occurs when lattice vacancies make interchanges with adjacent atoms. The sequence of events whereby a vacancy moves through the lattice is shown in Fig. 4-1*a*; clearly, as the vacancy moves from site 1 to 2 to 3 to 4, the atoms on sites 2, 3, and 4 make one jump each. The overall process obviously depends on two factors, generation of lattice vacancies and their subsequent motion. *Interstitial* diffusion is also a two-step process. It involves the jumping into an interstitial site of a regular lattice atom, from point 1 to point 2 of Fig. 4.1*b*, and the subsequent jumping of this interstitial atom into adjacent interstitial sites 2 to 3 to 4 to 5 to 6, etc. Diffusion by an *interchange* mechanism may occur in one of several ways. Simultaneous interchanges of two, three, four, or more atoms may occur,

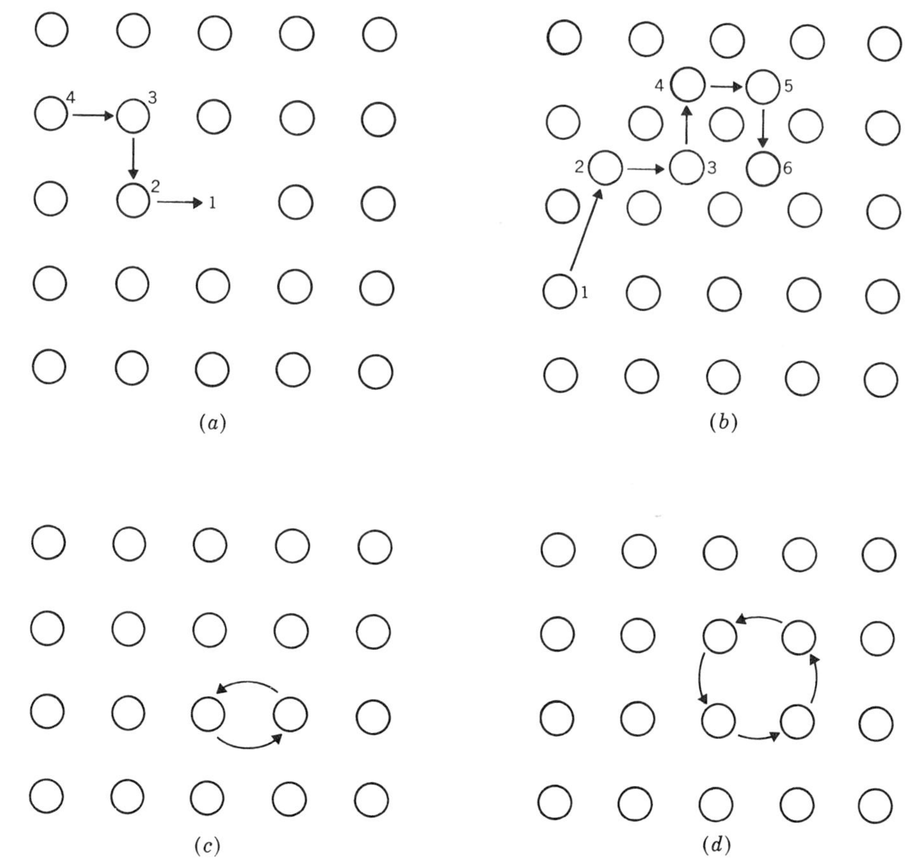

Figure 4-1 Atomic motions which result in diffusion. (*a*) Vacancy; (*b*) interstitial; (*c*) interchange of two; (*d*) ring of four.

and the interchanges of two and of four atoms in the cubic lattice are sketched in Fig. 4-1*c* and *d*.

Detailed theoretical calculations of the energy required for atomic motion to occur by the mechanisms cited above show that diffusion by vacancy motion should predominate in pure metals and substitutional alloys. Many experiments bear out this choice. Hence, succeeding sections examine the vacancy picture in detail. For one group of alloys, however, interstitial motion certainly predominates. This motion is the diffusion of small atoms dissolved in interstices in the parent crystal (such alloys are called *interstitial alloys*). Since diffusion of these interstitial atoms is the simplest to analyze in detail, it is considered first.

4-3 Diffusion in Interstitial Alloys

The rate of diffusion of the small interstitial atoms in alloys is controlled by the energy of motion of the interstitial atoms in the structure.

The elementary diffusion step in such alloys is illustrated in Fig. 4.2a. Since the interstitial sites at points a and b are equilibrium positions, they are energy minima. From symmetry considerations, one sees that a maximum in energy exists at point c midway between the equilibrium sites. The potential energy of the atom as a function of position is sketched in Fig. 4-2b.

The increase in energy E_m of the crystal when the atom is at the midpoint is mainly elastic-strain energy which results when the diffusing atom squeezes through the narrow space between the lattice atoms. For many of the alloys, E_m is about 1 ev. Since the average thermal energy of an atom at reasonable temperatures is no greater

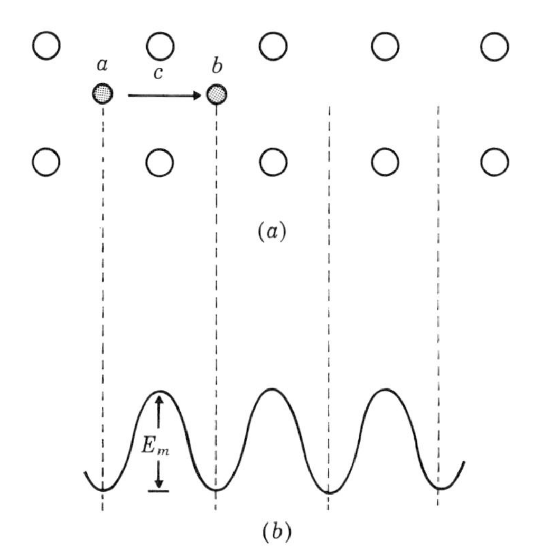

Figure 4-2 Interstitial diffusion. Potential energy of crystal as a function of the position of a diffusing atom.

than 0.1 ev, motion over the barrier requires a large fluctuation in energy. Thus, the frequency of jumping of an atom is controlled by the frequency with which energy fluctuations of sufficient magnitude occur.

An expression for the frequency of jumping f_m can easily be written. In its normal thermal vibration about the equilibrium site the atom strikes against the potential barrier ν times per second. Most of the time its energy is too low for it to pass over the barrier, but occasionally a fluctuation brings its energy to E_m, and it does cross the barrier. Since the fraction of the time which it has this requisite energy is $e^{-E_m/kT}$, the frequency of jumping to a neighboring lattice site is given, in order of magnitude, by $\nu e^{-E_m/kT}$. Since the atom can jump into Z equivalent neighboring sites, the total frequency of jumping out of the initial site is

$$f_m = Z\nu e^{-E_m/kT} \tag{4-1}$$

Recall that ν is about 10^{13} cps. For interstitial diffusion in bcc crystals, $Z = 4$; for fcc crystals, $Z = 12$. Since f_m depends exponentially on temperature, it changes rapidly with temperature. For example, for carbon diffusing interstitially in iron, E_m is about 0.9 ev; consequently at room temperature an atom of carbon makes about 1 jump in 25 sec, whereas at the melting point of iron (1545°C), it makes about 2×10^{11} jumps per second.

4-4 Diffusion by Vacancy Motion

The details of diffusional motion are a little more difficult to understand in substitutional alloys, where vacancy motion is responsible for the atomic jumps. Nevertheless the general principles are clear, and for this case an expression can also be derived for the frequency of jumping of a given atom.

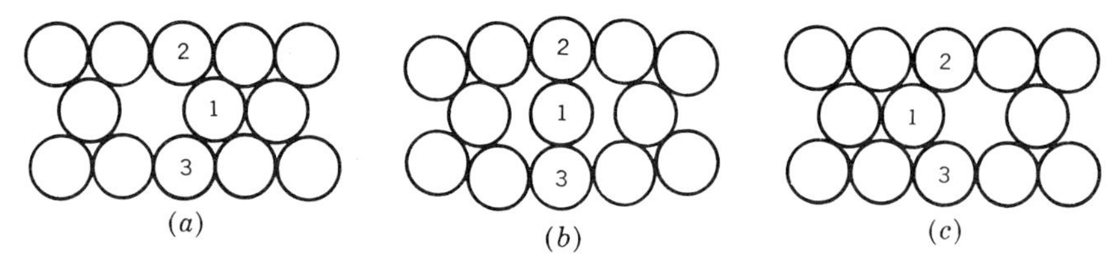

Figure 4-3 Motion of vacancy constrained to move in a close-packed plane.

The geometry of the vacancy-atom interchange gives the clue to the method of attack. Consider the successive steps in this interchange as they are portrayed in Fig. 4-3. Atom 1 in the configuration shown in (*a*) can exchange places with the adjacent vacancy in the close-packed plane and arrive in the configuration shown in (*c*). These two arrangements of atoms are geometrically, and hence energetically, equivalent. To go from the arrangement of (*a*) to (*c*), atom 1 must squeeze by atoms 2 and 3 in this plane; in the intermediate position (*b*) atoms 2 and 3 are displaced from their normal position, and the lattice locally has considerable strain energy. Hence the intermediate position (*b*) is a higher energy arrangement than (*a*) and (*c*).

The planar motion of an atom-vacancy interchange does not show everything. Above and below the plane of atoms of Fig. 4-3, other atoms also impede the interchange. Two other atoms in particular, one in the plane above and the other in the plane below, also squeeze against the jumping atom in the same way. The atomic arrangement around the vacancy in the fcc lattice (see Fig. 4-4) is such that the exchange of the vacancy, □, with atom 1 requires the diffusing atom to squeeze through the channel between atoms 2, 3, 4, and 5, pressing them apart as it passes. The potential energy as a function of position is sketched in Fig. 4-5. Again the height of the barrier is called E_m.

The motion of an atom next to a vacancy is then similar to the motion of the diffusing interstitial treated in Sec. 4-2. The atom strikes the barrier ν times per second and has enough energy to surmount it a fraction of the time given by $e^{-E_m/kT}$. An additional factor enters this calculation, however, for an atom has a vacancy adjacent to it in a given lattice site only a small fraction of the time, a fraction given by a second exponential $e^{-E_v/kT}$, where E_v is the formation

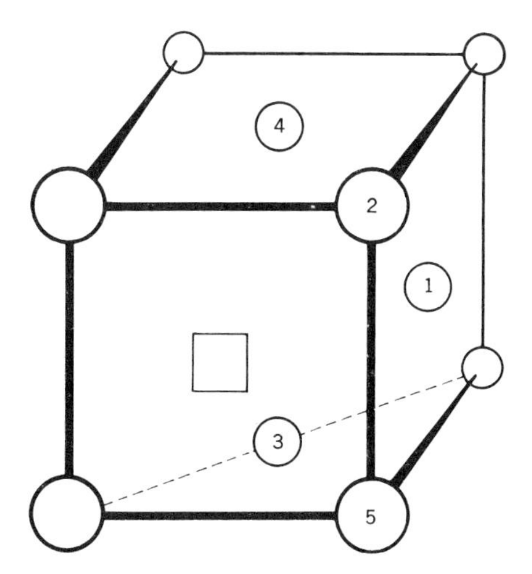

Figure 4-4 Exchange of a vacancy with an atom (atom 1) in an fcc structure. Four other atoms bear directly on the diffusing atom at the midpoint.

energy of a vacancy. The frequency of jumping, f, is proportional to the product of these factors and is, to an order of magnitude,

$$f = Z\nu e^{-E_m/kT} e^{-E_v/kT} \qquad (4\text{-}2)$$

where Z is again the number of equivalent neighboring sites.

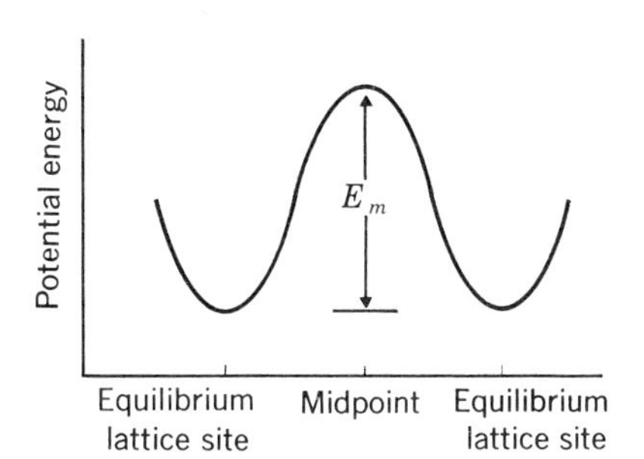

Figure 4-5 Potential energy of diffusing atom as a function of position.

Again f is extremely temperature dependent. E_v and E_m are both about 1 ev for metals such as Cu, Ag, Fe, etc., so that f for them is much smaller than the corresponding jump frequency for a typical impurity interstitial. For example, although an interstitial carbon atom in iron at room temperature jumps about once in 25 sec, a substitutional zinc atom in copper at room temperature makes only one jump in about 1,000 years.

4-5 Long-distance Motion

The calculations made in the preceding sections describe only the nature of the elementary jump process; they do not in themselves describe the long-range motion of atoms which have made many jumps. This latter problem is solved by consideration of the statistics of adding successive jumps.

The displacement of an atom after it has made a definite number of jumps will now be calculated. The calculation is made easier by the fact that the jumps are all equal in length (the atomic spacing) and that they take place in a highly symmetrical lattice. However, the motion of an atom is presumed to be completely random in the various permissible crystallographic directions; it may go forward,

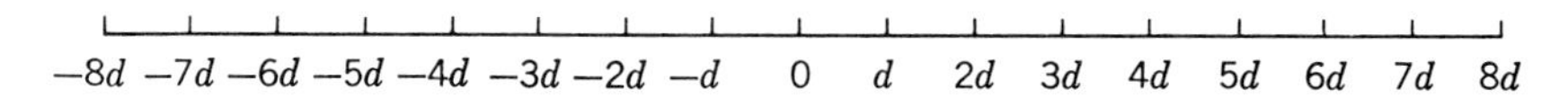

Figure 4-6 Coordinates of jumping for one-dimensional random jumping.

backward, or up and down so that its net motion after a certain number of jumps is not at all predictable. Hence, only the displacement averaged over a large number of diffusing atoms can be specified with acceptable precision.

The calculation can be made most simply by considering the motion of an atom in only one dimension. Suppose that an atom starts from point zero and makes successive jumps of length d, Fig. 4-6. The direction of each jump is completely random; it does not depend on anything that has occurred before. The net distance X covered by the atom after n jumps is simply the algebraic sum of all the individual jumps,

$$X = d_1 + d_2 + d_3 + d_4 + d_5 + \cdots + d_n \qquad (4\text{-}3)$$

where d_1 is the first jump, d_2 the second, etc. X is clearly a distance which is some multiple of length d, but it may be positive, negative, or zero. In fact, while X has the range $+nd$ to $-nd$, its average for many atomic jumps is exactly zero. This *simple* average is an algebraic average, and the fact that it is zero tells us only that the positive direction of jumping is not preferred over the negative direction. Other kinds of averages exist, however, which are not zero. For example, the root-mean-square (rms) average is a measure of the total nonalgebraic distance the atom travels from its starting point.

The mean value of X^2 (i.e., the average value of X^2 for many atoms each making n jumps) is not difficult to calculate. X^2 itself can be written

$$\begin{aligned} X^2 &= (d_1 + d_2 + d_3 + \cdots + d_n)(d_1 + d_2 + d_3 + \cdots + d_n) \\ &= d_1^2 + d_2^2 + d_3^2 + \cdots + d_n^2 + 2d_1d_2 + 2d_1d_3 + \cdots \qquad (4\text{-}4) \\ &\quad + 2d_1d_n + 2d_2d_3 + \cdots + 2d_{n-1}d_n \end{aligned}$$

The average value of X^2 (call it $\overline{X^2}$) is given by the sum of the averages of the individual terms. Clearly, each of the squared terms is equal to d^2, since $|d_1| = |d_2| = \cdots = |d_n| = d$. The terms $2d_1d_2, \ldots, 2d_{n-1}d_n$ are all equal to zero when averaged over the motion of many atoms, since each of the $d_1, d_2, \ldots, d_n$ has an equal chance to be positive or negative. Hence

$$\overline{X^2} = d_1^2 + d_2^2 + d_3^2 + \cdots + d_n^2 \tag{4-5}$$

This equation may be rewritten for convenience

$$\sqrt{\overline{X^2}} = \sqrt{nd^2} \tag{4-6}$$

This square-root expression shows that many jumps are required for the rms values of X to be appreciable. To recall an earlier example, consider the average number of jumps required for a Cr atom to diffuse 0.025 cm into the steel of an automobile bumper. The jump distance d is about 2.5 A, so that $\sqrt{\overline{X^2}}$ is about $10^6\, d$. Hence n must be about 10^{12}, and nd, the total linear distance, is about 250 m.

Equation (4-6) may be expressed in a somewhat different form for convenience. The number of jumps, n, may be expressed as a product of the frequency of jumping, f, and the time t required for the atom to make the n jumps, that is, $n = ft$. Then Eq. (4-6) may be rewritten

$$\sqrt{\overline{X^2}} = \sqrt{ftd^2} \tag{4-7}$$

Note that the quantity fd^2 is a property of the material and the temperature. Commonly a one-dimensional *diffusion coefficient* D is defined as

$$D = \frac{fd^2}{2} \tag{4-8}$$

Substitution in Eq. (4-7) yields

$$\sqrt{\overline{X^2}} = \sqrt{2Dt} \tag{4-9}$$

The rms distance covered by a diffusing atom varies as the square root of the time. The factor $\frac{1}{2}$ is inserted into Eq. (4-8) to make Eq. (4-9) agree with an equation which is derived in Sec. 4-6.

The preceding calculations are basically sound, but unfortunately they are not complete, since the motion of atoms is rarely limited to motion along a line. Ordinarily the atoms have access to three directions of jumping. Similar calculations can be made for the rms radial displacement $(\overline{R^2})^{\frac{1}{2}} = (\overline{X^2} + \overline{Y^2} + \overline{Z^2})^{\frac{1}{2}}$ in any direction from the initial point. The result is

$$(\overline{R^2})^{\frac{1}{2}} = (fd^2t)^{\frac{1}{2}} \tag{4-10}$$

where f is the frequency with which the atom changes position in the structure. In many problems in three-dimensional diffusion, interest is confined to the resultant diffusion displacement in one particular coordinate direction, although the atom jumps in other directions as well. Since, by symmetry, $\overline{X^2} = \overline{Y^2} = \overline{Z^2}$, the displacement $\overline{X^2}$ is given by

$$(\overline{X^2})^{\frac{1}{2}} = \left(\frac{\overline{R^2}}{3}\right)^{\frac{1}{2}} = \left(\frac{d^2ft}{3}\right)^{\frac{1}{2}} \tag{4-11}$$

From this equation the "three-dimensional" *diffusion coefficient* D is defined as

$$D = \frac{fd^2}{6} \tag{4-12}$$

again in terms of the jump frequency f and the atomic diameter d.† These two equations, (4-11) and (4-12), are adequate to describe diffusion in most of the cases to be discussed.

Values of D vary greatly for different solids at a given temperature. This variation is only in slight part caused by variations in d but is almost entirely a result of greatly different values of f. For the common metals (and for Ge and Si) at room temperature, D commonly lies in the broad range 10^{-20} to 10^{-50} m²/sec.

The expression for f in Sec. 4-2 can be used to show the explicit temperature dependence of D. It can be written (for cubic crystals)

$$D = Z\frac{d^2\nu}{6}e^{-(E_v+E_m)/kT} \tag{4-13}$$

Commonly the factors ahead of the exponential are grouped into a term called D_0 and the quantity $E_v + E_m$ is set equal to Q. Equation (4-13) then becomes

$$D = D_0e^{-Q/kT} \tag{4-14}$$

D_0 is often called the *frequency factor*, and Q is always called the *activation energy*.

Several features about this expression are of interest. First, it predicts that D depends on temperature exponentially whenever a single diffusion mechanism is operative. Such temperature dependence is found experimentally to be true for most solid substances (see, for example, the experimental data for diffusion of N in Fe, Fig. 4-7). Second, it predicts that D_0 should be a constant, independent of temperature, and that it should have a value of about 10^{-5} to 10^{-7} m²/sec. Actually D_0 does appear (from experiments) to be nearly independent of temperature, but experimental values of D_0 for many metals and semiconductors (and their alloys) are somewhat larger than 10^{-5}.

† Note that f in Eq. (4-12) is not the same as f in Eq. (4-8).

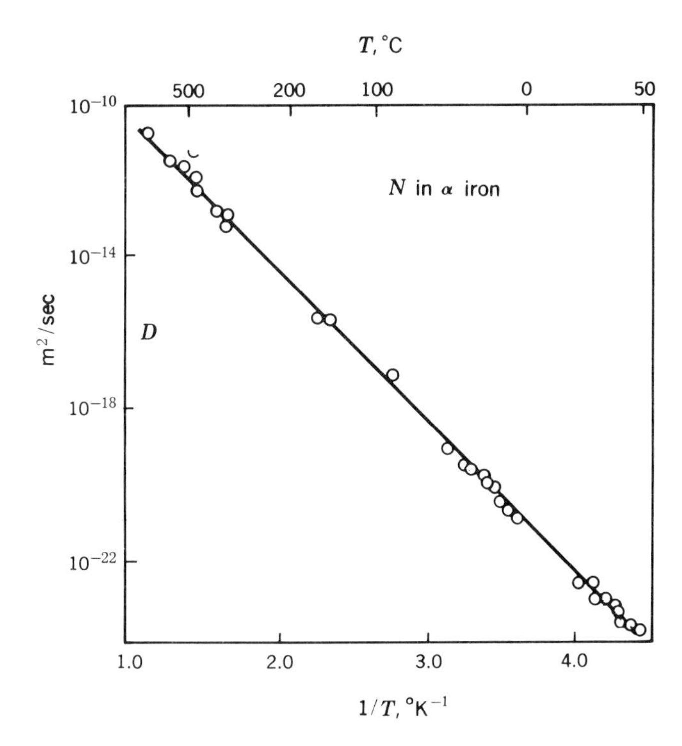

Figure 4-7 The diffusion coefficient of N in bcc Fe as a function of temperature. The data points for several experimental methods are shown. The data fit a straight line when plotted as log *D* versus 1/*T* over 14 cycles of 10.

The quantity Q is about 1 to 4 ev per atom for many materials of interest. About half the value of Q is the formation energy of the vacancy, E_v; the other half is the motion energy E_m. Experimental values of D_0 and Q for a number of materials are listed in Table 4-1.

Diffusion of impurities in the elemental semiconductors Ge and Si deserves some detailed comment. They divide into two general groups with respect to activation energy for diffusion. One group consists of such elements as Sb, As, In, P; they diffuse with an activation energy about equal to that of self-diffusion of Ge and Si in themselves. Another group including Li, Au, Ag, Cu, Fe, have an activation energy much lower—lower by at least a factor of 2. These elements may be interstitial in the semiconductor, in which case they would not require the presence of vacancies to allow atom movement. Cu, in fact, is anomalous in Ge even among these fast diffusers, having two separate domains of particular behavior. For additional comment on these points the reader is referred to the reference book *Diffusion in Semiconductors*, B. Boltaks, Academic Press, Inc. New York (1963).

4-6 Macroscopic Diffusion

The previous discussion has been concerned with the average motion of individual atoms. Historically the treatment of the individual atom motions has been preceded by another type of analysis, that which treats the bulk mass transport of atoms. This method, which might be called the *macroscopic approach*, is basically that of determining how fast atoms diffuse past a certain plane in the lattice. Fortunately, for the simpler types of geometry, an accurate calculation

Table 4-1a

Metals	D_o, m^2/sec $\times$ 10^4	Q, cal/mole
Fe in Fe (bcc)	1.4	56,500
Cu in Cu	0.69	50,200
Ta in Ta	2.0	110,000
Nb in Nb	1.3	95,000
U in U	1.8	27,500
Pt in Pt	0.14	66,100
Pb in Pb	0.28	24,200
Ag in Ag	0.4	44,100
Au in Au	0.09	41,700
Na in Na	0.24	10,500
Ni in Ni	1.30	66,800
Li in Li	0.23	13,200
Al in Al	0.18	30,100
Sb in Sb	1.05	39,500
Mo in Mo	0.5	97,000
Zn in Cu	0.34	45,600
Fe in Nb	1.5	77,700
C in Fe (bcc)	0.02	20,000
C in Fe (fcc)	0.15	34,000
Co in Mo	18.0	107,000
Nb in Mo	14.0	108,000
W in Mo	1.7	110,000
Re in Mo	3.2	113,000
N in Nb	0.01	34,900
H in Ni	0.005	9,600
H in Fe	0.0004	1,760
N in Fe (bcc)	0.005	18,350
Au in Ni	0.02	35,000

of this rate can be made in terms of the same atomic factors which were used previously.

For simplicity, atom movements are considered in the simple cubic lattice of Fig. 4-8, where a cut is shown along a cube face. Consideration is given to two adjacent planes of atoms, say, those which are indicated by numbers 1 and 2. Let the crystal be divided up into thin slabs of cross-sectional area L^2 and an atomic diameter d thick. Let some of the atoms in each plane be impurity atoms, say, N_1 in a plane 1 and N_2 in plane 2 (these might be, for example, As in Ge). When diffusion occurs, atoms move past plane A in both directions. Under certain circumstances a net flow exists in one direction or the other; it is this net flow which we wish to calculate.

For convenience several conditions and two new variables are

Table 4-1b

Semiconductors	D_o, $m^2/sec \times 10^4$	Q, cal/mole
Ge in Ge	7.8	69,000
Si in Si	32.0	98,000
As in Ge	6.0	57,500
Sb in Ge	12.0	57,500
P in Ge	2.0	57,500
Zn in Ge	0.4	57,500
In in Ge	0.06	57,500
Li in Ge	0.002	11,800
Ag in Ge	0.04	23,000
B in Si	10.5	85,000
Al in Si	8.0	80,000
Sb in Si	5.6	91,000
Ag in Si	0.002	36,800
Li in Si	0.0094	18,100
Ga in Si	3.6	81,000
In in Si	16.5	90,000
As in Si	0.32	82,000
Bi in Si	1036.0	107,000
Ni in Si	1000.0	97,500
Cu in Ge (above 750°C)	0.0002	4,100
Cu in Ge (below 750°C)	0.04	23,000
Au in Si	0.0011	25,800

introduced. C_1 is defined as the volume concentration of atoms in plane 1 and C_2 that in plane 2. Then $C_1 = N_1/L^2d$ and $C_2 = N_2/L^2d$, where L^2d is the volume per atom plane. The jump frequency f is also reintroduced. Each impurity atom makes one jump in $1/f$ sec on the average. The atoms are first restricted to jumping either to the right or to the left; i.e., after one jump an atom in plane 2 is either in plane 1 or in plane 3. (This restriction is relaxed later to show the effect of crystal type.) In a single jump half the N_1 atoms in plane 1 jump in either direction and so, during the time $1/f$, $\frac{1}{2}N_1$ atoms pass

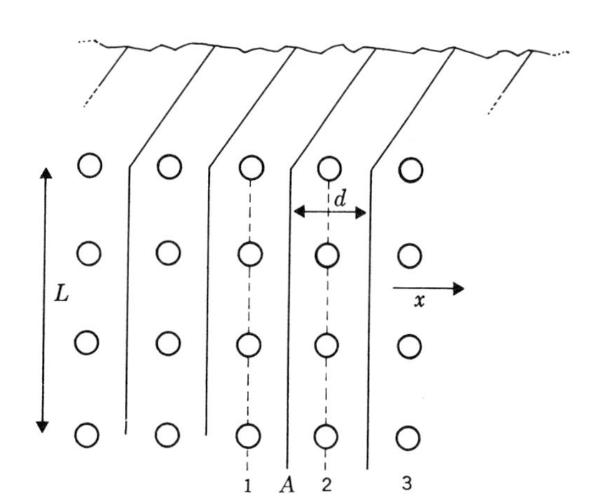

Figure 4-8 Geometry of diffusion in a bulk simple cubic crystal.

the plane A to the right. Similarly $\frac{1}{2}N_2$ atoms go past plane A to the left in that period of time. The net flow of atoms per second, dN/dt, over plane A is then just the difference between these two.

$$\frac{dN}{dt} = -\frac{f}{2}(N_2 - N_1)$$

$$= -\frac{f}{2}L^2 d(C_2 - C_1) \qquad (4\text{-}15)$$

where dN refers to a net flow of atoms to the right and $C_1 > C_2$. (Any consistent scheme of signs could be used.)

The quantities C_1 and C_2 can be replaced by using the concentration gradient dC/dx. Then

$$C_2 - C_1 = \frac{dC}{dx}d$$

so that

$$\frac{dN}{dt} = -\frac{f}{2}L^2 d^2 \frac{dC}{dx}$$

or

$$\frac{1}{L^2}\frac{dN}{dt} = -\frac{fd^2}{2}\frac{dC}{dx}$$

Again let

$$\frac{fd^2}{2} = D$$

Then

$$\frac{1}{L^2}\frac{dN}{dt} = -D\frac{dC}{dx} \qquad (4\text{-}16)$$

In Eq. (4-16), the left-hand side is the net flow of atoms per second over a plane of unit area at point A; this is the flux of atoms per second. Observe that the constant D has just the same form as it did in the "statistical" approach earlier in the chapter for unidirectional flow [Eq. (4-8)]. Observe also that Eq. (4-16) is exactly analogous to the well-known formula for heat conductivity,

$$\text{Heat flux (cal/sec unit area)} = -K\frac{dT}{dx}$$

where dT/dx is the temperature gradient and K is the thermal conductivity.

Taking account of three-dimensional jumping of the diffusing atoms is a simple matter. Consider again the simple cubic structure. Each atom may change places with any one of the six nearest neighbor atoms (see Fig. 4-9). Only one of these six jumps of an atom in plane

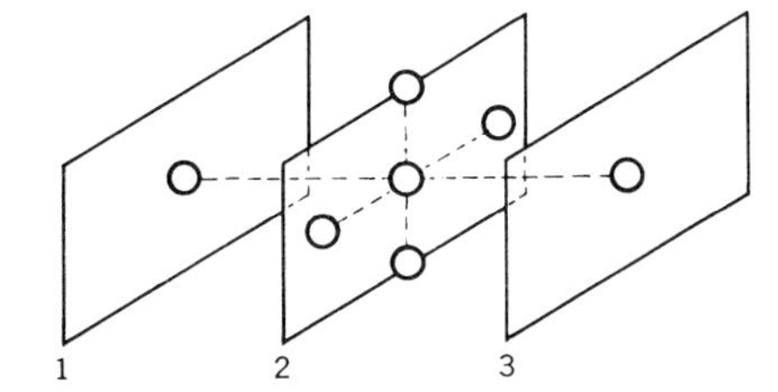

Figure 4-9 Geometry of six nearest-neighbor exchanges for simple cubic lattice.

2 carries the atom into plane 1. Similarly only one of six possible jumps of an atom in plane 1 carries the atom into plane 2. Thus Eq. (4-15) is modified to be

$$\frac{dN}{dt} = -\frac{f}{6} L^2 d^2 (C_2 - C_1) \tag{4-17}$$

and

$$\frac{1}{L^2}\frac{dN}{dt} = -D\frac{dC}{dx} \tag{4-18}$$

where

$$D = \frac{d^2 f}{6}$$

in agreement with Eq. (4-12). Equation (4-18) is commonly known as *Fick's law.*

In Sec. 4-2 the statement was made that the motion of vacancies is the primary mechanism of diffusion in pure metals and substitutional alloys. The reason for the correctness of this statement can now be made more clear by using the concepts of macroscopic diffusion. A classical experiment by Smigelskas and Kirkendall† provided the proof. They welded a piece of pure Cu to a piece of brass (70 per cent Cu–30 per cent Zn) and permitted interdiffusion of Cu and Zn to occur at high temperature. They found that Zn atoms diffused past the welded interface in one direction faster than Cu atoms diffused in the other direction and that the original interface moved into the brass. In addition, voids formed in the alloy on the brass side of the interface. They interpreted these observations to mean that a net flux of vacancies from the Cu to the brass caused the creation of extra lattice planes on the Cu side of the interface and destruction of atom planes on the brass side. In addition the presence of voids was interpreted by them to result from a condensation of excess vacancies in the brass.

This measurement and many others like it show that several parameters are needed to specify atom movements in the presence of concentration gradients. Except on very special planes or in very special systems, the flux of atoms of one kind in one direction is not numerically equal to the flux of atoms of the other kind in the other

† Smigelskas and Kirkendall, *Trans. AIME,* **171:** 130 (1947).

direction. In binary systems for one-dimensional diffusion, these fluxes, call them J_A and J_B, are related to the vacancy flux J_v by the equation

$$J_A + J_B = -J_v$$

If the fluxes of the diffusing species are related to the local concentration gradients through Eq. (4-18) that is, $J_i = -D_i\, \partial C_i/\partial \chi_i$, then separate diffusion coefficients exist for each species.

As an example, consider Cu and Ni diffusing at a 50-50 concentration at 1000°C. The two diffusion coefficients are found to be

$$D_{Ni} = 2.8 \times 10^{-15}\ \text{m}^2/\text{sec}$$

and

$$D_{Cu} = 9 \times 10^{-15}\ \text{m}^2/\text{sec}$$

Thus the flux of Cu atoms is numerically a little more than three times that of Ni atoms; the difference is made up by an accompanying flow of vacancies.

Prior to 1945, most determinations of diffusion coefficients were

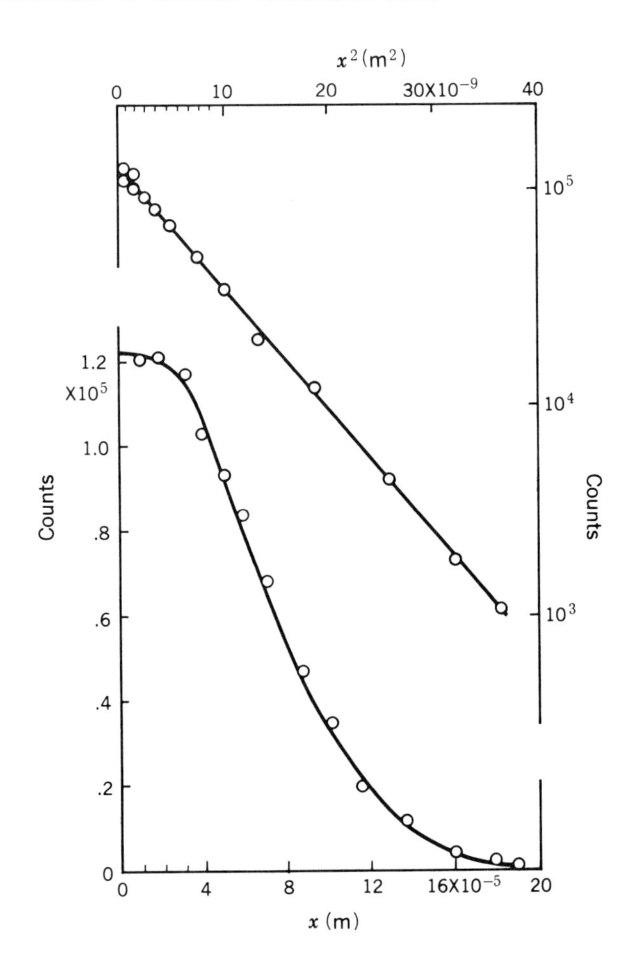

Figure 4-10 Radioactivity as a function of distance for diffusion of Fe^{59} in an equiatomic alloy of FeCo. Diffusion of the radiotracer was carried out for 18 x 10^5 seconds at a temperature of 1129°K. The data are drawn both as a function of x and of x^2. From the slope of the line in the upper part of the figure, the value of D (1129°K) can be calculated to be 1.04 x 10^{-15} m^2/sec. Data of Dr. Stephen Fishman, University of Illinois at Urbana-Champaign.

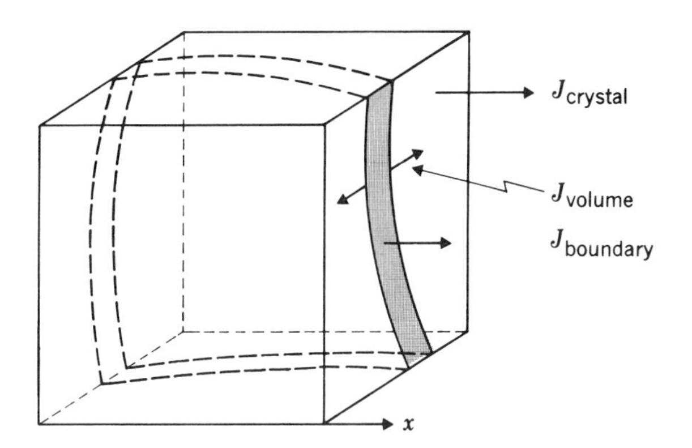

Figure 4-11 In the presence of a boundary, the total diffusional flux is the sum of the crystallographic and the boundary flux. The effective area for volume diffusion is nearly the area of the solid, that for boundary diffusion is the linear extent of boundary times some effective thickness t, usually considered to be 5 to 10 Angstrom units.

made by chemical analysis of interdiffusing pure metals or selected alloys. Usually a single diffusion coefficient was determined. Such determinations of values of D are now known to be imprecise, and almost no measurements of that type are now made. Rather, use is made of radiotracers, and separate values of D are determined for each of the diffusing species. In a typical experiment, a layer of radioactive atoms is evaporated or electroplated on the clean, flat surface of the solid for which knowledge of the diffusion coefficient is desired. The solid is then heated to the temperature at which D is to be determined for a time t long enough for the radioactive atoms to diffuse into the surface an appreciable distance (say, $X_{\text{rms}} \approx 0.5$ mm). The concentration of radioactive atoms is then measured as a function of depth into the solid. This is done by counting the radioactivity of thin slices of the solid, which are removed by a machining operation.

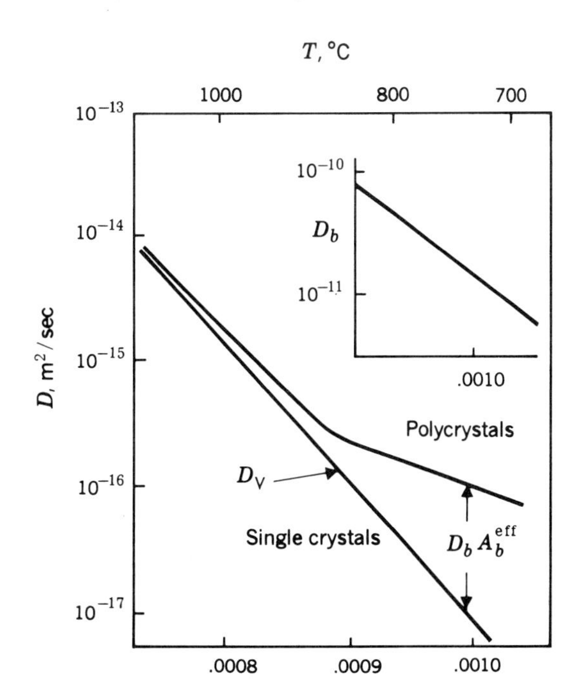

Figure 4-12 Lattice and grain boundary diffusion of gold in nickel. The polycrystals had an average grain diameter of 5×10^{-6}m. After Chatterjee and Fabian (*Jour. Inst. Metals*, **96**, 186 L968).

The activity as a function of distance in an actual experiment is shown by the data points in Fig. 4-10. These data should fit the equation

$$a = \frac{A}{\sqrt{t}} e^{-X^2/4Dt} \tag{4-19}$$

where a is the specific activity at depth X from the surface after a diffusion time t (see Prob. 4-3 for discussion). The constant A is proportional to the number of radioactive atoms initially deposited on the surface. Equation (4-19) fits the data of Fig. 4-10 for a diffusion time of 18×10^5 sec if D has the value of 1.04×10^{-15} m²/sec. A may be selected so that the data fit the curve at $X = 0$.

Diffusion by means of crystal vacancies predominates in most instances of diffusion in pure elements and substitutional alloys. Nevertheless, atom flux by other mechanisms may predominate under certain conditions. In particular, crystal surfaces, both internal and external, provide continuous paths through which atoms can diffuse readily. The total macroscopic flux of diffusing atoms is then the sum of two terms:

$$J_{\text{total}} = J_{\text{crystal}} + J_{\text{boundary}}$$

$$D\frac{\partial c}{\partial x} = -D_v \frac{\partial c}{\partial x} - D_b A_b^{\text{eff}} \frac{\partial c}{\partial x} \tag{4-20}$$

Here D_v and D_b represent the volume diffusion coefficient in vacancies and the boundary diffusion coefficient, respectively. A_b^{eff} is the effective cross-sectional area of a boundary (see Fig. 4-11).

Results of an experiment showing the effect of boundary diffusion are given in Fig. 4-12 for Au diffusing into single and polycrystals of Ni. In the single crystals, only the first term, D_v, of Eq. (4-20) is important, and a single straight line on this logarithmic plot is seen. This line has the equation

$$D_v = 2 \times 10^{-6} e^{-55{,}000/RT} \text{ m}^2/\text{sec}$$

In the polycrystals one sees the effect of both of the fluxes: the term D_v predominates at high temperatures and the term $D_b A_b^{\text{eff}}$ at low. Analysis of the grain boundary term gives the result for D_b given in the inset of Fig. 4-12, namely

$$D_b = 3 \times 10^{-4} e^{-30{,}000/RT} \text{ m}^2/\text{sec}$$

An approximate value of the effective width of the grain boundary can be calculated using the data in Fig. 4-12. For example, at 725°C, $D_b A_b^{\text{eff}} \approx 10^{-16}$ m²/sec. For the grain diameter of 5×10^{-6} m, a simple calculation gives the effective width to be about 10 angstrom units (see Prob. 4-13).

In actual fact, the simple treatment considered here overlooks one important feature of mixed boundary and crystallographic diffusion. Lateral diffusion in and out of the boundary occurs readily since lateral concentration gradients develop as diffusion proceeds. The solution to the problem is thus not as simple as Eq. (4-20) implies, and several assumptions and approximations must be made to extract D_v and D_b from measured plots of isotope penetration. The reader is referred to the paper by Chatterjee and Fabian for discussion of these points.

Other types of paths of high diffusivity exist, such as free surfaces, low-angle boundaries and individual dislocations. With appropriate experiments and interpretation, both the diffusion coefficients and effective areas may be determined.

4-7 Diffusion in Liquids

Diffusion usually occurs in liquids more rapidly than in the solid phases of the same substances. Experimental observation of its occurrence is difficult, because it is easily masked by massive convection currents, which are not diffusion at all. The importance of liquid diffusion has unfortunately been underestimated in the past. It is of great importance in the purification of Ge and Si and is essential both in zone refining (a process used in purification of Ge and Si) and in several processes used for making *p-n* junctions and *p-n-p* transistors, as we shall see in later chapters.

Two aspects of diffusion in liquids are strikingly different from diffusion in solids. First, the geometry is not so simple, and, second, diffusion is very rapid in liquids compared with that in solids at the same temperature. These two points are considered in the following paragraphs.

The geometry of diffusion in liquids is not nearly so well understood as it is for solids, because the detailed geometry of atomic arrangement in liquids is imperfectly known. Experiments do show that some geometrical order exists in liquids as far as nearest neighbors are concerned; i.e., a given atom has approximately the same number of nearest neighbors, and they are at about the same distance as nearest neighbors of a crystal of the same substance. Longer-range correlations between atoms in liquids are much lower than those for solids, however. This lack of complete understanding of the geometry of liquids makes impossible the definition of a unique path length for the elementary diffusion jump in the same way as was done for solids in Fig. 4-1. Nevertheless, by supposing that a range of path lengths exists and by carrying through an analysis of the rms distance covered by atoms in a certain time one can write an equation analogous to Eq. (4-6) (for one-dimensional motion),

$$\overline{X^2} = 2D_L t \tag{4-21}$$

where D_L is a quantity again called the diffusion coefficient. The important fact here is that the rms distance covered is proportional to $t^{\frac{1}{2}}$, as it is for one-dimensional solid diffusion. Similarly, the macroscopic diffusion of atoms in a concentration gradient obeys an equation of type (4-16). Hence it is reasonable to suppose that an adequate description of the rate of diffusion in liquids can be made by specifying a diffusion coefficient D_L.

The fact that diffusion in liquids is not so well understood geometrically as that in solids does not mean that Eq. (4-20) is less valid than the corresponding equation (4-9). The derivation of Eq. (4-21) does not, in fact, depend on exact knowledge of the geometry of liquids. It does require that the individual jumps be random in direction and also not biased in length in either the one direction or the other. They may be of equal length, as is true for solids, or of random length as well. The first derivation of Eq. (4-21), in fact, was made by Einstein to describe the Brownian motion of fine particles in water. Both the direction and the jump length were presumed to be random. The diffusion coefficients of Brownian particles can be calculated by fitting the experimental data to the theory [using Eq. (4-18) and somewhat similar equations for two and three dimensions]. They are rather low compared with those for diffusion of atoms in liquids.

The difficulty of ascribing a detailed geometry to the individual jumps in diffusion in liquids is accompanied by a second difficulty: an explicit expression for the jump frequency cannot be written. In fact it seems certain that no unique jump frequency exists. Hence D is not written in terms of an activation energy, since a clear meaning cannot be assigned to it. Experimental data do show some temperature variation of D with apparent values of Q much less than 1 ev. One interesting feature of all the data for simple liquids is that all values of D are in the range 1×10^{-9} to 10×10^{-9} m^2/sec near the melting point of the liquid. This fact is apparent from observation of the data in Table 4-2. One might, apparently, not be far wrong in supposing that for simple liquids and for simple atoms or small molecules diffusing in them (i.e., not high polymers in either case) the diffusion coefficient at the melting point is of the order 3×10^{-9} m^2/sec. This value is much higher generally than that of solid diffusion just below the melting point, as can be demonstrated from the data of Table 4-1.

Table 4-2 Diffusion coefficients of some elements and compounds in a number of liquids

Diffusing substance	Diffusion medium	T, °C	D, 10^{-9} m^2/sec
Mg	Al	700	7.5
Si	Fe	1,480	2.4
Au	Pb	500	3.7
Ag	Sn	500	4.8
$AgNO_3$	KNO_3	360	4.6
TlBr	KI	770	4.3
Methanol	H_2O	18	1.4
Glucose	H_2O	18	0.5
CO_2	H_2O	18	1.5
I	CCl_4	20	1.2

4-8 Diffusion of Electric Charges

The concepts of the preceding section can be extended to diffusion of charged particles. Not only do charged ions diffuse in solids (say Na^+ ions in NaCl), thereby transporting charge, but also electrons (and holes) themselves diffuse as well. The same general expressions governing random walk and transport of atoms in solids apply also to the diffusion of charge, and again a quantity called the diffusion coefficient can be defined.

Although diffusion of charges in all types of solids is amenable to analysis, that of charges in semiconductors is the most interesting. Consider first the motion of electrons. They move about in the crystal with a certain average velocity v, making collisions with the lattice ions after travelling some average distance λ. The quantity f in Eq. (4-12) is then the reciprocal of the mean time between collisions and is given by v/λ and the quantity d is just λ. Thus the equation equivalent to (4-12) is for this case

$$D_n = \frac{v\lambda}{b} \tag{4-22}$$

Unfortunately neither v nor λ can be measured directly, so D_n cannot be calculated explicitly with any great accuracy. In this we are no worse off, however, than we were for diffusion of atoms in crystals, for D cannot be calculated with precision for that case either.

Experimental values of D_n can be determined by observing drift phenomena in semiconductors and applying Eq. (4-11). One finds the value of D_n for electrons in Ge at room temperature to be about 10^{-2} m^2/sec. Since the average value of the velocity of the electrons at room temperature is about 10^5 m^2/sec, the mean path between collisions, λ, must be about 10^{-7} m (about 1000 angstrom units).

Measured values of D for electrons and holes in several semiconductors at room temperature is given in Table 4-3. Variation with temperature is not a simple function. In some temperature regions the Ds vary as some small negative power of T, say T^{-2}, in others as some more rapidly varying function of temperature. The scant data available for several of the more useful semiconductor materials are

Table 4-3 Diffusion coefficients for several elemental and compound semiconductors at room temperature.†

	D_n, m^2/sec	D_p, m^2/sec
Ge	0.0095	0.0045
Si	0.0035	0.0012
GaSb	0.01	0.002
InSb	0.18	0.003
GaAs	0.02	0.001

† For Ge and Si, the values are direct drift measurements; for the compounds, they are calculated from measured mobilities.

listed material by material in the reference book *Materials Used in Semiconductor Devices*, C. A. Hogarth, Editor, Interscience, New York, 1965.

REFERENCES

1. P. G. Shewmon, *Diffusion in Solids*, McGraw-Hill Book Company, Inc., New York, 1963.
2. B. Boltaks, *Diffusion in Semiconductors*, Academic Press, Inc., New York, 1963.
3. C. A. Hogarth, *Materials Used in Semiconductor Devices*, Interscience Publishers, a division of John Wiley & Sons, Inc., New York, 1965.
4. C. Hilsum and A. E. Rose-Innes, *Semiconducting III-V Compounds*, Pergamon Press, New York, 1961.
5. A. S. Grove, *Physics and Technology of Semiconductor Devices*, John Wiley & Sons, New York, 1967.

PROBLEMS

4-1 Suppose that the potential function of Fig. 3-1 is approximated by the curve (Fig. P4-1) for a two-dimensional close-packed structure. The equilibrium separation of atoms is 3 A. The Einstein frequency of the atoms is 2×10^{12} cps. Suppose that the atoms have a mass of 6×10^{-26} kg (i.e., about the mass of an atom of K).

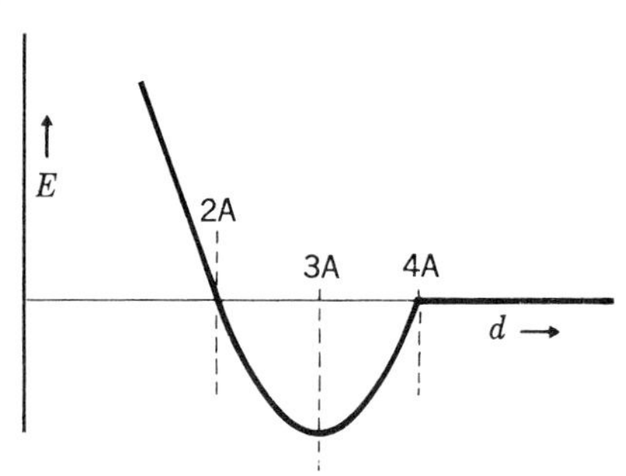

Figure P4-1

Calculate the following, assuming no relaxation of interatomic distances:

(*a*) The energy to form a vacancy in the structure

(*b*) The fraction of vacant sites at 1,000°K

(*c*) The motional energy of a vacancy

(*d*) The diffusion coefficient at 1,000°K

(*e*) The rms distance that an atom would travel in a given coordinate direction, say, X, in 1 year.

Explain any assumptions you use in the calculations.

4-2 Suppose that impurity atoms move in one-dimensional diffusion in the x direction in a solid of unit cross-sectional area. Assume that Fick's law holds. Show that the time rate of change of concentration c in an element of thickness dx is given by

$$\frac{\partial c}{\partial t} = D \frac{\partial^2 c}{\partial x^2}$$

4-3 Let a thin evaporated layer of atoms diffuse into a thick solid on which they have been deposited.

(*a*) Verify that, after a time t, the concentration c, as a function of distance, X, into the solid is given by

$$c = \frac{A}{\sqrt{t}} e^{-X^2/4Dt}$$

(*b*) Make a sketch of this equation.

(*c*) Find the value of X at which the curvature of c versus X is zero. How does this value compare with the rms value of X of Eq. (4-9)?

(*d*) Estimate the fraction of atoms that diffuse at least as far as 2 rms distances.

4-4 Following is a tabulation of data made for the diffusion of radioactive Ag in an alloy of Ag–16.7 per cent In at 728.5°C. The time of the diffusion anneal was 5.9×10^4 sec.

Calculate D for these conditions. (After A. Schoen.)

Penetration distance, 10^{-4} m	Specific activity, arbitrary units
0	600
0.84	540
1.32	450
1.83	360
2.30	250
2.79	160
3.29	88
3.76	50
4.25	25
4.70	12
5.20	5
5.68	2

4-5 From data similar to those of Prob. 4-4, the following diffusion coefficients were obtained for the diffusion of Zn in single crystals of pure Cu. From these values, calculate Q and D_0. (These diffusion coefficients are the original data from which the values of Q and D_0 listed on line 16 of Table 4-1 were obtained. Data of J. Hino.)

T, °K	D, m^2/sec
1,322	1.0×10^{-12}
1,253	4.0×10^{-13}
1,176	1.1×10^{-13}
1,007	4.0×10^{-15}
878	1.6×10^{-16}

4-6 Steel parts subject to wear are usually given a wear-resistant surface by heating them for a period of time at high temperature in a carbonaceous atmosphere. Carbon diffuses into a layer near the surface during this time; this carburized layer responds to later heat-treatments at lower temperatures to form a hard, wear-resistant surface layer called a *case*. The diffusion coefficient of carbon in the steel is given by the formula

$$D = 0.12 \times 10^{-4} \exp\left(\frac{-32{,}000 \text{ cal/g-mole}}{RT}\right) \qquad \text{m}^2\text{/sec}$$

About how much time is required to produce a case 0.5 mm thick on a piece of steel of radius 2 cm at a diffusion temperature of 925°C?

4-7 The thickness of an oxide layer formed on a piece of flat metal increases with time. Suppose that the atmosphere keeps the oxygen dissolved at the surface at a level C_0 and that the oxygen level at the oxide-metal interface remains at a lower level C_i. (See Fig. P4-7.) If the rate

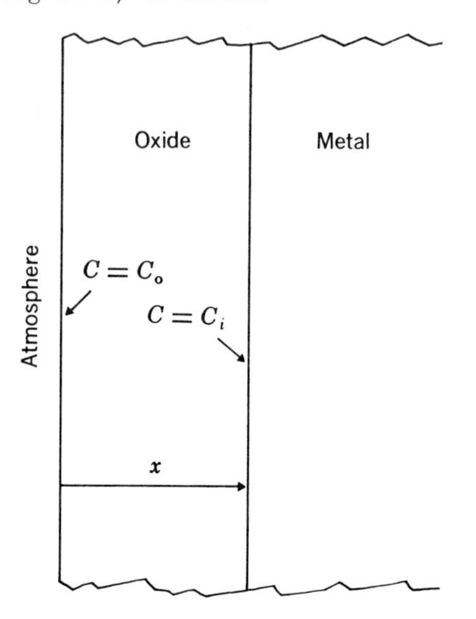

Figure P4-7

at which the oxide layer thickens is controlled by the diffusion rate of oxygen through the layer, show that x varies with time according to the equation

$$x = Bt^{\frac{1}{2}}$$

where B is a constant. Disregard the oxygen which forms a solid solution in the metal.

4-8 From the value of Q and D_0 in Table 4-1, calculate the temperature at which an atom of Li dissolved in solid Ge will make a single jump in 1 sec. Calculate the same temperature for B in Ge.

4-9 The formation energy of a vacancy in Al is about 0.76 ev. The energy for diffusion of Al in Al is 1.48 ev. Suppose that, in a certain piece of Al rapidly cooled to room temperature from near the melting point, the excess vacancies disappear (to about $1/e$th the initial excess) in about 100 hr at room temperature (i.e., the average lifetime of a vacancy is about 100 hr). About how many jumps does a vacancy make on the average before it disappears? About what rms distance does it travel in this period of time? (Assume that the vibrational frequency ν of a vacancy is about 10^{13} cps, that is, about that of an atom.)

4-10 The lifetime of an electron (i.e., period of existence of an electron in its excited state) in a particular piece of doped Ge might be of magnitude 10^{-4} sec. Calculate the mean diffusion distance of an electron in such material at room temperature. Calculate also the total path length during its lifetime.

4-11 N. Nachtrieb has shown that the activation energy for self-diffusion of cubic metals is related to the latent heat of fusion, H_m, by the equation

$$Q = AH_m$$

From data on self-diffusion of the metals in Table 4-1 and from handbook values of H_m, deduce the value of A.

5

Metallic solutions and compounds

5-1 General

Almost no elements are used in pure form; they are nearly always combined with each other to improve one or more properties. Steel, for example, is basically an alloy of iron with carbon, although many other metals are commonly added, metals such as manganese, nickel, chromium, tungsten, or cobalt. The alloy duralumin has, as its chief constituent, aluminum, which accounts for its light weight; the aluminum is strengthened by additions of metals such as copper, silicon, and magnesium. Solid electric-power rectifiers may be made from silicon alloyed with extremely small amounts of elements such as arsenic and indium.

The ways in which elements are combined to produce useful alloys form a large body of engineering knowledge. Most alloys used commercially have been developed over a long period of time by trial-and-error methods. Development of an alloy is often exceedingly laborious, and methods are constantly sought to shorten the development time. In this effort, a large amount of information about various combinations of elements has been accumulated. To tabulate certain aspects of this information concisely, engineers and scientists depend in large part on a scheme called the *phase diagram.*

5-2 The Phase Diagram

Several variables control the product obtained when two or more elements are combined. Both temperature and pressure may be varied, as well as the relative concentration of the constituent elements. To describe the product of such a combination, we use the concept of a *phase*, which is defined as being a portion of a system, bounded by a surface, having a characteristic arrangement of atoms and possessing uniform intensive properties (e.g., temperature, pressure, composition, electric polarization, etc.). Of course, several such portions having the same intensive properties are the same phase. Ice, for example, is a phase of H_2O; liquid water, another. A mixture of the two at the same temperature is a two-phase mixture. Water containing dissolved air is a single phase because a change in an intensive variable, temperature or pressure, is required to effect a separation of air and water. A *phase diagram* is a graphical description of the phases which are present in equilibrium at the many combinations of the variables composition, pressure, and temperature.

The simplest kind of equilibrium phase diagram is one which shows the structure of a pure homogeneous material as a function of

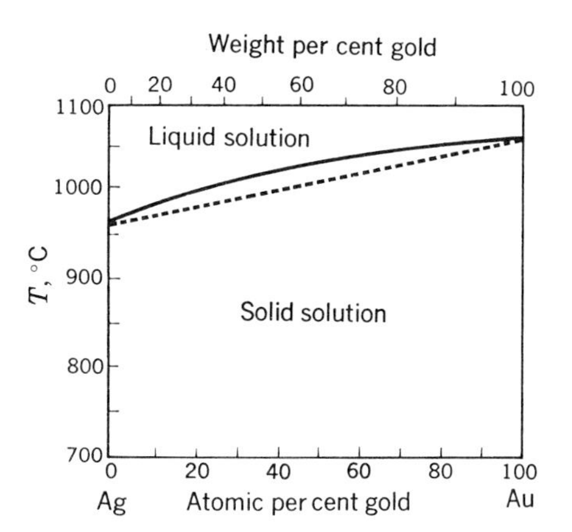

Figure 5-1 The Ag-Au phase diagram.

pressure and temperature. Perhaps the most commonly known diagram of this sort is the P-T diagram of H_2O. Such one-component systems are not so important to us, however, as systems of two or more components; hence we proceed immediately to a study of the latter. The simplest of these are the two-component systems, the binary systems.

A *binary* phase diagram has three major variables: pressure, temperature, and composition (others, such as polarization and magnetization, are irrelevant to this discussion). These three variables are entirely independent of each other; so a binary diagram is three-dimensional. For convenience of presentation on a two-dimensional surface, any one of these three may be held constant. A simple example of such a diagram is shown for the system Ag-Au in Fig. 5-1. Here the phases are plotted as a function of temperature and composition for a pressure of 1 atm. (Since those parts of phase diagrams involving solid reactants are usually insensitive to moderate changes of pressure near 1 atm, almost all phase diagrams are plotted at 1 atm.) This system has two components, Ag and Au. The system has a liquid phase at high temperatures and a solid phase at low temperatures. A gaseous phase, which exists at very high temperatures, is not shown. No compound is formed at any temperature or composition. Notice that the phase diagram of Fig. 5-1 has the composition expressed in atomic per cent. By changing the abscissa scale appropriately, it may also be expressed in weight per cent.

The phases present in this system are typical of those found in many other systems. The liquid phase is a *liquid solution;* in it the atoms of Ag and Au are intimately mixed. Such solutions are well known to everyone, since their use is common. Sugar in tea, CO_2 in soda pop, antifreeze in water—these are all liquid solutions. The solid phase is a *solid solution;* the Ag and Au atoms in it are intimately mixed on the face-centered cubic crystal lattice. Solid solutions, too, are common, though they are not usually recognized as such. Some brasses, 14-karat gold, some stainless steels—these are solid solutions. They have the characteristic, in common with the solid solution of the Ag-Au alloys, that the chance of a given lattice site being occupied by either the one type of atom or the other is random and thus depends only on the composition of the alloy. For example, at the composition 25 atomic per cent Ag–75 atomic per cent Au, each lattice site has one chance in four of being occupied by an Ag atom. This type of solid solution, in which the two atoms substitute freely for one another on the lattice sites, is called a *substitutional solid solution.*

5-3 Ordered Solutions

In the strict sense implied above, probably there are no true substitutional solid solutions (except for very dilute solutions), because the placement of atoms of one type in the structure is always correlated to some degree with the placement of atoms of the other type. Suppose that, in a Cu-Au solid-solution alloy, a certain site is occupied by an Au atom. Occupancy of neighboring sites is not mixed in the average ratio of Cu to Au atoms, but the chance of a neighboring site being occupied by a Cu atom is slightly higher than this ratio. This phenomenon, called *ordering,* exists because of slight differences in interatomic forces between Au-Au, Au-Cu, and Cu-Cu neighbors. In some alloys such as the 50–50 Au-Cu solution the tendency toward ordering is so large that the structure at low temperatures is almost like that of a chemical compound (Fig. 5-2). At higher temperatures the order tends to disappear, but it does not completely disappear even near the melting point.

5-4 Changes in Phase

A phase which is stable at one set of values of pressure and temperature may change to a different phase if either or both of these variables are changed. All elements can be melted and vaporized under proper conditions. These changes of state of aggregation are changes in phase. In addition, many elements change crystal type under appropriate changes of pressure and temperature; these changes, too, are changes in phase. For example, the body-centered cubic structure of iron (α Fe) is the most stable structure from absolute zero to 910°C (for low pressures), where it changes to a face-centered cubic structure, γ Fe. The face-centered cubic structure in turn changes back to the body-centered cubic form at 1,402°C, δ Fe. Finally δ Fe melts at 1,539°C. Such changes of crystal type (termed *allotropic* phase changes) are common among the elements.

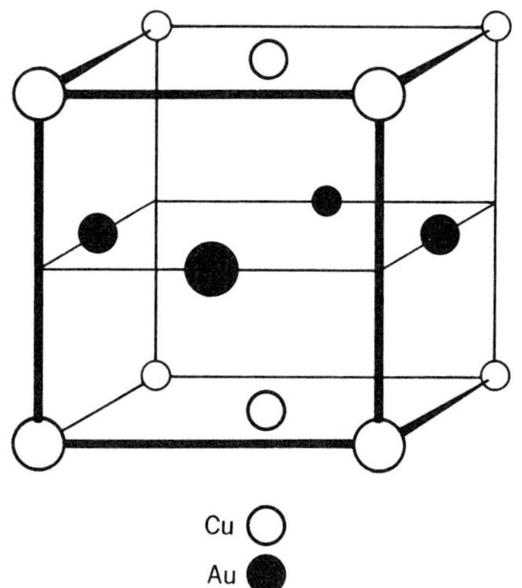

Figure 5-2 The completely ordered 50 per cent Cu-Au structure. The copper and gold atoms alternate on (001) layers of the fcc structure. The ordering also slightly distorts the crystal so that the vertical edge of the cell is slightly different from the horizontal edges. (*After Barrett.*)

Phase changes of pure elements occur at sharply defined temperatures. The heat evolved or absorbed during the change in phase is the latent heat. Phase changes of alloys often do not occur at a sharply defined temperature; an example is shown in Fig. 5-1. At any composition away from the pure constituents, the melting point is broadened into a melting range (which in this system is rather narrow—it never exceeds a few degrees).

The details of changes of phase for an alloy are important. They can be demonstrated through the use of Fig. 5-3, which shows an enlarged section near the liquid-solid region for a simple alloy system. Let the composition of a piece of the alloy be C_0; at a temperature below T_1 it is a homogeneous solid solution. If the temperature is raised just above T_1, some of the alloy melts. The first liquid has the composition C_1; it is more rich in A than the original alloy. As the temperature of the alloy is raised, more liquid forms. At each temperature the composition of the liquid moves along the line bd, the *liquidus* line. The entire mass of liquid at each temperature is homogeneous in composition. It is always richer in element A than the original solid. Since matter must be conserved, the solid remaining at any temperature is more lean in A than the original solid. As the temperature is raised, its composition moves along the line *ac*, the *solidus* line. When the temperature reaches T_3, the last bit of solid has the composition C_2. Above T_3, the entire liquid has the initial composition C_0.

The region of the phase diagram between the solidus and liquidus lines is a two-phase region; i.e., two phases coexist at equilibrium in this region. To specify completely the two phases at any temperature, say T_2, both the composition of each phase (i.e., the weight per cent B) and the relative quantity of the two phases must be known. The compositions are read directly from the abscissa; they are the values

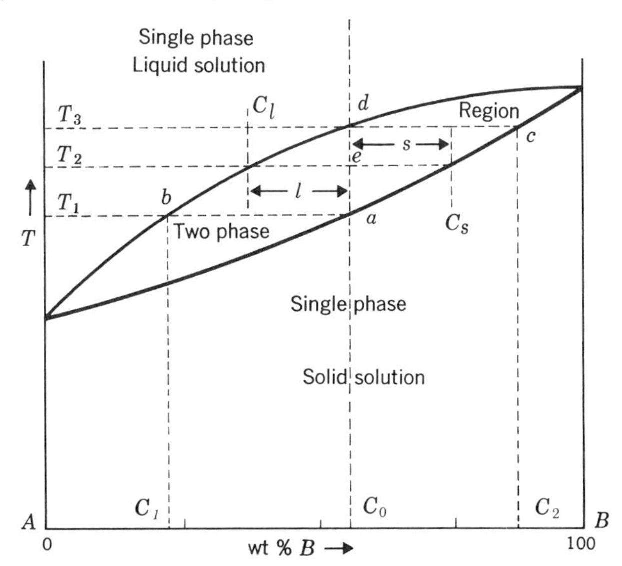

Figure 5-3 Hypothetical liquid-solid region of a phase diagram.

along the liquidus and solidus lines which correspond to intersections of these lines with a horizontal (constant-temperature) line through *e*. The quantity of each phase present may also be determined from the phase diagram by use of a formula called the *lever rule*.

The lever rule is important enough in the quantitative use of phase diagrams to warrant a derivation, which can easily be done with reference to Fig. 5-3. Again consider the alloy at temperature T_2 in the two-phase region. The two phases which coexist are (1) a liquid of composition C_l; (2) a solid of composition C_s. The overall concentration of B is C_0. Clearly the amount of B in the liquid plus the amount in the solid must equal the total amount.

$$L \cdot C_l + S \cdot C_s = (L + S) \cdot C_0$$

where L is the total weight of liquid and S the total weight of solid. This equation can be rearranged to yield the ratio

$$\frac{S}{L} = \frac{C_0 - C_l}{C_s - C_0}$$

or

$$Ss = Ll \tag{5-1}$$

where the composition differences s and l are the lever arms for which the rule is named. The lever rule yields the relative amount of the two phases present in terms of the composition of each and the average composition of the alloy.

Two things must be noted about phase diagrams in general.

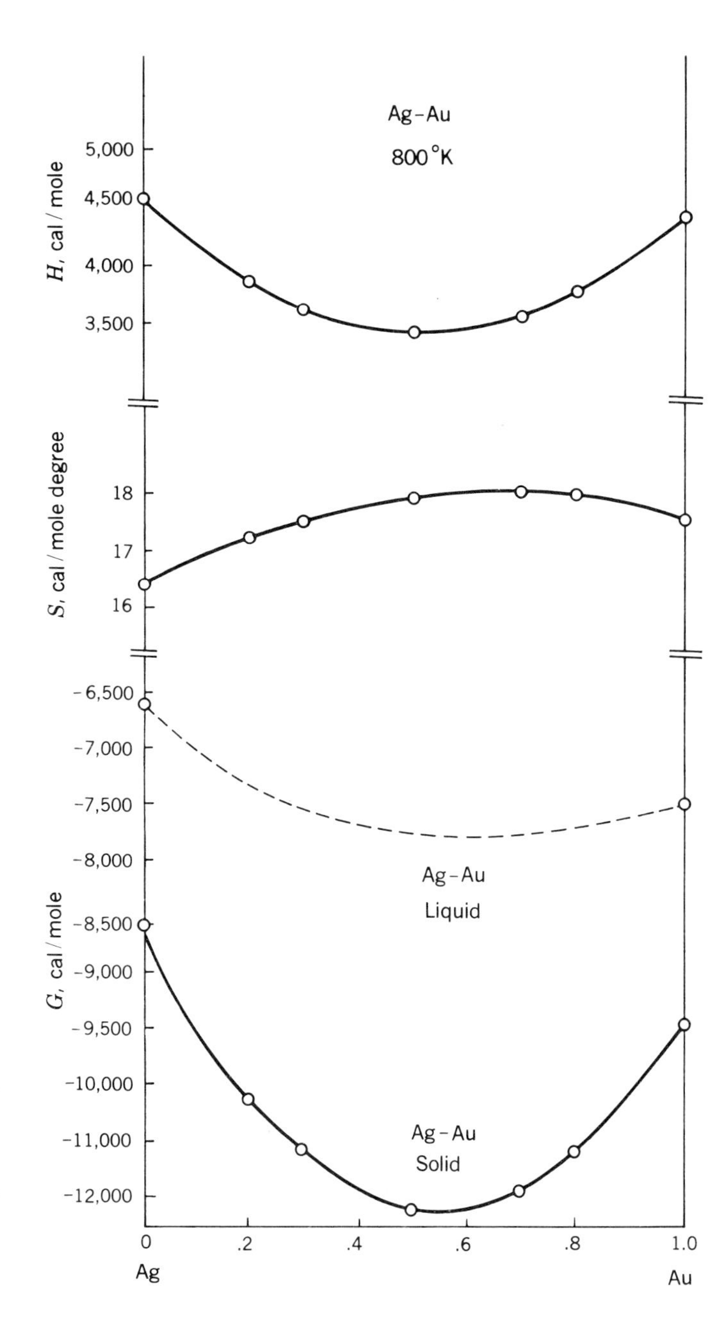

Figure 5-4 Enthalpy, entropy, and free energy of Ag-Au alloys at 800°K. The circles are numbers obtained from experiment. The dashed curve for the free energy at 800°K of possible liquid Ag-Au alloys is extrapolated from measurements made on true liquids above the melting temperature.

First, the ideas of this example of Fig. 5-3 can be extended to any phase change, including solid-solid and liquid-liquid transformations. Second, the construction of the phase diagram is based on equilibrium thermodynamics, and it gives no information at all about the rate or method of approach to equilibrium. Since time is required for the atom rearrangements necessary to achieve equilibrium, changes in phase should be made slowly if equilibrium is desired at all times. Ideal, infinitely slow changes are never achieved in practice. At the other extreme is quenching, in which temperature changes are made rapidly enough to suppress phase changes. Again this state is difficult to achieve completely, since infinitely rapid temperature changes are not possible.

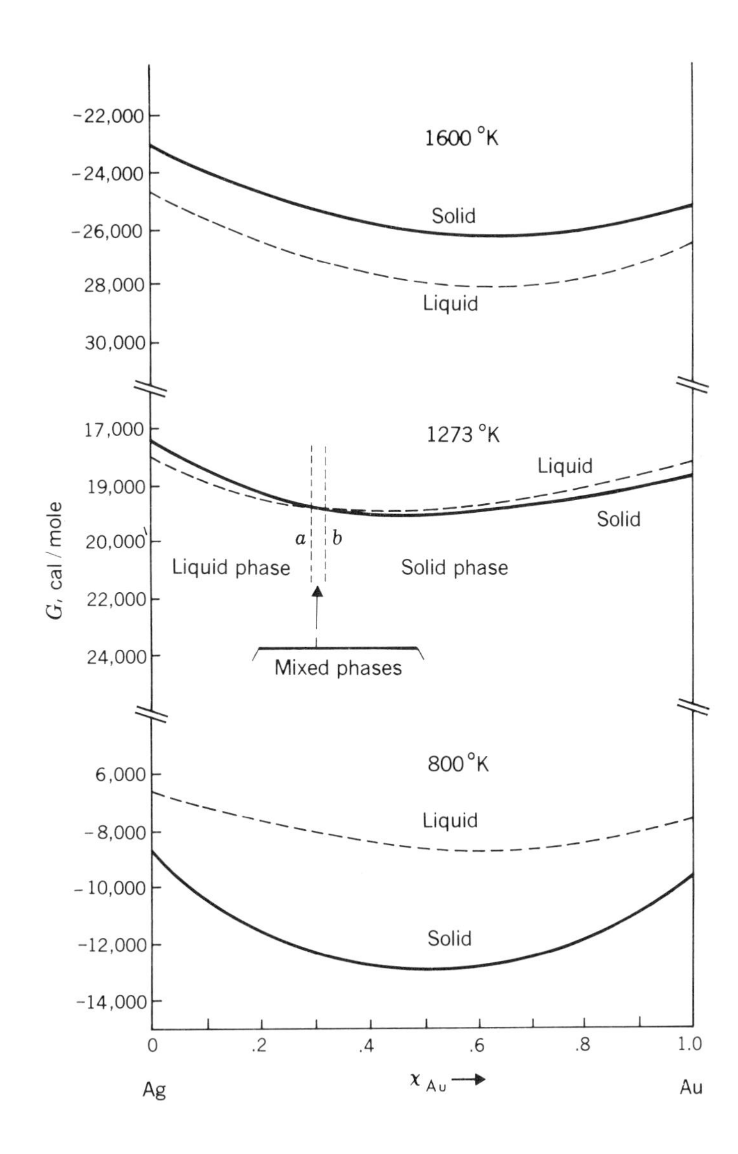

Figure 5-5 The free energy of Ag-Au alloys at three temperatures—one above, one below, and one in the central part of the melting range.

5-5 Thermodynamics of Phase Equilibria

The phase relations described in the preceding sections are controlled by equilibrium thermodynamics; the overall free energy of the system must be a minimum. Although detailed thermodynamic descriptions of all alloy systems described in this chapter are not necessary for adequate understanding, some discussion of a few of the simpler systems is desirable.

The PV term in the Gibbs free energy of solid and liquid systems at pressures near 1 atm is small compared with the internal energy at room temperature, so the Gibbs and Helmholtz free energies are about the same:

$$G \approx F = E - TS \tag{5-2}$$

The internal energy of a solid or liquid pure element under no external constraints is the heat content. That of an alloy is both the heat content and an additional term called the heat of solution. At 800°K, for example, the internal energy of Ag is 4,565 cal/mole and of Au is 4,615 cal/mole; the values for solutions of these elements in each other are plotted in Fig. 5-4. One sees that the heat of solution is small and is indeed negative.

The entropy of a pure element at temperature T is given by the well-known expression

$$S_{\text{pure}} = \int_0^T \frac{C_v^e(T)\,dT}{T} \tag{5-3}$$

Here $C_v^e(T)$ is the specific heat of the pure element. For Ag and Au at 900°K, S_{pure} is 16.44 and 17.52 cal/mole degrees, respectively. For solutions of these two elements an additional term is necessary:

$$S_{\text{solution}} = \int_0^T \frac{C_v^s(T)\,dT}{T} + S_{\text{mixing}} \tag{5-4}$$

In this expression C_v^s is the specific heat of the solution. The term S_{mixing} is the additional entropy arriving from the mixing of atoms of different types on the crystal lattice. It is always a positive term and has the value for random mixing in a binary solution of

$$S_{\text{mixing}} = -R(X_A \ln X_A + X_B \ln X_B) \tag{5-5}$$

Here X_A and X_B are the atom fractions of A and B in the solution. Values of S_{solution} for alloys of Ag and Au are also plotted in Fig. 5-4 for a temperature of 800°K.

The overall free energy of the solid solution is then given by

$$G_{\text{solution}} = E_{\text{solution}} - T\int \frac{C_v^s(T)\,dT}{T} - TS_{\text{mixing}} \tag{5-6}$$

This free energy is plotted at the bottom of Fig. 5-4. Notice that the curve is concave upward for all compositions. Complete solid solution results at 800°K, as the phase diagram shows. Sketched in is the estimated free energy of a liquid Ag-Au alloy at 800°K; since it is higher at all compositions than that of the solid, no liquid forms at this temperature.

The relationship of the free energy of the Ag-Au alloy system to the phase diagram in the region of melting is shown in Fig. 5-5. At 1600°K, the free energy of the liquid phase is lower at all compositions than that of the solid phase, hence only a homogeneous liquid phase is present. At an intermediate temperature, say 1273°K, the two curves cross, so that a liquid phase is stable over a certain temperature range and a solid phase stable over another. In range of compositions between a and b, the two phases coexist; this region is determined by the common tangent of the two curves. The fact that the two free-energy curves are not tangent at any composition is the reason that the alloys melt over a *range* of temperatures and not at a *sharp* temperature.

The free-energy curves at 800°K are repeated in the lower part of the figure. Only the solid solution forms at this temperature.

The principles used in elucidating the nature of this most elementary of binary phase diagrams may easily be extended to much more complicated systems. Some simpler extensions of these arguments to more complicated systems are made in later sections. Measurements of thermodynamic parameters of phases have proved to be more difficult than determination of phase boundaries themselves, so thermodynamic understanding has, in general, lagged behind the knowledge of what the phase diagram looks like.

5-6 Eutectic Phase Diagrams

A different type of phase diagram results if the pair of metals chosen are dissimilar instead of similar, as Ag and Au are. In particular, if the metals are completely miscible in the liquid state but completely insoluble (or nearly so) in the solid state, a type of diagram called the *eutectic* often results. A typical example is the diagram of alloys of the metals Bi and Cd, Fig. 5-6 The addition of Cd to Bi (right side) or Bi to Cd (left side) causes the melting point to be spread into a

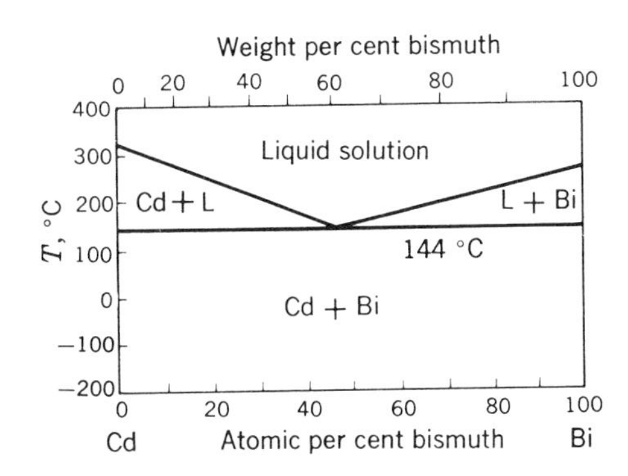

Figure 5-6 The Cd-Bi phase diagram.

melting range which, in this case, extends to increasingly lower temperatures as the departure from the pure metals is greater.

In the solid state nearly pure Cd and nearly pure Bi are present merely as a mixture of the two. At a particular composition called the *eutectic composition*, a mixture of 45 per cent Cd and 55 per cent Bi melts at a sharply defined temperature (144°C) called the *eutectic temperature*. At compositions on either side of the eutectic composition, melting occurs over a range of temperatures, pure Bi or pure Cd remaining solid to higher temperatures, depending on which is in excess of the eutectic composition. For this system, just as for the Ag-Au system, the compositions of the phases in equilibrium are those at the end of a horizontal line drawn across a two-phase region. Again, the amount of each phase is found by use of the lever rule.

Many pairs of metals show neither the complete mutual solid solubility of the Ag-Au system nor the nearly complete solid insolubility of the Bi-Cd system. Often the metals are partially soluble in each other. For such pairs of metals a eutectic system may also exist. At either extreme a solid solution exists, and the two solid solutions are separated by a two-phase region. Ag and Cu alloy in this way (Fig. 5-7*a*). The solid solution α is a face-centered cubic crystal of Ag containing a small amount of dissolved Cu, while the solid solution β is a face-centered cubic crystal of Cu containing a small amount of dissolved Ag. A second eutectic of this type is the Sn-Pb system (Fig. 5-7*b*). This important commercial alloy system includes many of the solders. A 60 weight per cent Sn–40 weight per cent Pb solder is used commonly for joining metals, since it has a low melting temperature. A solder with a lower Sn content is used as a wiping

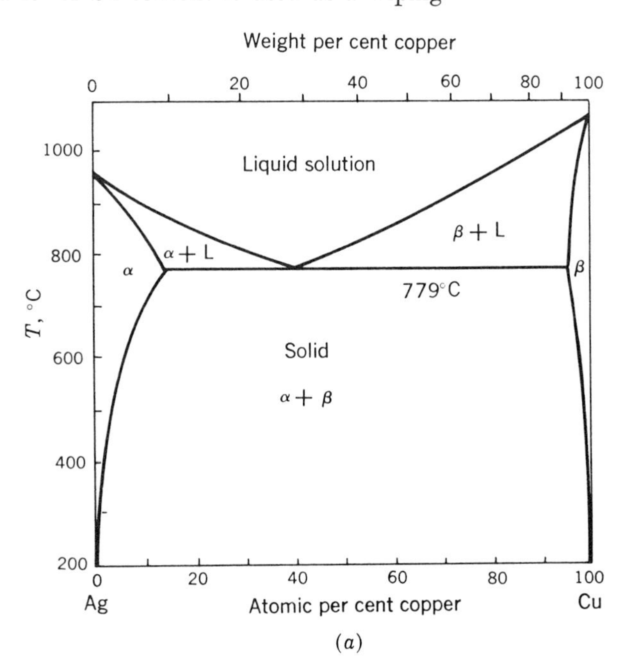

Figure 5-7 (*a*) Partial solid solubility. Eutectic type of phase diagram.

solder or as an automobile-body solder; the broad melting range ensures a long, mushy stage of solidification so that time is available for fitting and shaping.

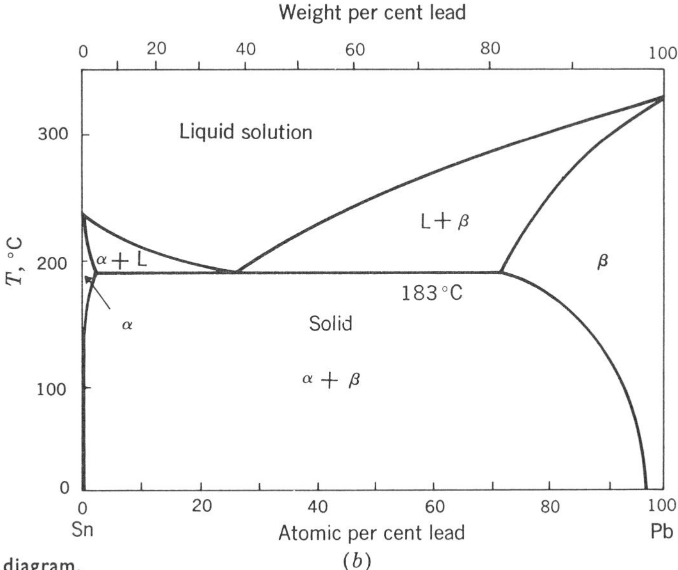

Figure 5-7 (*b*) Sn-Pb phase diagram.

The thermodynamic factor which accounts for the particular character of the eutectic phase diagram is a large positive enthalpy of solution. The Ag-Cu system has limiting slopes for $|\partial H/\partial \chi|$ of more than +5,000 cal/mole. Isomorphous systems such as Ag-Au, Cu-Au, Nb-Ta, Ge-Si possess values of $\partial H/\partial \chi$ ten times less, or values which may even be negative. Thus the difference.

To see this effect more explicitly, consider the curves of thermodynamic parameters for the Ag-Cu system plotted in Fig. 5-8. The top curve shows the enthalpy, including both the heat content and the heat of solution. At this temperature of 1000°K, the limiting solubilities of the α and β phases are about 14 and 5 per cent, respectively. Hence the data are only known over that range (dark lines); the extrapolations estimated into the two-phase region are given by the dashed lines.

The entropy and free-energy curves are also plotted in Fig. 5-8. The free-energy curve shows two minima, one near each extreme of the phase diagram. The common tangent gives the extremes of the solubility in each terminal solution.

At lower temperatures, the effect of the large enthalpy of solution becomes even more pronounced, since it remains about constant even though the term $-TS$ becomes smaller as temperature goes down. Hence the minima in the free-energy curves move even more toward the extreme edges of the phase diagram. The equations of the lines of maximum solubility (see the lines called solvus lines of Fig. 5-7) are

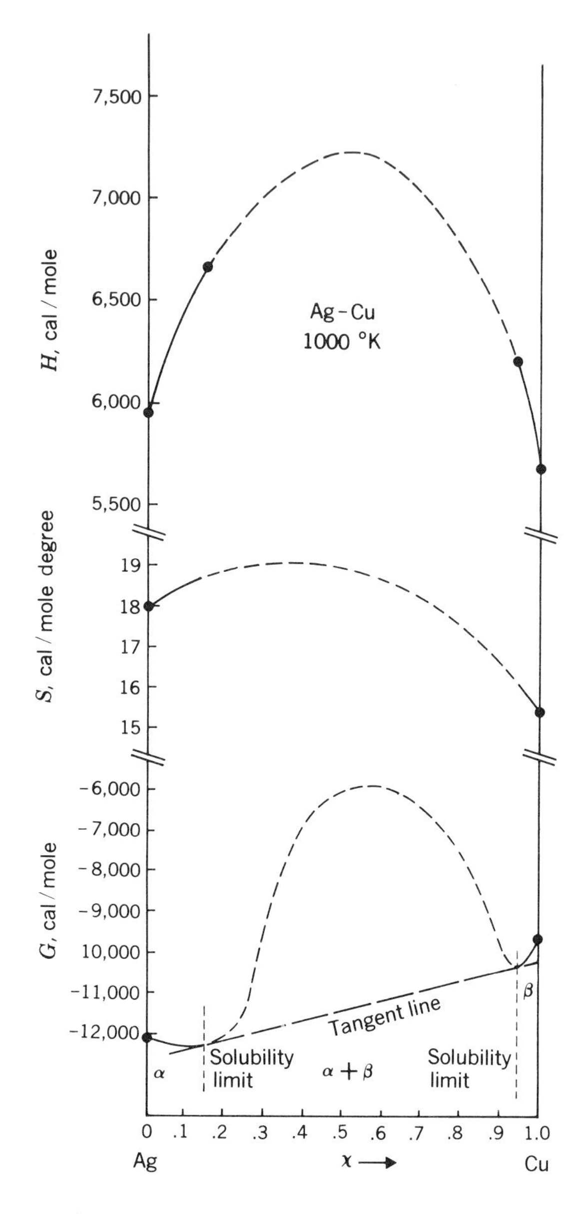

Figure 5-8 The enthalpy, entropy, and free energy plots for the Ag-Cu system at 1000°K (just below the eutectic temperature). The solid lines are plots of experimentally determined parameters; the dashed lines are estimates in the composition range where phase separation occurs. The values of χ at the points of common tangency of the two curves are the points on the solvus lines at 1000°K.

in fact given with good accuracy by the expressions

$$\chi = \chi_0 e^{-\Delta H/RT} \tag{5-7}$$

where ΔH is the limiting value of $\partial H/\partial \chi$ at the edge of the phase diagram and χ_0 is a constant. (See Prob. 5-6.)

The free-energy curves of the liquid and solid for this system are sketched in Fig. 5-9. Part (*a*) shows the normal behavior for a liquid solution with no solid present, and part (*d*) the behavior for a solid with no liquid present (this is a sketch of part of the data in Fig. 5-8). Part (*b*) shows the situation for a mixed liquid and solid alloy, and part (*c*)

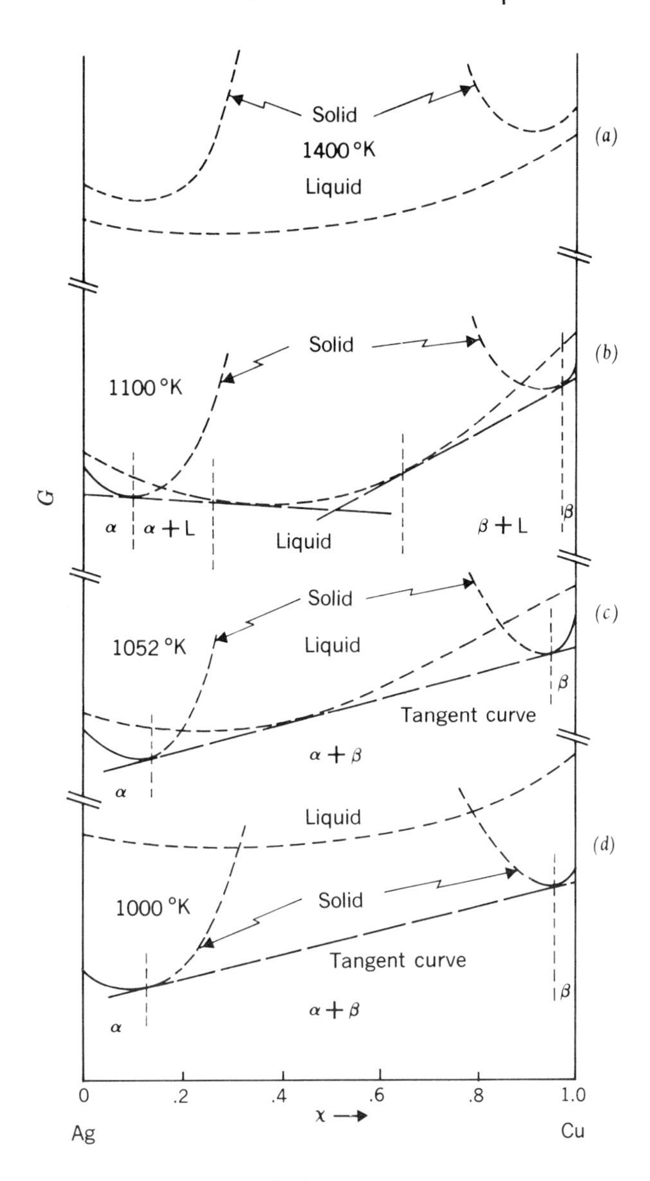

Figure 5-9 Free energy plots at several temperatures in the Ag-Cu system. Many of these data are extrapolations of values measured in other temperature regions, so they may not correspond exactly to values on the appropriate lines on the phase diagram. They are correct in principle, however.

the situation at the eutectic temperature where a single line is tangent to the free-energy curves of the two solid phases and the liquid phase of the eutectic composition as well.

For other details of the thermodynamics of this and similar systems, the reader should consult the excellent treatments by P. Gordon and R. Swalin listed at the end of the chapter.

5-7 Peritectic Phase Diagrams

Not all alloy systems whose components have limited mutual solubility form eutectics. Some of them have a phase diagram called the *peritectic*, an example being the Cu-Co diagram, Fig. 5-10. The peritectic has some features similar to those of the eutectic. It has extensive two-phase regions, and it has a composition for which a temperature change results in a single phase transforming into two phases. The peritectic point, however, is the highest melting point of a particular solid phase, whereas the eutectic point is the lowest freezing point of a particular liquid phase. Since not many alloys to be considered later have the peritectic type of phase diagram, no detailed discussion of it is given here.

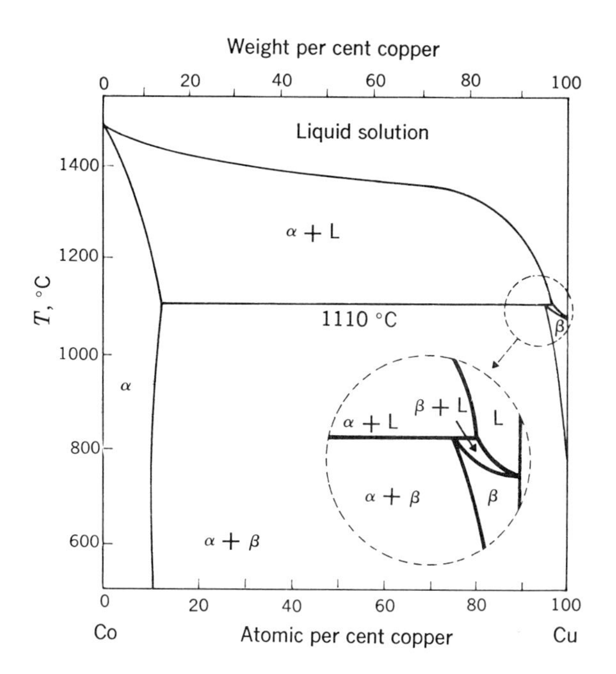

Figure 5-10 Partial solid solubility. Peritectic type of phase diagram.

5-8 Systems with Compound Formation

Only alloy systems between elemental metals have been considered so far. The same features of phase diagrams are found when one or both of the constituents is a compound. The systems Pb-Mg_2Pb and Mg_2Pb-Mg are shown in Fig. 5-11*a* and *b*. Each is a eutectic system with partial solubility on one side and nearly zero solubility on the other. These two diagrams are really just halves of the Pb-Mg system and may be placed together as in Fig. 5-11*c*.

In such a manner a complicated phase diagram may be made up of simple systems which are joined together by compounds or solid solutions. While some significant diagrams are made of simple eutectics like that of Fig. 5-11, others are more complicated. For all, however, the two-phase fields of the complicated diagrams are separated by single-phase solutions or compounds. The phases present at equilibrium, their compositions, and the proportion of each phase are found in the same way as is described for the elementary systems.

Alloys of definite composition are usually made by melting together known weights of the constituent metals. For this reason compositions are frequently desired in terms of *weight* per cent, not *atomic* per cent. The formula for converting from the one to the other is derived in Prob. 5-1.

5-9 Alloys of Semiconductors

The two important elemental semiconductors, Si and Ge, alloy in a simple way with each other and with other elements. That they do so is important to the electronics industry, for the purification of Ge and Si and their subsequent alloying with appropriate impurities depend on the simplicity of the phase diagrams. A few typical phase diagrams are described in this section.

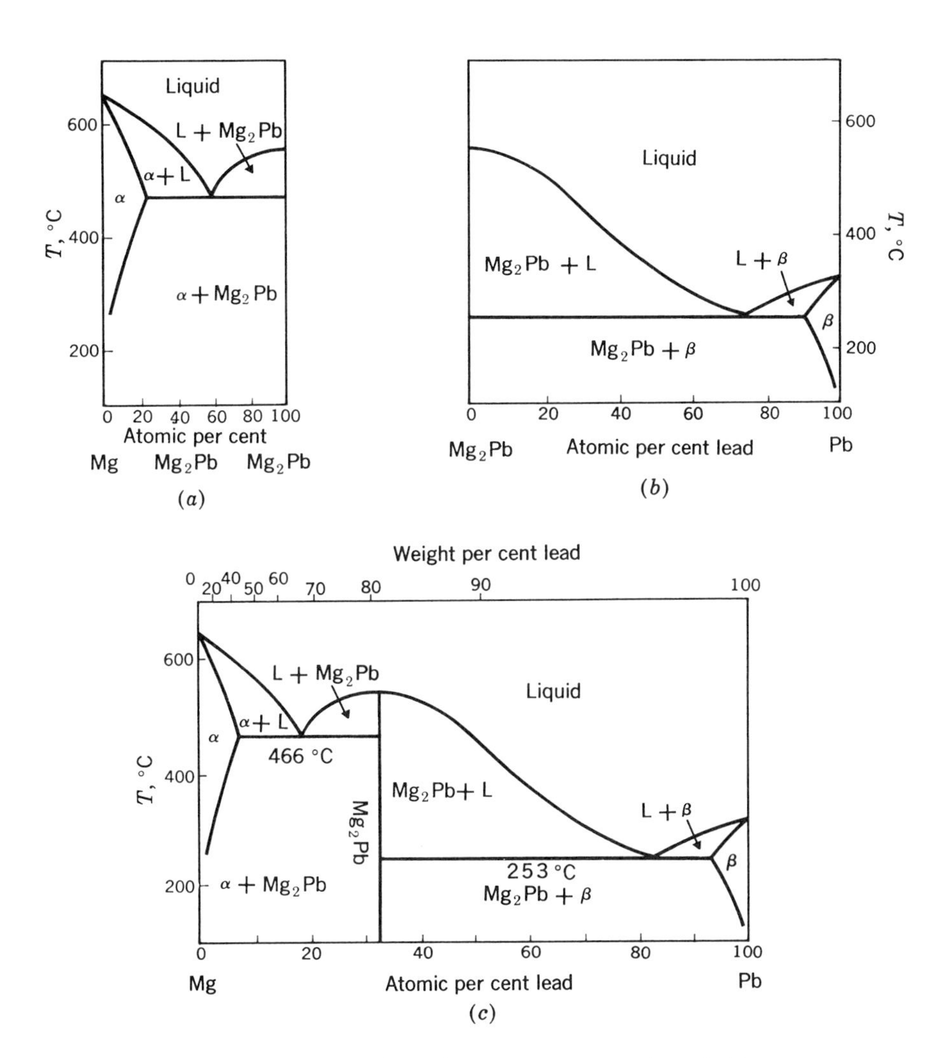

Figure 5-11 (*a*) Metal-compound system; (*b*) metal-compound system; (*c*) the Pb-Mg system.

Both Ge and Si solidify in the diamond cubic structure. As a consequence, the possibility of their forming complete solid solutions with other elements is limited to the few elements which also have the diamond structure. Because of other factors, most of these complete solid solutions do not form, and the only phase diagram containing Ge or Si which is known to exhibit complete solid solubility is the binary Ge-Si alloy system (Fig. 5-12*a*). These alloys show the same general features as the Ag-Au system (Fig. 5-1). Melting and solidification of alloys of Ge and Si follow the same principles as were outlined in Sec. 5-4.

The alloys of Si or Ge with other elements are, almost without exception, eutectics. Some of these phase diagrams are simple eutectics; an example is the Ge-Sb system (Fig. 5-12*b*). The solid solubility of Sb in Ge is not completely zero as is indicated in Fig. 5-12*b*, although

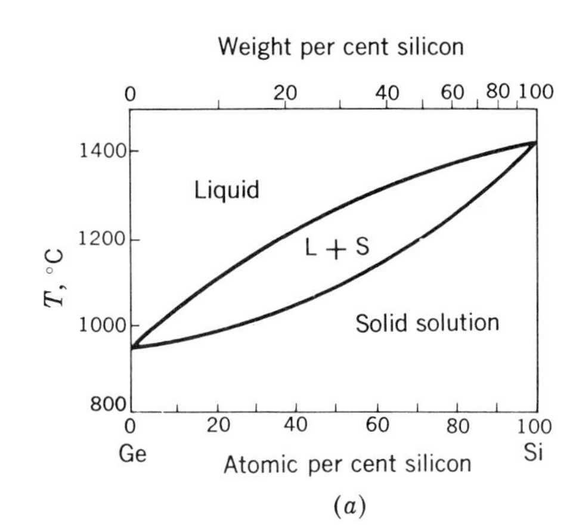

Figure 5-12 Typical binary phase diagrams of semiconductors. (*a*) The Ge-Si system.

it is small. An expanded view of this side of the diagram is sketched in the inset of Fig. 5-12*b*. Another example of a eutectic is the system Ge-In, an alloy system of great practical importance in the manufacture of transistors. These elements are virtually insoluble in one another; furthermore the eutectic composition is extremely close to the pure In side of the diagram. The general features of this phase diagram are sketched in Fig. 5-12*c*; the details of the important solid-solution regions are sketched in the insets. Not all diagrams of the Ge or Si binary alloys are this simple; one of the more complicated is the Cu-Ge system (Fig. 5-12*d*). Alloys rich in Cu have a variety of structures. For alloys of more than 30 per cent Ge, however, the phase diagram consists of a eutectic between the Cu_3Ge phase and pure Ge. For purposes of semiconductor manufacture, this is the only important feature. Again, the solubility of Cu in Ge is not zero; it does dissolve slightly in Ge (to a maximum of about 10^{-4} atomic per cent at about 850°C). The few phase diagrams discussed here are but a small number of the many which are known. The eutectic nature of most of the diagrams and the extremely small solubility of

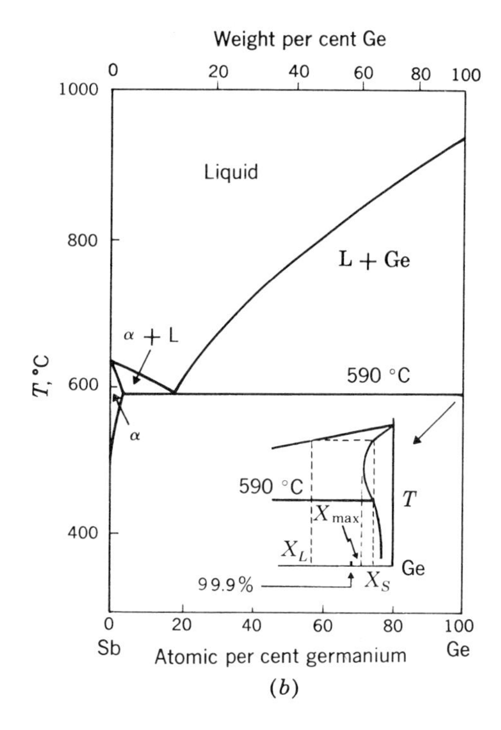

Figure 5-12 (b) The Ge-Sb System.

elements used for alloying are important for the manufacture of rectifiers and transistors.

The low solubility of the alloying elements in Ge and Si is directly a result of the high heat of solution of elements in the elemental semiconductors.

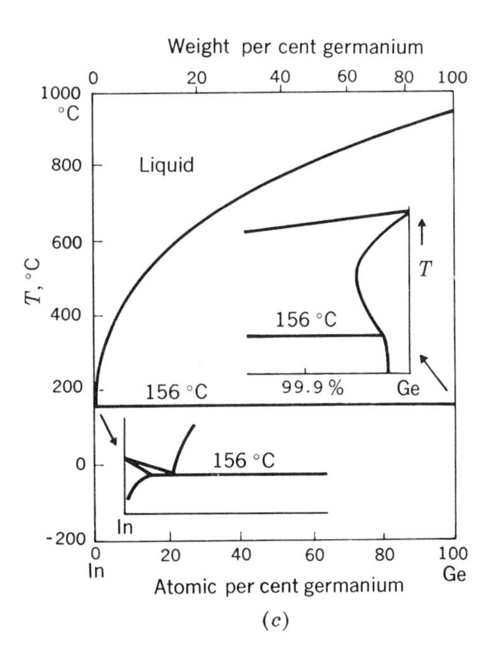

Figure 5-12 (c) The In-Ge system.

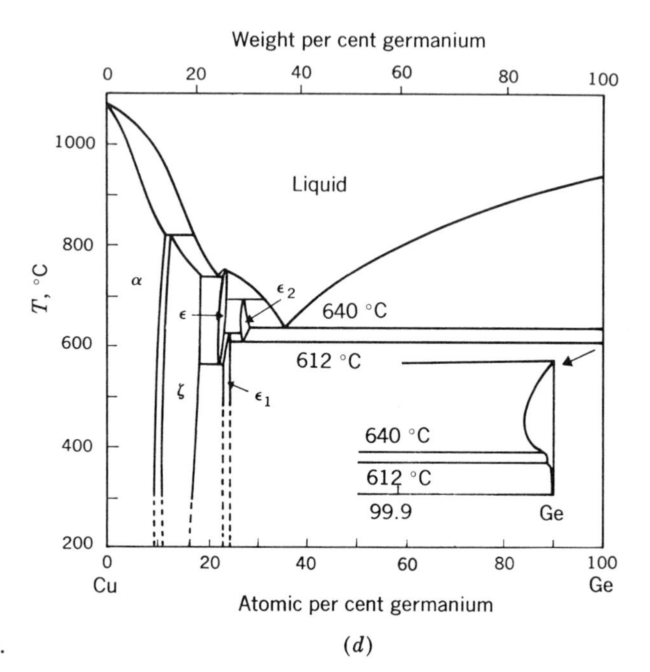

Figure 5-12 (d) The Cu-Ge system.

For example, Cu in Ge has a limiting slope of the heat of solution of about 35,000 cal/mole atom-fraction. Its maximum solubility in Ge is much less than 0.1 atom per cent—see Fig. 5-12*d*. In contrast, Cu in Ag has a limiting heat of solution of less than 10,000 cal/mole atom-fraction; hence, the maximum solubility is much greater, more than 10 per cent (Fig. 5-8). Furthermore, the limiting slope of the curve of heat of solution of Cu in Ni is less than 1000 cal/mole atom-fraction; Cu and Ni exhibit complete solubility in the solid state. The reader should be well aware of the enormous importance of the heat of solution in controlling both the general and also the detailed shape of phase diagrams.

Determination of the thermodynamic parameters which characterize alloying is difficult. Even for simple systems such as Ag-Au, collection of data like those in Fig. 5-4 requires painstaking measurement. For semiconductor alloys the problem is even more difficult.

An important term used to characterize one feature of the phase diagram is the *distribution coefficient k*. It is defined as the ratio of the slopes of the liquidus and solidus lines where they intersect at T_m (for example, see Fig. 5-12*b*). This definition is equivalent to forming the ratio of X_s/X_L in the inset of this figure.

$$k = \frac{X_s}{X_L} \tag{5-8}$$

In general k is a small number, varying between 0.5 and 10^{-7} for most important alloying elements in Ge and Si.

An interesting correlation can be made between k for a given element and its maximum solubility $X_{\max}$ (Fig. 5-12*b*). If $X_{\max}$ be

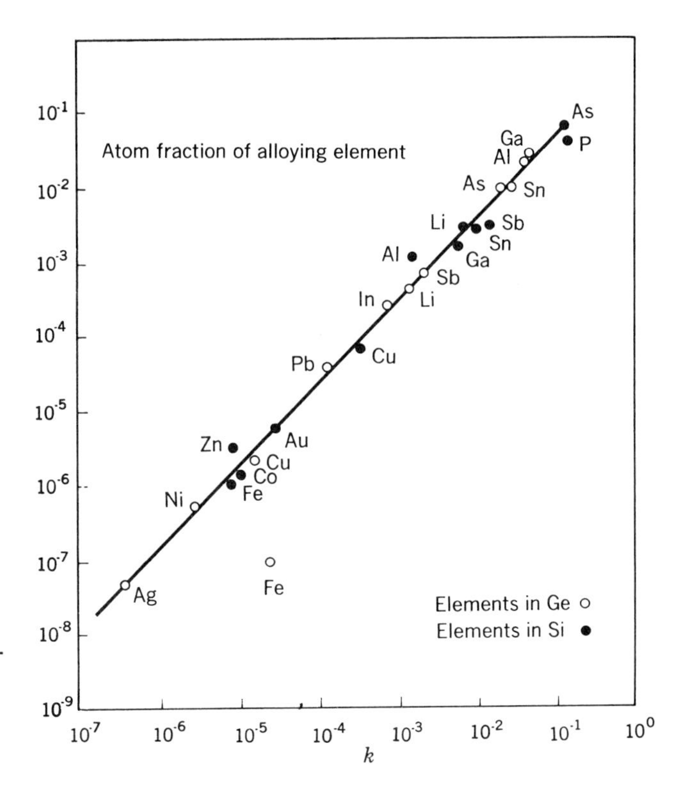

Figure 5-13 The maximum solid solubility of a given alloying element in Ge or Si is a simple function of the distribution coefficient. [*See S. Fischler and R. Brebrick, Bull. Am. Phys. Soc.*, (II) **7:** 235 (1962).]

expressed in atom fraction, then to a rather good approximation

$$X_{\max} = 0.1k \tag{5-9}$$

The validity of this expression for a number of alloying elements is shown in Fig. 5-13. A notable discrepancy exists for Fe in Ge, but the correctness of this rule to within a factor of about 2 for most elements is apparent.

Many other features of the phase diagrams are technologically important. Several of the books and review articles listed at the end of the chapter discuss them in detail.

5-10 Alloys of Three or More Elements

So far only phase diagrams of pairs of elements have been considered, the binary systems. If three elements are present, call them A, B, and C, three binaries, AB, AC, and BC, may be formed, as well as combinations of the three elements not found in the binaries. Such a ternary system is represented at constant pressure in a three-dimensional plot consisting of a triangular-based prism having the compositional variables plotted on the sides of the triangle and the temperature plotted vertically. The phase diagrams AB, AC, and BC are binaries plotted on the vertical faces as a function of temperature, as before. In the space enclosed by the three vertical faces are the alloys of all three components. The simplest kind of system of

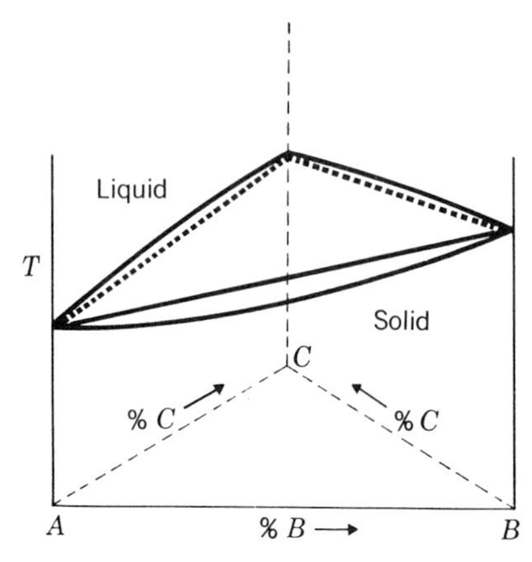

Figure 5-14 Isomorphous ternary system.

this type is one which has complete solubility for all compositions in both the solid and liquid states. Such a system, termed *isomorphous*, is sketched in Fig. 5-14. The liquidus and solidus lines of the three binaries are the lines of intersections of the liquidus and solidus surfaces with the faces of the equilateral prism. Few, if any, ternary phase diagrams are this simple. A ternary system a little more complicated, but still relatively simple, is shown in Fig. 5-15. This is a system in which each binary is a eutectic with small solid solubility. The ternary diagrams in general may exhibit all the features described for binaries: solid solutions, compounds, eutectics, and peritectics. In addition they possess other features not found in binary systems (see Ref. 1 and 2 at the end of the chapter).

Systems of more than three components cannot be represented in their entirety by a model. Various simplifications have been intro-

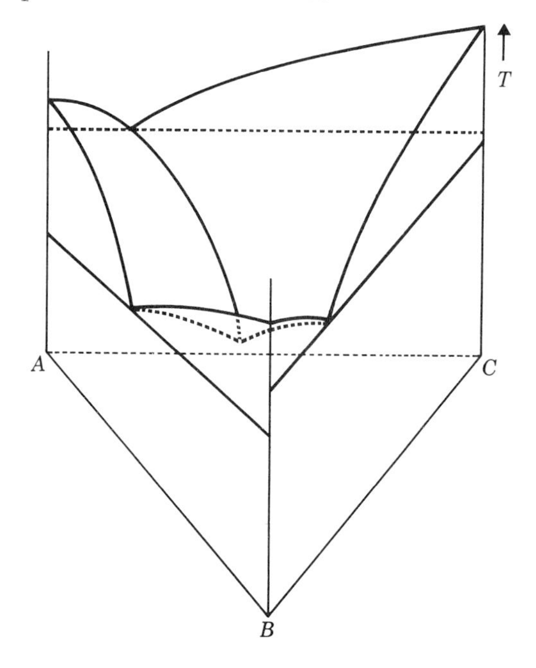

Figure 5-15 A eutectic-type ternary.

duced to represent such systems adequately. One of the most common is to hold the compositions of enough of the components constant so that the remainder can be represented in two or three dimensions. Since many commercial materials (e.g., steels, duralumin, brasses, bronzes, to name a few) have more than two elements, occurring intentionally or accidentally, their phase diagrams must be presented by using grossly simplifying assumptions.

5-11 Factors Governing Phase Equilibrium

Several factors are important in determining the nature of phase diagrams. Two of these deserve consideration here: (1) the *relative sizes* of the constituent atoms and (2) the *compound-forming tendency* of the elements involved. The effect of each of these factors can be conveniently expressed as a rule: (1) The existence of large regions of solid solution usually requires that the atoms be nearly the same size. (2) Strongly bound compounds usually dominate the phase diagram, restricting regions of solid solution severely.

The relative size effect is a mechanical effect, and its importance may be seen by considering the geometry of lattices packed with atoms of different sizes. Recall, for example, the crystal structure of Ag; it is face-centered cubic, with 12 neighbors in the close-packed configuration at a distance of 2.88 A. Atoms of Au substituted for atoms of Ag in this structure fit perfectly, because Au atoms also have an atomic diameter of 2.88 A. The phase diagram of Ag-Au reflects this ease of substitution, since the two metals form a complete solid solution (Fig. 5-1). Atoms of Cu substituted for atoms of Ag in the Ag lattice fit poorly because Cu atoms have a smaller diameter, 2.56 A. Ag atoms which are neighbors of a Cu atom must collapse a bit around the Cu atom; bonds in the vicinity of the Cu atom are thereby stretched a little, producing a local increase in lattice strain energy. Many of these local effects added together when numerous Cu atoms are substituted for Ag atoms cause the internal energy of the structure to increase appreciably. The Ag-Cu phase diagram reflects this energy increase. As Cu is added to Ag, the mechanical (internal) energy increases but the entropy also increases because of the nearly random way in which the Cu atoms substitute for Ag. Thus the increased internal energy is more than offset by the increased entropy and the free energy decreases when the solid solution is formed. Above a certain Cu concentration, however, the internal energy becomes intolerably high, and the crystal reduces its free energy by segregating further additions of Cu into a structure having more nearly the character of the Cu lattice. The same arguments apply to Ag additions to Cu-rich alloys. The result is that of Fig. 5-3: alloys in the middle of the diagram consist of particles of Ag-rich structure (with average atomic diameter near 2.88 A), mixed with particles of a Cu-rich structure (with average atomic diameter near 2.56 A). A photomicrograph

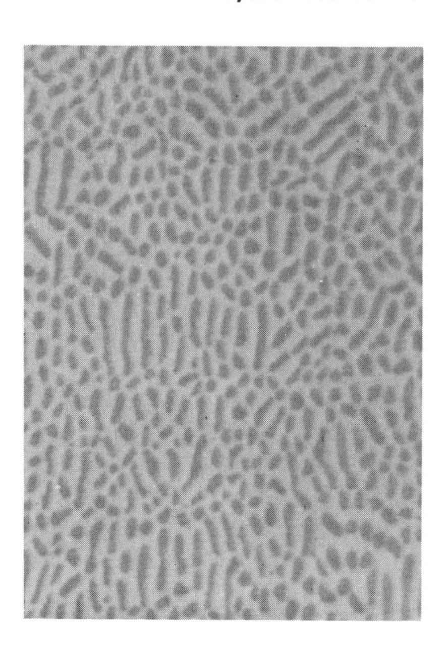

Figure 5-16 The structure of the Ag-Cu eutectic as seen through a microscope at a magnification of about 500 times.

of a polished and chemically etched surface of this phase mixture is shown in Fig. 5-16 for the Ag-Cu eutectic.

The solid solutions at the sides of the phase diagram are called *terminal* solid solutions. They clearly must have the crystal structure of the corresponding element and also have nearly the same unit-cell size. The extent of solid solubility is generally smaller the larger the size difference between the two atoms. A rule for this, first stated by the metallurgist Hume-Rothery, describes the behavior found for many alloy systems. Large terminal solid solubility is not possible if the two atoms differ in size by much more than 15 per cent. A size difference less than 15 per cent is said to be "favorable" for large solubility; a size difference more than 15 per cent "unfavorable." As an example of the correctness of this rule, data for several alloys of Cu are listed in Table 5-1. The reader should be aware that the rule must be applied with care. A favorable size ratio does not guarantee

Table 5-1

Alloy system	Size ratio	Solubility in Cu, at.%
Cu-Be	Favorable	16.4
Ga	Favorable	19.9
Ge	Favorable	11.8
Si	Favorable	11.0
Zn	Favorable	38.4
Cd	Unfavorable	1.7
Li	Unfavorable	Near zero
Mg	Unfavorable	7.0
P	Unfavorable	3.5
Tl	Unfavorable	Near zero

large solubility, because chemical factors may play a more decisive role; a favorable size ratio is a *necessary* but not *sufficient* reason for large solubility. An unfavorable size ratio, on the other hand, is usually a sufficient reason for small solubility, but not a necessary one.

The second factor, the tendency toward compound formation, is a chemical effect. An alloy may be able to achieve a low energy state by forming a definite chemical compound. Since the stability of compounds is often high, the effect on the phase diagram is generally large. Again, extreme cases may be used to illustrate the effect. No compounds are formed in the Ag-Au system, since the valence of each is +1. However, Br, with valence of −1, forms with Ag a stable salt AgBr, which dominates the phase diagram of Ag and Br. The solubilities of Br in Ag and of Ag in Br are both vanishingly small. Furthermore, only a little excess Ag or excess Br can be dissolved in the salt itself; so the compound always has nearly the ideal ratio of atoms of Ag to Br.

The more electropositive the one element and the more electronegative the other, the greater is the tendency toward compound formation, with a corresponding restriction in the ranges of the terminal solid solutions. Since the elements at the top left of the periodic table are the most electropositive and those at the upper right are the most electronegative, the relative behavior of phase diagrams of elements from different parts of the periodic table can be estimated.

These general rules and some others of less importance enable one to retain a feeling for the nature of phase diagrams. A concise statement of these rules follows, with several examples of each:

1. Alloys of atoms having the same valence, the same crystal structure, and distinctly favorable size ratio tend to form a continuous solid solution over the entire range of composition (Ag-Au, Cu-Ni, Nb-Ta).
2. Alloys of atoms having about the same electronegativity (or electropositivity) and unfavorable size factors have limited terminal solid solubilities, but no compounds usually form. Such phase diagrams are usually simple eutectic or peritectic, though additional intermediate solid solutions may form (Ag-Cu, In-Ge, Pb-Sn).
3. A metal of lower valency tends to dissolve a metal of higher valency to a greater extent than the reverse (Si-Cu, Mg-Au).
4. Alloys of atoms having a great difference in electronegativity usually form stable compounds. Phase diagrams of such elements usually show small terminal solubility (Na-Cl, Si-O, Mg-Pb).

5-12 Interstitial Alloys

The preceding sections are concerned with phase diagrams for elements which form solid solutions by substituting for each other on the lattice sites. These substitutional solid solutions, be the solubility high or low, are usually formed with the metals. Another type of solid solution is possible, one in which the dissolved atoms insert themselves in the holes between the solvent atoms, in the so-called "interstices" of the structure. These solid solutions are termed *interstitial solutions.* The solvent structure is usually a metal such as Fe, Ni, V, or Ti, and the dissolved atoms are the small atoms H, N, C, or O. The resultant solid solution always has the crystal structure of the metallic element. The lattice of the solid solution, however, is always expanded compared with the pure metal because the diameter of the dissolved interstitial atom is larger than the hole into which it is inserted (perhaps this is not so for all H solutions). The local lattice strain around the interstitial atom is usually quite large, and so the maximum solubility is generally rather small.

The phase diagrams of the interstitial type have not been determined for many combinations of elements: in fact there is not one which has been completely determined for all compositions. The experimental determinations of these phase diagrams are rather difficult: X-ray analysis is difficult for alloys of light elements, chemical analysis is often tedious, and several compounds often exist. One system which has been determined accurately over a part of the composition range is the system Zr-H (Fig. 5-17). The α phase is an interstitial solution of H in the hexagonal-close-packed form of Zr. The maximum solubility (at 547°C) is about 6 atomic per cent. The

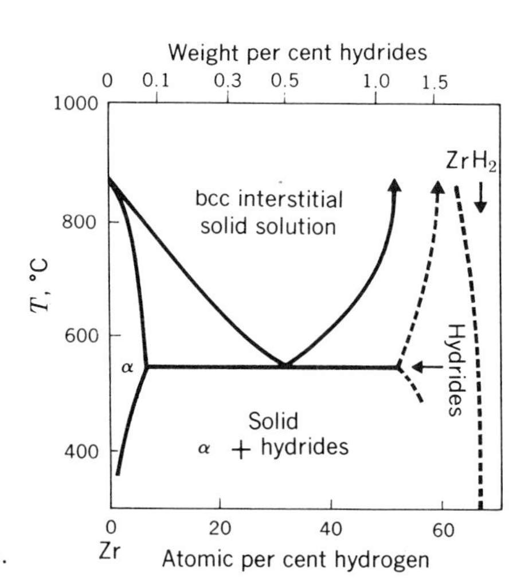

Figure 5-17 Part of the Zr-H phase diagram.

β phase, based on the body-centered cubic form of Zr, has an enormous composition range; at 850°C the solubility of H is nearly 50 atomic per cent.

Many of the common structural materials are interstitial alloys. Much of their strength comes from the dissolved interstitial atoms and from the stable compounds formed in the alloy.

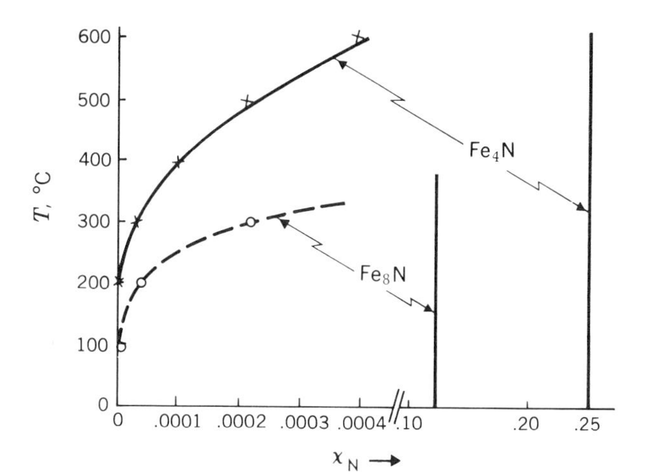

Figure 5-18 Part of the phase diagram of the Fe-N system, showing the solubility of N in bcc iron in equilibrium with the two nitrides Fe_4N and Fe_8N. Note the change in scale of the abscissa between the two halves of the phase diagram.

5-13 Nonequilibrium States

Phase diagrams represent equilibrium states. One assumes that sufficient time has been granted at all temperatures so that rearrangement of atoms gives, overall, the lowest free energy. This situation may occur in liquids of small volume (say a few cc), but it almost never occurs in macroscopic solids. Indeed, almost all usable alloys are in nonequilibrium states or at least in metastable states. Steels, cast iron, duralumin, magnesium alloys, alnico magnet steel, transistors, *p-n* junctions—almost all materials are deliberately left in some desired nonequilibrium state. For example, a properly tempered knife blade contains metastable carbides to give it hardness: heat it too hot on an emery wheel and it loses its ability to stay sharp.

One can calculate rather readily the temperature below which nonequilibrium states will exist if he knows the rate at which alloys are cooled or heated, the size of inhomogeneous regions which form, and the diffusion coefficients of the component atoms. Alloys of aluminum with a few per cent total of Cu, Ag, Mg, or Fe (these are the duralumins) can be quenched from some 500°C to room temperature to retain the solid solution characteristic of the high-temperature state. This solid solution partially decomposes at about 250°C to give a fine precipitate with a distance of separation of some 1000 angstroms. Copper and the other alloying elements diffuse in Al with a diffusion coefficient of about 10^{-18} m^2/sec at 250°C. Hence, the time of heat treatment to produce this precipitate is about

$$t = \frac{\bar{x}^2}{2D} = \frac{(5 \times 10^{-8})^2}{2 \times 10^{-18}} = \frac{25 \times 10^{-16}}{2 \times 10^{-18}} = 10^3 \text{ sec}$$

Once this metastable precipitate has formed, it is stable at room temperatures for a long time. At room temperature $D \approx 10^{-30}$ m^2/sec. Consequently, degradation of the metastable precipitate into large, stable precipitates of $CuAl_2$ (which don't give much strength to Al) does not occur rapidly during use at room temperature, since the time for even a single place change of Cu is

$$t = \frac{\bar{x}^2}{2D} = \frac{(2.5 \times 10^{-10})^2}{2 \times 10^{-30}} = \frac{5 \times 10^{-20}}{2 \times 10^{-30}} \approx 10^{10} \text{ sec}$$

Thus (at room temperature), the duralumin wing structure of aircraft is stable for many years. High-speed aircraft in which the structure may reach several hundred degrees centigrade need a more stable metal than duralumin, however; for them stainless steels or metals such as titanium must be used.

A second example of metastability directly related to a phase diagram is afforded by the iron-nitrogen system. A partial phase diagram is sketched in Fig. 5-18. A phase Fe_8N exists with the microstructure shown in the upper inset of Fig. 5-18. This phase is in equilibrium with a solid solution having the dashed solvus line. This phase is not stable relative to Fe_4N, the microstructure and equilibrium solubility of which are also given in that figure.

A great many metastable situations exist in metal and semiconductor alloys. Some are troublesome and should be avoided, but many of them form the backbone of practical materials usage.

REFERENCES

1. F. Rhines, *Phase Diagrams in Metallurgy*, McGraw-Hill Book Company, New York, 1956.
2. G. Masing, *Ternary Systems*, Dover Publications, Inc., New York, 1960.
3. M. Hansen and K. Anderko, *Constitution of Binary Alloys*, 2d ed., McGraw-Hill Book Company, New York, 1958.
4. R. Elliott, *Constitution of Binary Alloys, First Supplement*, McGraw-Hill Book Company, New York, 1965.
5. Paul Gordon, *Principles of Phase Diagrams in Materials Systems*, McGraw-Hill Book Company, New York, 1968.
6. F. A. Trumbore, "Solid Solubility of Impurity Elements in Ge and Si," *Bell System Tech. J.*, **39**:205 (1960).
7. William Paul and Douglas Warschauer, *Solids under Pressure*, McGraw-Hill Book Company, New York, 1963.

PROBLEMS

5-1 Phase diagrams are drawn with the relative composition of the constituents expressed in terms either of weight per cent or of atomic per cent. Derive an expression for binary systems to convert from one to the other.

5-2 Use the expression derived for Prob. 5-1 to calculate what fraction of the weight of NaCl is contributed by the Na ions.

5-3 From the phase diagram of the Sn-Pb binary (Fig. 5-7*b*), estimate the composition of the two phases present just below the eutectic temperature for a eutectic alloy. Approximately what fraction of the total weight of the solid is contributed by each phase?

5-4 Suppose that an alloy of 75 atomic per cent Pb and 25 atomic per cent Sn is maintained at 100°C until equilibrium is achieved. What fraction of the total number of atoms are in each phase?

5-5 Phase diagrams are usually determined by experimental methods.

(*a*) From the following experimental data construct a phase diagram for the system *A*-*B*:

1. Pure metal *A*: fcc, atomic radius 1.28 A, melting point 1083°C.
2. Pure metal *B*: fcc, atomic radius 1.25 A, melting point 1453°C.
3. Complete liquid and solid solubility.
4. For an alloy of 50 atomic per cent *B*, cooled from the liquid state, the first solid phase appears at 1305°C. This phase contains 64 atomic per cent *B*.
5. With further cooling the last liquid phase exists at 1240°C. This phase contains 34 atomic per cent *B*.

(*b*) Suppose that an alloy of 25 atomic per cent *B* is held at 1175°C until equilibrium is established. State the composition of the phases present and the quantity of each.

(*c*) Estimate the size of the unit cell for a solid solution containing 75 atomic per cent *B*.

5-6 (*a*) Dilute solutions of one metal in another are described, in general, by the same thermodynamic equations which describe dilute solutions of vacancies in a metal [Eq. (3-20)]. Using the same reasoning that was used in the solution of Prob. 3-12, develop an equation for the maximum concentration (in dilute solution) of one metal in another as a function of temperature,

$$C = C_0 \, e^{-\Delta H/RT}$$

where ΔH is the heat of solution.

(*b*) Show that the solvus line separating the silver-rich α phase of the Ag-Cu phase diagram obeys this equation and that the maximum concentration C of Cu in the α phase is given (over the region $T = 200°C$ to $T = 779°C$) by

$$C = 900 \, e^{-9{,}000/RT} \qquad \text{atomic \% Cu in Ag}$$

The energy 9,000 cal/g-mole is the heat of solution necessary to dissolve one mole of Cu atom taken from phase β in phase α at concentration C, all at constant temperature. This energy difference is about 0.39 ev per atom of Cu.

(*c*) The energy difference of 0.39 ev should be almost completely elastic energy if the arguments of Sec. 5-10 are valid. Suppose that the force constant between atoms is about 25 to 50 newtons/m (see Sec. 3-2) and that the smaller Cu atoms stretch their bonds with the Ag atoms when they go into the α-phase by the difference in radii between the two. Show that, for the coordination number 12 of the fcc lattice, the elastic-strain energy around a Cu atom is (with no relaxation of the α phase) about $\frac{1}{4}$ to $\frac{1}{2}$ ev per Cu atom, i.e., of the same order of magnitude as was calculated in Prob. 5-6*b*.

(*d*) What can you say about the solubility of 0.35 atomic per cent Cu at 200°C, listed by Hansen and Anderko in the lower left-hand corner of Fig. 5-7*a*?

5-7 (*a*) Show that the value of $\partial S_{\text{mixing}}/\partial \chi_A$ tends to ∞ as $\chi_A \to 0$.

(*b*) The slope $\partial H/\partial \chi_A$ remains finite as $\chi_A \to 0$ although it may become very large. Find the value of

$$\lim_{\chi_A \to 0} \frac{\partial G}{\partial \chi_A}$$

(*c*) What do parts (*a*) and (*b*) demand of the shape of the free-energy plot near the sides of the phase diagram?

5-8 Your great-grandfather may have "purified" applejack by freezing part of the water out of the original solution. Sketch a possible phase diagram of the water-alcohol system which would permit this process to occur.

5-9 (*a*) Sketch the free-energy curves of the several phases in the Mg-Pb system at 200°C (see Fig. 5-11).

(*b*) What can you say about the variation of the free-energy curve or χ_{Pb} for a compound like Mg_2Pb, which has an extremely narrow composition range?

5-10 The solubility of C in Mo has the following values at several temperatures:

T, °C	χ, atom fraction
1795	0.0023
1900	0.0037
1940	0.0042
2005	0.0059
2060	0.0066
2110	0.0087

(Data from Gebhardt, Fromm and Roy, *Z. Metallk.* 57:732, 1966)

(*a*) Are these values consistent with each other?

(*b*) What is the indicated solubility at room temperature? About how large a piece of Mo would be required to contain just 1 carbon atom at room temperature?

5-11 Consider an alloy of average composition χ_B° at temperature T where the free-energy or composition curve has the shape shown.

(*a*) Show that a mixture of phases having compositions χ_B' and χ_B'' has free energy lower than that of a possible homogeneous phase χ_B° (or than

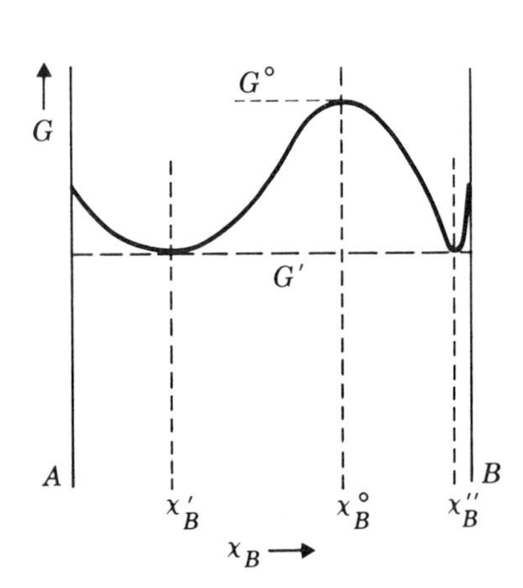

Figure P5-11

that of any other possible combination of compositions).

(*b*) How would the situation change if a small additional amount of pure A were added to the system and equilibrium were to be reestablished?

5-12 The internal energy of Cu at room temperature relative to that at absolute zero is about 1200 cal/mole. If a mole of Cu were subdivided into a fine powder, this energy would increase because of the increase in the free surface area.

(*a*) Calculate the absolute energy increase and the fractional increase accompanying subdivision into spherical particles of diameter 2 microns.

(*b*) The internal energy is also increased if the particle is put under pressure. Suppose a mole of Cu were lowered into one of the deepest parts of the ocean, the Mariana trench (depth about 35,000 feet). How much would its internal energy change?

5-13 The limiting slope of the heat of solution of Cu in Ge, $\partial U/\partial \chi_{\text{Cu}}$, is about 35,000 cal/mole atom-fraction (i.e., about 350 cal per atom per cent of Cu added to a mole of Ge). The extra entropy added to Ge by small additions of Cu is almost completely the entropy of mixing given by Eq. (6-4). Calculate the shape of the free-energy curve for small additions of Cu to Ge at the entectic temperature 640°C. Assume a value of free energy for pure Ge of G_0

5-14 The Clausius-Clapeyron equation equates the derivative with respect to pressure of the temperature T at which a phase change occurs to a function of the latent heat of the transformation, ΔH, and the volume change ΔV accompanying the transformation.

$$\frac{dT}{dp} = T\frac{\Delta V}{\Delta H}$$

(*a*) Cu has a 4.2 per cent volume change upon melting and has a latent heat of melting of 3120 cal/mole. Calculate the change in melting point accompanying an increase in pressure to 10 kilobars.

(*b*) The $\alpha \rightarrow \gamma$ transformation in iron occurs at 910°C for a pressure of one atm. The volume change is about -1.0 per cent, and the enthalpy change is 215 cal/mole. How does the calculated value of dT/dp compare with the measured initial slope of this parameter of -9.6°K/kilobar?

6 Dislocations

A dislocation is a more complicated imperfection than any of the point defects. The discussion in this chapter is therefore more qualitative than in the preceding chapters. We first define the dislocation and discuss it in a general way as a possible type of imperfection. At the end of the chapter its properties are related to the mechanical behavior of solids, which is the principal application of dislocation concepts.

Two extreme types of dislocations are possible, the edge dislocation and the screw dislocation, and any particular dislocation is usually a mixture of these two extremes. The geometry of a general dislocation is difficult to describe, but the extreme edge and screw types can be understood by means of sketches, and we discuss them separately.

6-1 Edge Dislocations

The arrangement of atoms characteristic of an *edge* dislocation is sketched in Fig. 6-1*a*. The distortion of the crystal structure may be considered to be caused by the insertion of an extra plane of atoms part way into the crystal. Most of the distortion is concentrated about the lower edge of the "half plane" of extra atoms, and so we speak of the dislocation as being that line of distortion along the end of the plane. Thus, the dislocation is a "line" imperfection, in contrast to the "point" imperfections discussed earlier. A sketch in depth of the atom positions about an edge dislocation in a simple cubic crystal is seen in Fig. 6-1*b*.

All dislocations have the important characteristic that the distortion is severe in the immediate vicinity of the dislocation line. In fact, along the dislocation line the atoms may not even possess the correct number of neighbors, while, not more than a few atom distances away from the center, the distortion is so small that the crystal is locally nearly perfect. The region near the dislocation line where the distortion is extremely large is called the *core* of the dislocation; here the local strain is quite high. Far away from the core, the strain is small enough (a few per cent or less) to be described by small-strain elasticity theory, and so it is called the *elastic region*.

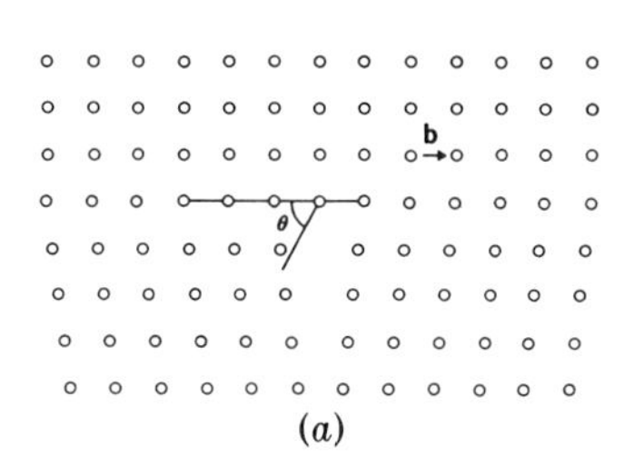

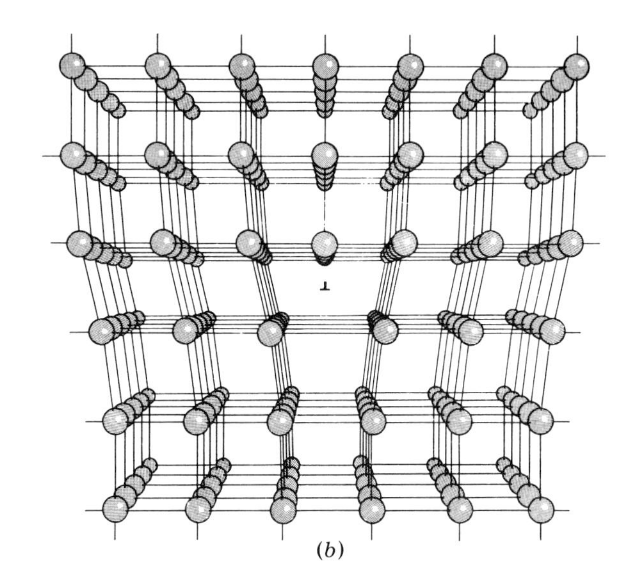

Figure 6-1 (*a*) Atomic arrangement of atoms in an edge dislocation (cross-section view); (*b*) perspective of atom arrangements about an edge dislocation in a simple cubic crystal.

An important characteristic of the distortion of the atoms around an edge dislocation is visible in the figure. The distortion at the end of the extra half plane is caused by the attempt of the atoms to accommodate themselves to the sudden end of the plane. Thus, the atoms just above the end of the extra half plane are in compression; i.e., the two rows of atoms to the right and left of the extra plane are squeezed together. However, just below the half plane, the same two rows of atoms are farther apart from each other than is normal, and the structure is expanded.

This local expansion, termed the *dilatation*, around the edge dislocation can be described by a simple analytical expression. The dilatation Δ at a point near an edge dislocation is given by

$$\Delta = \frac{\Delta V}{V} = \frac{b}{r} \sin \theta \qquad (6\text{-}1)$$

The factor b, the magnitude of a vector called the *Burgers vector*, is a measure of the strength of the distortion caused by the dislocation (**b** is defined exactly in Sec. 6-3). The quantity r is the radial distance from the point to the dislocation line, and θ is the angle defined in Fig. 6-1*a*. The sign of the dilatation, positive for expansion and negative for compression, is proper if θ is measured as shown. In addition to the simple expansion and contraction of the lattice in the vicinity of the dislocation, the structure is also sheared, but this shear distortion is somewhat more complex than the dilatation.

6-2 Screw Dislocations

The features of a screw dislocation can best be seen by considering the following operation on a perfect crystal: Let a sharp cut be made part way through a perfect crystal; further, let the material on one side of the cut be moved up relative to that on the other by one atomic spacing and the rows of atoms then placed back into contact. The resultant external character of the crystal is that of Fig. 6-2. Clearly a line of distortion exists along the edge of the cut; this line is called a *screw* dislocation.

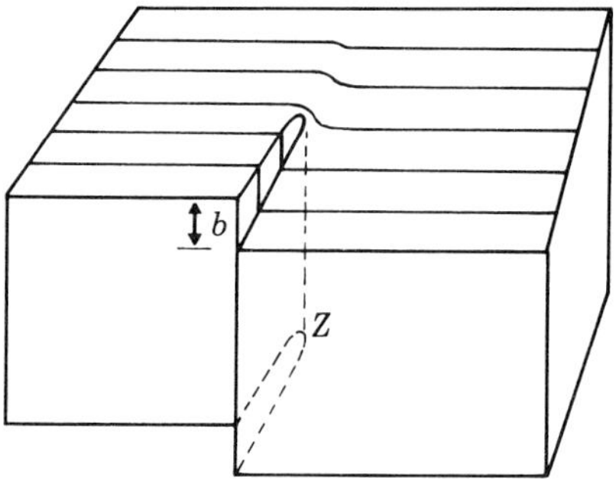

Figure 6-2 Structure of a crystal containing a screw dislocation. The height of the step on the top surface is usually one lattice spacing. The atom rows perpendicular to the dislocation are on a spiral ramp.

The most important feature of the screw dislocation is the new character of the atom planes. Complete planes of atoms normal to the dislocation no longer exist. Rather, all atoms lie on a single surface which spirals from one end of the crystal to the other, hence the word screw. The pitch of the screw may be either right- or left-handed, and it may also be one or more atom distances per revolution. The simple dislocation just described has a pitch of one atom spacing per revolution.

Again, the degree of distortion varies with distance from the center of the dislocation. Atoms far away from the screw dislocation are in regions of very little local distortion, while the atoms near the center are in regions highly distorted and the local crystalline symmetry is destroyed.

In contrast to the type of distortion surrounding the edge dislocation, the atoms near the center of the screw dislocation are not in dilatation but are on a twisted or sheared lattice. These atoms, being on a spiral ramp, are displaced from their original positions in the perfect crystal according to the equation which describes the spiral ramp,

$$u_z = \frac{b}{2\pi}\theta \tag{6-2}$$

In Fig. 6-2 the z axis lies along the dislocation, and u_z is the displacement in this direction. The angle θ is measured from some axis perpendicular to the dislocation; note that, as θ increases by 2π, the displacement increases by the factor b. The vector **b** is again the Burgers vector and is the measure of strength of the dislocation.

6-3 Dislocations in General

The geometry of dislocations can be described in a more general manner which shows how the edge and screw dislocations are related to one another. Consider a perfect crystal within which a closed curve C is drawn (Fig. 6-3a). The only restriction on the curve is that it be closed on itself or run out to the surface of the crystal. In addition, let a surface be drawn in the crystal which has the curve as a boundary. Of course, many surfaces with the same boundary can be drawn in a crystal, and no importance is attached to the particular one chosen. Let the atoms which face each other across the surface be disconnected from each other and the material on one side of the surface displaced from the material on the other side *rigidly* by the distance **b**, where **b** is a vector giving both the magnitude and the direction of the displacement. The vector **b** is the Burgers vector of the dislocation which will result. In general, the displacement may leave a cavity in the crystal, or it may make the material overlap, depending upon the direction of the displacement. If a cavity is left, it is filled up with new atoms; if overlap occurs, atoms are removed from the crystal until it fits together again. At this time the atoms are all reconnected to their new neighbors across the surface, and the crystal is allowed to relax into that configuration which minimizes its total energy. The dislocation line is defined to be the curve which forms the boundary of the surface. The distortions of the structure are concentrated along this boundary.

The general case just cited has special cases which are simple to see. The special case for the screw dislocation can be demonstrated by using Fig. 6-3b. The closed curve is the rectangle $ABCD$, consisting of a straight line through the crystal and three other lines along the surface. A shearing displacement of the material on the two sides of the plane $ABCD$ generates a screw dislocation along the line AB. The special case for the edge dislocation may be seen by using Fig. 6-3c. The defining curve is again a rectangle, $ABCD$. Let the material on the two sides of the plane be separated so that a cavity is produced. The cavity is filled with one or more planes of atoms to eliminate the void, and all atoms are allowed to relax to equilibrium positions. This sequence of events puts one or more extra planes of atoms the size of the original rectangle in the crystal, and the result is an edge dislocation.

The Burgers vector for these simple cases and for the general case, too, is a vector which describes both the direction and the magnitude

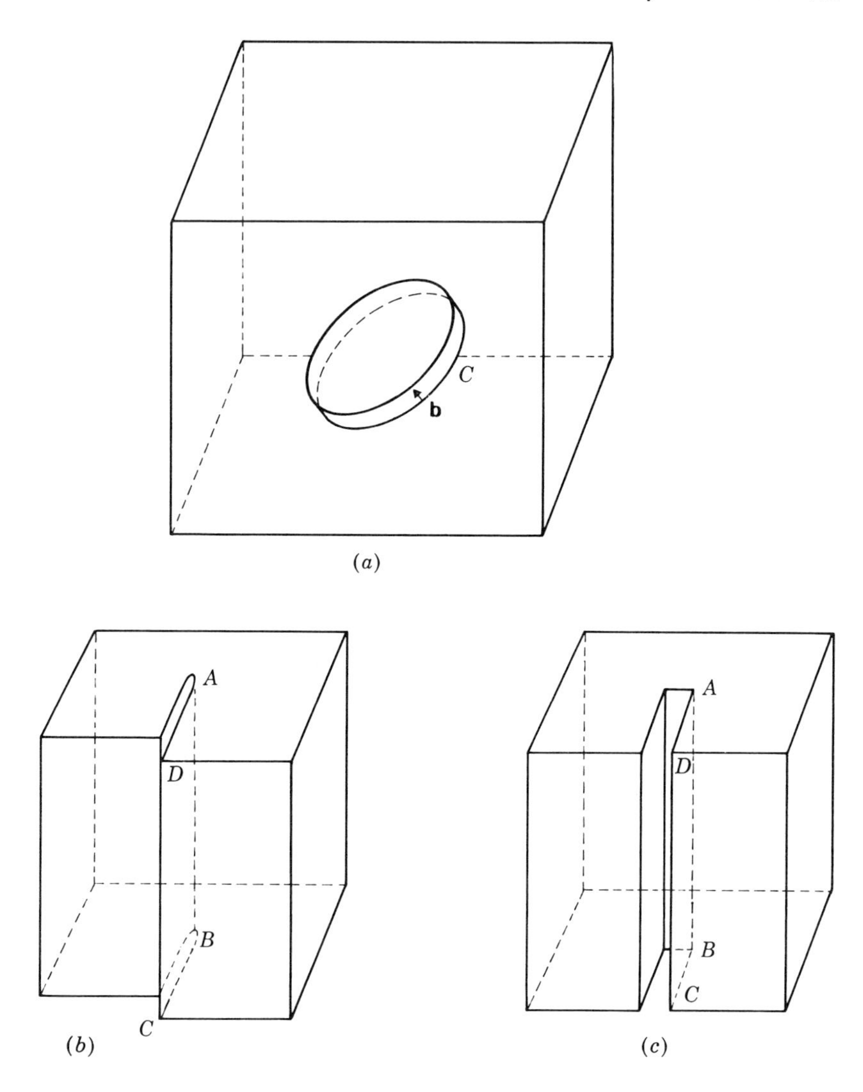

Figure 6-3 (*a*) General construction of a dislocation line; (*b*) special construction for a screw dislocation; (c) special construction for an edge dislocation.

of the rigid displacement. For the screw, the shear displacement, and hence the Burgers vector, lies along the dislocation line. For the edge, **b** is perpendicular to the line. For the general case, the Burgers vector may have other directions with respect to the dislocation, and for these cases the dislocation is a mixture of edge and screw types. The mixed dislocation is defined in terms of the direction of the Burgers vector. The geometry of these types is somewhat complicated, and no attempt is made to describe them here beyond the comment that they are simply the geometrical composite of both types.

So far, the size of the Burgers vector has not been specified. However, certain restrictions on its size and direction are necessary.

In Fig. 6-1 the Burgers vector is exactly one lattice spacing, because, when the material on the two sides of the cut is displaced, a chasm just wide enough for a single plane of atoms is made. If the Burgers vector was instead only half a lattice spacing, not enough room would be made in which to insert the extra plane of atoms and a simple dislocation could not be produced. *The Burgers vector therefore must be some multiple of the lattice spacing* or in general must be a "lattice vector" of the crystal so that when the new material is introduced, it fits exactly into the empty region made for it in regions far from the dislocation line.

The screw is easily described in the same terms. If the Burgers vector of the dislocation in Fig. 6-2 were a fraction of the lattice spacing, when a point makes a traversal of the dislocation it does not come back to a new plane of atoms but returns between two planes. Such a Burgers vector does not describe a simple dislocation. In general, in all instances where the Burgers vector is not a lattice vector, the cut plane of the construction of Fig. 6-3 is a plane where the atoms are out of registry with one another, and the result is a dislocation coupled to a plane of misfit. Although there are cases in nature where dislocations have "partial" Burgers vectors coupled with planes of misfit, they are not discussed here.

Clearly, the particular surface used in the general construction of a dislocation is not important. After the atoms are reconnected across the cut surface, almost complete accommodation of the atoms exists over the surface of the cut in regions far from the cut boundary and the fact that the cut was made in a particular area is obscured. The only important factors are (1) the Burgers vector whose relative direction specifies the general type of dislocation and whose size is a measure of the strength of the dislocation and (2) the curve bounding the cut which defines the position of the dislocation in the crystal.

6-4 Dislocation Energy

The energy of a dislocation can be estimated by assuming that the crystal behaves as an elastic solid during the process of creation of the dislocation. Calculation of the energy can easily be performed for a straight screw dislocation; however, the results are similar for an edge dislocation. Starting with a perfect crystal, let a cut be made in the way described in Sec. 6-2 for a screw dislocation, and let the two sides of the cut be displaced with respect to each other by the distance b in the prescribed manner. For the displacement to be made, a distribution of forces must be exerted over the surface of the cut, and the work done by the forces in making the displacement $\mathbf{b}$ is equal to the energy of the dislocation, U_D.

$$U_D = \int \mathbf{F} \cdot \mathbf{b} \, dA \tag{6-3}$$

The integral is evaluated over the area of the surface of the cut. The force **F** is the average force per unit area at a point on the surface during the displacement. The average value must be used because the force at a point builds up linearly from zero to a maximum value as the displacement is carried out.

To calculate the energy, we must have a quantitative estimate of the magnitude of the force **F**. The magnitude of **F** can be obtained by the following stratagem. Let the crystal be considered to be a series of concentric cylindrical shells, with the dislocation in the center.

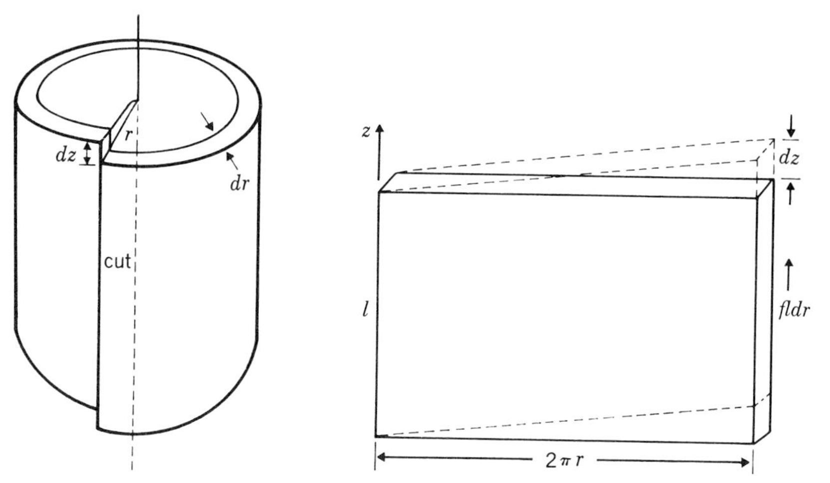

Figure 6-4 The cylindrical shell around a dislocation is opened out into a platelet of length l and width $2\pi r$, where r is the radius of the shell. The thickness is dr. The shell is sheared in the z direction as shown. The force $\mathbf{f}l\ dr$ is applied to the edge of the plate, where $l\ dr$ is the area of the edge and **f** is the force per unit area. It produces a displacement dz.

Each shell is cut along its cylindrical length where the cut surface intersects the shell (see Fig. 6-4). The shell on one side of the cut is then displaced with respect to the other side by the distance b. If the thickness of the shell is small, the geometrical configuration of the shell is not important for calculating the resistive force during the displacement. In particular, the force will be the same if the shell is opened out into a flat plate. Then the problem is reduced to consideration of the shearing displacement of a plate as shown in Fig. 6-4.

For small shearing displacements, Hooke's law surely holds. Therefore the force per unit area, **f**, necessary to shear by an amount **b** a cylindrical shell a distance r from the center of the dislocation into the final configuration is just

$$\mathbf{f} = \frac{\mu \mathbf{b}}{2\pi r} \tag{6-4}$$

where μ is the shear modulus of the material (values of μ and Y for some of the common solids are listed in Table 6-1). In the calculation of the energy needed to displace the cylindrical shell, the average force during the complete displacement to **b** is used.

Table 6-1 Elastic moduli and dislocation energies for various materials

	Y, Young's modulus, 10^{10} newtons/m²	μ, shear modulus, 10^{10} newtons/m²	U_D, ev/atom length
Al	2.5	2.85	3.1
Cu	6.0	7.56	5.3
Ag	12.0	4.4	4.5
Diamond	95.0	43.0	29
Ge	12.9	6.7	18
KCl	4.1	0.6	9.3
Si	16.7	7.9	19
W	50.0	15.1	13

The average force $\langle \mathbf{f}_{av} \rangle$ is just half the final value when the displacement is **b**. Thus $\mathbf{F} = \langle \mathbf{f}_{av} \rangle = \frac{1}{2}\mathbf{f}$. Hence

$$\mathbf{F} = \frac{\mu \mathbf{b}}{4\pi r} \tag{6-5}$$

Then the energy of the dislocation is

$$U_D = \int \frac{\mu b^2}{4\pi r}\, dA$$

Since $dA = dz\, dr$, U_D for a dislocation of length l is

$$U_D = \int_{r_0}^{r_1} \int_0^l \frac{\mu b^2}{4\pi r}\, dz\, dr \tag{6-6}$$

Thus

$$U_D = \frac{\mu b^2 l}{4\pi} \ln \frac{r_1}{r_0} \tag{6-7}$$

The energy calculated by using Eq. (6-7) depends on the values taken for the integration limits on r. For a crystal of infinite size, the energy of a single dislocation is infinite. However, crystals of ordinary size (say, 1 cm on an edge) contain many dislocations. These dislocations in a random distribution cancel out each other's strain fields at distances approximately equal to the mean distance between them. Experimental observations show that the mean distance of separation of dislocations in crystals is about 10^4 atom spacings; hence r_1 is of this order.

As $r \rightarrow 0$ in Eq. (6-7) or (6-5), the expression becomes divergent. However, because of the finite size of the atoms, we cannot consider any region of atomic dimensions as an elastic continuum, and elasticity theory ceases to be correct. It is reasonable therefore to consider the region within about one lattice spacing of the geometric center of the dislocation to be a void and to delete it from consideration. In the region within about two atom spacings of the center of the dislocation, the core, the discrete nature of the structure is important and displacements are sufficiently large to bring in nonlinear terms in the force law. Fortunately, the region outside the core where the strain does obey linear elasticity theory is so large that for many purposes the core effects can be neglected. For example, the energy of the dislocation [Eq. (6-7)] resides primarily in the strained region well outside the core, because the logarithm is not very sensitive to the size of r. If r_0 is about one or two atom spacings, the ratio r_1/r_0 is about 5×10^3 in Eq. (6-7).

Once this ratio r_1/r_0 is determined, the energy U_D can be calculated easily. The Burgers vector $\mathbf{b}$ is about 2.5 A; μ is about 10^{11} newtons/m^2. Then U_D is about 4×10^{-9} joule/m of dislocation length, equivalent to about 10×10^{-19} joule (or about 6 ev) per atom length (about 2.5 A) of dislocation line. Considerable variation in energy can exist for different solids, but the range of energy for screw dislocations should be about 3 to 10 ev per atom length. Values calculated by using Eq. (6-7) for several solids are tabulated in Table 6-1.

The displacement around the edge dislocation is more complicated than for the screw; so the energy calculation is more involved. However, the energy in this case is about the same as Eq. (6-7).

One of the important characteristics of dislocations is the fact that they are athermal. Unlike vacancies, which at equilibrium exist in significant numbers given by the Boltzmann factor $e^{-E_v/kT}$, dislocations in solids seem to have nearly zero equilibrium densities. The reason is that the formation energy of a dislocation is very high. With an elastic energy of about 8 ev per atom length, the Boltzmann factor is insignificant at all normal temperatures. Thus the dislocation densities experimentally observed depend upon the previous treatment of the crystal, the method of growth, previous mechanical treatment, previous thermal history, etc.

6-5 Dislocation Motion: Slip and Climb

Just as point defects can move about in the lattice, so can dislocations. The dislocation is obviously more constrained in its motion since it must always be a continuous line. Two general types of motion are possible, termed climb and slip.

Climb is the name given to the motion of a dislocation with an edge component when the extra half plane is extended farther into the crystal or withdrawn partially. Since the dislocation is defined as the end of the extra plane, as the plane is extended or withdrawn, the dislocation moves with it. Climb is a process which is observed in real crystals, but a dislocation in an actual crystal does not climb, of course, by moving the extra half plane bodily farther into the

Figure 6-5 Climb of a dislocation by (*a*) the addition or (*b*) subtraction of atoms from the extra half plane.

crystal. The same effect is realized, however, if extra atoms are added individually from the crystal to the end of the plane (Fig. 6-5*a*). These atoms may have existed within the crystal in the form of interstitials, for example. The reverse process is also called climb and occurs when vacancies from the lattice are annihilated at the bottom of the extra half plane, causing it to recede from the interior of the crystal (Fig. 6-5*b*).

A more analytic way of describing the motion of the dislocation in climb is in terms of the slip plane of the dislocation. The slip plane is defined for a straight dislocation as that plane which contains both the Burgers vector of the dislocation and the line of the dislocation. It is shown in cross section for an edge dislocation in Fig. 6-6. Climb of the dislocation corresponds to motion of the dislocation either up or down from this plane. When the motion of the dislocation is downward in Fig. 6-6, the dislocation absorbs additional atoms from the crystal, and when it is upward, the dislocation absorbs vacancies.

Four different mechanisms can cause dislocations to climb: (1) Vacancies from the lattice can be annihilated at the bottom of the extra plane. (2) Vacancies can be produced in the lattice from the bottom of the extra plane when a neighboring atom jumps to the end

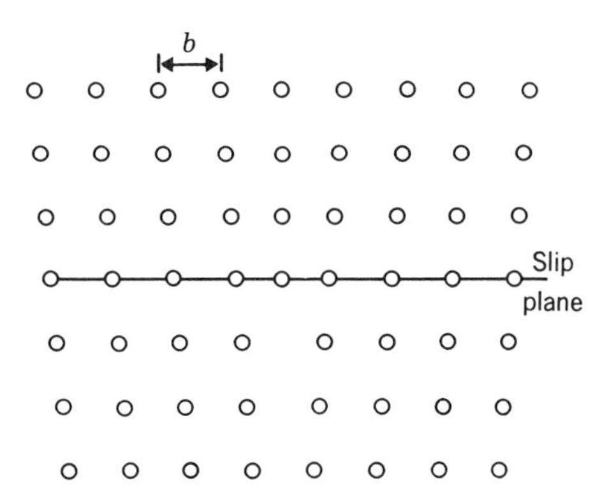

Figure 6-6 Slip plane of an edge dislocation.

of the extra plane, extending it into the crystal. The vacancy thus formed is free to diffuse into the bulk of the crystal. (3) Interstitial atoms can attach themselves to the bottom of the plane. (4) Interstitial atoms can be created within the crystal when atoms jump off the end of the extra half plane into neighboring lattice interstitial positions.

The climb of the edge dislocation just described is reasonably straightforward and is valid for a general dislocation containing an edge component. The screw dislocation, however, has no unique slip plane, because the Burgers vector is parallel to the dislocation line. Thus no rigid motion of the screw dislocation corresponds to climb.

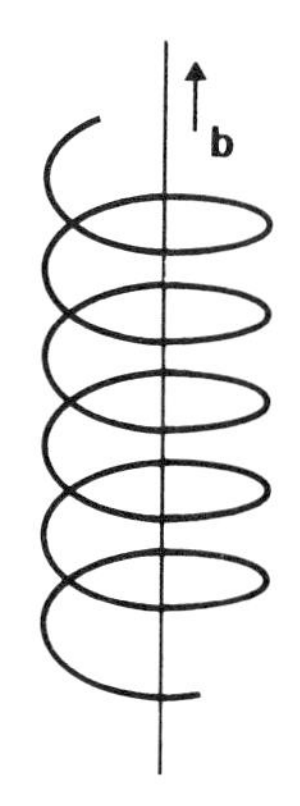

Figure 6-7 Screw-dislocation climb. The screw dislocation generates an edge component in climb by formation of a helix. An example of such a helix is shown in Fig. 6-17.

This result is also consistent with the fact that the screw dislocation has no extra half plane. A screw dislocation does have, however, a quite complicated motion which has been observed and which does correspond to climb. If the screw-dislocation line twists itself into a spiral, then the spiral must have an edge component and can climb. The climbing configuration is shown in Fig. 6-7. Climb causes the spiral to expand radially.

Slip is a result of motion of a dislocation of quite another sort. Motion of an edge dislocation normal to the slip plane corresponds to the addition or subtraction of atoms from the extra half plane; motion of the dislocation within the slip plane does not involve the diffusion of atoms to or from the line. This latter motion is much easier since it does not have to await the slow rearrangement of atoms attendant upon diffusion. A careful study of the atom arrangement in Fig. 6-8 shows that, when an edge dislocation moves from one lattice site to another on the slip plane, the atoms in the core move in a small, shuffling motion so that the extra half plane in the configuration at one lattice position becomes connected to a plane of atoms below the slip plane and the neighboring plane of atoms becomes the new extra half plane. This type of motion of the dislocation line is called slip, or glide.

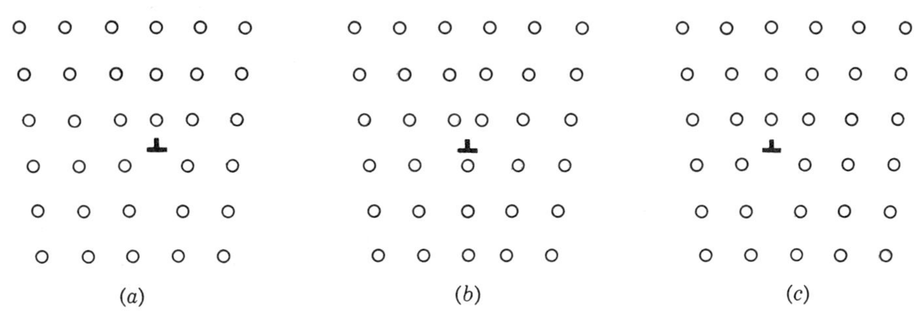

Figure 6-8 Slip of a dislocation showing that small, shuffling rearrangements of the atoms of the core of the dislocation are equivalent to motion of the dislocation from one lattice position to the next.

The importance of slip of dislocations is easily seen from the series of figures in Fig. 6-9. When the dislocation begins on one side of the crystal and moves through to the opposite surface, the top half of the crystal moves one lattice spacing with respect to the bottom half of the crystal. The glide motion of a dislocation line is thus induced when a shearing couple is applied to the surface of the crystal.

Glide motion is equally possible for the screw. However, as already remarked, all planes which contain the screw dislocation also contain the Burgers vector, and so the pure screw dislocation can move by glide in any direction. Thus all planes containing the screw

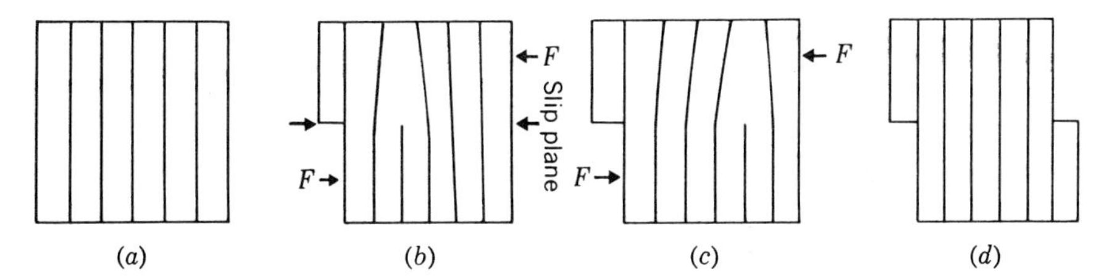

Figure 6-9 Motion of an edge dislocation through a crystal along its slip plane. When the dislocation moves completely from one surface to the other, the top of the crystal slips with respect to the bottom by a distance of one Burgers vector.

dislocation are glide planes. (All dislocation lines which are not exactly parallel to their Burgers vectors have only one unique slip plane.) Passage of a screw dislocation through a crystal alters the shape of the crystal in a somewhat different way from the passage of an edge dislocation through it. A study of the arrangement of Fig. 6-2 shows that the ledge does not form parallel to the screw dislocation as it does for the edge (Fig. 6-9); rather the ledge appears on the surface of the crystal at the two ends of the dislocation, and as the dislocation moves through the crystal, the ledge trails behind it (Fig. 6-10).

The slip plane has been defined only for the special case of straight dislocations. If the line of the dislocation is curved, the slip surface

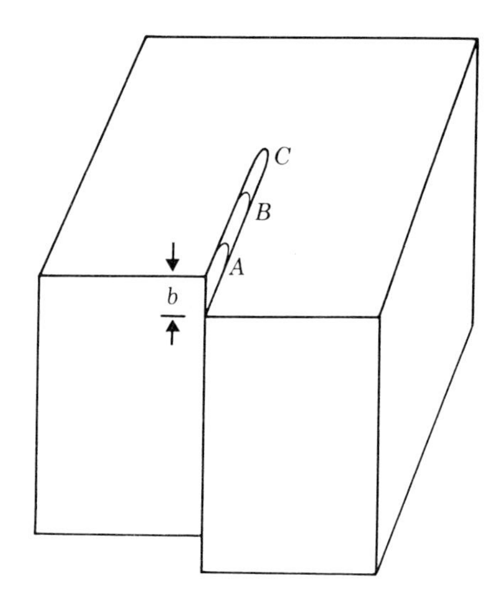

Figure 6-10 Moving screw dislocation. As the dislocation moves from A to B to C, the ledge of height b trails behind it, causing slip in the direction of $\mathbf{b}$.

is that cylinder generated when the Burgers vector moves around the dislocation (Fig. 6-11).

6-6 Plastic Deformation

The previous discussion has treated the dislocation only as a logically possible type of defect in a crystal lattice. However, the dislocation has enormous practical importance because it is the determining factor in the mechanical strength of a solid. This role of the dislocation is most easily brought out by the consideration of a stress-strain diagram.

Suppose that a long, uniform wire is placed in an arrangement where it can be stretched under a force. If the stretching force is plotted as a function of the wire extension, a curve results which typically has the form plotted in Fig. 6-12. (Actually, in normal practice, the stress σ, defined as the force divided by the cross section

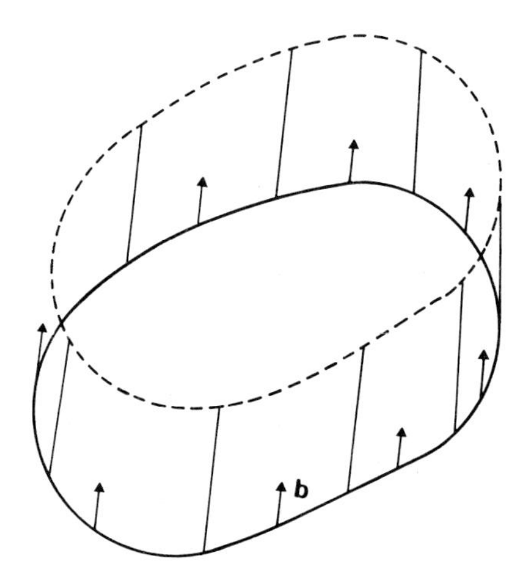

Figure 6-11 The slip surface of a curved dislocation.

of the wire, is plotted against the strain ϵ, defined as the extension of the wire divided by the total length. In these units, the plot is characteristic of the material and not of the particular geometry of the wire.) Figure 6-12 shows several characteristic features. First, the plot for small stresses shows a linear response which is in agreement with Hooke's law. In this elastic region, the stress and strain are proportional to each other. Besides linearity, the elastic range of stress also has the property that, when the stress is removed, the solid reverts to its original shape. When the weight is removed, for

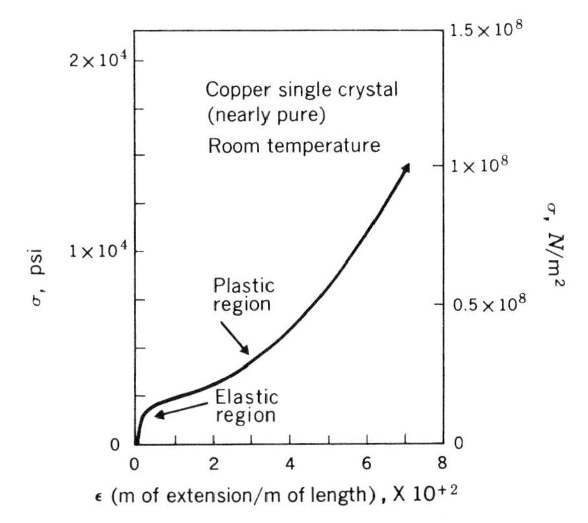

Figure 6-12 Stress-strain diagram for a typical crystal. The initial linear region is followed by a plastic region with smaller slope. As plastic deformation increases, the slope increases likewise in the phenomenon of work hardening. *(Data obtained by H. Birnbaum.)*

example, the wire returns to its original length. Beyond the elastic range, the curve bends over in a rather abrupt manner to the *plastic* range. In this region the wire does not return to its original length when the force is removed. In the plastic region the slope of the curve is much smaller than before but gradually increases as the strain increases. Finally, of course, as the stress is increased further, the crystal breaks.

The maximum range of elastic strain is normally very small, commonly of the order of 10^{-3} or less. Although the elastic range is small, some crystals, notably metals, are very ductile and can be deformed by large amounts without fracture. Some soft metals, in fact, can sustain plastic strains of 10 or more.

The most important feature of the plastic range is its nonreversibility. Once the crystal has been extended into the plastic range, it does not return to its original shape or length when the stress is removed. A permanent deformation has been introduced. This difference between an elastic deformation of the crystal and a plastic deformation is a fundamental one which is reflected on the atomic scale. In the elastic range, every atom of the crystal is displaced slightly from its lattice site, and the deformation throughout the crystal is uniform for a uniform stress. On the other hand, plastic deformation is very heterogeneous, often so much so that slip bands appear on the surface of the crystal (see Fig. 6-13). The slip bands are approximate planes in the crystal upon which large shear motion occurs. The planes are generally planes of high symmetry in the crystal, and the direction of shear displacement in the planes is also a direction of high symmetry. Characteristic slip planes and slip directions for a few simple crystals are listed in Table 6-2.

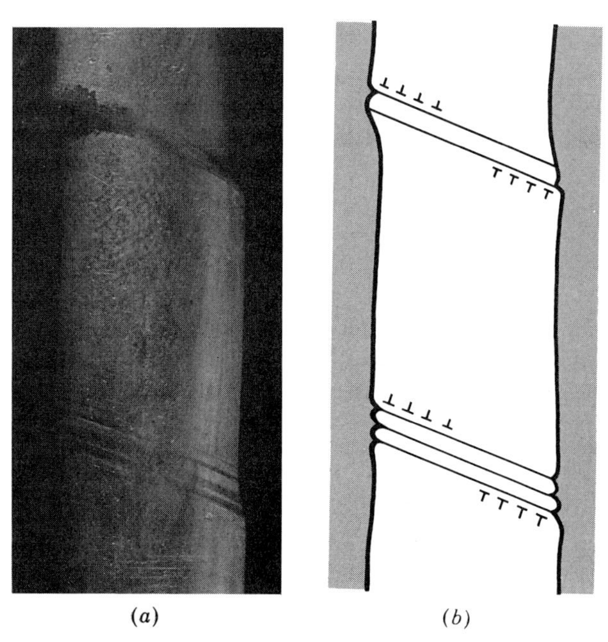

Figure 6-13 (*a*) Slip bands on a deformed crystal. (*This photograph first appeared in A. Guy, "Physical Metallurgy," Addison-Wesley Publishing Company, Inc., Reading, Mass. Reproduced with the permission of the publisher.*) (*b*) Schematic drawing showing representative motions of some of the dislocations which produce the slip bands in (*a*).

The connection between plastic slip and dislocations is easily guessed from a reexamination of the sequence in Fig. 6-9. The result of the passage of a dislocation through a crystal is a slip step of atomic scale. The Burgers vector gives the direction of the slip, while the slip plane of the dislocation is the same as the plane of the slip band of the crystal. Thus slip bands can be the result of the passage of a large number of identical dislocations on closely neighboring slip planes. Observations bear out this supposed cause of plastic flow: (1) Dislocations can be visually inspected in crystals (see Figs. 6-14 to 6-18), and above the critical shear stress they multiply and move in the proper planes. (2) The production of slip bands is observed to be caused by the increase in numbers of dislocations on the same slip plane or on closely spaced slip planes. Direct observation of dislocation multiplication is observed in Fig. 6-18.

Striking techniques have been developed to enable one to study single dislocations and groups of dislocations. Figures 6-14 to 6-18 serve to illustrate these techniques for several kinds of crystals. The captions to these figures give further details.

Table 6-2 Slip directions and slip planes

Crystal	Crystal type	Slip plane	Slip direction
Al	Face-centered cubic	(111)	[110]
Cu	Face-centered cubic	(111)	[110]
Ag	Face-centered cubic	(111)	[110]
Au	Face-centered cubic	(111)	[110]
Fe	Body-centered cubic	(110) (112) (123)	[111]
W	Body-centered cubic	(112)	[111]
Mg	Hexagonal close-packed	(0001)	[1120]
Zn	Hexagonal close-packed	(0001)	[1120]
Cd	Hexagonal close-packed	(0001)	[1120]
NaCl	NaCl	(110)	[110]
NaF	NaCl	(110)	[110]
KCl	NaCl	(110)	[110]
KBr	NaCl	(110)	[110]
KI	NaCl	(110)	[110]
Ge	Diamond	(111)	[110]
Si	Diamond	(111)	[110]

6-7 Dislocation Mobility and Multiplication

The crucial factor in the role played by dislocations in plastic deformation is the relatively small stress needed to cause dislocation motion and dislocation multiplication. Both motion and multiplication are important, and each is discussed in turn.

Consideration of the relative weakness of crystals leads to the conclusion that a defect like the dislocation is a necessity. We pointed out that crystals normally sustain only a very limited elastic shear strain before the crystal either flows plastically or fractures. A crystal which contains no dislocation should deform elastically to a strain of approximately $\frac{1}{2}$ before the onset of plastic flow. Instead plastic flow is observed at a strain of only 10^{-4} to 10^{-3}. Why can dislocations trigger plastic flow at the observed small strains and associated small stresses? On an atomic scale, the only way a deformation can be made permanent is for the atoms involved to move permanently from one lattice site to another. If an atom is deformed with respect to its neighbors by only a small fraction of a lattice distance, then when the force is withdrawn the atom will simply fall back into its original lattice position and the deformation is elastic.

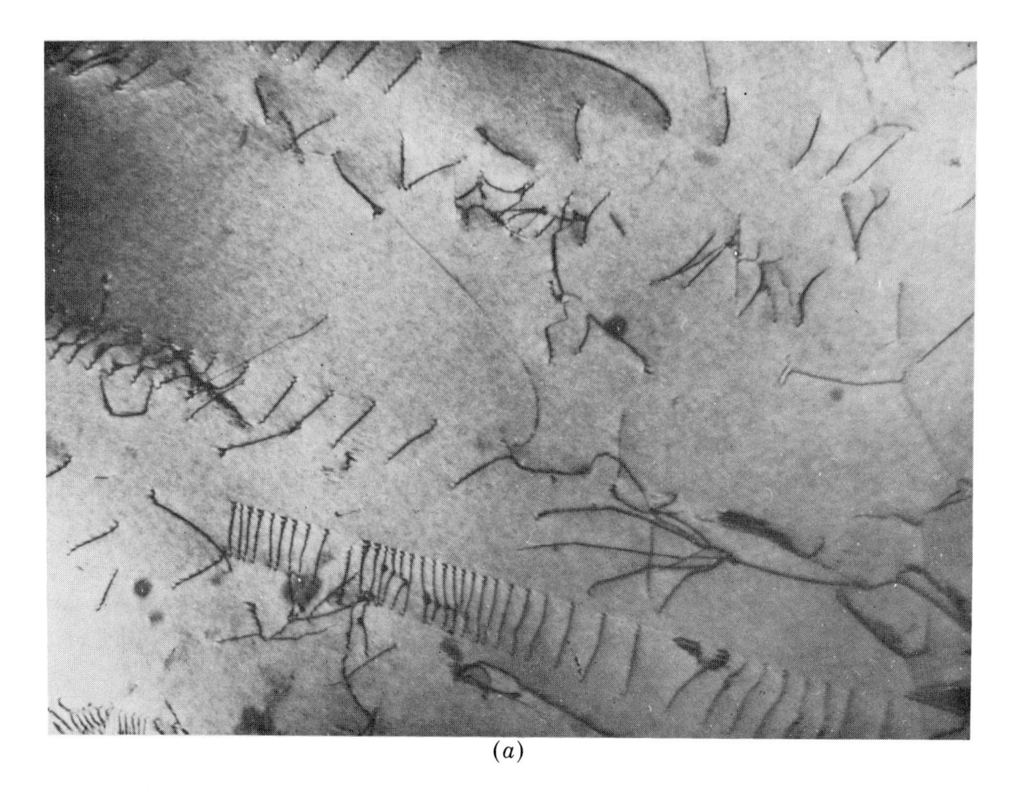

(a)

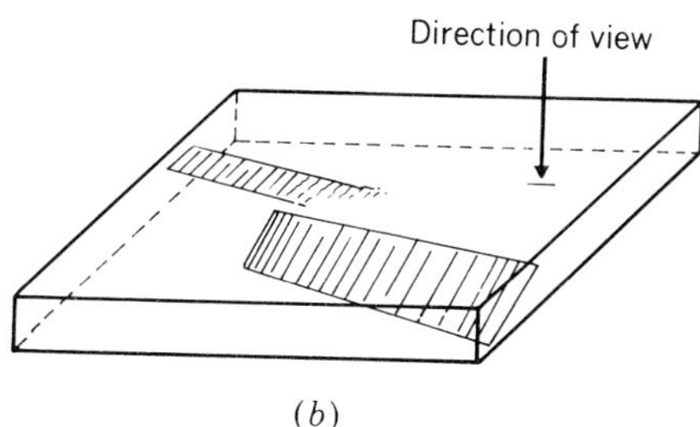

(b)

Figure 6-14 (*a*) Electron-microscope picture of dislocations in a thin foil of Cu + 4 per cent Al. Magnification about 30,000 times. The regular array of dislocations diagonally across the picture from the lower right has apparently originated from a common source. (*Unpublished work of A. Arts.*) (*b*) Oblique sketch of the foil shown in Fig. 6-14*a*, showing the geometry of some features seen in the photograph.

Consider these arguments in detail. Suppose that a uniform shear stress is induced throughout the crystal, that no dislocations are present, and that the stress is increased until the atoms ride over their neighbors into the next lattice positions. Fig. 6-19 shows, in general, that restoring forces are present on the atoms for relative displacements between the atoms of approximately one-half the lattice spacing. When the lattice deformation becomes approximately this value, then the atoms are in a position of unstable equilibrium: they fall back if the stress is decreased but ride into the next atom position

Figure 6-15 Electron microscope picture of a dislocation network in a thin foil of the semiconducting compound 67 per cent Bi_2Te_3–33 per cent Sb_2Te_3. Magnification 20,000 times. These dislocations have coalesced into a hexagonal network through dislocation interactions between the original dislocations in the crystal. This phenomenon is a frequent occurrence when a crystal containing many dislocations is heated to a high temperature for a long period (annealed). (*Courtesy of J. N. Bierly, Franklin Institute Research Laboratories.*)

if the stress is increased a little more. The restoring force during such a large displacement cannot be described exactly in terms of Hooke's law; nevertheless, the order of magnitude of the maximum restoring force during the deformation will be the same as if the forces were assumed to be Hookean up to deformations of the order of one-half lattice spacing. Consider the following crude estimate of critical shear stress: Hooke's law for the deformation is given by $\sigma = \mu\epsilon$. The strain ϵ in this case is $\frac{1}{2}$ when an atom is displaced one-half lattice spacing relative to its neighbor. Thus the stress necessary to produce elastic strains of this magnitude is about half the shear modulus, which itself is of order 10^{11} newtons/m^2. More refined estimates place the theoretical shear stress perhaps an order of magnitude lower. Hence the stress necessary to cause plastic deformation without dislocations is about 10 per cent of the shear modulus μ, or about 10^{10} newtons/m^2.

The experimental facts are that pure single crystals with only a few dislocations deform plastically at stresses of order 10^5 to 10^6 newtons/m^2, or at least four or five orders of magnitude less than their shear moduli. Hence dislocations must be important, not only in

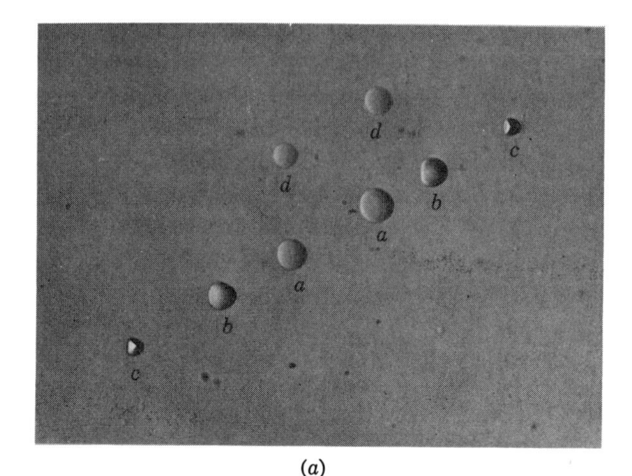

(a)

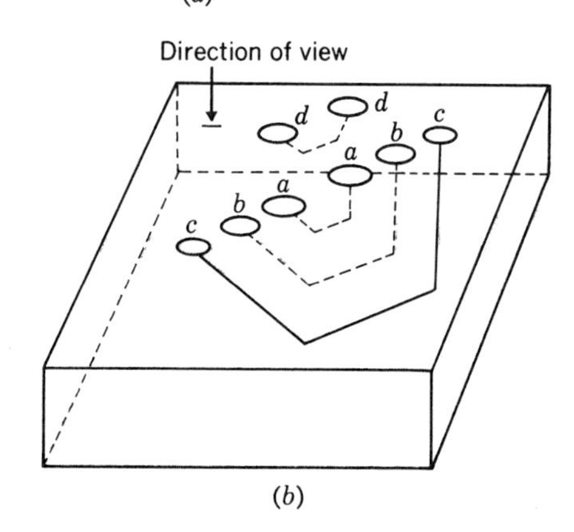

(b)

Figure 6-16 (a) Dislocation etch pits on a highly polished surface of Ge. Where a dislocation intersects the surface, the highly strained region near the core is rapidly attacked by particular chemical etchants and a depression, or pit, is formed. These pits show the successive positions of a dislocation loop which moved successively from position *a* to *b* to *c* under successive increments of stress. The loop at *d* has apparently moved completely out of the field of view. (*After Kabler.*) (*b*) Sketch of dislocation half loops which end in the etch pits shown in Fig. 6-16*a*.

producing deformation within solids, but also in being mobile under stresses which are much less than the shear modulus. Further inspection of the shuffling motion of the core atoms depicted in Fig. 6-8 as the dislocation moves shows the reason. When a dislocation moves from one lattice row to the next, only a very minor atomic rearrangement is necessary. Since the atoms in the core of the dislocations are highly distorted from their normal lattice positions, some of them are in relative positions where they can move past one another with a very small shear force. It happens that the shear force necessary to move a dislocation from one lattice site to the next is related to the size of the core region of the dislocation. If the core is large compared with a lattice spacing, the shear force necessary to displace the dislocation becomes vanishingly small; if the dislocation core is only one lattice distance, the shear force approaches the shear modulus.

As suggested at the beginning of this section, most crystals in the pure-single-crystal state with low dislocation content are soft and ductile, showing that normally dislocations are easily moved under small stress. As a group, the metals are good examples of ductile crystals. Some crystals, however, like diamond, Ge, and Si do not

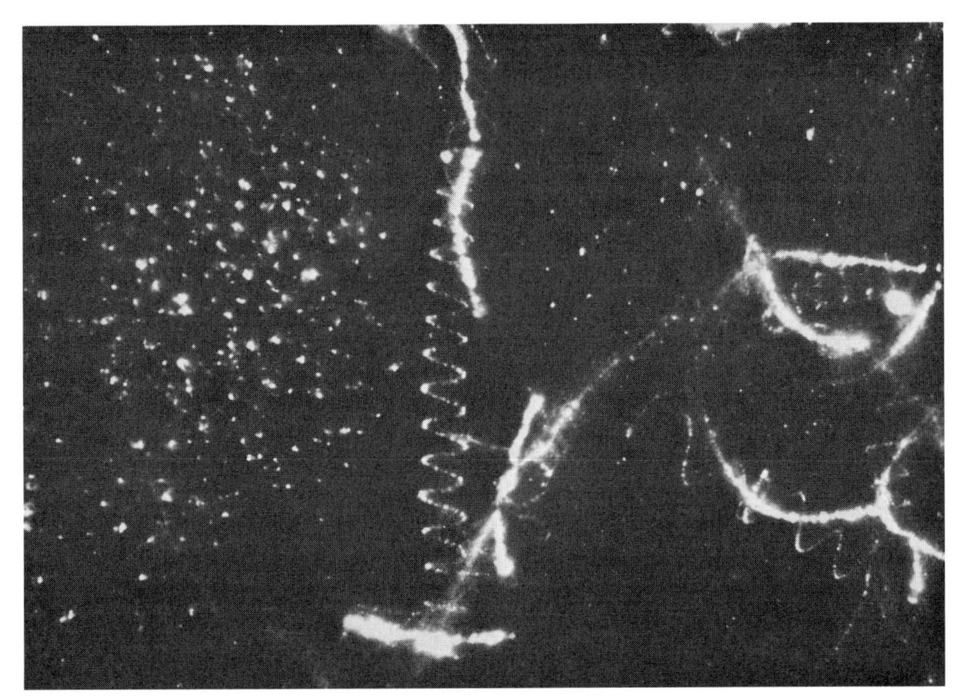

Figure 6-17 Observation of dislocations in CaF_2 by decoration. Dislocations in transparent ionic crystals can sometimes be made visible by precipitating impurity atoms along their cores; the precipitate is viewed through an optical microscope. This picture shows several spiral screw dislocations. [*W. Bontinck and S. Amelinckx, Phil. Mag.*, **2**, 94 (1957).]

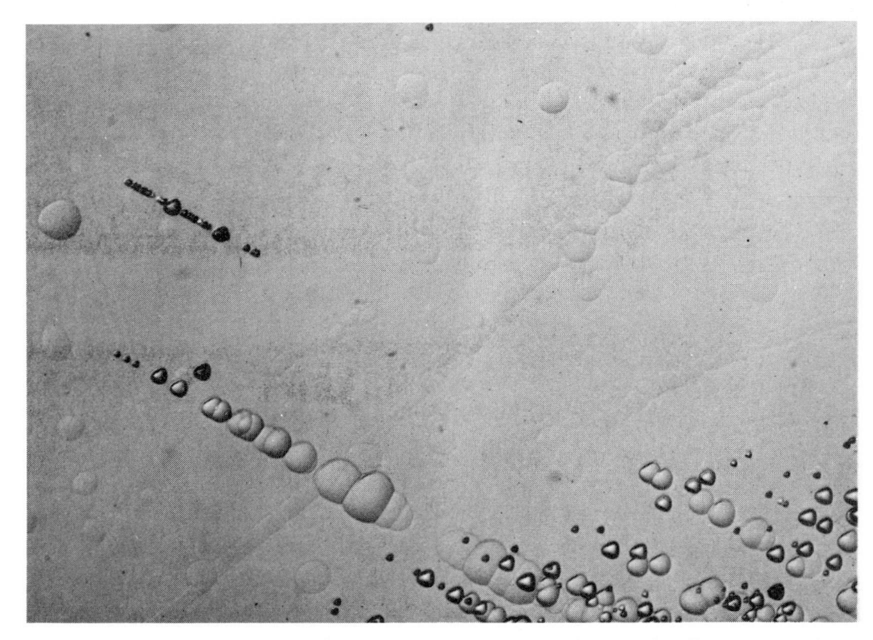

Figure 6-18 Dislocation multiplication in Ge as shown by increase in density of etch pits along a slip plane at upper left. (*After Kabler.*) The crystal has been etched twice in the manner explained in Fig. 6-16. In the first etch, only a single dislocation loop was present; in the second etch, the original loop has multiplied manyfold.

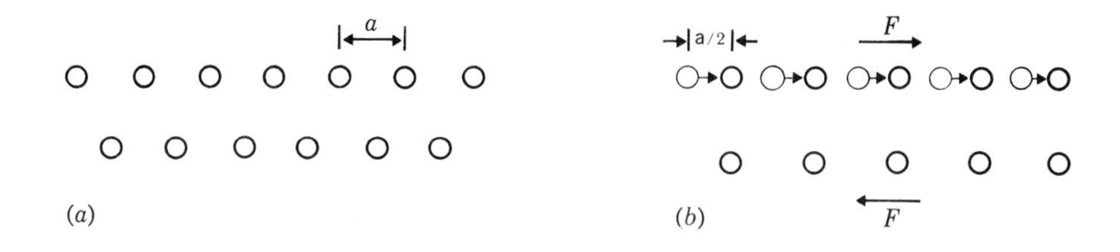

Figure 6-19 The perfect crystal lattice (*a*) will become plastic when adjacent rows of atoms roll over one another as shown in the figures. The critical point of instability (*b*), when the restoration forces are overcome, comes when the upper row has become displaced by one-half lattice spacing.

deform plastically at room temperature but are brittle and fracture when the stress is raised sufficiently. The brittle behavior of these crystals is caused by fracture processes being initiated before the dislocations can cause large-scale slip. Dislocations in these materials are probably very narrow compared with those of metals. Even for Ge and Si, however, the crystals become ductile as the temperature is raised. The increased ductility at high temperature is not only a property of these particular materials but also a general property of nearly all solids.

If dislocations are to explain the plastic deformation of crystals during plastic flow, dislocation multiplication must occur: crystals do not initially contain enough dislocations to permit the large plastic strains observed. Examination of deformed crystals provides the experimental proof that multiplication must take place but does not suggest a mechanism. F. Frank and W. Read have suggested a possible mechanism, depicted in Fig. 6-20. The original dislocation segment is at 1 and under the stress bows out through the stages 2–4. With the collapse of the cusp at 5, it returns to the original configuration after the creation of an expanding loop. The Frank-Read generator is only one of many possible mechanisms for dislocation multiplication. Real crystals usually behave in somewhat more complex ways.

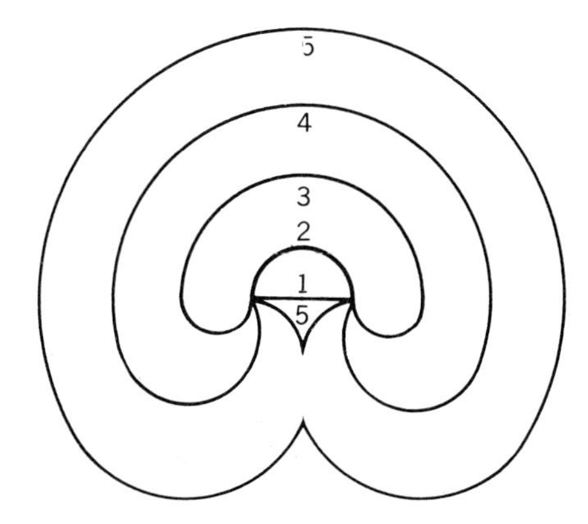

Figure 6-20 Frank-Read source. Under the stress the dislocation bows out, with a healing reaction at the rear, leaving the source in its original configuration at 5 with the production of an expanding loop.

6-8 Dislocation Interactions

The previous sections have demonstrated that dislocations provide the practical mechanism by which crystals deform plastically. They have also indicated that most pure single crystals are relatively soft compared with the so-called "theoretical" shear strength, $\simeq \mu/10$. As a practical matter, the mechanical strength of solids is actually extremely structure-sensitive. That is, the strength depends very strongly upon the way in which the crystal has been prepared. From the discussion of the previous sections it would seem that the critical stress for plastic flow of any crystal could be found by simply calculating the mobility of the dislocation. Predictions made on this basis

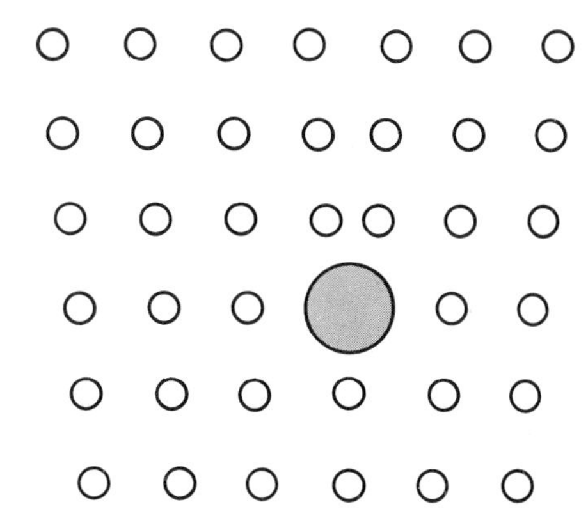

Figure 6-21 The interaction of impurity atoms at an edge dislocation. The large impurity fits more easily in the expanded region under the dislocation than in the perfect lattice.

are incorrect, however, since motion of dislocations is strongly perturbed by interactions with other imperfections (including other dislocations) in the crystal. In fact, the critical stress for plastic deformation of the crystal depends not so much on the intrinsic mobility of the dislocation in a perfect lattice as it does on the forces necessary to make the dislocation overcome other imperfections and obstacles that are also present in the crystal. This property is verified by the earlier statement that metal crystals grown with special care to eliminate all impurities, grain boundaries, etc., do in fact deform plastically with great ease.

The reason for strong dislocation interactions is that the severe displacements of atoms in the core of a dislocation can be greatly reduced when other imperfections are nearby to partly cancel the distortion. Impurities, in particular, can reduce the local distortion energy of the dislocation by a large amount. Figure 6-21 is a schematic picture showing the interaction of a large impurity with the expanded region of an edge dislocation.

In commercial practice the goal often is the fabrication of metals and alloys which are as strong as possible. Steel is a dramatic example. Pure single crystals of iron are difficult to grow but when obtained are very soft. They begin to deform plastically at about 5×10^5

newtons/m^2. On the other hand, steel, an alloy of iron, can be made to resist plastic deformation to stresses more than 100 times this amount. The alloy additions necessary for this strengthening—carbon, nitrogen, and metals such as Mn, W, Cr, Mo, and V—interact strongly, singly or in compounds, with the dislocations and impede their motion. Steel also has small grains so that a moving dislocation can glide only a relatively small distance before encountering a grain boundary which further impedes its motion.

Dislocations also interact strongly with other dislocations. In the beginning of plastic flow a nearly perfect crystal flows easily. However, as the deformation increases, the stress necessary to continue the deformation becomes larger. The reason is that the dislocation content increases greatly during the process of deformation, and the dislocations become tangled together in such a disordered fashion that further dislocation motion becomes difficult. This phenomenon, called *work hardening*, is one of the standard methods for obtaining a hard material. Thus a slab of steel might be hardened by passing it between rollers which roll the slab under great pressure into flat sheets. If the steel is maintained at a low temperature, the sheets retain many dislocations and are much more resistant to plastic deformation than the original slab. Work hardening is demonstrated by the fact that the curve in Fig. 6-11 rises sharply for large plastic deformations.

In summary then, the plastic deformation of solids depends upon the properties of dislocations in the solid. The ultimate strength of the solid, however, is not a simple matter but depends upon complex interactions between dislocations with themselves and with other imperfections in the solid. These interactions are so complex that predictions about the strength of solids cannot be made on the basis of simple theories concerning dislocations. However, dislocation theory does provide a qualitative understanding of the mechanical properties of solids.

PROBLEMS

6-1 The local strain at a point is defined in terms of a set of quantities like

$$\frac{\partial u_x}{\partial x}, \frac{\partial u_x}{\partial y}, \text{etc.}$$

where $u_x(r)$ is the displacement at the point r in the x direction. Compute the strain component

$$\frac{\partial u_z}{\partial y}$$

for a screw dislocation as a function of distance. The dilatation is given by

$$\frac{\Delta V}{V} = \frac{\partial u_x}{\partial x} + \frac{\partial u_y}{\partial y} + \frac{\partial u_z}{\partial z}$$

Prove that $\Delta V/V = 0$ for the screw.

6-2 Suppose that a prismatic dislocation is constrained to have a constant area and to have a rectangular shape, with the dislocations running in the cubic directions in the xy plane. If the dislocation lying along the x axis has twice the energy of that lying along y, what is the equilibrium ratio of the x and y lengths?

6-3 If there are 10^6 straight-edge dislocations passing through every square centimeter of a crystal, how much would each dislocation climb if the crystal were heated from 0 to 1,000°K? Assume $E_v = 1$ ev and $a = 2A$.

6-4 It has been speculated that one of the sources of dislocations in a crystal is the formation of prismatic loops from these vacancies. Suppose there is 10^{-3} atom fraction of vacancies in a crystal at the melting point. If it takes 10 sec to cool from the melting point to room temperature, calculate the density and radius of prismatic loops of dislocations in the crystal at room temperature. For the purposes of this calculation, assume that the "effective temperature" during which the vacancies diffuse to the loops is 1,200°K, that the crystal is divided into cubes, with one loop in each cube, and that the size of the cube is given by the "diffusion length" of vacancies for this temperature. The crystal is cubic, with lattice spacing 3 A, $E_m = 1$ ev, and $\nu = 10^{13}$ per second. (A prismatic dislocation loop is a pure edge loop which might be formed by the collapse of a platelet of vacancies.)

7

The physics of submicroscopic particles: quantum mechanics

In previous chapters, classical mechanics and Newton's laws could be used with little inconsistency because these chapters were concerned exclusively with the placement of atoms on lattices. Only in discussing the relatively subtle effects of atomic vibrations were quantum ideas necessary. Atoms, when considered as whole atoms, are large enough so that macroscopic Newtonian mechanics is usually adequate. However, the internal structure of atoms does not follow classical laws. Electrons, in particular, are so small in size and mass that ordinary macroscopic physics is not adequate for the description of their behavior; quantum mechanics is the result of the attempt to understand these electronic processes. Since subsequent chapters are concerned primarily with the electronic properties of solids, an elementary introduction to the ideas of quantum mechanics must be presented. Since this book is devoted to the properties of solids, no attempt will be made to give a complete treatment of the subject of quantum mechanics. The reader should seek this elsewhere. Some appropriate treatments of quantum mechanics are cited in the References at the end of the chapter. The purpose of the next two chapters is only to review and emphasize those aspects of quantum mechanics and atomic physics which are particularly useful in the study of solids.

7-1 Heisenberg's Uncertainty Principle

The most fundamental physical principle of quantum mechanics is the uncertainty principle. The fundamentals of this law can be understood by examination of a simple problem in Newtonian mechanics. Consider a projectile which starts from the origin with a given velocity and falls under the influence of gravity.

Newton's laws for the projectile are

$$F_y = m\ddot{y} = -mg$$
$$F_x = m\ddot{x} = 0 \tag{7-1}$$

where m is the mass of the particle and g is the acceleration due to gravity. The solutions of the differential equations (7-1) are simple and familiar.

$$y(t) = -\tfrac{1}{2}gt^2 + \dot{y}_0 t + y_0$$
$$x(t) = \dot{x}_0 t + x_0 \tag{7-2}$$

The future positions of the projectile, $x(t)$ and $y(t)$, are completely determined if the initial conditions x_0, $\dot{x}_0$, y_0, and $\dot{y}_0$ are known and a unique trajectory is thereby completely specified.

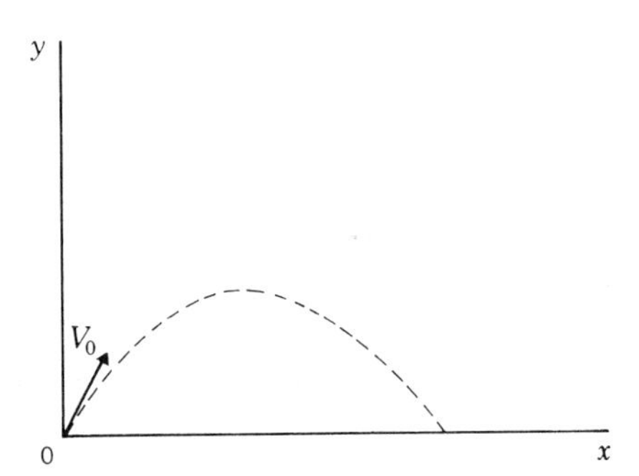

Figure 7-1 The trajectory of a projectile shot from the origin of the coordinate system with velocity v_0 is an inverted parabola.

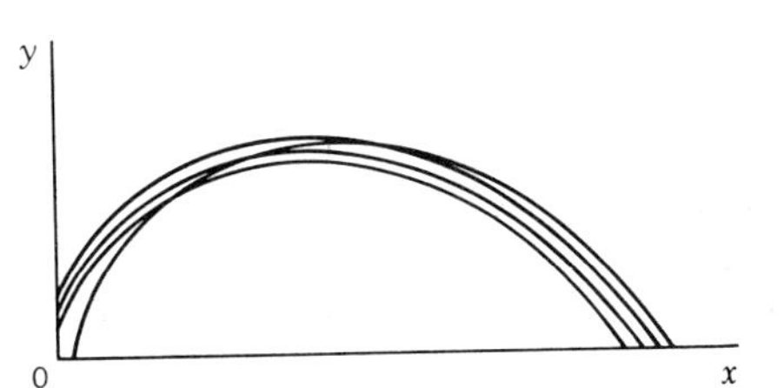

Figure 7-2 Possible trajectories of a projectile with somewhat inexact initial conditions.

Quantum mechanics makes a qualifying statement about the boundary conditions of this problem. The Heisenberg uncertainty principle states that simultaneous measurements of both the initial position and the velocity of any particle cannot be made with exactness. Hence, the unique trajectory of Fig. 7-1 cannot be determined according to the uncertainty principle because the initial conditions cannot be specified exactly. The trajectory becomes somewhat "fuzzy," more or less as pictured in Fig. 7-2. Hence, the most complete knowledge one can obtain concerning the projectile is found by solving for all the trajectories which lie within the probable range of the initial conditions. At any time during its flight the most that can be said is that the particle lies within some region of space and has a velocity in a certain range.

This example shows the basic meaning of the uncertainty principle and its result, but a more precise statement is possible. The Heisenberg uncertainty principle states that pairs of variables exist in physics which have an ultimate inherent error imposed upon them jointly; it is usually written quantitatively as

$$(\Delta p_x)(\Delta x) \geq h \tag{7-3a}$$

Here the momentum p_x and the position x are the two parameters of the system which are connected by the uncertainty principle; Δp_x and Δx are the uncertainties in p_x and x, and h is Planck's constant ($h = 6.62 \times 10^{-34}$ joule-sec). The two variables, when they are connected by a relation like (7-3a), are called *conjugate variables*. The most common conjugate variables are position and momentum, and energy and time. Hence, for all these variables,

$$\begin{aligned} (\Delta p_x)(\Delta x) &\geq h \\ (\Delta p_y)(\Delta y) &\geq h \\ (\Delta p_z)(\Delta z) &\geq h \\ (\Delta E)(\Delta t) &\geq h \end{aligned} \tag{7-3b}$$

Note that the uncertainty principle is stated in terms of a product of uncertainties, and the product of the uncertainties is Planck's constant h. The error imposed by Eq. (7-3b) is thus a joint, or reciprocal, property of the two variables taken together. For example, one can make an indefinitely accurate measurement of p, but such a measurement would imply a correspondingly large uncertainty in the measurement of the position of the particle.

An important feature of the uncertainty principle is that it is indeed a principle of physics: it has nothing to do with the details of experimental apparatus. For any real experimental measurement, of course, the crudity of the measuring apparatus gives rise to a range of error in the measurement. In measuring a length, the measurement cannot be made to more than a certain number of significant figures, because the scales on rulers can be read by eye with only finite accuracy. In weighing an object, friction in the bearings of the balance imposes a limit on the sensitivity of the balance. This type of error, however, can be improved upon by building better measuring devices. In principle one can always imagine an apparatus, in classical or macroscopic physics, which can make a measurement to any finite accuracy. In fact, in classical physics, construction of a measuring apparatus of any finite precision is only a matter of care and expense. The uncertainty principle, on the other hand, is a different kind of statement. It states that any measurement contains a "built-in" error which cannot be improved upon, no matter what the sophistication of the measuring apparatus: the uncertainty relation is thus a theorem concerning the limitations of measuring devices. To repeat, one cannot build a measuring device which will make a simultaneous measurement of position and velocity to any greater accuracy than is implied in the uncertainty principle.

From the standpoint of macroscopic physics, the uncertainty principle is a curious statement, and its meaning is fully realized only by considering the basic act of measurement. When the position of a particle is measured, in the very process of measurement one must "grab hold" of the particle by some means. One can bounce a light photon off the particle in order to see it, or one can place it in some kind of little, rigid box whose position is known. Obviously such procedures disturb the particle so much that its momentum is not what it was before the position measurement. Hence, if two consecutive measurements are made, the first of position and the second of momentum, the accuracy of the earlier measurements of position is destroyed by the crude handling to which the particle is subjected in the momentum measurement. The complete subtleties of measurement are beyond this brief introduction, but the interested reader is referred to the beautifully written discussion by Heisenberg.†

Of course, the uncertainty principle imposes significant uncertainties only in the microscopic realm. For example, the uncertainty in measurement of position of a mass of 1 kg whose velocity has an uncertainty of 1 m/sec is about 10^{-34} m, a distance some 10^{21} times smaller than the radius of the electron. However, for an electron in orbit in a hydrogen atom, the uncertainty is of the order of 10^{-10} m, or about the size which is usually ascribed to atoms themselves.

7-2 Wave Functions and Schrödinger's Equation

Quantum physics could proceed in just the way described in connection with the uncertain trajectories of Fig. 7-2. A map could be constructed of all the possible trajectories consistent with the range of initial conditions given by the uncertainty relations, and the approximate position of the particle at future times could be predicted. However, a more satisfactory method is to introduce at the start the concept of the probable behavior of the particle by means of some function which represents the probability of position of the particle. Then the equations for the probability function are developed directly. Historically, this latter method is the one which has actually been pursued, and through use of the Schrödinger equation the probability function is obtained.

Customarily a function $\Psi(x,t)$ is defined by a relation giving the probability density that the particle be at the position x at the time t,

$$\text{Probability density} = \Psi^*(x,t)\Psi(x,t) \tag{7-4}$$

† W. Heisenberg, *Physical Principles of Quantum Theory*, University of Chicago Press, Chicago, 1930.

In general, Ψ is a complex quantity. Since the probability must be a real quantity, Ψ must be multiplied by its complex conjugate $\Psi^*(x,t)$ to give the probability density. Since $\Psi^*(x,t)\Psi(x,t)\,dx$ is the probability that the particle is in the interval between x and $x + dx$ at time t and since the probability is equal to unity that the particle be somewhere in space at any time t, then

$$\int_{-\infty}^{\infty} \Psi^*(x,t)\Psi(x,t)\,dx = 1 \tag{7-5}$$

With the decision to forgo the description of the particle in terms of trajectories obtained from Newton's law and to define a probability function instead, an equation equivalent to Newton's laws must be introduced to give a prescription for finding Ψ in a particular physical problem. As mentioned previously, Schrödinger's equation is the equation sought. In one dimension, it is

$$\frac{-h^2}{8\pi^2 m}\frac{\partial^2\Psi(x,t)}{\partial x^2} + V(x)\Psi(x,t) = \frac{-h}{2\pi i}\frac{\partial\Psi(x,t)}{\partial t} \tag{7-6a}$$

where h = Planck's constant
m = mass of the particle
$V(x)$ = potential energy of the particle at the point x

Equation (7-6a) is written in only one dimension for simplicity. In three dimensions, the unknowns are functions of all three coordinates, and $\partial^2\Psi/\partial x^2$ is replaced by $\partial^2\Psi/\partial x^2 + \partial^2\Psi/\partial y^2 + \partial^2\Psi/\partial z^2$; so

$$\frac{-h^2}{8\pi^2 m}\left(\frac{\partial^2\Psi}{\partial x^2} + \frac{\partial^2\Psi}{\partial y^2} + \frac{\partial^2\Psi}{\partial z^2}\right) + V(x,y,z)\Psi = \frac{-h}{2\pi i}\frac{\partial\Psi}{\partial t} \tag{7-6b}$$

The reader might well be disturbed by the *ad hoc* introduction of the Schrödinger equation without any derivation. Historically, of course, a long period of development led up to the final enunciation of the Schrödinger equation, but it was never derived. It cannot be derived from more basic concepts, just as Newton's laws cannot be derived from something more basic. Schrödinger's equation is therefore simply a "law" of physics which explains physical phenomena. To be considered valid, it must lead to a correct prediction of experimental findings. Schrödinger's equation cannot be derived from Newton's laws, since Schrödinger's equation is supposed to be valid in a region where classical concepts are not entirely valid. Quantum theory requires, not that Newton's laws be entirely relinquished, but only that the complete specification of initial conditions be dropped. Therefore, Schrödinger's equation must be consistent with Newton's laws *on the average*. Furthermore, when the particle size and mass become macroscopic in size, the predictions of quantum theory and classical theory are consistent with one another because the uncertain path becomes more closely like a unique trajectory.

If the reader still has reservations about the proper meaning of a "basic law of physics," let him recall the historical development of

Newton's laws. They were the culmination of experiments by Galileo with balls on inclined planes. Galileo discovered the principle that balls dropping a certain vertical distance (even though they might move variable distances horizontally) always reached the same velocity; he was even able to formulate Newton's second law for this case in terms of the force on the particles. Newton later realized that the law discovered by Galileo in the special case actually applied to all cases. Thus, Newton's law was discovered by induction from experiments.

Likewise, in the first years of this century, scientists realized that various discrepancies existed between the predictions of classical theory and experiments on atomic structure. The most notable of these discrepancies was the discovery that physical systems are quantized. Gradually, through a period of thirty years, the full significance of the experiments became realized. After pieces of the puzzle were discovered by many persons, Schrödinger proposed his equation as the final explanation of atomic structure in terms of the probability function $\Psi(x,t)$. The various quandaries raised by the experiments of the period and the attempts of physicists to explain them make fascinating reading, but since this book is primarily devoted to the specific topic of solids, the reader is referred for further discussion to any of the several excellent books on atomic physics cited in the References.

7-3 Time-independent Schrödinger Equation

The form of Schrödinger's equation suggests that, so far as the time is concerned, the solutions should be simple, since the time enters only as a first derivative on the right-hand side. In fact a particular solution for the special case where V is not a function of t can be written in the form

$$\Psi(x,t) = \psi(x)e^{(-2\pi i/h)Et} \tag{7-7}$$

$\psi(x)$ must then satisfy (in one dimension) the equation

$$\frac{-h^2}{8\pi^2 m}\frac{\partial^2\psi(x)}{\partial x^2} + V(x)\psi(x) = E\psi(x) \tag{7-8}$$

as substitution of Eq. (7-7) into Eq. (7-6a) requires. Note that Eq. (7-8) does not contain time at all; hence it is called *the time-independent Schrödinger equation*. While Eq. (7-7) is only a particular solution of the time-dependent Schrödinger equation (7-6), the general solution is the sum of all the various particular solutions which have the form (7-7). Whereas the time dependence of $\Psi(x,t)$ is simple, the spatial dependence is not at all elementary, since Eq. (7-8) for one choice of the potential function $V(x)$ is completely different from that for another. Actually Eq. (7-8) can be solved analytically only for a few particular types of functions $V(x)$.

The interpretation of E in Eq. (7-7) is important; it is explained in the following way: The time dependence of the function $\Psi(x,t)$, given by Eq. (7-7), is an exponential in t, where the coefficient of t in the exponential is chosen so that the form of the right-hand side of Eq. (7-8) is a simple multiplying constant E. On the left-hand side of Eq. (7-8), ψ is multiplied by the potential energy $V(x)$. Therefore, for consistency, E must also have the dimensions of energy. The only important quantity with the dimensions of energy which is a constant in mechanics is the total (conserved) energy of the system; so one would suspect that E is the total energy. The physical interpretation of Schrödinger's equation is that E *is* in fact the total energy of the particle in the motion described by $\Psi(x,t)$.

7-4 Free-particle Waves

The time-independent Schrödinger equation (7-8) is therefore the equation of primary interest, and a solution can be attempted when the force field implied in $V(x)$ is known. Just as with Newton's laws, the simplest situation is that where the force is zero. Let $V(x) = 0$, and investigate the solution. Then Eq. (7-8) becomes

$$\frac{\partial^2 \psi}{\partial x^2} + \frac{8\pi^2 mE}{h^2}\,\psi = 0 \tag{7-9}$$

This equation is very simple, since it is the same one which describes the harmonic oscillator when x is substituted for the time. The solutions are thus the sine and cosine functions, but the exponential form is more convenient. A particular (not the most general) solution is $\psi = \exp\,[ix\sqrt{8\pi^2 mE}/h]$, and the time-dependent Ψ function becomes

$$\Psi(x,t) = \exp\left[\frac{-2\pi i}{h}\,(Et - x\sqrt{2mE})\right] \tag{7-10}$$

This function is a traveling wave.

To confirm the traveling wave character of Eq. (7-10), consider a string which is vibrating with the displacement in the y direction given by

$$y = e^{i(kx-\omega t)} \tag{7-11}$$

The displacement of the string at time $t = 0$ is given by e^{ikx}, which is a sinusoid with the wavelength $\lambda = 2\pi/k$. The motion of the particle at the origin, $x = 0$, is given by $e^{-i\omega t}$, which is also a sinusoid with the period $t = 2\pi/\omega$. The constant k is called the propagation vector, or wave vector, and ω is the angular frequency. Of course, at other times and places, the vibration is still sinusoidal, but its phase is shifted from the previous cases. The wave is transported down the string with the velocity $v = \omega/k$, since at time t the displacement (or phase) which was $x = 0$ at $t = 0$ has traveled to $x = \omega t/k$. (See the example of Fig. 7-3.)

The Ψ function of Eq. (7-10) has the same form as the wave on a string, and so the probability function for the particle is a traveling wave. This information is of use in itself, but two highly important relations can be obtained from Eq. (7-10) by a little manipulation. First write this equation as

$$\Psi = e^{i(kx-\omega t)}$$

where

$$E = \frac{h\omega}{2\pi} \tag{7-12}$$

Here the frequency ω is expressed in terms of the energy of the system. The constant k is also a function of the energy, $k = \sqrt{8\pi^2 mE}/h$.

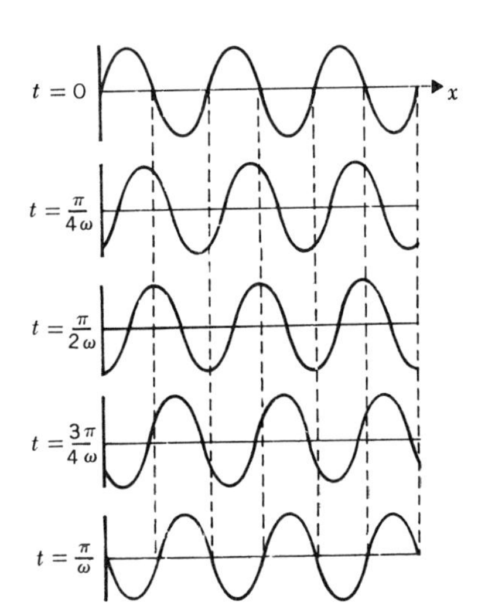

Figure 7-3 A traveling wave, $y = \sin(kx - \omega t)$, showing how, as time proceeds, the phase at a given point travels uniformly with phase velocity ω/k to the right.

Instead of leaving k in this form, we note that the energy is completely kinetic $[V(x) = 0]$. Therefore $E = p^2/2m$, where p is the momentum of the particle. Thus

$$k = \frac{2\pi p}{h} \tag{7-13}$$

The relations (7-12), called the *Einstein relation*, and (7-13), called the *de Broglie relation*, state that the particle motion has a wavelike nature. Together they represent one of the most important concepts of quantum mechanics, namely, the wave-particle duality of matter. The wave nature of matter is exceedingly difficult to understand from the standpoint of classical physics because waves and particles are treated there as fundamentally different entities. That real particles actually do possess a wavelength given by the de Broglie relation has an undeniable experimental basis, however. For electrons, neutrons, protons, and atoms, diffraction experiments have demonstrated beyond any doubt the wavelike character of matter. The first demonstration of this behavior was made for electrons in 1927 by Davisson and Germer; their work has been followed by a myriad of experiments on many kinds of particles.

To what extent are particles which exhibit wave characteristics still to be considered real particles? The answer to the paradox that particles, say, electrons, are both particles and waves is that the wave function of the electron is a traveling sinusoidal wave (when it is free of forces) but that, when an electron is actually detected, it is detected as a real and completely localized particle. For example, if a beam of electrons is incident on a crystal under proper conditions of energy and diffraction geometry, the wave function for any individual electron results in diffraction of the electron as though it were a wave. However, further suppose that the electron is sufficiently energetic to excite other electrons on particular atoms of the crystal. The particle nature of the electron is demonstrated by the fact that the electron does not partially excite a large number of atoms as the wave function for the electron traverses the crystal. Rather, the electron appears suddenly at a particular atom and fully excites that atom with its entire energy, just as would be expected if the electron were a real classical particle colliding with the atom. The only difference is that, before the electron strikes a particular atom, no prediction can be made of the particular atom along the wavefront of the wave function which will be struck by the electron. However, to repeat, when an atom is struck, it is struck with the complete electron. Thus the probability function of the particle describes a real wave and shows the diffraction effects of a wave, but the particle is always detected as a complete particle with the use of suitable detection instruments.

7-5 Energy Quantization: A Particle in a Box

One more concept fundamental to quantum mechanics can be derived simply from the Schrödinger equation, namely the *quantization of energy*. Consider a particle moving in one dimension; further let it be restricted to the "box" region $0 \leq x \leq L$. Let the potential energy be infinite for $x < 0$ and $x > L$ and zero for $0 \leq x \leq L$ (Fig. 7-4). The Schrödinger equation for the region $0 \leq x \leq L$ is still Eq. (7-9), and the solution is still the sinusoidal solution obtained earlier. However, now the ψ function must be zero at $x = 0$ and for all negative values of x, since the particle is not allowed to climb over the walls of the box. The function must also be zero for all values of x greater than $x = L$. The function desired is some combination of sines and cosines,

$$\psi(x) = \begin{cases} \sin kx \\ \cos kx \end{cases} \tag{7-14}$$

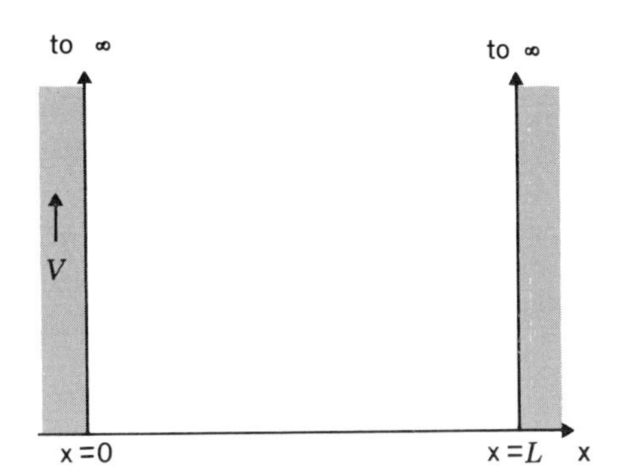

Figure 7-4 Potential energy of a particle in a one-dimensional box.

Since the cosine function is not zero at the origin, whereas the sine function is, the ψ function must be a pure sine function. Also, the solution must be restricted to just those sine functions which are also zero at $x = L$. Some ψ functions satisfying these conditions are those depicted in Fig. 7-5. Analytically, the requirement at $x = L$ is that

$$\psi(L) = \sin kL = 0 \tag{7-15}$$

According to Eq. (7-15), not all values of k are allowed; those which are allowed are only those for which

$$k = \frac{n\pi}{L} \qquad n = 1, 2, \ldots, \infty \tag{7-16}$$

If the wave vector k is restricted to only a discrete set of values, the energy of the particle is also restricted and the energy is quantized.

$$E = \frac{h^2k^2}{8\pi^2 m} = \frac{h^2n^2}{8mL^2} \qquad n = 1, 2, \ldots, \infty \tag{7-17}$$

The meaning of Eq. (7-17) is that the particle in the box can possess, not an arbitrary amount of energy, but only energy values in a discrete set. The energy then has a discrete spectrum as shown in Fig. 7-6. The reason that the completely free particle does not have quantized energy can easily be seen by observation of Eq. (7-17), for

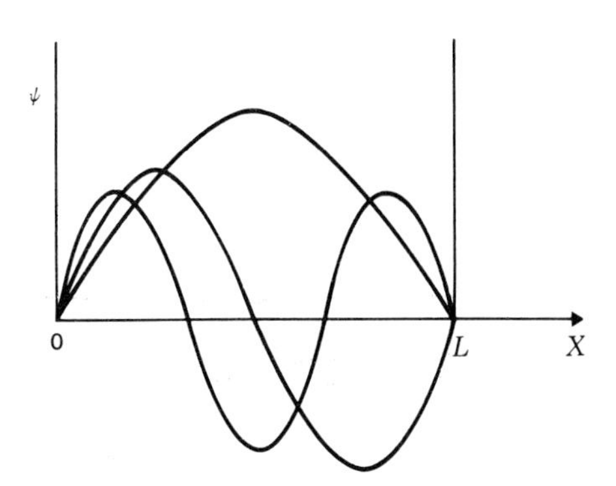

Figure 7-5 Some of the wave functions of the particle in a one-dimensional box.

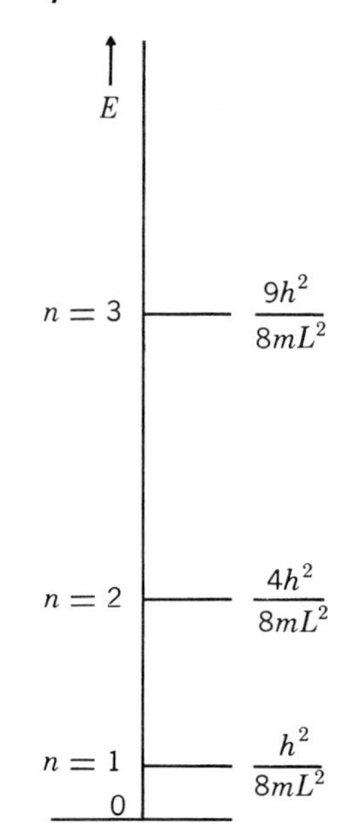

Figure 7-6 The spectrum of energy of the particle in a one-dimensional box.

if the length of the box is infinite, the spacing of the "energy levels" becomes infinitesimal. In general, quantum mechanics requires that completely unbound particles which are nearly free have continuous energy spectra, while particles which are bound by forces to finite regions of space possess discrete energy spectra.

One aspect of the discussion of the allowed solutions in Fig. 7-5 might be questioned. The wave function was assumed to go to zero at $x = 0$ and $x = L$, because, outside these values, the wave function must be zero. The functions thus are continuous through the walls of the box, but their first derivatives are not. The question therefore arises: Why can the function itself not be discontinuous at the walls in the manner of Fig. 7-7? The reason for not choosing such functions is not apparent until further calculations of the consequences of such a choice are made. Such calculations show (they are carried out in detail in the References listed at the end of the chapter) that, for a ψ function which is discontinuous at a point, the velocity of the particle

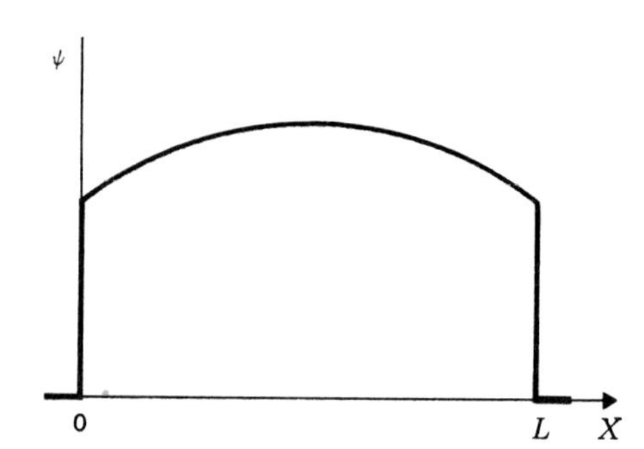

Figure 7-7 Discontinuous wave function for the particle in a box.

at that point is infinite; therefore the kinetic energy is infinite. With the assumption that the total energy of the particle is not infinite, such a solution is not physically admissible. The energy, then, is quantized for the reason that only certain ψ functions behave in a physically sensible manner. In short, the energy would not be conserved during the motion in a box of a particle with the wrong energy.

The discontinuity in slope of the wave function is allowed because such behavior of ψ can be shown to correspond to an infinite force on the particle. Since the wall of the box is impenetrable, the instantaneous force exerted by the box when the particle strikes the wall is infinitely sharp, but also infinitesimally short in duration.

The quantization of energy is seen to be the result of requiring physically meaningful solutions for the Schrödinger equation. Since the Schrödinger equation is required by the uncertainty principle, it is instructive to show that the minimum energy of the spectrum of Fig. 7-6 is in fact predicted by the uncertainty principle. If a particle is confined to a box, the largest possible value of the uncertainty in the position of the particle is the size of the box. Thus $\Delta x = L$. Let the particle have the momentum p. Since the particle must bounce back and forth from one side of the box to the other, the uncertainty in the momentum of a particle localized within the box must be $2p$, since the momentum constantly changes back and forth from $+p$ to $-p$ as collisions with the walls occur. Hence, the uncertainty relation for a maximally nonlocalized particle (and also minimized energy) is

$$(\Delta x)(\Delta p) = (L)(2p) = h \tag{7-18}$$

With the relation that $E = p^2/2m$, the energy becomes

$$E = \frac{h^2}{8mL^2} \tag{7-19}$$

which is consistent with Eq. (7-17) for $n = 1$.

7-6 A Particle in a Three-dimensional Box

Since all physical problems involve three dimensions, to make the previous calculations complete the Schrödinger equation must be solved in the three-dimensional form shown in Eq. (7-6*b*). If the time dependence is assumed to be of the form given in Eq. (7-7), the three-dimensional time-independent equation has the form

$$\frac{-h^2}{8\pi^2 m}\left(\frac{\partial^2\psi}{\partial x^2} + \frac{\partial^2\psi}{\partial y^2} + \frac{\partial^2\psi}{\partial z^2}\right) + V(x,y,z)\psi = E\psi \tag{7-20}$$

The potential of a three-dimensional box is the natural extension of the potential shown in Fig. 7-4.

$$V = 0 \qquad \begin{cases} 0 < x < L \\ 0 < y < L \\ 0 < z < L \end{cases} \tag{7-21}$$

$$V = \infty \qquad \text{everywhere else}$$

Here the box is assumed to a cube with side L. As for the one-dimensional box, the ψ function must vanish at the surface of the box and remain zero everywhere outside. Thus,

$$\psi = 0 \qquad \begin{cases} x = 0,\ L \\ y = 0,\ L \\ z = 0,\ L \end{cases} \tag{7-22}$$

The problem then reduces to solving the equation,

$$\frac{\partial^2\psi}{\partial x^2} + \frac{\partial^2\psi}{\partial y^2} + \frac{\partial^2\psi}{\partial z^2} + \frac{8\pi^2 mE}{h^2}\,\psi = 0 \tag{7-23}$$

with the boundary conditions given by Eq. (7-22).

This equation is one of the simplest partial differential equations, and its solutions appropriate to the boundary conditions are useful in many types of problems. The procedures for its solution are similar to those used to derive the time-independent Schrödinger equation (7-8). The ψ function is assumed to be a product function of form

$$\psi(x,y,z) = \psi_1(x)\psi_2(y)\psi_3(z) \tag{7-24}$$

Upon substitution of the assumed ψ function into the Schrödinger equation the result is

$$\left(\frac{1}{\psi_1}\frac{\partial^2\psi_1}{\partial x^2} + \frac{1}{\psi_2}\frac{\partial^2\psi_2}{\partial y^2} + \frac{1}{\psi_3}\frac{\partial^2\psi_3}{\partial z^2}\right) + \frac{8\pi^2 mE}{h^2} = 0 \tag{7-25}$$

This equation has been multiplied by the factor $1/\psi_1\psi_2\psi_3$.

The partial differential equation is now easy to solve because it has been "separated"; i.e., each of the terms in the parentheses is a function of only one of the independent variables. For example, the term $\dfrac{1}{\psi_1(x)}\dfrac{\partial^2\psi_1}{\partial x^2}$ is a function of only the x coordinate. Therefore, since x, y, and z are all independent quantities, each of the terms in the parentheses of Eq. (7-25) must be a constant. Thus, Eq. (7-25) becomes a system of equations,

$$\begin{aligned} \frac{1}{\psi_1}\frac{\partial^2\psi_1}{\partial x^2} &= -k_1^2 \\ \frac{1}{\psi_2}\frac{\partial^2\psi_2}{\partial y^2} &= -k_2^2 \\ \frac{1}{\psi_3}\frac{\partial^2\psi_3}{\partial z^2} &= -k_3^2 \\ k_1^2 + k_2^2 + k_3^2 &= \frac{8\pi^2 mE}{h^2} \end{aligned} \tag{7-26}$$

The differential equations in Eq. (7-26) are now only ordinary differential equations with exponential or sinusoidal solutions depending upon whether the constants k are real or complex.

$$\begin{aligned} \psi_1 &= e^{\pm ik_1 x} \\ \psi_2 &= e^{\pm ik_2 y} \\ \psi_3 &= e^{\pm ik_3 z} \end{aligned} \tag{7-27}$$

If any of the k's is purely imaginary, the solution has a real exponential. In this case the boundary conditions [Eq. (7-22)] at $x =$ zero and at $x = L$ cannot be satisfied. Hence, sinusoidal solutions are the only possibility. As for the one-dimensional problem, the boundary conditions of a three-dimensional box are satisfied by the set of functions

$$\begin{aligned} \psi &= A \sin k_1 x \sin k_2 y \sin k_3 z \\ k_1 &= \frac{n_1 \pi}{L} \qquad n_1 = 1, 2, \ldots \\ k_2 &= \frac{n_2 \pi}{L} \qquad n_2 = 1, 2, \ldots \\ k_3 &= \frac{n_3 \pi}{L} \qquad n_3 = 1, 2, \ldots \\ E &= \frac{h^2}{8mL^2}(n_1{}^2 + n_2{}^2 + n_3{}^2) \end{aligned} \tag{7-28}$$

One of the important results of this three-dimensional solution is that a quantum number exists for each dimension of the problem. Thus, while in one dimension only one quantum number is necessary, in three dimensions the ψ function depends on the three quantum numbers n_1, n_2, and n_3. Three quantum numbers result because the partial differential equation in three dimensions is separated into three independent equations, each equation leading to a new quantum number. Notice that the energy also depends upon all three quantum numbers in Eq. (7-28).

7-7 Quantum State and Degeneracy

The energy-level scheme of the three-dimensional particle in a box is somewhat more complicated than that for the one-dimensional box and is sketched in Fig. 7-8. In this figure a notation is used which is very useful in later chapters. The ψ function is labeled by the quantum numbers associated with that wave function. Thus, the ψ function for a particular set of quantum numbers n_1, n_2, and n_3 is written as $\psi(n_1,n_2,n_3)$. It is one of the linearly independent solutions of the differential equation. Such a function represents a possible state of the system different physically from all other states or functions. The term *quantum state* is used to designate the system as it is realized by a particular linearly independent solution of the Schrödinger equation specified by the appropriate set of quantum numbers.

Figure 7-8 shows an effect which is characteristic of three-dimensional problems. According to the expression for the energy given in Eq. (7-26), a given energy level may be obtained by taking several combinations of the quantum numbers. For example, the second energy level corresponds to quantum states (112), (121), or (211).

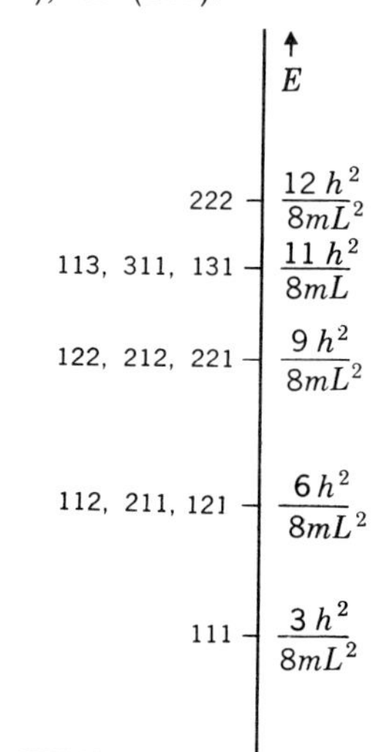

Figure 7-8 Energy levels of a particle in a three-dimensional box.

These ψ functions, and therefore the associated quantum states, are actually different from one another, as one can see when the ψ functions are written out explicitly.

$$\begin{aligned}
\psi_{112} &= A \sin\frac{\pi x}{L} \sin\frac{\pi y}{L} \sin\frac{2\pi z}{L} \\
\psi_{121} &= A \sin\frac{\pi x}{L} \sin\frac{2\pi y}{L} \sin\frac{\pi z}{L} \\
\psi_{211} &= A \sin\frac{2\pi x}{L} \sin\frac{\pi y}{L} \sin\frac{\pi z}{L}
\end{aligned} \tag{7-29}$$

In the quantum state (112) the particle bounces back and forth in the z direction with twice the momentum that it has in the x and y directions. In state (121) the particle has twice the momentum in the y direction as in the x and z directions, etc. The main point, however, is that each of these three quantum states has the same energy. When an energy level can be attained via more than one quantum state, the energy level is called *degenerate*. This particular energy level is threefold degenerate because it can be realized in three different ways. The lowest energy level in Fig. 7-8 is nondegenerate because it can be realized by only one choice of quantum numbers. Almost all energy levels to be considered in the band theory of solids in Chap. 9 are manyfold degenerate.

7-8 Recapitulation

The structure of submicroscopic physics is seen to be governed by a new law of physics known as Schrödinger's equation, which takes the place in the scheme of quantum mechanics that Newton's laws take in macroscopic classical physics.

Three fundamental concepts are peculiar to quantum mechanics. (1) At the root of all quantum theory are the uncertainty relations of Heisenberg. (2) Somewhat related to (1) is the wave character of matter expressed by the frequency and wavelength relations of Einstein and de Broglie. The wave characteristics are a natural outgrowth of the uncertainty relations, because waves by their very nature are unlocalizable and spread over all space when their wavelength and frequency are well defined. (3) Bound particles, as contrasted to free particles, have quantized energy levels. Only certain discrete energy values are allowed; the details of the spectrum of levels are different from one force field to another.

In the limit, as the size and mass of particles become macroscopic, the predictions of quantum mechanics become the same as those of classical mechanics.

REFERENCES

1. R. Sproull, *Modern Physics*, 2d ed., John Wiley & Sons, Inc., New York, 1963.
2. C. Sherwin, *Introduction to Quantum Mechanics*, Holt, Rinehart and Winston, Inc., New York, 1960.
3. R. B. Leighton, *Principles of Modern Physics*, McGraw-Hill Book Company, Inc., New York, 1959.
4. F. K. Richtmyer, E. H. Kennard, and T. Lauritsen, *Introduction to Modern Physics*, 5th ed., McGraw-Hill Book Company, Inc., New York, 1955.

PROBLEMS

7-1 Estimate the uncertainty in momentum of a 1-g mass whose center is restricted to a region 10^{-6} m in size. What is the lowest energy which such a particle could have? (This problem shows the quantum-mechanical limit of macroscopic measurements, since 10^{-6} m represents an optical limit of easy laboratory measurement.)

7-2 A 1-g mass is known to be at $x = 0$, with an error (quantum-mechanical in origin) of 10^{-10} m. This measurement implies an uncertainty in velocity. If the body has the lowest velocity commensurate with the

Heisenberg relation, how long will it take to move 1 cm from its original position?

7-3 We have described atoms in the crystal as analogous to balls on springs. Suppose that an atom of copper with spring constant equivalent to a vibration frequency of 10^{13}/sec is in its lowest possible state. Find the value of the "zero-point" amplitude of vibration. Atoms in a real crystal thus are never exactly at a lattice site, even at zero temperature, but execute random motions which have their origin in the uncertainty principle. (See Ref. 2.)

7-4 In the spirit of Prob. 7-3, find a relation between the spring constant and the mass of the atoms such that the zero-point oscillations are of the order of a lattice spacing of 4 A. A solid of this sort is no longer able to maintain its lattice structure, but even at 0°K remains a liquid. Helium is thought to be such a substance. It has low mass, weak interatomic forces, and a zero point oscillation of the interatomic spacing.

7-5 Electrons can be used in the same way as X rays to give diffraction patterns of crystals. What is the energy an electron must have to give the reflection from (100) planes at a Bragg angle of 45° for a crystal with lattice spacing 4 A?

7-6 For a three-dimensional box, derive an expression for the number of states whose energy is less than E^*. *Hint:* Let the three axes of a coordinate system be labeled as k_x, k_y, and k_z. In this k space, draw a sphere in the first octant of radius corresponding to E^*, and find the number of possible k values within the octant.

8

Atomic physics

8-1 Introduction

The internal structure of atoms can be studied by means of the concepts of quantum mechanics of the previous chapter. The hydrogen atom, with only one electron, is the simplest atom, and rather exact calculations can be made for it. In spite of its simplicity, it provides the key to all atomic structure, because the electronic structure of the more complicated atoms has important similarities to that of the H atom.

An analogy exists between such a quantum-mechanical calculation for hydrogen and the two-body problem of classical mechanics. In classical mechanics, no problem containing more than two particles which attract each other can be solved exactly, because the mathematics is too complex. The solar system, with the sun and numerous planets, is in this class: the motion of the earth cannot be specified by a precise analytical function. In spite of this restriction, however, much can be learned about complex systems like the solar system in Newtonian mechanics by use of various mathematical tricks which lead to approximations to the accurate and rigorous solutions. For example, the movement of the earth is governed mainly by its attraction to the sun, and the effects of the other planets and the moon can be treated as small perturbations upon this two-body motion. Similarly, the energy levels for the hydrogen atom can be determined exactly by using quantum mechanics, but those of more complicated atoms cannot be determined without some form of numerical approximations. By analogy to the solar system, moreover, the more complicated atoms can be discussed in terms of approximations built around the hydrogen atom.

In this chapter, therefore, the description of the hydrogen atom, including electron spin, is given, leading into a discussion of angular momentum in quantum mechanics. More complex atoms are then described, and the structure of the periodic table of elements is explained.

8-2 Electron Spin

As a preliminary to the discussion of the hydrogen atom, the concept of *spin*, an important property of all real particles, must be introduced. The most convincing demonstration of the spin of an electron is the experiment of Stern and Gerlach. In this experiment, a beam of silver atoms, all having the same energy, is projected down an evacuated

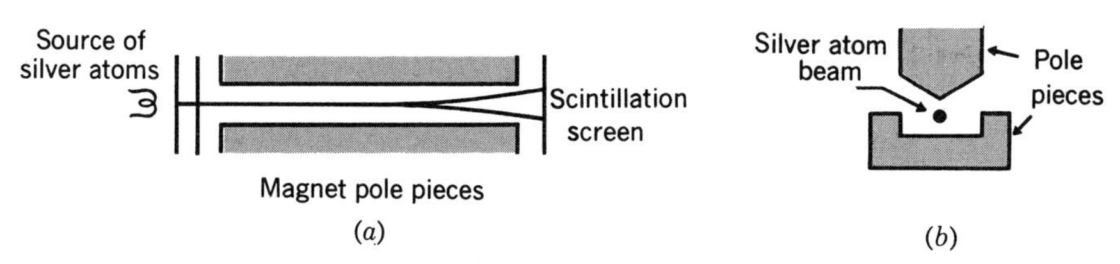

Figure 8-1 The Stern-Gerlach experiment: (*a*) Plan view; (*b*) section view. A collimated beam of monoenergetic silver atoms is shot between two pole pieces of a magnet. The pole pieces are designed to deflect the atoms if they possess a magnetic moment. A uniform magnetic field exerts no force on a magnetic dipole (see Chap. 19), but a nonuniform field does. The pole pieces thus are constructed with a "tongue-in-groove" arrangement to cause the nonuniform field. The beam is then split into two parts, corresponding to $+\frac{1}{2}$ or $-\frac{1}{2}$ spin on the valence electrons of the silver atoms, the amount of the deflection being a measure of the magnitude of the magnetic moment.

tube between the poles of a magnet whose field is strongly divergent (see Fig. 8-1). Some of these atoms are deflected up by the magnetic-pole pieces, while the others are deflected down with equal force. If a detecting screen is inserted at the point where the atom beam emerges from the magnetic-pole structure, two separate spots are observed. The two spots are displaced vertically by equal but opposite amounts. The experiment actually involves neutral silver atoms; however, the vertical deflection is a property of the single-valence electrons on the atoms, as will be clear from the discussion of the latter part of this chapter. The silver atoms serve only as an agent for transporting the electrons. The experiment shows that an electron is not a simple, structureless particle and that all electrons traveling with the same speed in the same direction are not alike.

The result of the Stern-Gerlach experiment can be understood if an electron is postulated to be a magnetic dipole. The dipole can be aligned only parallel or antiparallel to the external field. The diverging field of the magnet produces a force on the electron which deflects it either one way or the other: hence some electrons are deflected upward, some downward. A magnetic dipole, according to electromagnetic theory, is caused by a circulation of current in the electron "system." In a qualitative sense, then, an electron can be imagined to be a small sphere of charge which spins on its axis. The magnetic moment of the spin must be fixed in magnitude to agree with the amount of the deflection in the Stern-Gerlach experiment, and its direction must be either up or down in the field. The magnetic moment of the electron is 9.27×10^{-24} amp/m^2. This number is called *one Bohr magneton*, β.

For reasons which are evident later, the quantum number of the state with spin up is labeled $\frac{1}{2}$, while the quantum number of the state with spin down is labeled $-\frac{1}{2}$. Therefore, an electron as a particle in three-dimensional space has quantum states denoted by four quantum numbers. Three of the quantum numbers relate to the dimensionality of the space, while the fourth is the spin quantum number, taking values $\pm\frac{1}{2}$.

8-3 The Hydrogen Atom

The hydrogen atom consists of one proton in the nucleus and one electron in a valence shell. The Coulomb force between the positive charge of the nucleus and the negative charge of the electron binds the charges together. Thus the potential energy of the electron in the field of the proton is

$$V(r) = -\frac{e^2}{4\pi\epsilon_v r} \tag{8-1}$$

Of course, the Schrödinger equation for the H atom must be written in its full three-dimensional form. Since the potential $V(r)$ is spherically symmetric, the equation is most easily solved in spherical coordinates instead of the cartesian coordinates which have been used exclusively until now. The transformation of the Schrödinger equation into spherical coordinates and the solution of the resulting equation are perfectly straightforward but involve more mathematical detail than is appropriate for this book. Therefore, the results of the mathematical solution are simply stated, and the reader is referred to any text on quantum mechanics for the details.

Since the electron of the H atom is bound to a finite region of space, the uncertainty principle dictates that the energy spectrum be quantized. The energy levels of the H atom are given by the formula

$$E = -\frac{R_\infty}{n^2} \tag{8-2}$$

where $R_\infty = \dfrac{me^4}{8h^2\epsilon_v{}^2}$

$$n = 1, 2, 3, \ldots$$

In this expression, the constant R_∞ is called the *Rydberg*. The integer

n is called the *principal quantum number* and is associated with the radial degree of freedom of the electron. In the formula, m is the mass of the electron, e is the charge of the electron (-1.6×10^{-19} coulombs), and h is Planck's constant. The energy spectrum is depicted in Fig. 8-2. ϵ_v, the capacitivity of a vacuum, has the value 8.854×10^{-12} farad/m.

The energy of Eq. (8-2) is negative for the reason that the zero of energy corresponds to that state for which the electron is infinitely far from the proton and is at rest. Hence, when the electron is in a

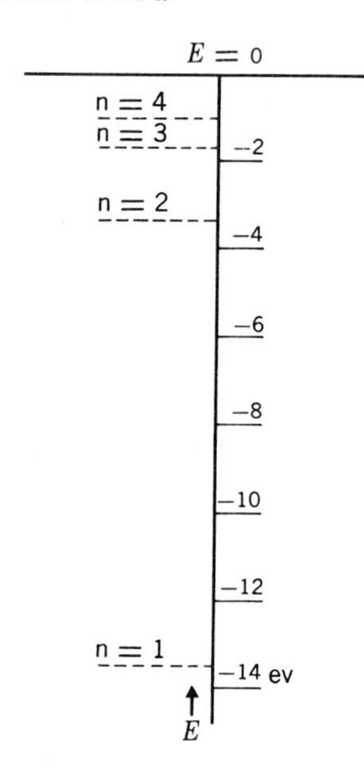

Figure 8-2 Part of the energy spectrum of the H atom.

bound state, energy must be supplied to the system to separate the particles. According to Fig. 8-2, there is a lowest energy given by $n = 1$. The energy of this state is -13.5 ev, and this number is the energy observed experimentally for the ionization energy of the H atom. The most striking characteristic of the energy spectrum of the H atom is that the levels converge rapidly for higher values of n so that the excited states become closer and closer together near the ionization limit.

The wave function of the H atom in the lowest state is simple, and it is worth discussing, since all atomic wave functions have some of its features.

$$\psi(r) = Ae^{-(r/a_0)} \qquad \text{for } n = 1 \tag{8-3}$$

In the lowest state, the wave function is spherically symmetric and is a simple exponential function. The constant A is found by the requirement that the probability that the electron is somewhere in space is equal to 1. The constant a_0 is more interesting, because it is a measure of the spatial region over which the wave function is appreciable. The value of a_0 is given by the theory to be

$$a_0 = \frac{\epsilon_v h^2}{\pi m e^2} = 0.529 \times 10^{-10}\ \text{m} = 0.529\ \text{A} \tag{8-4}$$

(see Prob. 8-3). Since the electron has appreciable probability for existing only within a radius of $r < a_0$, the constant a_0 is a convenient measure of the size of the H atom and is called the *Bohr radius.*

The result of the application of the Schrödinger equation to the H atom is seen to be that the atom has a region characteristic of its lowest energy state within which the electron is localized. This minimum size (which is associated with the lowest energy) is necessary because of the requirement of the uncertainty principle, in the same

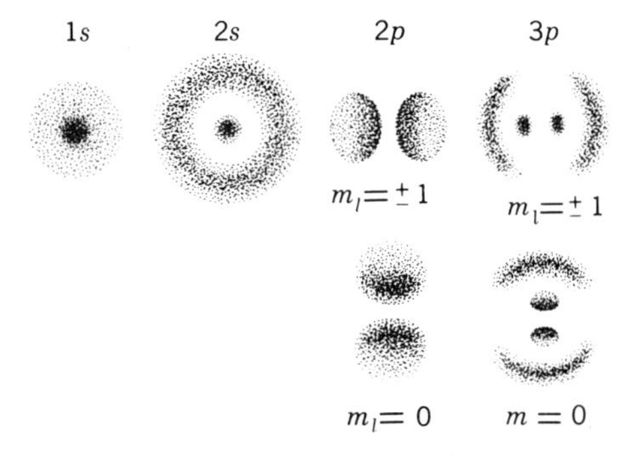

Figure 8-3 Pictograms of some of the electronic wave functions of atoms. *(From R. B. Leighton, Principles of Modern Physics,* McGraw-Hill Book Company, Inc., New York, 1959.)

way that the lowest energy was predicted by the uncertainty principle in the box problem of the previous chapter. Excited wave functions are, of course, larger, but the H atom normally exists most of its time in the lowest state for usual temperatures. The excited wave functions also differ from that of Eq. (8-3) in that some of them show an angular dependence as well as a radial dependence. However, the basic exponential dependence for large distance from the atom is always observed, so that the orbit has a rather well-defined maximum size. In Fig. 8-3, some pictograms of atomic wave functions are shown as examples.

According to Eq. (8-2), the energy of the H atom depends only upon the principal quantum number n. However, the development in Sec. 7-6 shows that each dimension of the problem leads to a distinct quantum number when the Schrödinger equation can be separated. Since V is a function only of r for the hydrogen atom, the Schrödinger equation can be separated in spherical coordinates into three ordinary differential equations; solution of these equations and application of appropriate boundary conditions again lead to three quantum numbers. These quantum numbers are designated n, l, m_l, and they are associated with the spherical coordinates r, θ, and ϕ, respectively. For the H atom, however, a complex relationship exists between the quantum numbers:

$$\begin{aligned} n &= 1, 2, 3, \ldots \\ l &= 0, 1, 2, \ldots, n-1 \\ m_l &= 0, \pm 1, \pm 2, \ldots, \pm l \\ s &= \pm\tfrac{1}{2} \end{aligned} \tag{8-5}$$

(A possible point of confusion in the notation lies in the fact that the letter m is used for both the quantum number and the mass. We shall always use the subscript form m_l for the quantum number.) All the quantum numbers except s, the spin quantum number, are positive or negative integers, but their magnitude is limited by Eq. (8-5). Thus, for $n = 1$ (the lowest energy level) the only values possible for l and m_l are $l = 0$, $m_l = 0$. However, $s = \pm\frac{1}{2}$. Thus there are two distinct states in the lowest level of energy of the energy spectrum of Fig. 8-2. It is therefore twofold degenerate in the sense of Sec. 7-7. For one of these states, $s = +\frac{1}{2}$ (spin up); for the other, $s = -\frac{1}{2}$ (spin down). The two wave functions in this case would be labeled $\psi\langle n = 1;\ l = 0;\ m_l = 0;\ s = +\frac{1}{2}\rangle$ and $\psi\langle n = 1;\ l = 0;\ m_l = 0;\ s = -\frac{1}{2}\rangle$ or, more simply, $\psi_{100\frac{1}{2}}$ and $\psi_{100(-\frac{1}{2})}$.

The degeneracy of the excited states can also be worked out from Eq. (8-5). For the first two excited states, a table can be given of all the different wave functions (these can be extended for higher excited levels by the reader).

$n = 2$		$n = 3$		
$211\frac{1}{2}$	8 states	$322\frac{1}{2}$	$322(-\frac{1}{2})$	18 states
$210\frac{1}{2}$		$321\frac{1}{2}$	$321(-\frac{1}{2})$	
$21(-1)\frac{1}{2}$		$320\frac{1}{2}$	$320(-\frac{1}{2})$	
$211(-\frac{1}{2})$		$32(-1)\frac{1}{2}$	$32(-1)(-\frac{1}{2})$	
$210(-\frac{1}{2})$		$32(-2)(\frac{1}{2})$	$32(-2)(-\frac{1}{2})$	
$21(-1)(-\frac{1}{2})$		$311\frac{1}{2}$	$311(-\frac{1}{2})$	
$200\frac{1}{2}$		$310\frac{1}{2}$	$310(-\frac{1}{2})$	
$200(-\frac{1}{2})$		$31(-1)\frac{1}{2}$	$31(-1)(-\frac{1}{2})$	
		$300\frac{1}{2}$	$300(-\frac{1}{2})$	

For a given value of l, $2l + 1$ different values of m_l are possible because $-l \leq m_l \leq +l$. The reader is asked to show in a problem that n^2 combinations of values of l and m_l exist for a given value of n, so that a total of $2n^2$ combinations of l, m_l, s exist for a given n. Thus, for the H atom, the degeneracy of an energy level, n, is given by

$$\text{Degeneracy of energy level } E_n = 2n^2 \tag{8-6}$$

8-4 Angular Momentum

In the previous section, the quantum numbers l and m_l were introduced simply as a consequence of the three-dimensionality of the problem. Actually, these parameters have important physical meaning. They are related to the total orbital angular momentum and the z component of the angular momentum, respectively. A more advanced treatment shows that

$$L^2 = \frac{l(l+1)h^2}{4\pi^2}$$
$$L_z = \frac{m_l h}{2\pi} \tag{8-7}$$

where L^2 is the square of the total orbital angular momentum of the electron on the atom and L_z is the component of the total angular momentum in the z direction. The quantum numbers l and m_l are given by Eq. (8-5).

The orbital angular momentum is defined in quantum mechanics, just as it is in classical mechanics, as the vector cross product,

$$\mathbf{L} = m\mathbf{r} \times \mathbf{v} \tag{8-8}$$

where $\mathbf{r}$ and $\mathbf{v}$ are the radius to the particle path and its velocity, respectively, and m is the mass of the particle. In quantum mechanics, since the particle does not execute a unique trajectory, the quantity $\mathbf{L}$ must be integrated over space, with the wave function as a weighting factor, to give the average angular momentum of the particle in its quantum-mechanical orbit. The result of such an integration for the hydrogen atom yields for the square of the angular momentum the value given by Eq. (8-7),

$$\langle L^2 \rangle_{\text{av}} = \int \psi^*_{nlm_l} L^2 \psi_{nlm_l}\, dv = \frac{h^2 l(l+1)}{4\pi^2}$$

$$\langle L_z \rangle_{\text{av}} = \int \psi^*_{nlm_l} L_z \psi_{nlm_l}\, dv = \frac{hm_l}{2\pi}$$

The use of the z axis to denote a direction in space along which the angular momentum can be measured requires additional explanation. As long as the atom is unable to distinguish the z axis from any other direction in space, states with differing values of m_l cannot be distinguished. However, let a magnetic field be applied along the z axis; then this direction is unique. The electrons in an atom, when placed in such a field, modify their random behavior in accord with the second of Eq. (8-7).

The relationship between the angular momentum of the electron in the hydrogen atom and the quantum numbers l and m_l suggests a very useful diagram for the angular-momentum vectors. If the z axis is drawn vertically, then the various states of orbital angular momentum can be represented as in Fig. 8-4. In Fig. 8-4, when the value of l is fixed, the possibilities for m_l are given by Eq. (8-5).

The striking feature of the angular momentum of the hydrogen atom in quantum mechanics is that it is quantized. For a given energy level, as specified by the quantum number n, only a discrete number of angular-momentum states are possible. In addition to this, when the magnitude of the total angular momentum is specified, only

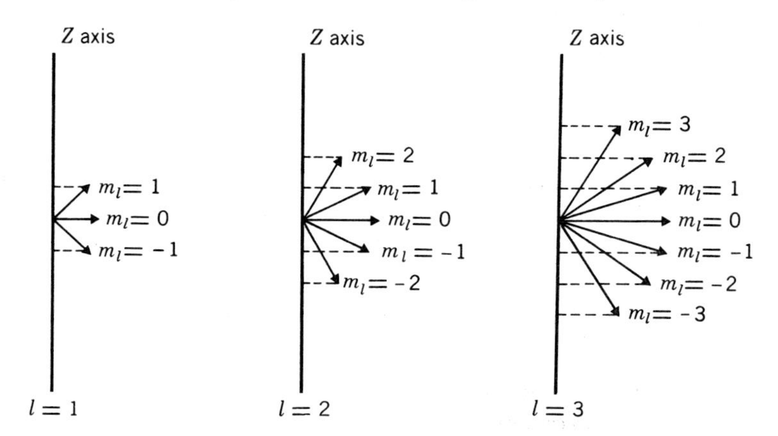

Figure 8-4 A representation of the angular-momentum states of the hydrogen atom. The angular momentum of each state is a vector which may have various values for its z component. The figure illustrates the cases of $l = 1$, $l = 2$, and $l = 3$.

a discrete set of directions is possible into which the angular-momentum vector can point. More strictly, the precise direction of the angular-momentum vector is never exactly specified, but only its component along the z direction. An important theorem concerning the angular momentum of the hydrogen atom states that the components in the other directions cannot be known precisely.

One of the peculiar characteristics of the angular-momentum vector is that it can never point exactly in the z direction. This is due to the fact that the magnitude of the angular-momentum vector, $(h/2\pi)\sqrt{l(l+1)}$, is slightly greater than its maximum component along the z direction, $hl/2\pi$.

A traditional notation is used to label the various states of an electron on the hydrogen atom, dating from the early days of spectroscopy. With this notation, an electron with different values of l is called by a letter according to the table shown on the next page.

In this scheme, an electron with $n = 1, l = 0$ is called a *1s electron*, an electron with $n = 2$, $l = 1$ is called a *2p electron*, etc.

The type of angular momentum associated with the quantum numbers l and m_l is called *orbital momentum* for the reason that the last quantum number s also relates to an angular momentum. Whereas the orbital angular momentum is just the angular momentum of a point particle with a certain mass in a rotating orbit, the spin relates to an intrinsic momentum possessed by the electron, analogous to its

l value	Spectroscopic notation
0	*s*
1	*p*
2	*d*
3	*f*
4	*g*

spinning about an axis. From the point of view of the early quantum mechanics, the spin of an electron is an anomalous effect which is introduced in an *ad hoc* way to satisfy the results of the experiment of Stern-Gerlach. Analysis has shown, however, that when relativistic effects are included in the Schrödinger equation, the spin of the electron is exactly predicted. It is a quantity whose magnitude is given analogously to L^2 in Eq. (8-7) by

$$S^2 = \frac{s(s+1)h^2}{4\pi^2}$$

Similarly

$$S_z = \frac{sh}{2\pi} \tag{8-9}$$

where $\mathbf{S}$ is the spin angular momentum and S_z is the z component. In this case, the quantum number s can take only the two values $s = +\frac{1}{2}$, $s = -\frac{1}{2}$, independently of all other quantum numbers. The total angular momentum of the electron is, of course, made up by the addition of all types of angular momentum, or, alternatively, the total specification of the angular momentum of an atomic state is given by giving all the values of the quantum numbers l, m_l, s.

8-5 The Pauli Exclusion Principle

All the previous discussion has concerned systems containing only one particle, i.e., one particle in a box, one electron in an orbit around a stationary nucleus, etc. However, most of the interesting problems to which one would like to apply the ideas of microscopic physics involve systems with more than one moving particle, for example, atoms with many electrons. On the basis of what has been said so far, one might suppose that, when more than one electron is involved, one need only treat each electron independently of the others and sum over the individual solutions to get a solution similar to that already found, say, for the hydrogen atom. Then the energy of the entire atom would simply be the sum of the energies of the electrons in their various states. The lowest energy in this picture is the one in which all the electrons are placed in the lowest energy state. However, an essentially new principle applies in quantum mechanics (unlike anything known in classical mechanics) and relates to those cases with more than one electron in the atom. This principle, the *Pauli exclusion principle*, states that only one electron can be in a particular quantum state at a given time.

This principle has important implications in discussing the energy states of more complicated atoms or the energy states of electrons in crystals. For example, if an atom has a system of energy states like the H atom, only two electrons can be accommodated in the lowest energy level, since the degeneracy of this level is 2. A third electron must be placed in the next higher level. After this level is full, additional electrons must go into even higher levels. To show how this affects placement of electrons in many-electron atoms, several of them are considered in detail.

8-6 Energy Levels of the Lithium Atom

The results of the quantum-mechanical solution of the hydrogen atom have been described in some detail. Not only is it the simplest atom, but the energy-level structures of the more complicated atoms have striking similarities to that of the hydrogen atom. The physical reason for the similarity is that, in the more complicated atoms, each electron finds itself approximately in the presence of a central force centered at the nucleus. A more exact statement is that a surprisingly good description of at least the qualitative energy-level structure of many-electron atoms is possible even if the noncentral components of the forces between the electrons themselves are completely ignored. In some atoms, notably the monovalent metal atoms, the central-force approximation is even quantitatively excellent.

In a later paragraph the physical reasoning which makes the central-force approximation plausible is examined. If, however, the assumption is made here that each electron of the multielectron atom is subjected to a central force, the relevance of the hydrogen solution is easy to understand. The reason is that the potential-energy function in the Schrödinger equation for a particular electron is then only a function of the radius from the nucleus. The Schrödinger equation is again separable in spherical coordinates, as was true for the hydrogen atom, and the same set of quantum numbers n, l, m_l, and s are found. More important, since the potential energy enters the radial part of the equation, the equations for the angular parts of the Schrödinger equation are exactly the same as for the hydrogen atom and the relations such as Eq. (8-5) and Eq. (8-7) for the quantum numbers

n, l, m_l, and s remain unchanged. However, the dependence of the energy of the particular electron on its principal quantum number may change. There is no reason why the energy might not depend upon the angular quantum numbers also, as it will be found to do. The energy does not depend upon m_l or s, for symmetry reasons that are not discussed, but it does depend upon the quantum number l as well as on n.

As an example of the many-electron atom, the structure of the lithium atom is discussed in detail. The lithium nucleus has an electronic charge of 3, and the neutral atom thus possesses three electrons. We proceed by first imagining one electron in an orbit about the bare nucleus, then a second, and finally a third, describing the changes which occur as the electrons are added.

The Li^{++} ion

One electron in an orbit about the bare nucleus is exactly like the hydrogen atom, of course, except that the potential energy function is $V = -3e^2/4\pi\epsilon_v r$ instead of simply $V = -e^2/4\pi\epsilon_v r$, as for hydrogen. If the appropriate potential function for lithium is used (Prob. 8-3), the lowest energy of the system is $E = -9R_\infty$, where R_∞ is again the Rydberg. In general, if the charge on the nucleus is Z, then the lowest energy level of a single electron in the field of the nucleus is $E = -Z^2R_\infty$. The second level has energy $E = -Z^2R_\infty/4$, etc. The relative energy level schemes are sketched in Fig. 8-5.

The Li^+ ion

When a second electron is added to the atom, the description is more complicated than that for a single electron. In addition to the force between the second electron and the nucleus, the first electron also exerts a force on the second. The total potential energy function of the second electron cannot be calculated, and an exact solution of the Schrödinger equation is impossible. The problem is the many-body problem alluded to earlier. Even though an exact analytic solution is not possible, an approximate one can be obtained by finding an approximate potential-energy function. First consider the excited states of the second electron. If it is put into an excited state with the first electron in its lowest energy state, an approximate value of the potential energy of the second electron due to the first is obtained by averaging over the positions of the first electron in its orbit (wave function). The total potential energy of the second electron is then the sum of (1) this average potential energy of the second electron in the field of the first and (2) the potential energy of the second electron in the field of the nucleus.

In computing the first of these we can conveniently use a theorem of electrostatics concerning force fields of spherical distributions of charge. According to the theorem, a spherically symmetric charge distribution has a force field at a point r within the distribution which is

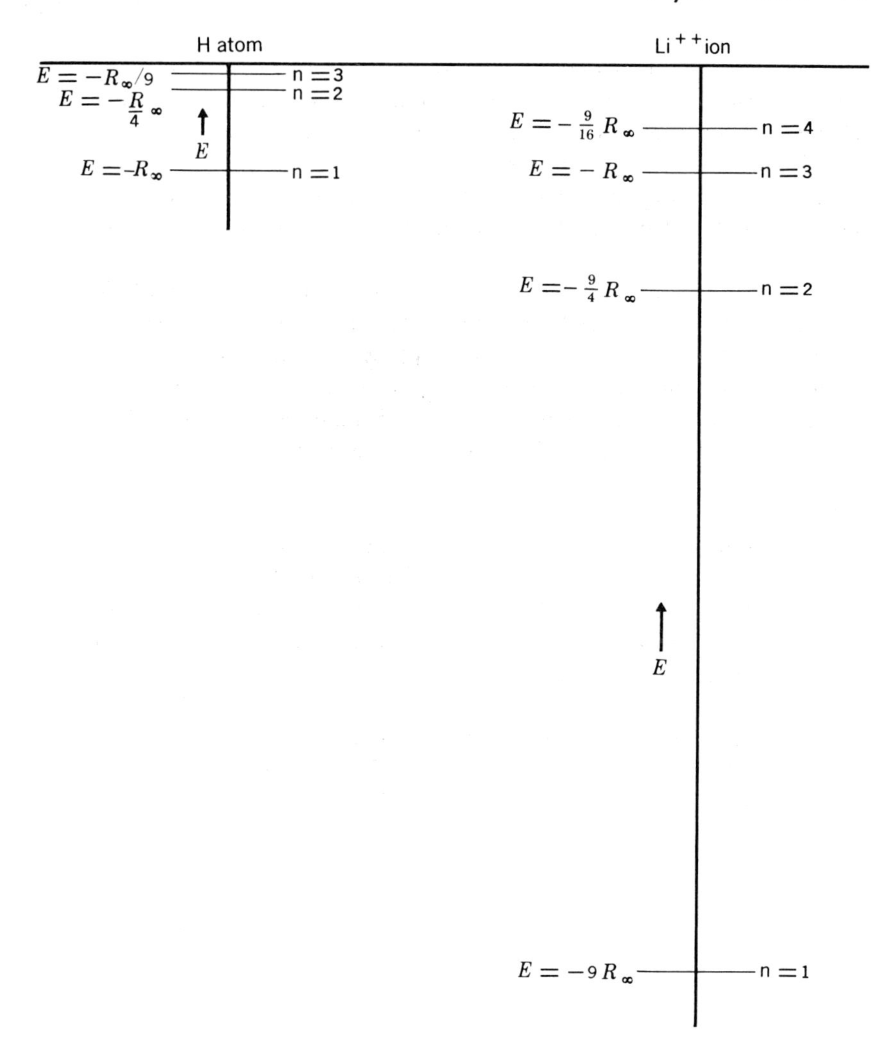

Figure 8-5 Energy-level schemes for a single electron in the field of a nucleus of a single charge and of a nucleus with a charge of 3, Li^{++}.

the same as if the charge within a sphere of radius r were all concentrated at the center of the sphere. The charge outside this sphere has no force at the point r. Thus the potential energy of a charge e at some point r in the field of a spherically symmetric distribution of charge is given by

$$V = \frac{eq(r)}{4\pi\epsilon_v r} \tag{8-10}$$

Here $q(r)$ is all the charge contained within the sphere of radius r. For the Li^+ ion, the second electron, when at the point r, interacts with a charge at the nucleus which is the actual charge on the nucleus

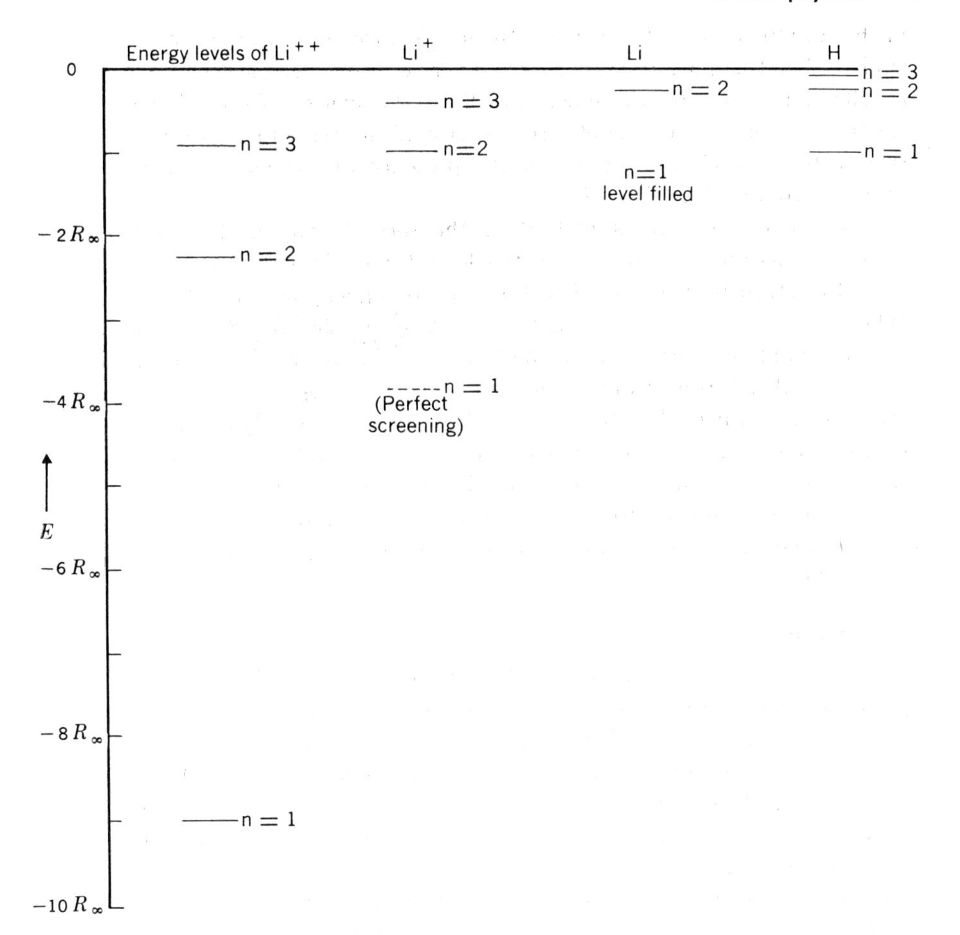

Figure 8-6 Comparison of relative energy levels of hydrogen and lithium. The levels shown are the levels for the last added electron on the assumption that previously added electrons are in their ground states. The energy levels are labeled with the approximate energies computed on the basis of complete screening. The correct experimental values of the energy are given on the energy scale.

minus the total charge of the wave function of the first electron which exists within a sphere of radius r. Thus the nucleus is partially "screened" from the second electron by the first electron. The effective charge on the nucleus is therefore something more than $2\,|e|$ but less than $3\,|e|$. Accordingly, the excited energy levels of the second electron are raised considerably above the corresponding levels of the first electron when it is present alone, as is plotted in the second column of Fig. 8-6.

To a crude approximation, this model holds even for the state where both electrons are in their lowest energy, i.e., in the $n = 1$ state. Then, of course, each electron is exactly like the other, and they have the same energy for removal from the ion, plotted as the lowest state in the second column of Fig. 8-6. In the plot, the energy of the state is plotted as though the screening were perfect: i.e., each electron sees an effective charge on the nucleus of $2|e|$. Of course, if one of the electrons is ionized completely, the energy of the remaining electron reverts to that of one electron in the presence of the bare nucleus shown in column 1 of Fig. 8-6.

In the excited states plotted in the second column, the first electron is assumed to remain in the lowest possible orbit; and the second electron is then placed in the various higher quantum states. When the second electron is in the $n = 2$ state, the $n = 2$ orbit is almost completely screened by the first orbit, since the $n = 2$ orbit is considerably farther from the nucleus than the first. The effective charge seen at the nucleus by the second electron is now only 2. Thus the first excited state of the second electron is very nearly at the position $E = -4R_\infty/4 = -R_\infty$. In the next higher quantum state, the orbit is even farther removed from the nucleus, the screening is even more perfect, and the state approaches even more closely to the value $E = -4R_\infty/9$.

The neutral Li atom

When the second electron is placed in its orbit, the Pauli exclusion principle must be satisfied. Two possible states exist for $n = 1$; the spin may be either up or down. Therefore, a maximum of two electrons may be placed in this lowest hydrogenlike energy level, and the lowest energy level available for the third electron is in the shell for which $n = 2$. The wave function for the third electron in this state again has a larger Bohr radius than that of the $n = 1$ state. Therefore, the third electron is almost completely screened from the nucleus by the two inner electrons, and the effective charge of the nucleus for the third electron (again on the assumption that the first two electrons remain in the $n = 1$ state) must be very nearly identical to the simple H atom with the $n = 1$ state deleted. The results are depicted in the third column in Fig. 8-7. The hydrogen spectrum is also given in column 4, for the sake of comparison.

So far, the dependence of the energy on l has not been considered. Of course, for $l = 0$, only spin degeneracy exists, and no splitting of the s energy levels occurs. However, for $n = 2$, p states as well as s states exist. For Li the $n = 2$ level (the ground state of the third electron) is not completely degenerate, as it was in the H atom. Instead, the s states are lowered somewhat below the p states. This dependence of the energy on l has as its primary reason the fact that the s states have a slightly smaller Bohr radius than the p states and so are slightly less screened. The result is approximately as shown in Fig. 8-7, which is a magnified picture of the Li energy spectrum already drawn in Fig. 8-6.

Screening is not the only effect which splits the levels; small magnetic forces within the atoms have the same effect, as do some

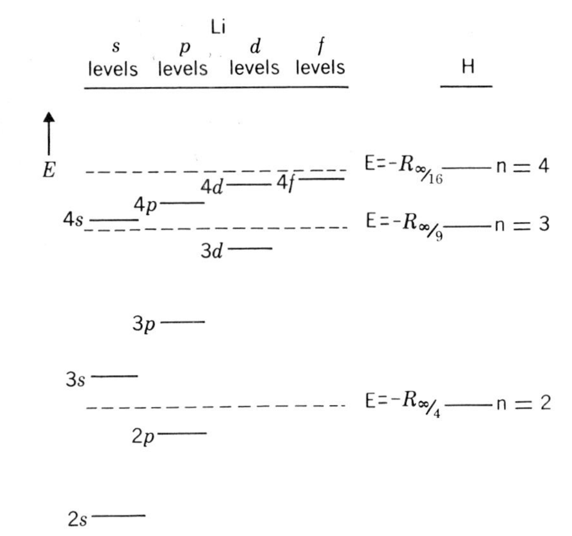

Figure 8-7 Magnified energy spectrum of Li, showing the splitting of the levels due to screening and other interactions. Only the 2, 3, and 4 states are shown.

specifically quantum-mechanical interactions which are not appropriately discussed here. In fact, even the levels shown in Fig. 8-7 are also split slightly, but this splitting is neglected here.

8-7 The Periodic Chart

The principles of the filling of the shells of other atoms are the same as those used for the Li atom. The level filling is performed on the basis of modified hydrogenlike energy-level schemes, the use of the Pauli exclusion principle, and electrostatic screening. Also, the degeneracy of the levels characteristic of the hydrogen atom is removed, as for Li, so that the energy of a quantum state of a multielectron atom depends upon the value of l as well as that of n. In fact, this tendency becomes very much greater than in Li as the shells fill, until for atoms of higher atomic number the splitting between the levels of a given value of n is enough to overlap the levels of a different n.

The spectroscopic notation already introduced in Sec. 8-4 is convenient for describing the energy levels of the complicated atoms. One additional modification is necessary in discussing many electrons, however. The number of electrons occupying a particular state is written as a superscript to the state. Thus an atom with two electrons in the $1s$ state, two electrons in the $2s$ state, and three electrons in the $2p$ state is written $1s^2 2s^2 2p^3$.

The electronic structure of many-electron atoms is easy to describe for the states with $n = 2$. As the atomic number increases above Li in the first period, first the $2s$ level is filled at Be with an electron configuration $1s^2 2s^2$ and further electrons must be placed in the $2p$ state. This occurs for B, C, N, O, and F. Finally, the $2p$ level is filled at neon when it contains six electrons; the neon configuration is $1s^2 2s^2 2p^6$. After neon, the $n = 3$ level begins to fill. Again, splitting of the $3s$ and $3p$ levels occurs so that the $3s$ level lies slightly below the $3p$ level. Sodium is the first atom in the second period, and the splitting is found to be larger than that for the $2s$ and $2p$ of Li. More important, however, the $3d$ level in sodium is not grouped with the $3s$ and $3p$ levels but lies among the higher $n = 4$ levels. The various neighboring levels to $3s$ are shown schematically in Fig. 8-8.

The $3s$ and $3p$ levels fill in the same fashion as before. For magnesium, the $3s$ level fills, and further electrons in Al, Si, etc., go into the $3p$ levels. For argon, the $3p$ level contains six electrons. For potassium, the first departure from the patterns above occurs.

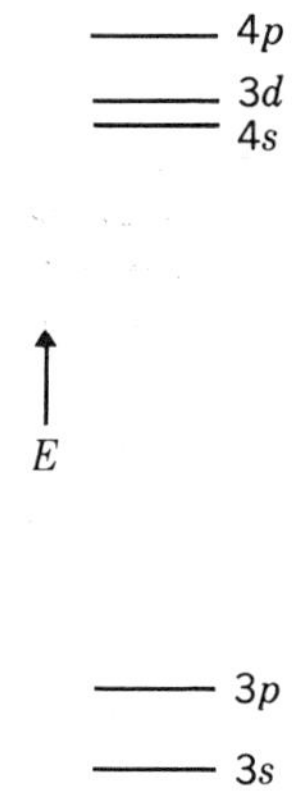

4p
3d
4s
E
3p
3s

Figure 8-8 The relative positions for Na of the $n = 3$ levels and the $4s$ level. The $4s$ level has dropped down below the $3d$ level, and the levels can no longer be grouped according to their principal quantum number n.

Its nineteenth electron does not go into the $3d$ level, but the irregularity noted above for the excited levels of sodium persists, and the $4s$ level lies below the $3d$. Thus the valence electron in the lowest state of potassium is a $4s$ electron. The second valence electron at calcium fills the $4s$ level, and at scandium the third electron goes into the $3d$ level. The group of elements which fill the $3d$ levels are nine more in number, ending at Cu, which has the structure $1s^2 2s^2 2p^6 3s^2 3p^6 3d^{10} 4s^1$. The third period of the chart is thus longer than the first two. This splitting becomes even more serious for states with $n = 4$. For these levels the overlap with other n values is large. Not only does the $4d$ level lie close to the $5s$ level, but the $4f$ level is badly split from the other $n = 4$ levels. Consequently it fills much later in the periodic table. A complete periodic chart is given in the Table 8-1, where the order of the filling is given, together with the electronic configurations of the valence electrons. The order of filling of the valence shells is shown in Fig. 8-9.

One of the peculiarities of the level filling as shown in the periodic chart is at Cr in the first long series. In the chromium atom, the d shell steals one of the electrons from the $4s$ shell, showing that the s and d levels are very close together. Further evidence is provided at copper, where the d shell fills at the expense of the s shell. These irregularities are repeated in the later transition-metal series.

Table 8-1 Periodic table, with the outer-electron configurations of neutral atoms in their ground states†

										55Cs	$5p^6 6s$	87Fr	$6p^6 7s$
										56Ba	$5p^6 6s^2$	88Ra	$6p^6 7s^2$
										57La	$5p^6 5d 6s^2$	89Ac	$6d 7s^2$
										58Ce	$4f^2 6s^2$	90Th	$6d^2 7s^2$
										59Pr	$4f^3 6s^2$	91Pa	$5f^2 6d 7s^2$
										60Nd	$4f^4 6s^2$	92U	$5f^3 6d 7s^2$
										61Pm	$4f^5 6s^2$	93Np	$5f^5 7s^2$
						19K	$3p^6 4s$	37Rb	$4p^6 5s$	62Sm	$4f^6 6s^2$	94Pu	$5f^6 7s^2$
						20Ca	$3p^6 4s^2$	38Sr	$4p^6 5s^2$	63Eu	$4f^7 6s^2$	95Am	$5f^7 7s^2$
						21Sc	$3d 4s^2$	39Y	$4d 5s^2$	64Gd	$4f^7 5d 6s^2$	96Cm	$5f^7 6d 7s^2$
						22Ti	$3d^2 4s^2$	40Zr	$4d^2 5s^2$	65Tb	$4f^8 5d 6s^2$	97Bk	$5f^8 6d 7s^2$
						23V	$3d^3 4s^2$	41Nb	$4d^4 5s$	66Dy	$4f^{10} 6s^2$	98Cf	$5f^9 6d 7s^2$
		3Li	$2s$	11Na	$2p^6 3s$	24Cr	$3d^5 4s$	42Mo	$4d^5 5s$	67Ho	$4f^{11} 6s^2$	99E	
		4Be	$2s^2$	12Mg	$2p^6 3s^2$	25Mn	$3d^5 4s^2$	43Tc	$4d^6 5s$	68Er	$4f^{12} 6s^2$	100Fm	
		5B	$2s^2 2p$	13Al	$3s^2 3p$	26Fe	$3d^6 4s^2$	44Ru	$4d^7 5s$	69Tm	$4f^{13} 6s^2$	101Md	
1H	$1s$	6C	$2s^2 2p^2$	14Si	$3s^2 3p^2$	27Co	$3d^7 4s^2$	45Rh	$4d^8 5s$	70Yb	$4f^{14} 6s^2$	102No	
2He	$1s^2$	7N	$2s^2 2p^3$	15P	$3s^2 3p^3$	28Ni	$3d^8 4s^2$	46Pd	$4d^{10}$	71Lu	$4f^{14} 5d 6s^2$	103Lw	
		8O	$2s^2 2p^4$	16S	$3s^2 3p^4$	29Cu	$3d^{10} 4s$	47Ag	$4d^{10} 5s$	72Hf	$5d^2 6s^2$		
		9F	$2s^2 2p^5$	17Cl	$3s^2 3p^5$	30Zn	$3d^{10} 4s^2$	48Cd	$4d^{10} 5s^2$	73Ta	$5d^3 6s^2$		
		10Ne	$2s^2 2p^6$	18A	$3s^2 3p^6$	31Ga	$4s^2 4p$	49In	$5s^2 5p$	74W	$5d^4 6s^2$		
						32Ge	$4s^2 4p^2$	50Sn	$5s^2 5p^2$	75Re	$5d^5 6s^2$		
						33As	$4s^2 4p^3$	51Sb	$5s^2 5p^3$	76Os	$5d^6 6s^2$		
						34Se	$4s^2 4p^4$	52Te	$5s^2 5p^4$	77Ir	$5d^9$		
						35Br	$4s^2 4p^5$	53 I	$5s^2 5p^5$	78Pt	$5d^9 6s$		
						36Kr	$4s^2 4p^6$	54Xe	$5s^2 5p^6$	79Au	$5d^{10} 6s$		
										80Hg	$5d^{10} 6s^2$		
										81Tl	$6s^2 6p$		
										82Pb	$6s^2 6p^2$		
										83Bi	$6s^2 6p^3$		
										84Po	$6s^2 6p^4$		
										85At	$6s^2 6p^5$		
										86Rn	$6s^2 6p^6$		

† Configuration assignments for the rare-earth and actinide elements are somewhat uncertain.

The $4f$ shell does not appear until atomic number 58, where it interposes in the filling of the transition series of the third long period. The group of rare-earth elements, where the f shell fills, is normally plotted separately on the periodic chart because of the peculiar properties of the rare earths. The separation chemistry of the rare earths is very difficult because of the great chemical similarity between all these elements. The $5f$ series, the actinide group, interposes again in the fourth long series.

A final comment concerning chemical reactions and the periodic table will complete this discussion of shell filling. Chemical reactions correspond to the exchange of electrons between elements. For example, Na reacts readily with Cl when the atoms approach one another

1s	2s	2p	3s	3p	(4s	3d)	4p	(5s	4d)	5p	(6s	4f	5d)	6p	(7s	6d	5f)
	2	6	2	6	2	10	6	2	10	6	2	14	10	6	2		
2	8		8			18			18				32				

No. of electrons to fill

Figure 8-9 Order of filling atomic levels. The parentheses represent close-lying shells with irregularities in the filling order. The braces enclose periods of the periodic table.

with the loss of the outermost electron of Na to the unfilled valence shell of Cl. Filling the shell of Cl corresponds to a stabilization of the structure of Cl at the expense of the ionization of Na. (Chapter 14 contains a more complete discussion of these ideas.) When the Cl is near the Na, the ionization of Na is not difficult, because the electron is not separated too far from its ion and because of the small binding energy of the electron to the Na atom. This rather minor energy loss is more than made up by placing the electron on the Cl atom in a lower energy level. If the valence electron had been in a deep-lying level like that in the ground state of helium, the ionization energy would have been tremendous compared with the energy to be gained by filling the valence shell of Cl, and the reaction would not be possible. Hence chemical reactions are possible between valence electrons in shells which are not too stable or in which electrons can be shifted to other close-lying or degenerate levels, so that small energies gained by stabilization of other atoms as shells are filled with extra electrons are sufficient to balance the slight energy losses. This discussion shows why an element at the end of a period of the chart is very inert chemically. No nearly degenerate levels lie near the valence electrons in atoms like He, Ne, etc., and excitation energies are too large for easy chemical reactions.

The plausible conclusion which can be drawn from this discussion is that atoms with the same number of electrons in their valence shell should behave similarly chemically. The success of the chemist's description of the chemical affinity of an atom solely in terms of valence (with the filled inner shells disregarded) attests to the truth of this conjecture. For this reason atoms which have similar valence shells are plotted vertically in columns in the periodic chart. The periodic nature of the periodic chart is, then, entirely a result of the successive filling of similar electron shells.

8-8 Recapitulation

The energy-level scheme of the hydrogen atom is the result of the solution of Schrödinger's equation for a central Coulomb-force law. The principal goal of the chapter, however, was the development of energy-level schemes for the many-electron atoms. The energy levels were described qualitatively by arguing from analogy to the central-force results on the hydrogen atom, together with the Pauli exclusion principle. These ideas led directly to the arrangement of atoms in the periodic chart.

PROBLEMS

8-1 Prove the "virial theorem," that PE = −2KE, for a particle in a circular orbit in a coulomb potential field.

8-2 Find the radius of the electron orbit in hydrogen if classical mechanics were valid and the binding energy of the electron around the proton were 1 Rydberg. Compute its angular momentum. Why does the electron in the lowest state in quantum mechanics not have angular momentum?

8-3 In spherical polar coordinates, when there is no angular dependence, the Schrödinger equation is

$$-\frac{h^2}{8\pi^2 m}\frac{1}{r^2}\frac{d}{dr}\left[r^2\frac{d\psi(r)}{dr}\right] - \frac{Ze^2}{4\pi\epsilon_v r}\psi(r) = E\psi(r)$$

Show that $\psi = Ae^{-r/a_0}$ is a solution if E and a_0 take certain values.

8-4 Assume that, in multielectron atoms, no splitting of energy levels occurs for different values of l, as is true for hydrogen. List atoms which would be inert gases for atomic number less than 90. Construct a periodic chart which would be valid in this case, with atoms of equal valence listed in columns as in the normal periodic chart. (Construct the chart for atoms up to atomic number 28.)

8-5 What is the frequency of light necessary to excite an electron of hydrogen (*a*) from the ground state to the first excited state; (*b*) to ionize the atom? What color would these light photons be?

8-6 Look up the value of the splitting of the *s-p* levels in the valence shells of Li and Na, and give an effective Z for the valence electron in each case.

8-7 Suppose that the dielectric medium between the nucleus and the electron in hydrogen had a dielectric constant $\epsilon = \epsilon_v\epsilon_r$. (*a*) Write expressions for the energy levels and the Bohr radius. (*b*) Compute and plot the energy levels for a dielectric medium for which $\epsilon_r = 16$. Compute the Bohr radius. (This problem shows approximately what the energy levels are for an electron or hole bound to an impurity in a semiconductor, as further discussed in Chap. 12.)

9

The electronic states in solids

The states of electrons in solids have an important similarity to the states of electrons of the free atoms, because the interactions between electrons on adjacent atoms can never completely destroy the basic structure of the electronic levels of the individual atoms. On the other hand, the interactions between atoms are strong enough to cause serious perturbation of the levels of the free atoms, and so some striking new phenomena occur in crystals. The most important is the splitting of the valence energy levels of the free atoms into nearly continuous energy bands. This splitting is the origin of a host of electrical, magnetic, and optical properties associated with one's everyday experience with metals and other solids.

The first part of the chapter is a discussion of the principles of electron interactions in solids. This discussion leads to equations which describe the bands of energy levels available for occupancy. The initial equations are derived from the simplest view possible for a system controlled by the laws of quantum mechanics, the application of the uncertainty principle. While this derivation leads to correct analytical expressions for the general nature of electron bands of levels, more sophisticated treatments are described to elucidate the finer details of band structure. Finally, discussion of the similarity and contrast between the nearly empty band and the nearly full band serves to point out both the properties of the electrons and the antithetical pseudoparticle, the hole.

9-1 Energy Bands

The formation of energy bands in a solid can be demonstrated easily by considering the changes in electronic energy levels which occur when a group of identical atoms with wide distances of separation are gradually brought together to form a solid. Let N identical atoms be placed on a lattice with interatomic spacing so large that the atoms have virtually no interaction with one another. For this system, the energy-level diagram of the solid is precisely the same as that of a single atom. However, while a single atom may have a level (say, the nth) with degeneracy g_n, the entire system of N atoms has degeneracy Ng_n at this level. If the lattice spacing of the ensemble of atoms is decreased, the atoms come within interaction range of one another and the free-atom energy levels change.

As an introduction to the behavior found for the real solid, first consider the changes in energy levels of a single atom when an outside, or perturbing, force is applied. If the perturbing force does work on the atomic electrons, the energy levels are shifted, of course, since the energy levels represent the total energy of an electron. An additional effect may occur: levels which were once degenerate in energy may split into levels with slightly different energies when the perturbing force is applied. The reason for the splitting is that an electron in one state may interact differently with the perturbing force from an electron in another. Only those orbits which are equivalent, or *symmetric*, to one another from the standpoint of the force field continue to have the same energy level, i.e., to be degenerate. An example of this kind of splitting is the Zeeman effect.

When the atoms are brought together in the solid, the interaction between atoms is a perturbing influence on the initial atomic levels. The atoms come so close together, in fact, that most of the symmetry possessed by the electron states in the isolated atom is destroyed, and hence the levels split. What was a single level for the solid with large lattice spacing broadens out into a large number of closely spaced levels for the solid with smaller lattice spacing. This *band of levels* retains some degeneracy, but not nearly so much as the Ng_n-fold degeneracy of the original group.

Some features of the band of levels are quickly apparent. First, the binding energy of the solid as a whole must be caused by shifts in the levels of the electrons, just as for chemical bonding. Thus, as the solid is formed, the energy levels must shift downward on the average. Second, the electrons which should be affected most are the outermost, or valence, electrons, since they are closest to the neighboring ions. Third, the equilibrium lattice spacing must be at an energy minimum, with the energy levels rising at distances of closer separation. These three requirements are taken into account in the sketch in Fig. 9-1 for a solid whose lattice spacing varies from large to small values. Fourth, as the atoms are brought together, as in Fig. 9-1, the states of the original system must deform continuously; the number of states in the solid must be the same as the number in the original separated atoms. Thus, since the "crystal" with large spacing has Ng_n states, the band at the distance of separation of the real crystal must also contain Ng_n

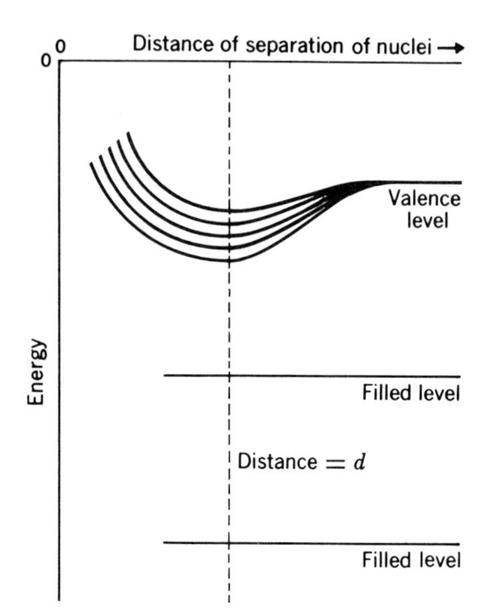

Figure 9-1 Energy of electrons in several levels as a function of the distance between atoms in a solid. The minimum in energy occurs at separation *d*, the atomic diameter.

states. An s band therefore holds 2 electrons per atom, a p band 6, a d band 10, etc.

The preceding argument is completely empirical except for the ideas from quantum mechanics. To understand the crystal band in any real physical sense, we must understand in detail at least these three things: (1) the basis for the forces which attract the atoms together, (2) the forces of repulsion when they are too close, and (3) the degree of splitting caused by the interactions.

The answer to the first point is complex because it is different for different crystals. It is therefore deferred to Chap. 10, where it can be discussed in the light of the ideas which will be formulated in later sections of this chapter.

The second point is more easily understood because it is the same for all solids. Atoms repel one another when they are too close, basically because the electrons of a given state must have a fairly well-defined volume to themselves. The Pauli exclusion principle states that the same wave functions from different atoms cannot occupy exactly the same space, for then they would be exactly the same state. When the atoms approach so closely as to compress the wave functions into smaller and smaller regions so that extensive overlap does occur and the exclusion principle cannot be satisfied, the energy rises because of the uncertainty principle. As the volume of the wave function decreases, Δx decreases. Then Δp and hence p must increase, thereby increasing the kinetic energy. The total energy thus rises as the atoms are pushed too close together. This is equivalent to a repulsive force.

The third point forms the subject of the most fundamental theorem concerning electrons in solids, namely, that the electrons in a band of energy levels are mobile and are not localized on individual atoms. Detailed calculations can be made which show explicitly the nature of the mobility, the topic considered in the next section.

9-2 Electron Mobility

The mobility of electrons in solids can be understood by considering the changes in wave function which occur when isolated atoms are brought close enough together so that appreciable overlap occurs. To be sure, some overlap occurs for any finite distance of separation; however, it becomes appreciable only when the separation distance becomes 10 A or less. The wave-function overlap at small separations for an s state is sketched in one dimension in Fig. 9-2. An electron which is at one

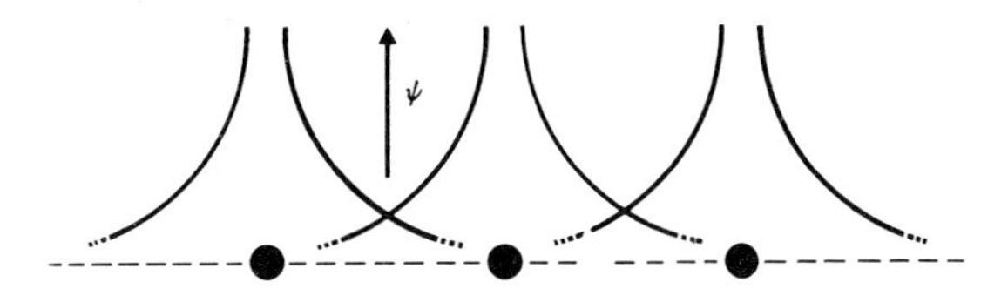

Figure 9-2 Wave-function overlap for nearest neighbors in the solid.

instant in orbit in a particular atom has a finite probability of being captured by a neighboring atom. The greater the amount of overlap, the greater is the probability that transfer occurs from atom to atom. At the distance between atoms corresponding to actual crystal-lattice spacings, the overlap is very high so that an electron hardly has time to circulate in its atomic orbit before it hops to the next atom site. Since this hopping process occurs rapidly, these electrons must be said to belong jointly to the entire collection of atoms of the crystal rather than to particular atoms.

The wave functions for electron states lying within the valence shell are more localized about the nucleus than are those of the valence electrons, so the overlap between these levels is less. Therefore the electrons in these states do not participate appreciably in the hopping process.

From the way in which the hopping features of the electrons have been introduced, the wave functions of the electrons seem to possess two characteristics. First, they have something of a free-atom character. The wave functions sketched in Fig. 9-2 are concentrated about the nuclei, even with overlap, and the individual wave functions about the atoms are primarily free-atom-like. Thus the energy of the electron in its orbit in a solid is nearly the same as the electron in a free atom. However, because of the overlap in the wave functions, a real transfer of the electron occurs from atom to atom in the crystal. This is an actual translation in space; hence there exists a translational momentum and a translational energy which are separate from, and in addition to, the strictly free-atom-like momentum and energy.

The total energy of the electron is the sum of these two factors,

$$E_{\text{total}} = E_{\text{atomic}} + E_{\text{trans}} \tag{9-1}$$

In this expression, E_{atomic} is the energy of the electron in its atomic orbit and E_{trans} the kinetic energy of hopping. Of course, E_{atomic} is not equal to the corresponding energy level of the free atom, because the wave functions (or orbits) of the electrons about the atoms in the solid are distorted from their free-atom shapes. In general E_{atomic} is *less than* the free atom energy; this is clear because, as mentioned earlier, the atoms do attract one another to form the solid. Fortunately, in many cases of interest, E_{atomic} for one state of translation is approximately the same as E_{atomic} for another. This results from the fact that the atomic orbit of an electron when it is hopping slowly is almost identical to that of an electron which is hopping faster. Therefore, for these states the energy levels of the various electrons can be sketched in a particularly simple way (see Fig. 9-3).

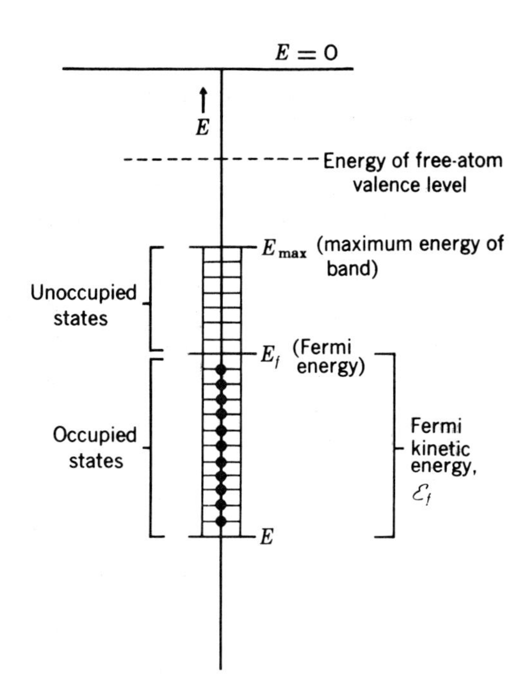

Figure 9-3 Energy-level scheme of a partly filled band.

In the figure several important features of the band have been labeled. The state labeled E_{atomic} must be the lowest energy state of the band, since E_{trans} must always be positive. Those electrons which do have translational energy have energies in a region above E_{atomic}. These electrons occupy the levels termed *occupied states* in the figure. The energy of the electrons with the highest velocity of translation is termed the *Fermi energy* E_f. Above E_f there exists a region of possible translational states up to some maximum in energy E_{max}, termed the *top* of the band. The fact that there is a maximum energy E_{max} follows from the fact that there are only Ng_n total states in the band. When Ng_n translational states are added to the purely atomic state at the bottom, the number of levels is exhausted and larger amounts of energy are not allowed in the band.

9-3 Fermi Energy

In a metal, the energy of the level with maximum energy, E_f, is a negative quantity (Fig. 9-3). The translational part of this energy, labeled $\mathscr{E}_f$ in Fig. 9-3, is a positive quantity since it is measured from E_{atomic}, the bottom of the band. The energy $\mathscr{E}_f$ will be called the *Fermi kinetic energy*. In many books, the distinction between E_f and $\mathscr{E}_f$ is not made, but such distinction is useful in discussion of both metals and semiconductors.

So far, nothing quantitative has been said about the translational motion. Fortunately, with a little help from quantum mechanics, the various translational levels can be described for the simplest metals. Corresponding to the translational motion there is translational momentum p_{trans}. The translational energy and momentum are, of course, related by the expression

$$E_{\text{trans}} = \frac{1}{2m} p^2_{\text{trans}} \tag{9-2}$$

where m is the mass of the electron.

The total band of levels contains exactly Ng_n states. Therefore the energy levels are a discrete but tightly packed set. For each of the (discrete) translational states to be distinguishable from all the others, the states must be resolvable from one another by an amount given by the uncertainty principle.

Consider, specifically, only one component of translation, say, the x direction. If an electron of the solid belongs to the entire solid, the uncertainty in position Δx is just the total extent of the crystal in that dimension, say, L. The uncertainty principle for the x components of momentum and position is written as

$$\Delta p_{x_{\text{trans}}} = \frac{h}{L} \tag{9-3}$$

where the minimum uncertainty in p_x is assumed. If Eq. (9-3) gives the minimum uncertainty in the momentum of any state, then the states are not discrete and cannot be distinguished from one another unless the momenta of all other states differ from the state under consideration by amounts which are greater than the basic uncertainty h/L. This momentum situation is sketched in Fig. 9-4 in a manner which suggests an analogy to the resolution of a microscope. If the optics of the microscope do not separate the diffraction patterns (uncertainty) of individual objects from one another, then the discrete set of objects is not resolvable under the microscope and a continuous blur results.

An actual crystal is three-dimensional, and the basic uncertainty of momentum in each direction is h/L (for a cube of edge L). The basic uncertainty of each state is, then, not a one-dimensional factor such as is sketched in Fig. 9-4, but a volume uncertainty in momentum space.

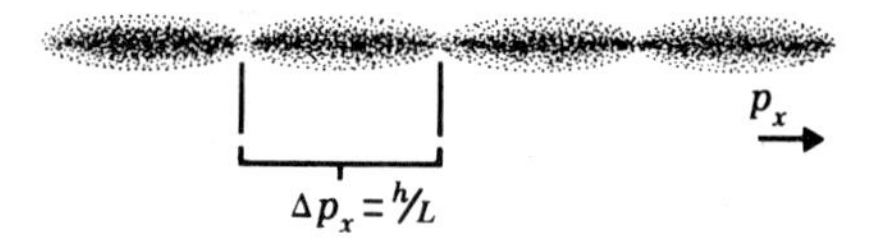

Figure 9-4 The allowed values of p_x are fuzzy regions whose distance of separation can be not less than that permitted by the uncertainty relation.

The minimum volume of uncertainty is

$$\Delta p_{x_{\text{trans}}} \Delta p_{y_{\text{trans}}} \Delta p_{z_{\text{trans}}} = \frac{h^3}{L^3} \tag{9-4}$$

This volume is sketched in Fig. 9-5.

Computation of the total number of states from the bottom of the band (that is, E_{atomic}) to the Fermi energy is a simple matter if Fig. 9-5

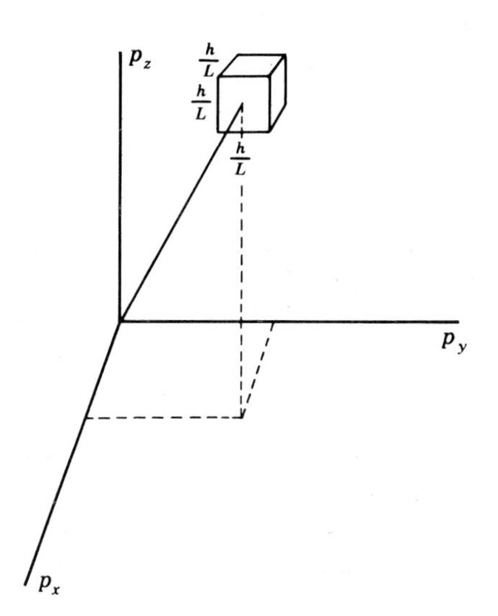

Figure 9-5 Uncertainty volume of a translational state of an electron in a band.

is used. The kinetic energy increases as the square of the distance p from the origin. The energy of all electrons whose momenta lie on a shell of radius p is the same. Those electrons with the Fermi energy have momenta on some spherical shell which has a certain value of p, say p_f (Fig. 9-6). All the electrons in the band are contained within the sphere so that the number of states within the sphere is equal to the total number of electrons, N. The total volume of the sphere in momentum space is simply given by $4\pi p_f^3/3$, and the volume in momentum space of each state is h^3/L^3. Since each momentum state holds two electrons, one with spin up and one with spin down, the total number of momentum states required to hold N electrons is $N/2$; this number is given by the quotient of the total volume of the sphere and the uncertainty volume h^3/L^3,

$$\frac{N}{2} = \frac{4L^3 p_f^3}{3h^3} \tag{9-5}$$

This equation does not contain the Fermi kinetic energy $\mathscr{E}_f$ ex-

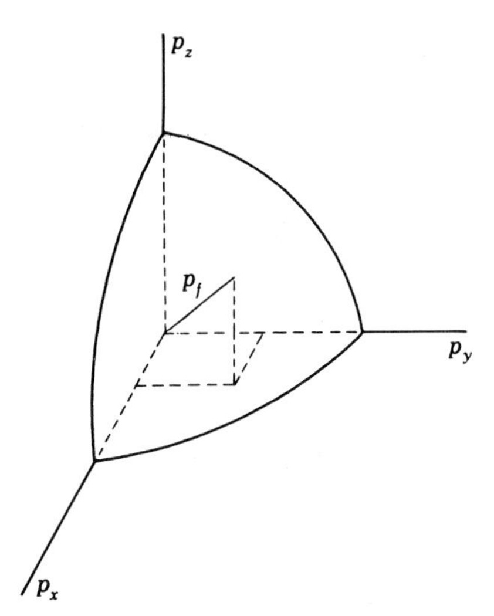

Figure 9-6 One octant of Fermi sphere in momentum space. All electrons in the band have momenta within this sphere.

plicitly, but p_f is related to the Fermi kinetic energy by the expression $\mathscr{E}_f = p_f^2/2m$. Therefore Eq. (9-5) can be written

$$N = \frac{8\pi(2m\mathscr{E}_f)^{\frac{3}{2}}L^3}{3h^3} \tag{9-6a}$$

Alternatively, $\mathscr{E}_f$ can be expressed in terms of the total number of electrons:

$$\mathscr{E}_f = \frac{h^2}{8m}\left(\frac{3}{\pi}\frac{N}{L^3}\right)^{\frac{2}{3}} \tag{9-7a}$$

The Fermi kinetic energy thus depends only on the number of electrons per unit volume, N/L^3.

Note that the Fermi kinetic energy given in Eq. (9-7*a*) represents the total translational energy of the electrons at the top of the distribution. Therefore it gives a quantitative expression for the width of a band in terms of the number of electrons in the band. For the simpler metals with a single *s* electron, such as the alkali metals Li, Na, K, etc., $\mathscr{E}_f$ is of order 3 to 5 ev (see Problems).

Treatment of the Fermi sphere as a smooth surface may be a little troublesome, since the momentum volume of each state has been considered to be a cube. This gives a fine, grainy structure to the surface of the sphere. This grainy structure, however, is very slight. $\mathscr{E}_f$ is of order 5 ev, so p_f is of order 10^{-24} kg m/sec. The uncertainty in each coordinate is, for a 1m cube of material, 6.6×10^{-34} kg m/sec. The granular structure of the surface is thus practically infinitesimal.

The picture of the band presented so far is appropriate only for bands made up of *s* electrons. Calculation of $\mathscr{E}_f$ for *p*, *d*, and higher states requires consideration of the basic degeneracy of these states. For example, a *p* state of a free atom contains six degenerate states. The application of the Pauli principle for the distinguishable translational states does not distinguish these six states from one another.

Hence, though Eq. (9-7a) does contain an explicit factor of 2 for the s band, this factor must be changed for the p and d bands. The general expressions for Eqs. (9-6) and (9-7) should be

$$N = \frac{4\pi(2m\mathscr{E}_f)^{\frac{3}{2}}gL^3}{3h^3} \tag{9-6b}$$

and

$$\mathscr{E}_f = \frac{h^2}{2m}\left(\frac{3}{4\pi}\frac{N}{gL^3}\right)^{\frac{2}{3}} \tag{9-7b}$$

where g for simple bands is the degeneracy factor for the atomic level given in the following table. We will see cases later (Ge and Si) where the degeneracy factor is more complex.

State	g
s	2
p	6
d	10
f	14

Equation (9-7b) predicts a dependence of bandwidth on atomic spacing. Qualitative arguments show that when atoms are very far apart the band is narrow, but as the atom spacing decreases, the band spreads apart because of the interaction between atoms. Equation (9-7b) shows this explicitly, since N/L^3 is proportional to the actual density of the solid. The interactions are complicated functions of atom separation; thus this equation does not predict accurately what happens at intermediate distances.

9-4 Electron Gas Approximation

There exists an alternative derivation of the results of the previous section which, because of its importance, is also given here. Suppose that the electrons in their translational motion form a free gas of particles in the solid. Then the model of a particle in a box is appropriate, and the results of Sec. 7-6 may be applied. We suppose that the solid amounts to the potential energy box shown in Fig. 9-7. Obviously, the zero of kinetic energy within the box is measured from E_{atomic} in Fig. 9-3, because this is the energy of zero translational motion. $E = 0$ corresponds to the electron outside the box as before and is the same as the zero of energy of Fig. 9-3.

Strictly speaking, if the solid is a box with a finite potential well, the results of Sec. 7-6 do not apply, since in that section the potential step at the edge of the box was supposed infinite. In that case, the wave function is totally excluded from the exterior of the box and must go to zero at the surface. In the present case, the wave function

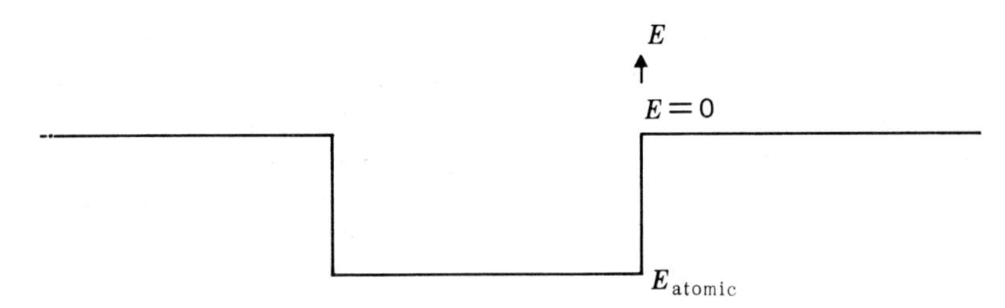

Figure 9-7 Potential energy of an electron in a solid considered as a particle in a box. The depth of the well is E_{atomic} in order to make this point the zero of kinetic energy.

merely decays exponentially outside the surface. However, the results to be derived are essentially the same in either case, provided surface effects are not considered specifically.

A more important problem, however, arises from the form in which we wish to express the solutions. We wish to use the pure exponential solutions, exp (ikx), instead of the sine functions which are, rigorously, the only true solutions to the box problem. The sine functions are standing waves, and as such correspond to particles with equal amounts of positive and negative momentum. That is, sin kx is written as a combination of exp (ikx) and exp $(-ikx)$. We have written of electrons in the last section as possessing either positive or negative momentum (not both at once), and it is a considerable convenience to continue to do so. But the exponential functions satisfy a different set of boundary conditions, the so-called periodic boundary conditions. That is, if

$$\begin{aligned} \psi(x,y,z) &= \psi(x + L, y, z) \\ \psi(x,y,z) &= \psi(x, y + L, z) \\ \psi(x,y,z) &= \psi(x, y, z + L) \end{aligned} \tag{9-8}$$

then the appropriate solutions of the Schrödinger equation satisfying these boundary conditions is

$$\psi(x) = e^{ikx} \tag{9-9a}$$

in which

$$\begin{aligned} k_x &= \frac{2\pi}{L} n_x \\ k_y &= \frac{2\pi}{L} n_y \\ k_z &= \frac{2\pi}{L} n_z \end{aligned} \tag{9-9b}$$

The solid with the boundary conditions (9-8) has no surface in the physical sense; it simply repeats itself in the three coordinate directions with period L. In one dimension, the analog is a string of length L

which is formed into the shape of a continuous loop. The string then has no end, but repeats itself with period L. Of course no physical solid has such a periodic character, but the true significance of the periodic boundary conditions (9-8) and the solutions (9-9) is that we have done away with the physical presence of the surface, while at the same time we have retained the concept of the size of the crystal lattice.

The allowed values of k given by the condition of Eqs. (9-9b) generate a cubic lattice of points in k space, with the energy of each point given by $E_{\text{trans}} = \hbar^2 k^2/2m$. Calculation of the number of such points within a sphere of constant energy in that space yields the result that

$$N = \frac{4\pi}{3} \frac{k_{\text{max}}^3}{(2\pi/L)^3} = \frac{4\pi}{3} \frac{p_{\text{max}}^3 L^3}{h^3} \tag{9-10}$$

which is, of course, the earlier result, Eq. (9-5). The rest of the results of Sec. 9-3 then follow automatically. The reader will note that the "unit cell" of the lattice in k space precisely corresponds to the uncertainty volume in momentum space, Eq. (9-4).

The physical picture of Sec. 9-3 emphasizes the underlying atomic character of the wave function of the electrons in a solid and only brings in the translational motion as a consequence of the overlap of wave functions on neighboring atoms. We have here dealt with the electron as if it were a free particle, and indeed have derived a plane wave function which makes no reference whatever to the atomic wave function which the electron must still largely possess. We are obviously in the presence of a basic duality: the electron is both atomic–like, and free-particle–like. We show in a later section how these two properties are combined in a product to give the total wave function.

The potential function in Fig. 9-7 (from which the free-electron approximation is obtained) can be considered to be the result of some form of averaging of the internal potential of the solid. This view, however, is not without its difficulties; one knows that the regions near the ion cores are very deep Coulomb potentials, so one would not expect to obtain very meaningful results from such a grossly averaged potential. However, modern theories have shown that, properly modified and interpreted, the free-electron picture comes very close to the truth for many metals and semiconductors. The reason is that an "effective potential" or "pseudo potential" for the electron can be defined which *is* very nearly the flat average value, E_{atomic}, and one obtains the true electron wave function from the simple plane wave solution above by gouging in it holes which look like atomic functions in the vicinity of the inner ion cores of the lattice atoms. The energy of this "orthogonalized" wave is then very nearly that of the free electrons described above.

9-5 Bloch's Theorem

The result of the past two sections demonstrates, on the one hand, that the wave functions of the valence shells simply overlap when the atoms come together to form the solid and, on the other hand, that the electrons in many solids may be thought to form almost a free gas of particles. We have hinted that these two attributes form a duality and that the wave function in the solid takes a special form which recognizes both aspects of this duality. The existence of such a wave function was first demonstrated by Bloch, and it bears his name. Bloch's theorem states that the wave function for a perfect, periodic lattice has the form

$$\psi(x) = e^{ikx}u_k(x) \tag{9-11}$$

The plane wave part of this function is precisely of the form given in Eq. (9-9*a*), and k has the allowed values given by Eq. (9-9*b*). The function $u_k(x)$ is a periodic function which is closely related to the free atomic functions. Since all atoms are alike, $u_k(x)$ is obviously the part of the wave function which is characteristic of the kind of atom involved. Also $u_k(x)$ becomes asymptotically the free-atom wave function if the lattice parameter is uniformly expanded until all the atoms are far apart from one another. A schematic representation of the total wave function is given in Fig. 9-8.

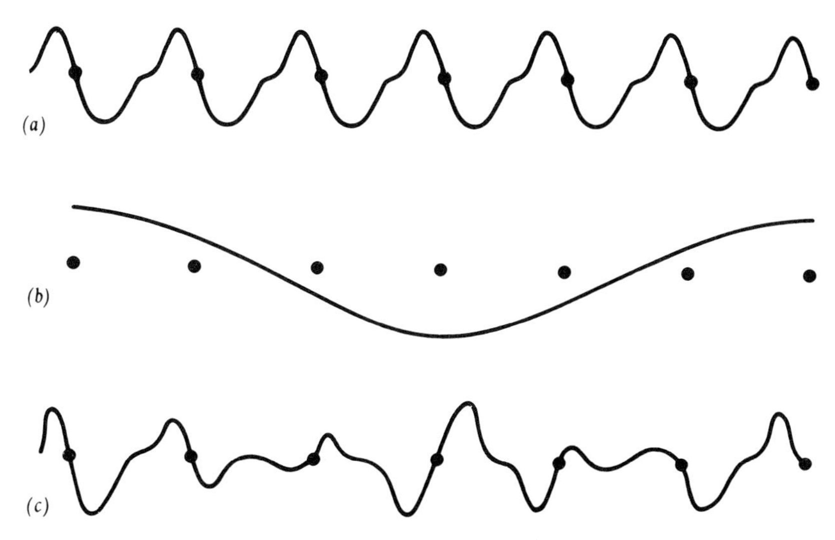

Figure 9-8 Representation of the Bloch wave function in the solid. (*a*) Atomic functions. (*b*) Plane wave. (*c*) Product wave function.

The Bloch wave represents a flux of particles, since the k vector is related to the momentum of a wave by the de Broglie relation. Further, the particle moves uniformly through the lattice completely unhindered by the atoms. This property is directly related to the perfection of the lattice, since, as we shall show, the validity of Bloch's theorem rests entirely upon the existence of a mathematically perfect crystal lattice. The completely uniform motion of the electron is reminiscent of the free-particle gas of the last section, for a perfect gas is composed of particles which interact only rarely with each other and the container walls. Similarly here, an electron in a Bloch wave starting on one side of the crystal moves in a newtonian-like line, straight through the crystal to the other side without either losing momentum or changing direction. This property is of central importance in the description of electron transport in electric fields (Chap. 11).

This preliminary discussion of the meaning and implications of Bloch's theorem has preceded the formal proof of the theorem, because the proof requires a somewhat deeper familiarity with quantum mechanics than has been reviewed in Chap. 7. Some readers may thus wish to skip on to the next section, rather than work through the development which follows.

Bloch's theorem follows from the fact that the potential function in the Schrödinger equation is completely periodic. The Schrödinger equation for two completely equivalent points of two different cells of the lattice is written (in vector form)

$$\frac{-\hbar^2}{2m}\nabla^2\psi(\mathbf{x}) + V(\mathbf{x})\psi(\mathbf{x}) = E\psi(\mathbf{x}) \tag{9-12a}$$

$$\frac{-\hbar^2}{2m}\nabla^2\psi(\mathbf{x}+\mathbf{R}_n) + V(\mathbf{x})\psi(\mathbf{x}+\mathbf{R}_n) = E\psi(\mathbf{x}+\mathbf{R}_n) \tag{9-12b}$$

Here $\mathbf{R}_n$ is a lattice vector connecting the two cells. Strictly speaking, we should have put $V(\mathbf{x}+\mathbf{R}_n)$ in place of $V(\mathbf{x})$ in (9-12*b*), but $V(\mathbf{x}) = V(\mathbf{x}+\mathbf{R}_n)$ since the lattice is completely periodic.

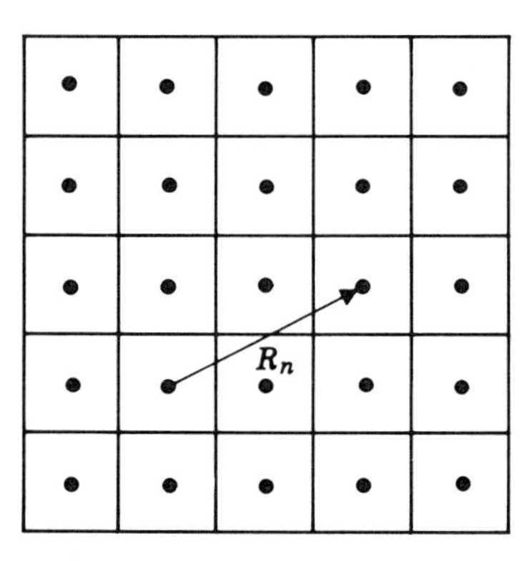

Figure 9-9 A lattice vector.

The two equations (9-12) are, of course, the same differential equation, except that the solutions, $\psi(\mathbf{x})$ and $\psi(\mathbf{x}+\mathbf{R}_n)$ are translated from one another by the vector $\mathbf{R}_n$. The two solutions must be the same *physical* wave function, therefore, since two solutions are equivalent in quantum mechanics if they are multiples of one another: since Ψ functions must be normalized, the multiple may be, in the most general case, a complex number of magnitude one. The most convenient representation of this multiple is $\exp(i\alpha_n)$, since $[\exp(i\alpha_n)]^2 = 1$.

$$\psi(\mathbf{x}) = e^{i\alpha_n}\psi(\mathbf{x}+\mathbf{R}_n) \tag{9-13}$$

Suppose two independent lattice translations, $\mathbf{R}_n$ and $\mathbf{R}_m$, are performed. The result must itself be a lattice translation, which we may call $\mathbf{R}_p$. Thus

$$\psi(\mathbf{x} + \mathbf{R}_n + \mathbf{R}_m) = e^{i\alpha_m}[\psi(\mathbf{x} + \mathbf{R}_n)] = e^{i(\alpha_m+\alpha_n)}\psi(\mathbf{x}) \tag{9-14}$$

But since

$$\psi(\mathbf{x} + \mathbf{R}_n + \mathbf{R}_m) = \psi(\mathbf{x} + \mathbf{R}_p) = e^{i\alpha_p}\psi(\mathbf{x}) \tag{9-15}$$

then

$$e^{i(\alpha_m+\alpha_n)} = e^{i\alpha_p} \tag{9-16}$$

or

$$\alpha_m + \alpha_n = \alpha_p \tag{9-17}$$

If Eq. (9-17) is valid for any translation, it is valid in particular for two primitive translations in the x direction. Let this be

$$\alpha_1 + \alpha_1 = 2\alpha_1 = \alpha_2 \tag{9-18}$$

Thus α_p is linear in the distance of translation, and we can write for a translation R_x in the x direction

$$\alpha = k_x R_x \tag{9-19}$$

Likewise for a translation which contains components in the other directions as well,

$$\alpha = k_x R_x + k_y R_y + k_z R_z \tag{9-20}$$

The solution to the Schrödinger equation must thus have the form,

$$\psi(\mathbf{x}) = e^{i\mathbf{k}\cdot\mathbf{R}_n}\psi(\mathbf{x} + \mathbf{R}_n) \tag{9-21}$$

If Eq. (9-21) is multiplied by exp $(i\mathbf{k} \cdot \mathbf{x})$ on each side, then

$$e^{i\mathbf{k}\cdot\mathbf{x}}\psi(\mathbf{x}) = e^{i\mathbf{k}\cdot(\mathbf{x}+\mathbf{R}_n)}\psi(\mathbf{x} + \mathbf{R}_n) \tag{9-22}$$

This equation is simply a statement that the product exp $(i\mathbf{k} \cdot \mathbf{x})\psi(\mathbf{x})$ is a periodic function which repeats itself within each cell of the lattice. If we call this function $u_k(x)$, then (in scalar form)

$$e^{ikx}\psi(x) = u_k(x) \tag{9-23}$$

In the more familiar form,

$$\psi(x) = e^{-ikx}u_k(x) \tag{9-24}$$

which is Bloch's theorem.

The fact that the k used in Eq. (9-24) is the same as that used in previous sections and has the allowed values (9-9*b*) follows immediately from the requirement that ψ as a whole must satisfy the periodic boundary conditions. (See the previous section.) Since $u_k(x)$ is periodic from one cell to the next of the lattice, it is already periodic with the period, L. Thus the total wave function ψ is periodic with period L in each of the coordinate directions only if the k vector satisfies the conditions of Eq. (9-9*b*).

The interpretation of the last two sections can now be unified. The Schrödinger equation for the solid is

$$E_k\psi = \frac{-\hbar^2}{2m}\nabla^2\psi + V(x)\psi \tag{9-25}$$

We have written a subscript, k, on E to emphasize that E_k is a function of k. After substitution of Bloch's theorem, Eq. (9-24), and division of the common exponential,

$$\frac{-\hbar^2}{2m}\left\{-k^2 + 2i\left(k_x\frac{\partial}{\partial x} + k_y\frac{\partial}{\partial y} + k_z\frac{\partial}{\partial z}\right) + \nabla^2\right\}u_t(x)$$
$$+ Vu_k(x) = E_k u_k(x) \tag{9-26}$$

In many cases, the terms in $k_x(\partial/\partial x)$ etc., can be neglected, for example if the ψ function is s-like, and the result then becomes

$$\frac{-\hbar^2}{2m}\nabla^2 u_k + Vu_k = \left(E_k - \frac{\hbar^2k^2}{2m}\right)U_k \tag{9-27}$$

In Sec. (9-3) and (9-4), the quantity $E_k - \hbar^2k^2/2m$ is just what we have called E_{atomic}. Thus (9-27) becomes

$$\frac{-\hbar^2}{2m}\nabla^2 u_k + Vu_k = E_{\text{atomic}}u_k \tag{9-28}$$

Within a cell of the solid, V is very nearly the atomic potential, so Eq. (9-28) has the same form as the Schrödinger equation for a free atom. Thus the solutions, u_k, are the solutions of the same differential equation as the free-atom ψ functions. The only difference is that the atom in the solid is not in free space, but is bounded by other atoms around it. This change in boundary condition does change the wave function, as we will demonstrate more fully in Sec. 10-1, and its energy is lower than the free-atom energy, as it must be. The important point for us here, however, is that the u_k is indeed an atomic-like function. Also, since the right side of Eq. (9-28) is just what has previously been called E_{atomic}, this approximation yields $E_{\text{atomic}} = E_k - \hbar^2k^2/2m$ (or $E_k = E_{\text{atomic}} + \hbar^2k^2/2m$). This latter equation represents precisely the free-electron approximation in the solid, and the duality mentioned above is clarified.

9-6 Effective Mass

The mass of the electron in the previous sections was assumed tacitly to be its ordinary value as measured in free space. However, the mass of electrons in solids can be measured directly (see Sec. 9-9), and such measurements, though generally giving values of the expected order of magnitude, normally do not yield values exactly the same as that of a true free electron. Further, the mass measured in one direction is often markedly different from that measured in another.

To explain this discrepancy, one must reflect upon precisely what is meant by the inertial mass of an electron in a solid. We define it through Newton's law of motion for an electron in an external electric field, $m^*\ddot{\mathbf{x}} = -e\mathbf{E}$. The free-space electron mass is that value of m^* when motion of the electron is completely free of any other influence. The motion of electrons in solids, however, can be greatly influenced by the ion cores. For example, suppose that the atoms in a hypothetical crystal are far apart. Then the hopping frequency of an electron from atom to atom is small because the overlap is small. In the electric field, the electrons are induced to jump from one atom to another at a faster and faster rate in the direction of the field. However, when the overlap is small to begin with, the electrons have only a limited ability to accelerate in the field. Consequently, since $\ddot{\mathbf{x}}$ is small, the inertial mass, defined by Newton's law, must be large compared to the free-electron mass.

The mass defined in this way, is thus a special type of quantity which not only includes the ordinary mass of the electron, but also includes the complex interaction between the electron and the atoms to which it is attached. As shown in this example, the degree of overlap is the most important factor which determines the ease with which the particle responds to forces tending to increase its translational motion through the crystal. To distinguish this type of inertial "mass" from the ordinary free electron mass, we speak of the mass of the electron in a solid as an *effective* mass, and denote it m^*.

When the atoms are as close together as they are in most solids, translational motion is easy, and electrons can be accelerated easily by an external field. Generally, then, the effective mass of valence electrons is of the order of magnitude of the free mass. In some crystals, the mass is even *less* than the free-space mass. In these cases, the electron in its atomic orbit has just the correct phase so that its response to a translational force is enhanced. One might say that the underlying atomic forces help kick the electron along when acted upon by an external electric field.

A more rigorous approach to the idea of effective mass is afforded through Eq. (9-26). The effective mass enters through the terms in $k_x(\partial/\partial x)$, etc., which were neglected in Eq. (9-27) and in later equations. In cases where the effective mass is not greatly different from the free mass, the neglected terms, being small, can be shown to be quadratic in k. When transposed to the right side of Eq. (9-26), they make a correction to the free electron term, $-\hbar^2k^2/2m$. This correction may then be interpreted as a deviation in the mass from its free-space value. Indeed, one can generalize to the effect that the mass is the coefficient of k^2 in $E(k)$. More precisely, for a simple scalar mass,

$$\frac{1}{2m^*} = \frac{\partial^2 E(k)}{\partial k^2} \tag{9-29}$$

The mass defined in this way can be shown to be exactly the translational inertial mass of the electron as measured by an electric field. It is obviously large for a narrow band, because in that case $E(k)$ is nearly flat and the "curvature" calculated in Eq. (9-29*a*) within the band is small.

In subsequent chapters, we deal with cases like the semiconductors, where $E(k)$ is not a simple quadratic function of k, but is a general quadratic function of k_x, k_y, and k_z. In that case, m^* is not a simple scalar, but is a function of direction, a tensor of second order. By contrast, most of the simple metals possess a simple scalar mass. Although the mass as defined in terms of the local curvature by Eq. (9-29) may vary through a band, one can derive an appropriate *average* over the band by applying Eq. (9-7), the formula for the Fermi energy, and by substituting m^* for m and the experimental value for $\mathscr{E}_f$. The effective masses derived in this manner for a few of the simpler metals are given by Table 9-1.

Table 9-1 Estimated effective mass and Fermi energy for some metals

Metal	m^*/m	$\mathscr{E}_b$, ev
Be	1.6	12.0
Li	1.2	3.7
Na	1.2	2.5
Cu	1.0	7.0
K	1.1	1.9
Ca	1.4	3.0
Al	0.97	11.8
Zn	0.85	11.0

† The mass is estimated from the width of the band.

As with Eq. (9-7), all the other equations of Secs. 9-3 and 9-4 should be rewritten in terms of the effective mass.

9-7 Electronic Density of States

The results of the previous calculation give only the bandwidth; they do not describe the details of the spacing of energy levels or the population of the levels. Determination of these latter quantities can, however, be made through extension of the previous ideas. Although the states are uniformly distributed in momentum space (see Fig. 9-6), they are not uniformly distributed on an energy scale. Furthermore, the population at a given energy level, called henceforth the *density of states*, must increase for the larger values of p, since the number of ways in which individual values of p_x, p_y, p_z can combine to form a given value of p increases as p increases. Quantitatively, the density of states $\rho(E)$ is defined as dN'/dE, where N' is the total number of states having energy between E_{atomic} and E,

$$N' = \frac{4\pi(2m^*)^{\frac{3}{2}}g\mathscr{E}^{\frac{3}{2}}L^3}{3h^3}$$

where $\mathscr{E} = E - E_{\text{atomic}}$.

Note the similarity to Eq. (9-6*b*). Carrying out a differentiation with respect to E gives for $\rho(E)$ the expression

$$\rho(E) = \frac{dN'}{dE} = \frac{2\pi(2m^*)^{\frac{3}{2}}gL^3\mathscr{E}^{\frac{1}{2}}}{h^3} \tag{9-30}$$

Thus $\rho(E)$ varies as the square root of $\mathscr{E}$, the translational energy.

The interpretation of $\rho(E)$ may be easily seen through the use of sketches of constant-energy contours in momentum space (Fig. 9-10).

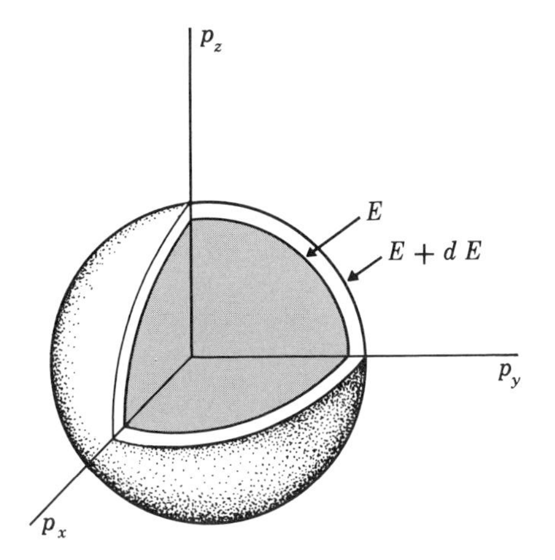

Figure 9-10 The number of states between E and $E + dE$ is given by the number of states in a spherical shell of thickness dE.

The inner sphere is a surface of constant p and hence of constant energy E. The outer sphere is also a sphere of constant energy $E + dE$. The number of electrons, dN', with energy between E and $E + dE$ is given by $\rho(E)\,dE$, hence by $A(E - E_{\text{atomic}})^{\frac{1}{2}}\,dE$, where $A(E - E_{\text{atomic}})^{\frac{1}{2}}$ is the area of the spherical shell.

The derivation of Eq. (9-30) does not depend on a given state being occupied by an electron, so it applies to the unfilled part of the band as well as the filled part. A plot of Eq. (9-30) from the bottom to the top of a band is given in Fig. 9-11. Calculation of representative values

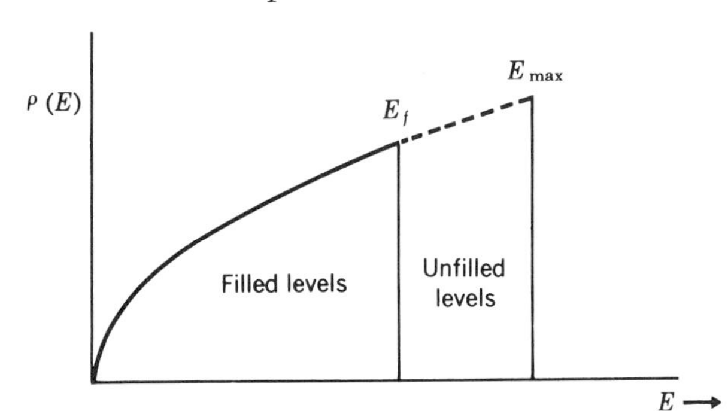

Figure 9-11 The density of states as a function of energy.

$\rho(E)$ at the Fermi level for various metals is made in several problems.

X-ray evidence confirms the plot of $\rho(E)$ versus E for the simpler metals. This confirmation gives credence to the physical concepts leading to a development of Eq. (9-30). For the simpler metals such as Na and Li the results are not only in good qualitative agreement but also in good quantitative agreement. More complicated metals and other crystals show deviations which indicate that the physical picture outlined here is too simple for them; these exceptions are noted and discussed as they arise.

One important application of Eq. (9-30) is the calculation of the *average* energy of the electrons in a band. In general the average of any quantity, say, r, which has a distribution function $\rho(r)$ is given by

$$r_{\text{av}} = \frac{\int_0^{r_{\text{max}}} r\rho(r)\,dr}{\int_0^{r_{\text{max}}} \rho(r)\,dr} \tag{9-31}$$

if the range of the variable is from 0 to r_{max}. For the energy of the electron distribution the average is

$$\mathscr{E}_{\text{av}} = \frac{\int_0^{\mathscr{E}_f} (E - E_{\text{atomic}})\rho(E)\,dE}{\int_0^{\mathscr{E}_f} \rho(E)\,dE} = \tfrac{3}{5}\mathscr{E}_f \tag{9-32}$$

The average translational energy of an electron in the ideal distribution is therefore three-fifths the Fermi kinetic energy. The total translational energy of N electrons in the band is thus $\frac{3}{5}N\mathscr{E}_f$.

9-8 The Top of the Band

The number of states in a band is finite; therefore the band of allowed states terminates at some maximum energy. The termination of the band has so far not appeared in the analysis. The most direct physical picture which describes the top of the band derives from the fact that the electrons, being waves, can be expected to be diffracted by the lattice for appropriate wave lengths just as X rays are. When one remembers that electrons at the Fermi velocity have wavelengths of the order of the atomic dimensions, diffraction becomes an important consideration.

We discuss the diffraction problem only for the case of the simple cubic crystal. As with X rays, the electrons are not diffracted if their wavelength is long compared to the distance between diffracting planes in the crystal; thus, electrons with energies near E_{atomic} are not diffracted. Electrons higher in the band have shorter wavelengths, though, and at certain values of energy, diffraction can occur. In the simple cubic crystal, the first planes off which diffraction can occur are the close packed [100] planes, since the distance between these planes is larger than that of any others. Bragg's law for the [100] planes Eq. (2-3) is

$$n\lambda = 2d \cos\theta \tag{9-33}$$

where d is the interplanar spacing, λ is the wavelength of the electrons, and n is the order of the diffraction. θ is the angle between the normal to the [100] plane and the incident wave. (This angular convention is different from that of Eq. (2-3) and Fig. 2-14, for reasons which will appear shortly.) This equation may be written in terms of the $\mathbf{k}$ vector, and it then takes the form

$$\pi n = \mathbf{k} \cdot \mathbf{d} \tag{9-34}$$

The vector $\mathbf{d}$ is normal to the (100) plane and has the value of the interplanar spacing. The geometry is sketched in Fig. 9-12. If the vector

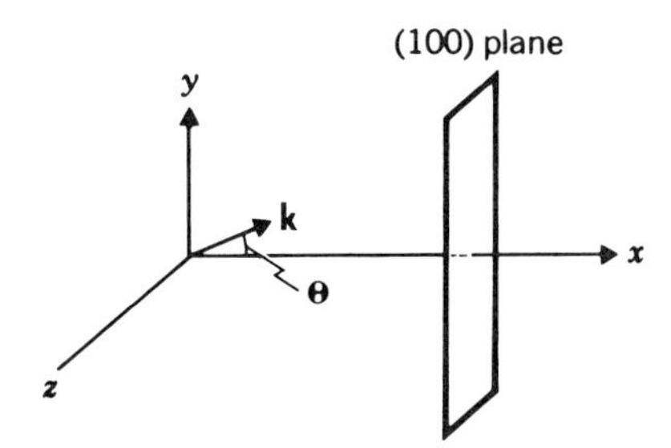

Figure 9-12 Sketch of the diffraction conditions in Bragg's law.

$\mathbf{d}$ is chosen as described, then Eq. (9-34) is the equation for a plane in k space over which diffraction will occur. (See Fig. 9-13.)

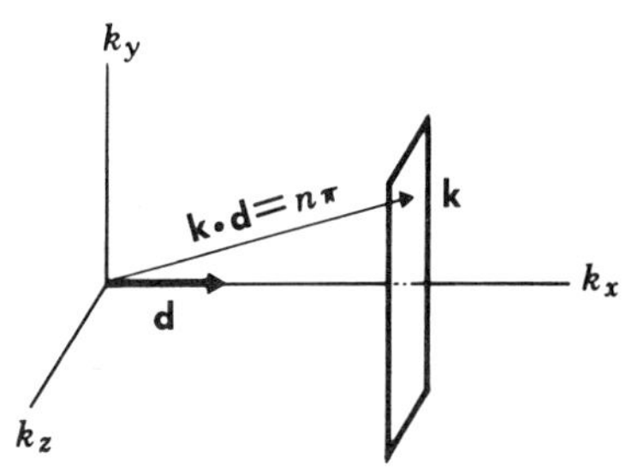

Figure 9-13 The diffraction condition in k space.

For diffraction of lowest order ($n = 1$), there are six equivalent diffracting planes in k space, corresponding to the six equivalent {100} planes. These are sketched in Fig. 9-14. The six diffracting planes enclose a cube in k space with edge $2\pi/d$ in length. The volume of the cube is $8\pi^3/d^3$, and if we divide this by the elementary uncertainty volume belonging to each allowed $\mathbf{k}$ vector [Eq. (9-4) or (9-9b)], we obtain the number of points allowed in the cube:

$$\frac{8\pi^3/d^3}{(2\pi/L)^3} = \frac{L^3}{d^3} = N \tag{9-35}$$

Here L^3/d^3 is the number of atoms in the simple cubic crystal. If the band is an s band, then spin degeneracy doubles each k point to give the total number of states, and the volume enclosed by the six diffract-

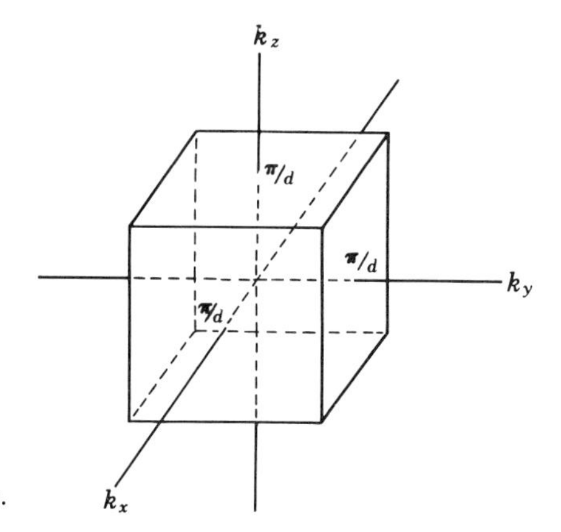

Figure 9-14 First Brillouin zone for simple cubic lattice.

ing planes contains exactly the total number of states in the band.

This argument demonstrates that as the energy of an electron increases through the band, it eventually reaches a value where the electron is strongly diffracted by the atoms of the crystal, and the free-electron picture breaks down. It also shows that the planes on which the diffraction occur completely enclose a volume of k space which just corresponds to the total band of the crystal. This volume is called a Brillouin zone, and the surfaces bounding it are called the Brillouin zone boundaries. The zone constructed in Fig. 9-14 is actually only the first Brillouin zone, since higher order diffractions give rise to other volumes in k space outside the first zone. These enclosed volumes are also called Brillouin zones.

This treatment does not show why the energy gap occurs when the diffraction condition is met. Indeed, a rigorous treatment of that fact is beyond the scope of this book. One can, however, show that inclusion of the periodic potential of the lattice yields not only the condition of diffraction but also the existence of a deviation in the energy of the diffracted wave relative to the expected parabolic value. Figure 9-15 shows how the energy of nearly free electrons varies as a function of k in a particular direction in k space. The parabolic form is closely

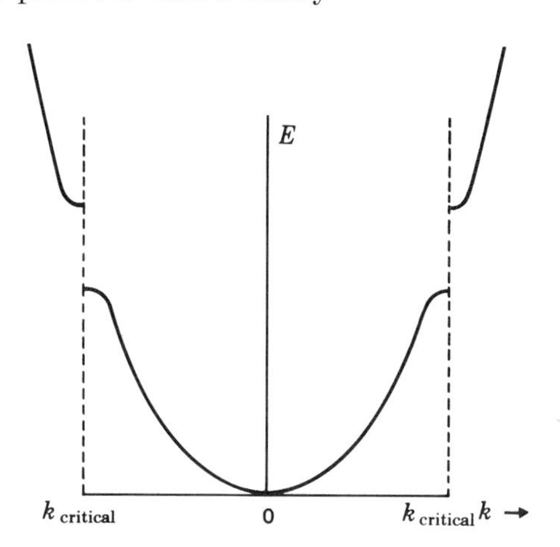

Figure 9-15 Variation of E with k for a particular direction in k space.

followed for nearly free electrons until the wavelength approaches the diffraction condition, where the curve bends over in a reverse parabola, leaving a gap in energy for values of adjacent **k** vectors on either side of the Brillouin zone boundary. Similar gaps appear as one approaches the Brillouin zone boundary along any other direction from the origin.

We will note, for future reference, that the bending over of $E(k)$ or k as the Brillouin zone boundary is reached corresponds to a lowering of energy relative to the free-electron value. If we recall that the Fermi surface corresponding to free electrons is a sphere, it follows that as the Fermi surface approaches the boundary, it is attracted to the boundary (adecrease of energy corresponds to an attractive force),and protuberances are formed adjacent to the boundary. A two-dimensional representation of how a zone fills as electrons are added is shown in Fig. 9-16. The consequences for the density of states near the top of the

Figure 9-16 As the Fermi sphere grows, dimples appear on the sections nearest the Brillouin zone boundary, and as overlap occurs, reverse curvature appears until the corners are finally filled in.

band are shown in Fig. 9-17. For small values of energy, the density of states is proportional to $\sqrt{E}$. As the Fermi sphere first reaches the Brillouin zone boundary, a peak appears in the density of states corresponding to the level portion of $E(k)$ as the Fermi surface first overlaps the boundary. Then the density of states rapidly decreases because the total area of the Fermi surface decreases as it recedes into the corners

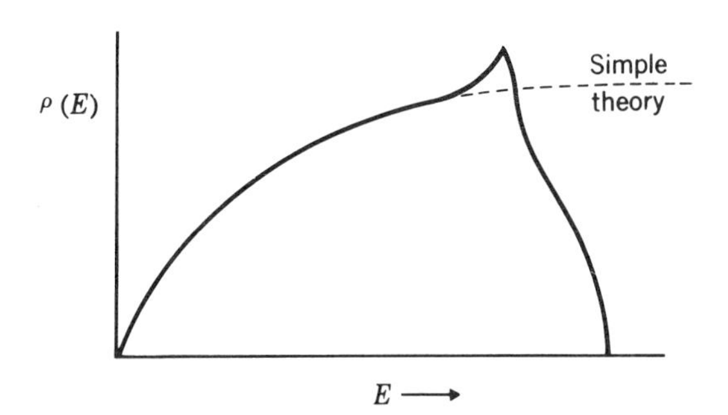

Figure 9-17 Density-of-states plot as a function of energy for the simple free-electron theory and the zone theory.

of the zone. (Remember that the density of states as it is defined in Sec. 9-7 is proportional to the free area of the Fermi surface.)

The zones for other more complex crystals are constructed in precisely the manner outlined for the simple cubic. The first Brillouin zones for fcc and bcc crystals are shown in Fig. 9-18. Calculations of their shapes is left for the problems.

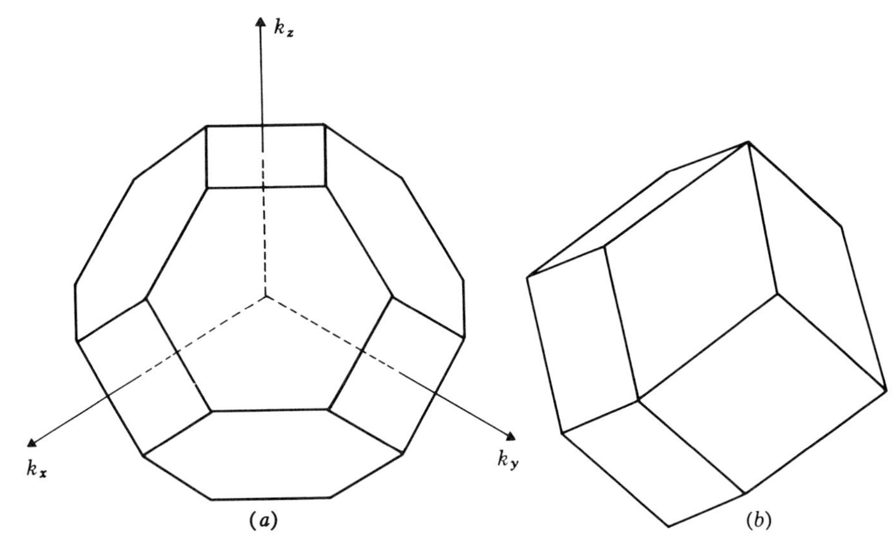

Figure 9-18 First-zone boundaries for fcc and bcc crystals. (*a*) Fcc, truncated octahedron; (*b*) bcc, regular dodecahedron.

9-9 Electrons and Holes

When one combines the comments of the last section to the effect that $E(k)$ near the top of a band has negative curvature (a negative parabola) with those of Sec. 9-6, which state that the effective mass is inversely proportional to the curvature of $E(k)$, a considerable paradox appears. The juxtaposition of these two facts predicts that an electron near the top of a band must have a negative mass! This prediction is entirely correct, and from the description already given of the concept of effective mass, we should not be surprised that the interplay between internal and external forces might make the electron behave occasionally as if it has a negative mass. Indeed, the fact that electrons near the top of the band have negative mass follows directly from the fact that the top of the band represents a diffraction condition. Diffraction, of course, means that the electron discontinuously suffers a complete reversal of velocity. For an electron near the Brillouin zone boundary, an external force bringing it closer to the diffraction condition will at the same time force it to experience some of the effects of diffraction. An increasing tendency toward diffraction is manifested as an acceleration in the opposite sense from the external force. This behavior follows, of course, from the ability of the quantum-mechanical system to sense the singular behavior at the Brillouin zone boundary somewhat before that boundary is actually reached.

The dynamics of a particle of negative mass are unnatural, however, so an alternative construct is adopted. The alternative is to focus attention not on the electrons of the nearly full band, but rather on the "holes" left over in the electron distribution at the top of the nearly full band. These holes can be treated as particles having opposite charge and mass from the electrons, whose absence they represent. Removal of one electron from a full band with consequent creation of a hole causes the charge of the crystal to be increased positively by the amount $|e|$ because a negatively charged electron is subtracted from the crystal. Thus the hole, considered as a particle, must be considered to have an effective positive charge of amount $+|e|$.

The mass of the hole is obtained by an application of Newton's law. This law, written for the electron which has been taken away, in the presence of an electric field, E, is

$$m_e^* \ddot{x} = -|e|\,\mathbf{E} \tag{9-36}$$

The hole must, of course, follow along in the electric field as the electron would have, had it been there. We rewrite the equation so that a positive charge of magnitude $|e|$ corresponding to the charge on the hole appears on the right side. Thus,

$$(-m_e^* \ddot{x}) = m_h^* \ddot{x} = |e|\,\mathbf{E} \tag{9-37}$$

We now interpret this equation as if it applied to a particle of charge $+|e|$, and mass $m_h^* = -m_e^*$, where the subscript h pertains to holes and e to electrons. The "particle" which results is obviously the hole. For a state near the top of the band, the hole has positive mass (and charge), and it can be considered as a conventional particle; this makes it preferable to the electron for these states.

A hole has a mirrorlike relationship to the electron. An electron is formed when a negatively charged particle is placed in a completely empty band. A hole results when one of the electrons of a completely full band is annihilated by addition of a particle of opposite charge and mass. The completely full band and completely empty bands have in common the property that each is a state of zero net charge flow, and thus each serves as a convenient "ground state."

Excitation of an electron to some higher state (e.g., a higher empty band) from a completely full band is equivalent to the formation of a "pair" of free but opposite particles. One is the electron in the higher state, and the second is the hole in the full band. The hole is free to move about spatially and to jump from state to state in the full band just as is the electron in the empty one.

As one might expect in the curiously inverted world of the hole language, the energy scale is inverted to that of the sea of electrons on which the hole floats. A hole at the top of the full band, for example, represents the *lowest* state of excitation of the full band. This property

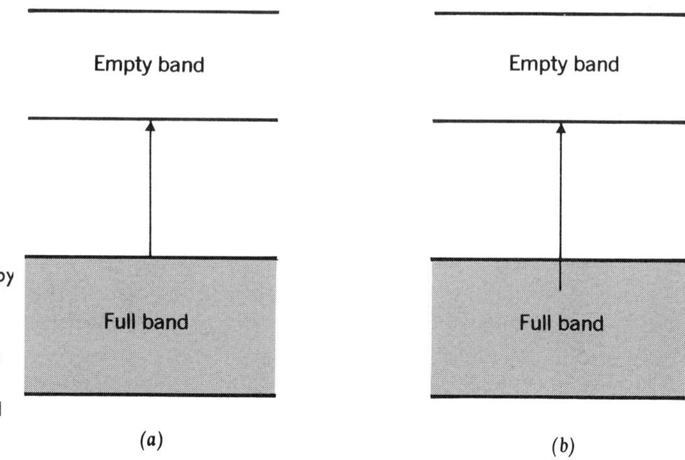

Figure 9-19 A crystal with one full and one empty band is excited to its lowest excited state by the transition represented by the arrow in (*a*). In (*b*) higher excitations are shown. Thus the energy scale for holes is downward from the top of the band.

can be demonstrated by creation of a pair in a crystal containing one completely full band and one completely empty band, Fig. 9-19. The first excited state of the crystal corresponds to taking an electron from the top of the full band to the lowest state in the empty band. States of higher excitation may be populated by taking electrons from deeper states in the full band. Alternatively, we can speak of exciting the hole from its ground state at the top of the band to excited states deeper in the band. Thus the energy scale for holes points downward from the top of the band.

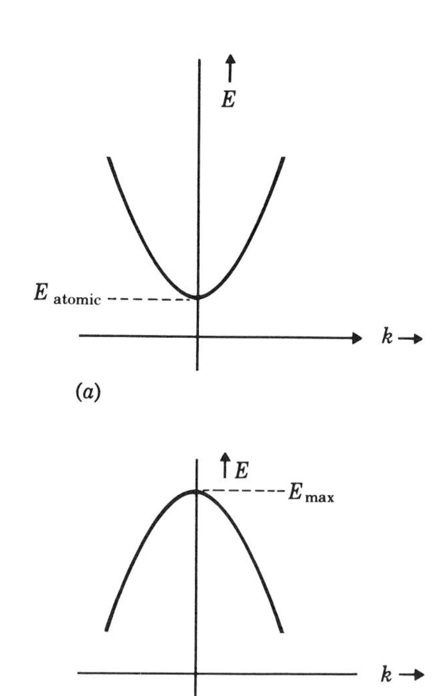

Figure 9-20 Energy plotted near the top of a band as a function of *k*. The energy at the top of a band is an inverted parabola.

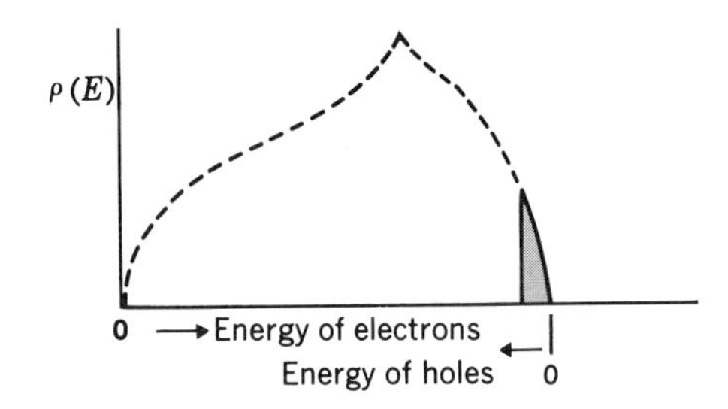

Figure 9-21 States occupied by holes in a nearly full band.

The arguments of this section are summarized in Figs. 9-20 through 9-22. If $E(k)$ near the top of the band is represented by a negative parabola,

$$E(k) = E_{\text{max}} - \frac{\hbar^2}{2m^*}(k - k_{\text{max}})^2$$

as illustrated by Fig. 9-20, its density of states is illustrated by Fig. 9-21. In k space, the energy contours at the top of the band are then spheres centered about the corners of the Brillouin zone, as shown, for example, for the cubic crystal in Fig. 2-22.

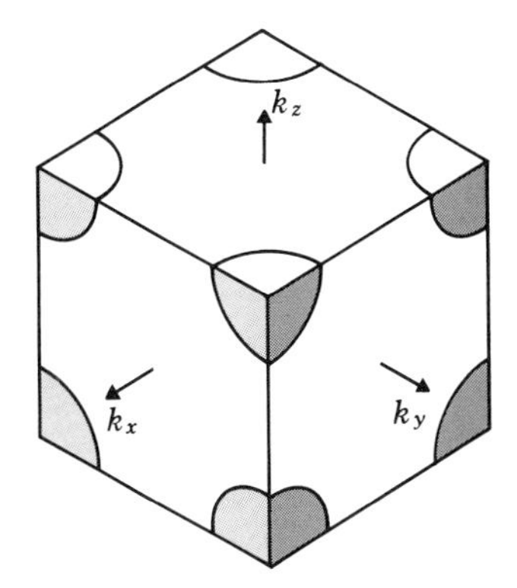

Figure 9-22 States occupied by holes in a nearly full zone of the simple cubic lattice. For convenience the k coordinates are plotted in terms of electron motion, as before.

9-10 The Experimental Study of Band Structure and Comparison with the Simple Theory

The treatment of band structure in the previous sections has made use of extremely simple physical ideas. The most important adjustable parameter in the theory is the effective mass, which we have made no attempt to calculate. However, the appearance of bands in real crystals can be studied by a variety of techniques discussed below. The picture obtained from the interpretation of experiments is that, although energy bands certainly exist and conform qualitatively to the theory presented here, the band structure of some materials shows important deviations from the simple picture. In particular, the energy as a function of k is usually not so simple as the expression $\mathscr{E} = h^2k^2/8\pi^2m^*$. (Recall that k was defined in Eq. (7-13) as being equal to $2\pi p/h$. Thus $\mathscr{E} = p^2/2m^* = h^2k^2/8\pi^2m^*$.) In fact, the bands

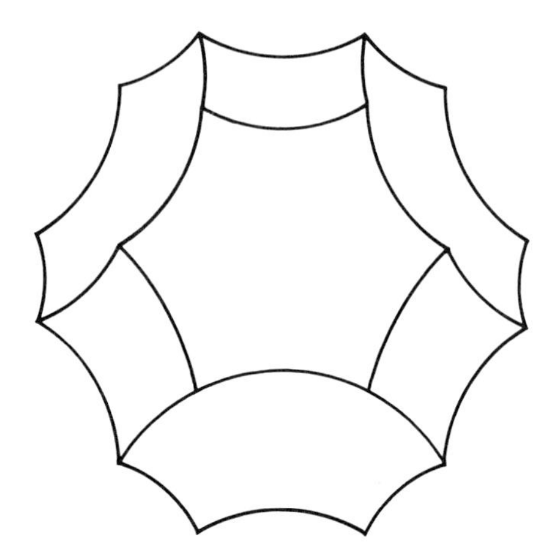

Figure 9-23 Fermi surface for Al.

of most solids deviate in varying amounts from the spherical energy surfaces shown in Fig. 9-6. Figure 9-23 represents, as an example, the shape of the Fermi energy surface as it is thought to be for aluminum. Of course, rigorous application of the Schrödinger equation to the electrons of the solid reveals many cases where the simple pictures can break down, and the interplay between the experimental studies on the shapes and dimensions of the electron bands and the theoretical predictions has led to a very active interest in the details of the band structures of various crystals. Discussion of the specific results is postponed to later chapters, but at this point we interject a short section on some of the experimental techniques which are used to study the details of band structure.

The techniques include (1) X-ray spectra of solids, (2) cyclotron resonance, (3) de Haas–van Alphen effect, (4) interaction of acoustic waves with electrons, (5) anomalous skin effect, (6) electronic specific heat, and (7) the magnetoresistive effect. Because of their relative simplicity, X-ray spectra are discussed rather fully and the other more complex methods more briefly.

X rays may be excited by bombarding atoms (either gaseous or

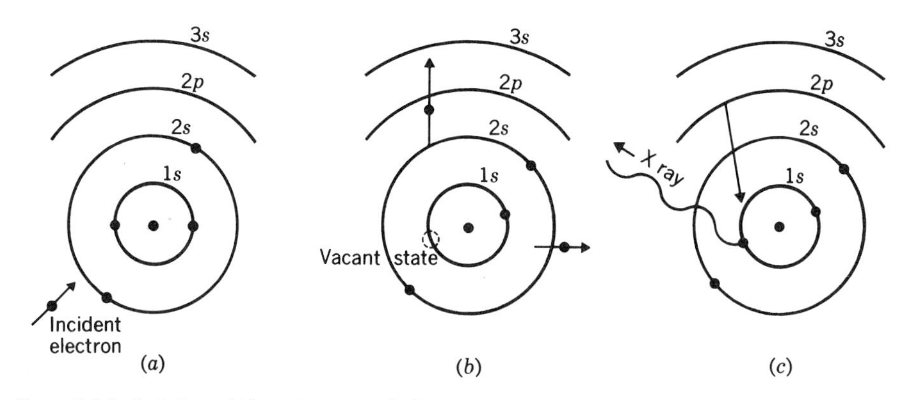

Figure 9-24 Emission of X ray by inner-shell transition in solid or gaseous Na.

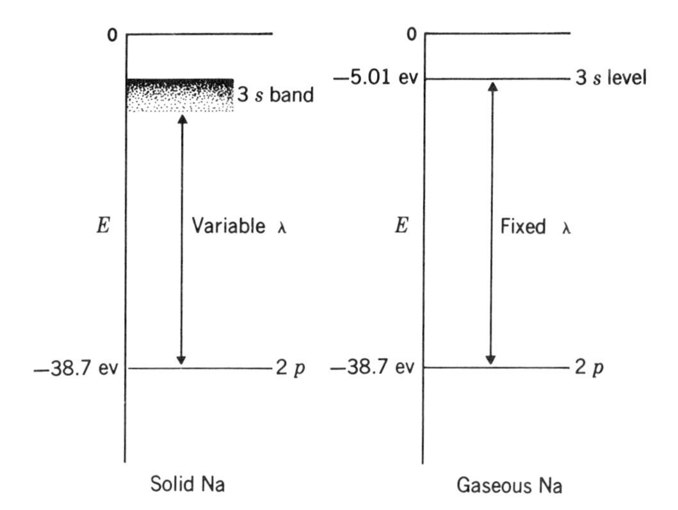

Figure 9-25 Soft X ray from 3s-2*p* transition in solid Na and Na vapor.

solid) by high-energy electrons. An incident electron knocks out of its orbit one of the inner electrons, say, a $1s$ electron. An X ray is emitted when another electron from one of the outer orbits, say, from the $2p$, falls into the vacated state. This sequence of events is illustrated in Fig. 9-24 with a free atom of Na as an example. The wavelength of the X ray emitted in this event is precisely determined; it is 11.909 A and corresponds to a transition in the lattice of energy 1,041 ev. All other X-ray lines of gaseous Na are also sharp.

This sharp line $2p$-$1s$ also exists for solid Na virtually unaltered. The $3s$-$2p$ line for Na, a sharp line for gaseous Na, of wavelength about 380 A, is spread into an emission band about 30 A wide for the solid. The reason for this width is clearly evident from observation of the energy-level diagram (Fig. 9-25). For gaseous Na, the transition is fixed and sharp. If transitions can occur between any of the levels in the valence band of the solid and the sharply defined $2p$ state, then the range of emitted wavelengths should correspond to the width of the valence band. Hence, by direct measurement of the width of the emitted X-ray band, the range in energy of the valence band can be determined. The values quoted for valence bandwidth (i.e., for the Fermi energy, Table 9-1) were determined in this way.

The density of states as a function of energy can also be determined from such measurements. If all transitions are equally likely, then the fraction of X-ray photons emitted at a given wavelength should be proportional to that wavelength. A sketch of the curve which has been obtained for solid Na is given in Fig. 9-26. First note that the bandwidth for Na is about that predicted by the theory (about 2.5 ev) and that its shape is approximately parabolic, as is also predicted. Data for other metals are given by Skinner, by Shaw, and by Hume-Rothery (see the references at the end of the chapter). They show that the bands of a number of the lighter metals are about as predicted.

Several departures from the simple theory can be noted in a de-

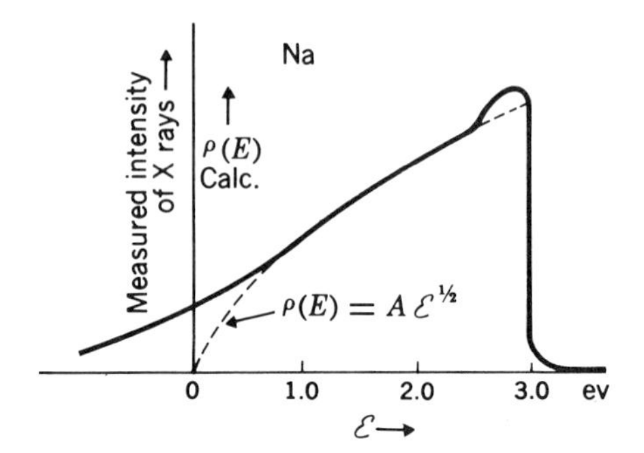

Figure 9-26 Density of states versus E for Na. (*Solid line from H. Skinner, Progress in Physics,* vol. V, 1938. *Dashed line plotted according to* $A\mathcal{E}^{\frac{1}{2}}$.)

tailed examination of the X-ray spectra, even for the simpler metals. The band trails off rather gradually at low energies (this may in part be due to experimental difficulties), and it also has some peaks and dips on the high-energy side. The latter deviations are significant, for they show that the top of the band needs more consideration than we have so far given it. A later section treats these complexities.

Cyclotron resonance is a technique in which the effective mass is measured. In this experiment, a few electrons are injected into the conduction band and are made to rotate in circles by use of a strong magnetic field. Such an electron rotates with the "cyclotron" frequency $\omega = |e|\ B/m^*$. The analogy to the cyclotron is complete, since the particles in a cyclotron also rotate in a magnetic field with a frequency given by the same expression. The frequency of rotation is measured by determining the response of the rotating electrons to an electromagnetic signal of variable frequency. When the frequency of the signal comes into coincidence with the rotational frequency of the electrons, a resonance occurs in the system. If the crystal is a part of a tuned circuit, the resonance causes an observable change in the parameters of the tuned circuit. Hence the resonance frequency can be accurately determined. Since the frequency depends upon the effective mass of the electrons, the mass is measured directly. More importantly, by changing the orientation of the magnetic field relative to the crystalline axes, the variation of the mass with direction in the crystal can be measured with considerable accuracy. This technique is one of the principal methods used in the study of the bands in insulators and in germanium and silicon.

The *de Haas–van Alphen effect* is a curious oscillation in the magnetic susceptibility of a crystal as a function of the magnetic field strength. When a metal is placed in a magnetic field, the electrons do not move in straight paths but rotate in circles. If the radius of the circle is such that the frequency of rotation of the electrons at the Fermi surface is just that which takes them completely around a circle on the Fermi surface, the magnetic susceptibility is enhanced. For a simple Fermi surface, the frequency is just that for going around a great circle of the Fermi sphere; however, for more complex surfaces, there may be several "great circles." As a function of direction in the crystal, these

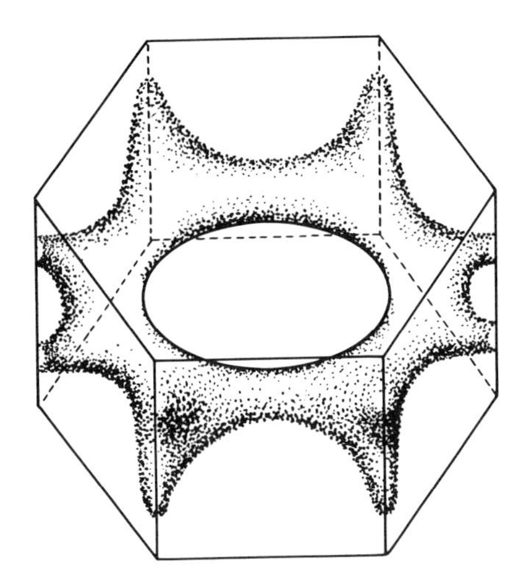

Figure 9-27 The Fermi surface of Mg metal. [*Reprinted with permission from W. Harrison and M. Webb (eds.), The Fermi Surface, John Wiley & Sons, Inc., New York, 1960.*]

de Haas–van Alphen oscillations have led to the prediction of some strange Fermi surfaces in metals. Figure 9-27 is the very surprising result for the Fermi surface of magnesium.

The *attenuation of sound waves by the conduction electrons* in metals provides a means of determining the Fermi energy and also the relaxation time τ (a quantity described in Chap. 11). In principle the phenomenon is simple. Acoustical waves cause the lattice ions to be displaced in a periodic fashion from their equilibrium positions. The motion of the Fermi electrons is perturbed by the displaced ions. The electrons gain energy from an acoustical wave, causing it to be attenuated in amplitude. Several parameters control the attenuation. It is extremely small for low frequencies (i.e., frequencies in the audio range) but becomes larger for ultrasonic frequencies. Optimum conditions for greatest attenuation occur, in general, for the following set of conditions: high frequencies (say, 10 Mc/sec or larger), pure metals, and low temperatures (less than 50°K). From measurements of attenuation under these conditions the Fermi energy can be calculated. Further information can be obtained by measuring the attenuation with and without a magnetic field; under certain conditions the quantity r, the mean time between collisions of the electrons with crystalline defects, can be measured. Details to supplement this brief account can be found in Ref. 4.

The *anomalous skin effect* measures the absorption of electromagnetic waves at the surfaces of metals. One of the basic properties of a metal is that it reflects light waves from its surface (see Chap. 10). This reflection takes place in a finite region near the surface, and a measure of the "skin depth" can be expected to yield information concerning the electronic states of the free electrons which are responsible for the reflection.

The *electronic specific heat* of a metal is directly proportional to the total density of states at the Fermi level. The theory of this effect is given in Chap. 11 and is not discussed here. This method has had particular use in comparing the specific heat of neighboring solids in the periodic table which have different numbers of electrons in the valence band but for which the overall band shape is similar.

The *magnetoresistance* of a crystal refers to the change of electrical resistance of a crystal when a magnetic field is imposed. This effect is due to the fact that, when the magnetic field is imposed, the paths of the electrons become curved and do not go exactly in the direction of the superimposed electric field. The *Hall effect* is a related phenomenon having very wide applicability; it is discussed fully in Chap. 11.

Most of the properties of a crystal depend in some way upon the band structure. In addition to the experimental techniques specifically mentioned, other effects have been used at one time or another to give information about the structure of bands. All the techniques mentioned, however, can be interpreted sufficiently straightforwardly so that they have been used extensively for this purpose.

REFERENCES

1. H. W. B. Skinner, *Phil. Trans. Royal Soc. London*, **A239** (801): 95 (1940).
2. W. Hume-Rothery, *Atomic Theory for Students of Metallurgy*. Institute of Metals, London, 1946.
3. C. H. Shaw, *Theory of Alloy Phases*, chap. 1, American Society for Metals, Cleveland, 1956.
4. D. F. Gibbons, *Resonance and Relaxation in Metals*, chap. 11, American Society for Metals, Cleveland, 1962.
5. C. Kittel, *Introduction to Solid State Physics*, 2d ed., John Wiley & Sons, New York, 1956.

PROBLEMS

9-1 Calculate the Fermi kinetic energy at $T = 0$ for valence electrons in Na and Li. Why does the Fermi energy not depend on the size of the piece of a given metal? (Assume $m^* = m$, $T = 0$.)

9-2 (*a*) Calculate the density of states for 1 m^3 of Na at the Fermi level for $T = 0$.

(*b*) Calculate the same quantity for 1 g-mole of Na.

(*c*) Why are these numbers different?

(*d*) What fraction of the 3*s* electrons of Na are found within an energy kT of the Fermi level?

9-3 Prove Eq. (9-32), that $\mathscr{E}_{\text{av}} = \frac{3}{5}\mathscr{E}_f$.

9-4 Calculate the linear velocity of the electrons at the Fermi level of Na at $T = 0$. Calculate the de Broglie wavelength of these electrons.

9-5 Since the spatial density of valence electrons in a solid is raised by compression of the solid, E_f and E_{av} are raised (E_{atomic} being assumed unchanged). The valence electrons thus resist being compressed.

(*a*) Show that the compressibility K, defined as $-\dfrac{1}{V}\dfrac{dV}{dp}$, for the valence electrons can be written

$$K_e = \frac{3}{2}\frac{V}{N\mathscr{E}_f}$$

where V is the volume (cubical) occupied by N atoms.

(*b*) For some metals, the compressibility of the electrons forms a major part of the entire compressibility of the solid. By inserting appropriate numbers in this expression, show that K_e is about equal to the actual compressibility of Na, for which $K \approx 15 \times 10^{-11}$ m²/newton.

9-6 The metal bonnevillium, which has a single electron in an s state, solidifies in a close-packed monolayer structure with atom spacing d.

(*a*) Show that, for N atoms per square meter,

$$\mathscr{E}_f = \frac{h^2 N}{4\pi m}$$

(*b*) Find the density of states as a function of energy.

(*c*) Determine the shape of the first Brillouin zone. How many electrons just fill this zone?

(*d*) Suppose that bonnevillium is alloyed with the divalent element Cd, which also has atomic diameter d. If no overlap of higher states occurs, at what Cd concentration is the zone just filled?

9-7 Show that the first Brillouin zone of the simple cubic structure (Fig. 9-14) contains just two electrons per atom.

9-8 Using the methods of Sec. 9-8, show that the Brillouin zone of an fcc metal and a bcc metal have the shapes drawn in Fig. 9-18.

9-9 The ratio of the number of electrons in a metal to the number of atoms is called the e/a ratio.

(*a*) Show that a spherical Fermi surface first touches the zone boundary of the fcc structure when the e/a ratio reaches a value of 1.36.

(*b*) Cu has a valence-electron structure $4s^1$, Zn, $4s^2$. When Zn atoms replace Cu atoms in the fcc brass structure, the e/a ratio increases from 1 for pure Cu to higher values. At what ratio of Zn to Cu does the e/a ratio reach the value 1.36? Where does an alloy of this composition occur in the phase diagram of Cu-Zn?†

(*c*) Show that a spherical Fermi surface touches the faces of the zone boundary of the bcc structure when the e/a ratio has the value 1.48.

(*d*) For alloys of Cu and Zn, at what composition does this e/a ratio occur? Where does this alloy occur in the Cu-Zn phase diagram?

9-10 The valence band of the simple cubic metal illium has the form

$$E = Ak^2 + B$$

where $A = 10^{-38}$ joules m²

$$B = -12 \text{ ev}$$

$$k = \frac{2\pi}{\lambda}$$

The ionization energy of the free atom is 1.0 ev, the band is a p band, and states up to $ka = (2\pi^2)^{\frac{1}{3}}$ are filled, where a, the cube edge of the structure, is 2.0 A.

† See M. Hansen and K. Anderko, *Constitution of Binary Alloys*, 2d ed., p. 650, McGraw-Hill Book Company, Inc, New York, 1958.

Calculate:

(*a*) m^*/m

(*b*) The number of valence electrons per atom

(*c*) The cohesive energy of the metal.

9-11 E_f for a metal is a function of temperature. Assuming that all the change is a result of thermal expansion through change in N/V with temperatures, write an expression for E_f as a function of temperature. What is the change in E_f for Na metal between 0°K and room temperature?

10

The basic solid types

The theory of bands developed in the last chapter forms the basis for our modern understanding of the physical properties of materials. As this is being written, a great deal of exploratory work of both an experimental and a theoretical type is under way. Some of the experimental techniques being currently used are listed in Sec. 9-10. A powerful synthesis is possible when the experimental results of a number of these sophisticated experimental techniques are used together with the theoretical results which can often be obtained with the aid of fast digital computers. This synthesis has already produced accurate data for the nature of the Fermi surface in such complex solids as the semiconductors Ge, Si, and GaSb, and the metals Cu, Al, and Bi. The Fermi surface of some of these materials is found to be much more complicated than the simple band structures suggested in the previous chapter. Detailed knowledge of the band structure of the transition metals has come more slowly, but a similar understanding should shortly be obtained for these metallurgically interesting metals.

Even before the present period, certain broad, general features of the band structure were recognized as being responsible for some of the characteristic qualitative differences between one crystal and another. Certain broad classes of solids can be distinguished and the existence of these classes is easily understood on the basis of their characteristic band structures. One distinguishing feature of these classes is the nature of the forces between atoms (or ions) in the solids. One type of crystal consists of ions of opposite sign held together by the Coulomb attraction of the ions for one another. Another type is held together by forces reminiscent of the carbon bond of organic chemistry. The atoms of other solids are held together by still different forces.

Associated with these distinctions are differences in physical properties. For example, crystals of one type conduct electricity in one way, crystals of another type in a different way. This difference in electrical conductivity is entirely a result of the nature of the band structure characteristic of each type.

This chapter describes in a comparative manner four properties of solids: electrical conductivity, optical properties, cohesion, and band structure. It serves as a terminal discussion for some of these topics, for example, the optical properties of metals, and as an introductory discussion for others, for example, the electrical properties of semiconductors.

No unique way of categorizing the various solids exists, since the particular grouping depends on the exact subject matter under consideration as well as on the viewpoint of the author. We make a choice, based on the type of force bonding the solid together, and list the following classes:

Metallic	Na, Al, Cu, Ag, Zn, Fe, etc.
Ionic	NaCl, MgO, LiF, etc.
Covalent	C, Si, Ge, GaSb, etc.
Molecular	Solid A, solid Kr, solid Xe, etc.

The lines of demarcation are not rigidly drawn between these four. Some solids exhibit properties of crystals of several types and they cannot be put with precision into any one of these four categories. An example is CuO, in which the bonds are partly ionic but which also behaves electrically like the covalent semiconductors.

10-1 Metals

The most important fact about the metals is that the valence band is not completely full of electrons. Put more precisely, the valence band of metals contains empty states above and immediately adjacent to the Fermi level into which electrons can be excited. These states require an almost infinitesimal energy excitation from the filled states at the Fermi level. [Metals in the superconducting state are an exception to this rule (Sec. 11-10).] For Na of volume 1 m^3, for example, an electron at the Fermi level can change states with energy absorption of only about 10^{-28} joules or 10^{-9} ev. Most of the properties of metals depend on this low excitation energy, among them thermal and electrical conductivity and opacity to electromagnetic radiation of all wavelengths longer than the short ultraviolet.

Electrical conductivity

The conduction of electricity in metals is a property of the valence electrons. These electrons have a mobility within the solid which comes from the overlap of the atomic state functions. These electrons can, in fact, traverse a perfectly crystalline metal with complete ease, suffering no collisions with lattice ions. In an electric field, they are accelerated into higher energy states with small energy absorption; their increase of velocity in the field gives rise to net charge transport and hence to electric current. In metals with high crystalline perfection the average velocity increase is relatively large so that the charge transport is rapid and the resistivity is low. In metals with low crystalline perfection, collisions with lattice ions keep the average velocity increase at a lower level, giving higher resistivity.

The problems of acceleration of charges in fields and of collisions with lattice ions have several complexities which need more extensive discussion. They are treated in the next chapter. Here we emphasize that the high conductivity of metals (relative to the insulators) is a basic property of the structure of the valence band—empty states are immediately adjacent to filled states.

Optical properties

The metals are opaque to all electromagnetic radiation from very low frequency to the middle ultraviolet, where they become transparent. In contrast the insulators are transparent over most of this region. Not only are the metals opaque to this large band of radiation, but they are also excellent reflectors. These two properties, high opacity and high reflectivity (which are related to each other by electromagnetic theory), have their basis in the nature of the electronic energy bands of metals.

Understanding what happens when radiation is incident on a flat solid surface is, in principle, not difficult. The Maxwell field equations are applied at the interface, and the boundary conditions on the field vectors of the wave, **E** and **H**, at the interface give the solutions. For all media the calculation shows that reflection, refraction, and polarization are given in terms of the geometry and the relative index of refraction of the adjoining media. In insulators the index of refraction is determined by interaction of the radiation with the bound electron clouds of the ions. No energy absorption occurs in the transparent medium; the index of refraction is a real number.

The index of refraction for a wave in a metal depends on the behavior of the electrons in the Fermi distribution. The electrons at the Fermi level can absorb energy readily according to the Einstein relation, $E = h\nu$, and in so doing, go into higher states in the band. The response of the electrons to the electric field of the radiation falls into two categories. They are, of course, accelerated in the field and hence gain energy. (1) If there is only a very small chance for a collision with the lattice ions during a period of the wave, the energy gained is re-radiated, there is no dissipation, and the crystal is transparent. (2) If the electrons make many collisions with the lattice ions during a cycle, their extra energy of translation is dissipated as heat, and the electromagnetic wave is absorbed. The second alternative applies to all metals in the optical band of frequencies. The response of the electrons is not in phase with the field vector of the radiation, however, and as is usual when the excitation and response are not in phase, a complex number is introduced to express the in-phase and out-of-phase components. The absorption is directly related to the imaginary part of the index of refraction in metals and both are very high. The real part has the conventional meaning of index of refraction.

The large imaginary index of refraction thus leads to high reflectivity in all metals, because high reflectivity at any surface is caused by a large relative change in index of refraction in crossing the surface. In the visible part of the spectrum (and on down in frequency to the radio waves) the reflectivity is nearly 100 per cent, and the attenuation is so strong that the wave is essentially damped within one wavelength of the surface. In fact, in the visible range, the damping is complete in the distance of a few hundred angstrom units, a distance considerably less than the wavelength. Even at radio frequencies the penetration is slight; the skin depth is only about $\frac{1}{10}$ mm. It is curious that in a highly absorbing medium like the metals the fraction of incident energy actually dissipated in the metal is very small because of the large reflectivity.

Note again that the important property of the metal is the ready availability of empty energy levels adjacent to the Fermi level. The Fermi electrons may thus absorb energy from the field of the wave. Were this not possible, metals would be transparent and would be much poorer reflectors.

Band structure of metals

The band structures of the several groups of metals have numerous small but important differences, so a short description of the band structures for typical metals follows. This description contains a brief statement of the nature of the bands but, more importantly, a description of the role of the bands in influencing physical properties, some of which are described briefly here and some of which are treated in detail in later chapters. This section is therefore important because it presents briefly many of the important properties of solids.

The monovalent metals, Li, Na, K, etc., are the simplest metals. In the past they have received more study, primarily theoretical, than any other group of metals. At the time of writing, modern experimental techniques are beginning to provide the same depth of understanding for somewhat more complicated metals. Of the monovalent metals, sodium is understood best. For sodium, the density of states is a simple function, since it varies almost exactly as the square root of the energy. The effective mass at the Fermi level is almost the rest mass of the electron, although it is somewhat larger down in the band: the effective mass averaged over the entire band is 1.2 m. The density of states plotted as a function of energy is therefore relatively squeezed at low energies but lengthened out near the Fermi energy.

For the alkali metals of higher atomic number, the band structure departs even more from that predicted by the simple theory, the bandwidth is narrower than that of Na (or than that given by the theory of Chap. 9), but the effective mass at the Fermi level is only slightly larger than the free-electron mass. For K, $m^*/m = 1.1$ at the Fermi level.

The monovalent alkali metals have been described as though only one free-atom level, an s level, and one band, an s band, exist. A free atom has excited levels, of course, and these levels split into bands for the solid even though they may be unoccupied. For free Na atoms the $3p$ level lies about 1 ev higher in energy than the $3s$ level. For solid Na the $3p$ level broadens into a band which lies just above the $3s$ band. The bottom of the $3p$ band just overlaps the top of the $3s$ band.

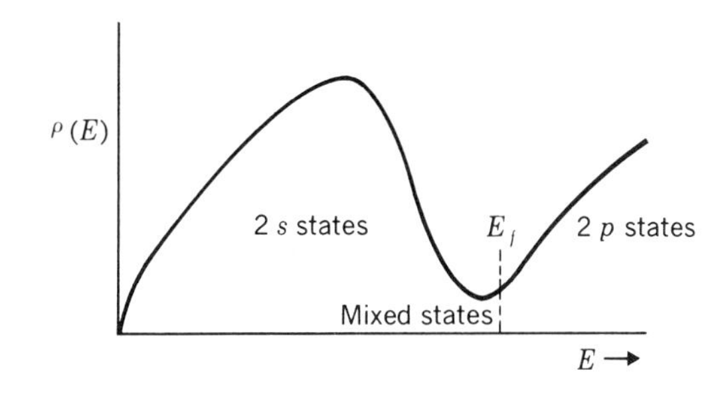

Figure 10-1 Density of states versus energy E for Be. Small overlap of the $2p$ and $2s$ states makes $\rho(E)$ low at the Fermi level.

For Na and the other alkali metals the Fermi level is well below this region of overlap, so the overlap is normally not important for them. [Note that the overlap could be important if it were sufficiently great so that the bottom of the $3p$ band were below the Fermi level. In such an event, the *total* density of states would be the sum of the densities of states of the $3s$ and $3p$ states and the simple $\rho(E)$ versus E dependence would be lost.]

The overlap of the s and p states is important for the metals of group II of the periodic table, the alkaline-earth metals. These elements have two electrons in an s subshell in the free-atom state; i.e., the s state is full. Hence these elements might be expected to have a *filled* band in the solid state and might be expected then not to fulfill the criterion we have set up for good electrical conductors, namely, that empty states be immediately adjacent to filled states. However, these elements are metals and are relatively good conductors, since the overlap of the p band adjacent to the corresponding s band provides empty states into which electrons can be excited.

Beryllium is one of the simplest of the group II elements. It is a metallic conductor, and its electrical conductivity is about the same as that of platinum. The density-of-states curve calculated for Be is presented in Fig. 10-1. The s band is about normal; it rises from zero as simple theory predicts, and falls off toward zero at an appropriate value of E. Instead of falling completely to zero, however, the density-of-states curve rises again as the p band begins. The Fermi level is about at the minimum point. This coincidence has a number of consequences: The specific heat of the electrons, to be discussed in Sec. 11-2, is lower than normal. The magnetic properties of Be are also anomalous. In short, Be barely qualifies as a metal. The other elements of group II have more overlap of the s and p bands; for them the Fermi level comes at a higher value of the density of states, and they are more normal metals.

The transition metals are characterized by extensive overlap between the s and d bands. The s band is thought to be very broad, with an effective mass approaching that of the free electronic mass,

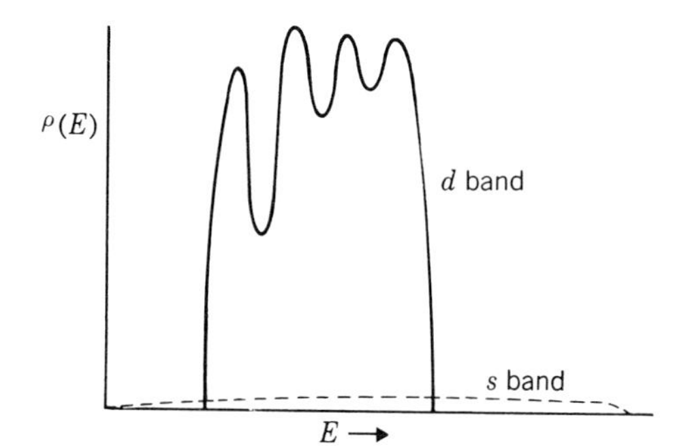

Figure 10-2 Complete overlap of the $(n + 1)s$ band with the nd band for transition metals. The exact shape of these bands is so far unknown.

while the d bands are much narrower, with a large effective mass. The general character of the bands is sketched in Fig. 10-2. The d-band wave functions are highly localized, with relatively little overlap between one atom and another. The result is that some of the properties of certain of the transition metals, e.g., ferromagnetism of Fe, Co, and Ni, can be treated by using a model of electrons localized on individual atoms. The exact nature of the d band is imperfectly known, partly because it is so thoroughly mixed with the s band, which overlaps it completely.

The transition metals, as a group, are not as good electrical conductors as the simpler metals, a fact which seems to be related to the high density of states in the d band. The current carriers in the transition metals are thought to be only the s electrons, since the d electrons have too high an effective mass to respond appreciably to an external field. However, the s electrons which are accelerated in the field can change their nature through collisions and become d electrons, i.e., occupy states in the d band. Scattering of s electrons by lattice vibrations is thus relatively easy compared with that for metals with no overlapping d band. Therefore the s electrons in transition metals experience relatively numerous collisions, and these metals have lower conductivities than the metals with filled d bands.

A final group of metals deserving individual comment is the group

Cu, Ag, and Au. They are like the alkali metals Na, K, Cs, etc., inasmuch as they have a single *s* electron outside completely filled inner shells. The Fermi level is well above the top of the *d* band. They are excellent conductors; while they do have a *d* shell, it is filled, so that scattering into it of the *s* electrons is not possible. The metal Cu has had more study than either of the others, and experiments show that the 4*s* band is well overlapped by the 4*p* band. While for many years the Fermi surface in momentum space was thought to be spherical, recent experiments show that it is considerably warped, with bulges in $\langle 111 \rangle$ directions. The effective mass of the electrons at the Fermi level in Cu is about 1.4 times the free-electron mass.

The *d* shells of these three metals are relatively large so that the wave functions of one atom overlap those of the adjacent atoms a

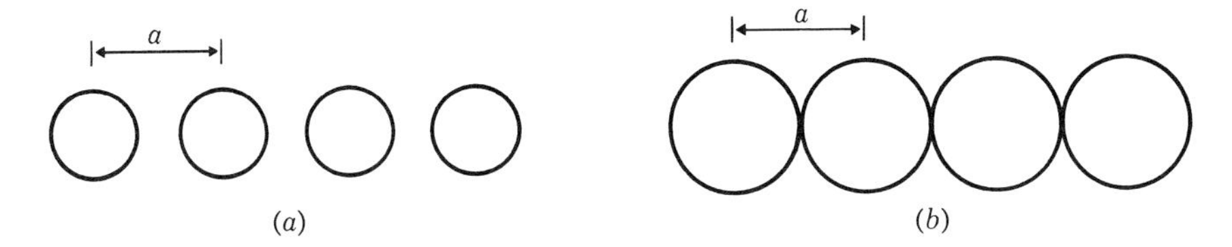

Figure 10-3 (*a*) Na metal; filled shells. For Na the inner 2*p* shell is small in size compared with the distance of separation of ion centers. Little repulsive force comes from overlap of these electrons. (*b*) Cu metal; filled shells. For Cu the inner 3*d* shell is large; it contributes a great deal to the repulsive forces between atoms.

little. Since closed shells repel one another—as is demonstrated by the monomolecular nature of the rare gases—a part of the repulsive energy of these atoms comes from the repulsion of the filled *d* shells. This situation is in contrast to the monovalent alkali metals, whose inner shells are small (Fig. 10-3), and whose repulsive energy comes mainly from the resistance to compression of the conduction electrons themselves (Prob. 9-5). The magnetic properties of Cu, Ag, and Au, discussed in a later chapter, reflect the presence of the relatively large *d* shells.

Cohesive energy

Work is required to separate the atoms in the solid into isolated neutral atoms. The energy expended in this way is termed the *cohesive energy*. Metals are about midway in the energy scale for solids, being much more tightly bonded than solid He, H, Ne, etc., and less tightly bonded than the covalent crystals diamond, Ge, and Si. For a typical metal, Cu, the cohesive energy is 81.2 kcal/g-mole (about 3.5 ev per atom). Values for many of the metals are listed in Table 10-1. The cohesive energy for a solid is somewhat temperature-dependent, and the data of Table 10-1 refer to the metals at room temperature.

Table 10-1 The heats of sublimation of monatomic metals† (In kcal/g-mole at room temperature)

		Monovalent Metals			
Li	36.5			Cs	18.8
Na	26.1			Cu	81.2
K	21.7			Ag	69.0
Rb	20.5			Au	82.3
		Divalent Metals			
Be	76.6			Zn	31.2
Mg	35.9			Cd	27.0
Ca	45.9			Hg	15.0
Sr	39.2				
		Trivalent Metals			
Al	74.4			Ga	66.0
Sc	93.0			In	58.2
La	88.0			Tl	43.3
		Tetravalent Metals (with Ge for comparison)			
Ti	112.0			Ge	78.0
Zr	125.0			Sn	72.0
				Pb	46.5
		Pentavalent Metals			
As	60.6				
Sb	60.8				
Bi	49.7				
		Transition Metals			
V	119.0				
Cr	80.0	Mo	155.5	W	201.6
Mn	68.1				
Fe	96.7	Ru	160.0	Os	174.0
Co	105.0	Rh	138.0	Ir	165.0
Ni	101.0	Pd	93.0	Pt	121.0

† Slightly modified from a table adapted by F. Seitz, *The Modern Theory of Solids*, McGraw-Hill Book Company, Inc., New York, 1940, from F. R. Bichowsky and F. D. Rossini, *The Thermochemistry of the Chemical Substances*, Reinhold Publishing Corporation, New York, 1936.

The valence electrons are responsible for the bonding of metals, and these electrons as a group occupy lower energy levels in the solid than they do in the free atom. The problem of understanding cohesion in solids is that of understanding why the average energy of the valence electrons is lowered by the formation of the metallic crystal.

The energy of a valence electron in a free atom is partly potential and partly kinetic, as we have seen. The wave function oscillates in the vicinity of the ion, then diminishes asymptotically to zero far away from the ion. This behavior is sketched in Fig. 10-4 for the $3s$

electron of Na. When a number of atoms are brought together to form a solid, the wave function is altered. Consider first the changes which occur at the bottom of the band, where $k = 0$ and translational motion from one atom to another can be neglected. Instead of dying off exponentially to zero at large distances, the wave function must connect smoothly with a similar function about an adjacent ion. Hence at a distance $d/2$ (where d is the radius of the ion) it is higher than it was (see Fig. 10-4). This changed wave function has a different energy; this is what we called E_{atomic}, and it is lower than the free-atom energy.

The energy E_{atomic} is partly potential and partly kinetic. The potential energy is lower than that for the free atom because the electron in a solid is never at a great distance from a positive ion. (The tail of the wave function for a free atom is concentrated close to the ion, as Fig. 10-4 shows.) The kinetic energy of the atomic part of the wave function (which is the orbital kinetic energy about the ion)

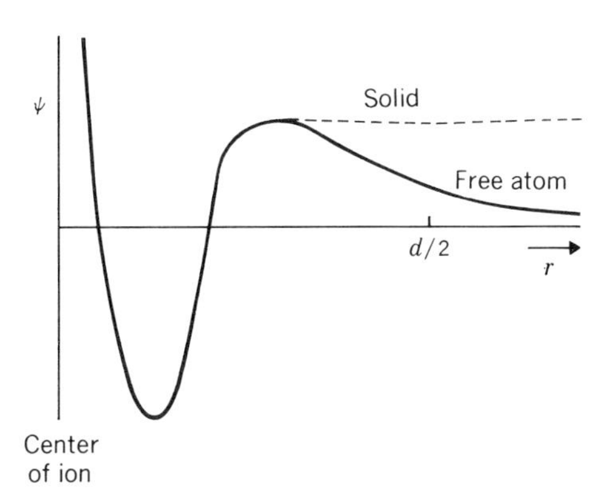

Figure 10-4 Because ψ for the solid must go smoothly from one atom to the next, the long asymptotic tail on ψ for the free atom is brought closer to the nucleus.

is also lower than it is in the free atom. The reason is that the changed wave function has lower curvature (is smoother) in the region of the dotted line of Fig. 10-4. The reader should recall that the kinetic energy of a wave function at a point is $-(h^2/8\pi^2m)\nabla^2\psi$. Thus the curvature and the kinetic energy are proportional to one another.

Changes in kinetic and potential energy, taken together, give a lower value for E_{atomic} than the free-atom energy of the valence electrons. This lowering of energy is the origin of the binding energy of metals.

The previous discussion is not complete, since the translational kinetic energy of the electrons in the band has been neglected. This energy, as an additional kinetic energy, reduces the binding energy. For the simple one-electron atoms the average translational kinetic energy has already been calculated in Eq. (9-32) and is equal to $\frac{3}{5}\mathscr{E}_f$ per electron.

The relation between the several energies just discussed is sketched in Fig. 10-5. The cohesive energy is the difference between the

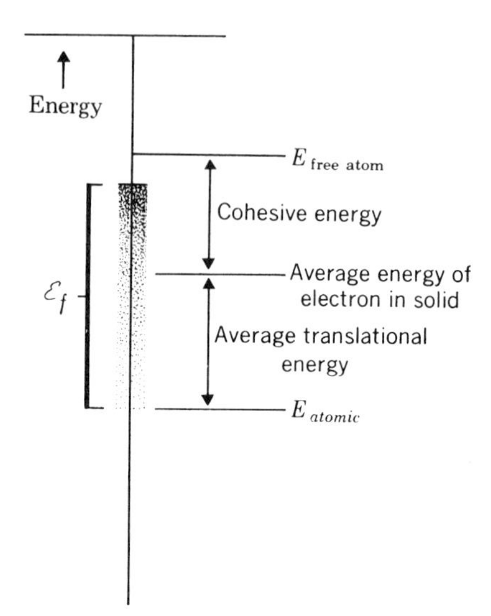

Figure 10-5 The cohesive energy of a metal is the energy necessary to bring all electrons in the solid up to the level for the free atoms.

free-atom energy and the average energy of an atom in the solid. It is simply

$$E_{\text{cohesive}} \text{ (per atom)} = E_{\text{free atom}} - E_{\text{atomic}} + E_{\text{av translational}}$$

E_{cohesive} is a positive quantity, i.e., bonding does exist for the metals, and is of order several electron volts per atom, as the data of Table 10-1 require.

10-2 Ionic Crystals

Most of the insulators are classified as *ionic crystals*. These crystals are never pure elements; they are always compounds. They may have a definite chemical formula, say, NaCl, KOH, or MgO, and in this sense they are an extremely narrow phase in a binary (or higher) phase diagram. On the other hand, they may have some variation in composition, such as the ferrite, $(FeO)_{1-x}(NiO)_x(Fe_2O_3)$. In this sense ionic crystals may have variation in stoichiometry, just as metallic solid solutions do. For ease of discussion, however, we consider only the simple two-constituent compounds of distinct stoichiometric composition.

The ionic crystals are mainly inorganic compounds and hence can usually be written with a definite chemical formula. This does not imply that a single molecule, say, a molecule of NaCl, commonly exists. A block of rock salt, for example, is not made up of many individual molecules of NaCl. Rather the entire block is one gigantic molecule in which the effect on a given ion of *all* other ions is significant. This significance becomes most apparent when cohesion and the band structure of ionic crystals (shown schematically in Fig. 10-6) are considered.

The properties of ionic crystals are intimately related to the chemical nature of the compounds. Consider again NaCl. The valence of Na is $+1$; that is, it has an electron which it can easily lose, the $3s$ electron. Cl, on the other hand, has a valence of -1; that is, it will accept an electron: This electron completes its $3p$ shell. By the transfer of the $3s$ electron of Na to the $3p$ of Cl, a change in energy occurs, since both ions have only filled shells. The ions, however, have net charge, the Na ion a positive charge and the Cl ion a negative charge. They therefore attract each other by Coulombic forces, and a lowering of potential energy occurs if they approach each other. The net energy of the crystal, calculated in detail in

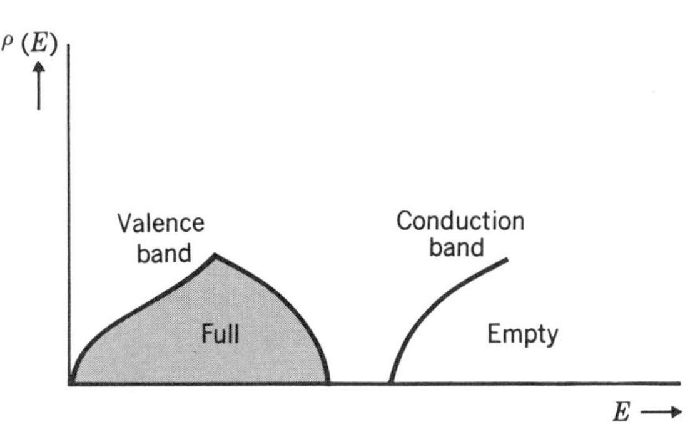

Figure 10-6 Schematic drawing for an insulator, showing full and empty bands separated by a region of forbidden energy values. In the ionic crystal, the full band corresponds to the full valence shell of the negative ion, and the empty band corresponds to the empty valence levels of the positive ion. The full band of the crystal is called the valence band; the empty band is called the conduction band.

Chap. 14, is lower than that of a gas of separated Na and Cl neutral atoms. The chief properties of these materials are discussed in the same order as for the metals. The introductory discussion of this chapter is followed more quantitatively by that in Chap. 14.

Electrical conductivity

The electrical conductivity of ionic crystals is much lower than that of the metals, at least 20 orders of magnitude lower at room temperature. Furthermore the conductivity of ionic crystals *increases* with increasing temperature, in contrast to metals, whose conductivity *decreases* with increasing temperature. These facts about ionic crystals indicate that for them the charge carriers are quite different from the charge carriers in metals. To be sure, some conduction occurs by electron flow, but the electrons are so tightly bound in the filled shells of the two types of ions that they cannot move appreciably in an external field. In fact one of the major characteristics of ionic crystals is that charge transport is by the charged ions themselves.

The fact that the ions are the current carriers is attested to by the fact that flow of current is accompanied by concurrent mass transport of ions. In NaCl, for example, free Cl gas is given off at the anode, and Na metal is deposited at the cathode during current flow. This implies that a directed diffusional motion of these ions occurs during conduction, the details of which are thought to be the following: Diffusion in ionic crystals occurs predominantly by lattice vacancies.

In the absence of an electric field, no *net* motion of ions occurs, since the individual jumps are random in direction. A field, however, causes the ions to jump somewhat more in one direction than the other, so a net ion flow occurs, and both matter and charge are thereby transported.

The diffusive nature of the conduction simultaneously accounts for both the smallness of the conductivity and its negative temperature dependence. The conductivity σ is low because diffusion (at reasonable temperatures) is low; the increase of conductivity with temperature occurs because diffusion becomes more rapid as temperature

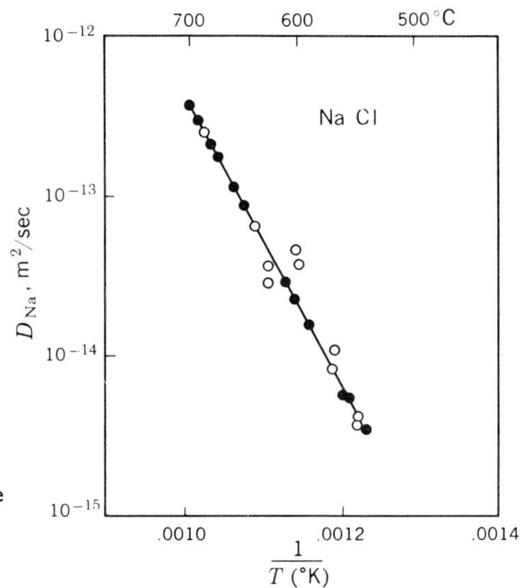

Figure 10-7 The diffusion coefficient of Na in NaCl at high temperatures as measured directly by radioactive tracers, -o-, and as calculated from conductivity measurements -●-. The diffusion of Na ions is given by the expression $D_{Na} \approx 10^{-3} \exp(-1.85 \text{ ev}/kT)$ m^2/sec. [*After D. Mapother, H. Crooks, and R. Maurer, J. Chem. Phys.*, **18:** 1231 (1950).]

increases. The conductivity and the diffusion coefficient D are related by a single equation known as the *Einstein mobility relation*,

$$\sigma = \frac{Nq^2}{kT} D \qquad (10\text{-}1)$$

where N is the number of Na ions per cubic meter and q is the charge on the ion. Since D varies with temperature according to an expression $D = D_0 e^{-Q/RT}$, where Q is usually 1 ev per atom or more, the main temperature dependence of σ is also exponential. Data for Na diffusion in NaCl measured by radioactive tracer and calculated from conductivity measurements using Eq. (10-1) are shown in Fig. 10-7. The agreement between the two methods of determining D is excellent.

The absence of conductivity by electrons gives a clue to the band structure of ionic crystals. The filled states of the topmost level are separated in energy from the nearest empty states, and no empty states are immediately adjacent to the Fermi level, as is true for metals. Furthermore the separation in energy between the filled and unfilled states, the energy gap, must be large, several ev or more, since even at high temperatures few electrons are excited into empty states. NaCl, for example, has an energy gap of nearly 7 ev.

Some ionic crystals have only modest energy gaps, 2 or 3 ev, and at high temperatures electrons are thermally excited into empty levels. These crystals are ionic semiconductors. The compound Cu_2O is an example.

Optical properties

The ionic crystals are transparent for almost all frequencies up to the point called the *fundamental absorption frequency*. At frequencies higher than this they are opaque. The transparency shows again that

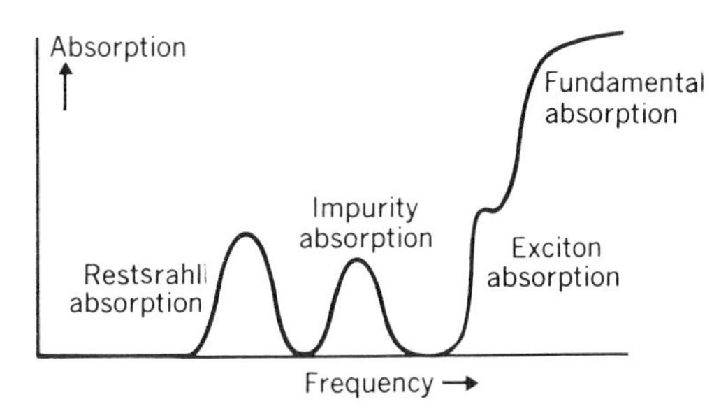

Figure 10-8 Absorption of electromagnetic radiation by ionic crystals as a function of frequency.

the electrons in the topmost occupied band do not have empty levels immediately adjacent to them; the topmost occupied band of levels is full. At the fundamental absorption frequency, however, photons have energy just sufficient to excite electrons completely across the gap into empty states, and photons with energy larger than this minimum are highly absorbed. A representative absorption curve is shown in Fig. 10-8. The fundamental absorption occurs for wavelengths in the ultraviolet, for which photons have an energy exceeding 4 ev. The gap width in the insulating crystals is thus larger than some 4 ev.

The sketch of Fig. 10-8 shows that the crystal is largely transparent below the fundamental absorption frequency. Several peaks of absorption exist at lower frequencies, but these are in no way caused by excitation across the complete gap. The peak at lowest frequency is caused by vibration of the ions themselves. The electric field of any electromagnetic wave exerts forces on the ions, causing them to vibrate. For the most part these forces cause vibrations of large amplitude only when the frequency of radiation is the same as the natural frequency of vibration of the ions in the crystal. At this resonant frequency, the oscillations become large, and during collisions with their neighbors the ions dissipate some of this energy as heat. Hence, incident radiation is strongly attenuated. The frequency of vibration of ions in a solid is about 10^{13} cps (Sec. 3-2); hence these frequencies, called *restrahl* frequencies, are in the infrared. For several crystals the measured frequencies are listed in Table 10-2.

Another absorption peak is that labeled impurity absorption in Fig. 10-8. This peak is caused by an impurity or a lattice defect. Since many different impurities may be found in ionic crystals, as well as several types of defects, the number of peaks of this type is variable. They are caused by electrons that are bound to the impurity or lattice defect somewhat more loosely than they are to the negative ion; these electrons are more easily excited into states in the empty band. The peaks thus correspond to localized absorption of radiation by electrons in special sites. Details of this rather complicated subject are left to Chap. 14.

Table 10-2

Crystal	Wavelength of maximum absorption (restrahl wavelength), A
NaCl	611,000
KCl	707,000
KBr	883,000
KI	1,020,000
TlCl	1,170,000
ZnS	330,000

The impurity absorption bands commonly occur in the visible part of the spectrum. They are, in fact, responsible for the color of many ionic crystals. The mineral corundum, crystalline Al_2O_3, is colorless when pure. One type of metallic impurity colors it blue to make it the precious gem *sapphire*. A different impurity colors it green to make it *Oriental emerald*. A third impurity colors it red; it is then the gem *ruby*. NaCl is colorless when it is a pure and perfect crystal, but negative-ion vacancies color it yellow. Innumerable examples such as these can be found.

The exciton absorption peak is always associated with the fundamental absorption process but is usually distinguishable from it. It corresponds to electron excitation into a set of discrete levels which exist just below the empty band in ionic crystals. The nature of these levels is discussed in detail in Chap. 14.

In addition to the regular ionic conduction in ionic crystals an electronic contribution to conductivity exists. This conductivity sheds further light on the band structure of these crystals. When light of a frequency corresponding to the fundamental absorption is incident on the crystal, electrons are excited, of course, into higher empty states. After excitation into the conduction band, they are capable of being accelerated by any external electric field and produce some electronic conduction. This conduction is possible only in the presence of the exciting photons, and so it is called *photoconduction*. The photoconductivity of ionic crystals is much less than metallic conductivity, because the number of electrons participating is much less; nevertheless it has the same general features as metallic conductivity.

Band structure of insulators

The conductivity and the optical properties of the ionic crystals permit one to deduce the general features of the band structure. The approximate energy levels are pointed out here; detailed discussion of the energy-momentum relationships is given in Chap. 14.

The ionic crystals are always made up of at least two elements; therefore the band structure of the solid has, at low energies, the inner electronic levels of both atoms. The only levels, in fact, which need

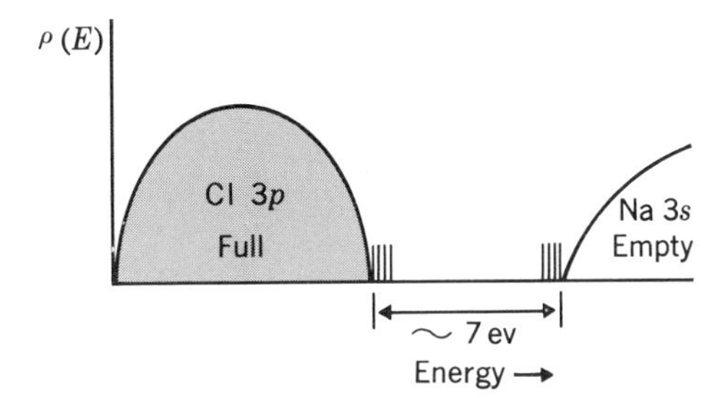

Figure 10-9 $\rho(E)$ versus E for NaCl. The discrete levels within the gap are associated with crystalline defects, impurities, and excitons.

to be considered as being broadened are the outermost levels of each ion. As an example, again consider NaCl. The $1s$, $2s$, and $2p$ levels for Na^+ are full and have low energies; even the $2p$ level has an energy lower than -30 ev. The $1s, 2s, 2p$, and $3s$ levels for Cl^- are also full and are also low in energy; even the $3s$ level has an energy of about -25 ev. None of these levels is important in the band structure. The $3s$ level of Na, the $3p$ level of Cl, and higher levels of each are the important levels for the solid.

The band structure for NaCl which fits with the conductivity and optical properties known for the solid is sketched in Figs. 10-6 and 10-9. The $3p$ level of Cl broadens into a band, and it contains its full complement of six electrons. The next higher band is the $3s$ of Na; it must be empty. The band gap, as given by fundamental absorption edge measurement, is nearly 7 ev. At ordinary temperatures virtually no electrons are excited from the Cl $3p$ band, the *valence* band, into the Na $3s$ band, the *conduction* band; so NaCl is an excellent insulator. Even at the melting point of NaCl, 801°C, the band gap is about $70kT$; so electronic conduction in NaCl is virtually nonexistent at any temperature.

Impurity, defect, and exciton levels are relatively narrow bands of levels within the gap. From detailed knowledge of the conductivity, optical, and thermal properties, they can be determined with reasonable exactness.

This example of NaCl is typical of the good insulators. The basic feature is a filled valence band separated by a large energy gap from a normally empty conduction band. Ionic crystals with a smaller band gap, say, only 1 or 2 ev, are poorer insulators, since, for them, electronic conduction is possible at reasonable temperatures. They are more like semiconductors; in fact, a gradual blending of properties from the good insulators to the semiconductors occurs as the gap decreases from 5 ev or more to less than 1 ev.

Cohesive energy

Any ionic crystal is a regular array of positive and negative ions. These ions exert both attractive and repulsive forces on each other which balance at the proper ion spacing in the solid. The attractive force is the Coulomb force between the ions. The repulsive force which keeps the crystal from collapsing on itself is the repulsion of filled shells. For a simple salt, say, NaCl, the attractive potential between any pair of ions with charges q_1 and q_2 is simply

$$V = \frac{q_1 q_2}{4\pi\epsilon_v a_0} \tag{10-2}$$

where a_0 is the distance between the centers of the ions (which are assumed to have spherical symmetry). Unfortunately for calculational purposes, this potential law must be summed over all pairs of ions in the crystal to get the total cohesive energy; the sum over nearest-neighbor ions is not a good approximation. That part of the binding energy contributed by Coulombic terms is then found by summing over each pair of ions. While this appears to be a formidable calculation, methods have been found to perform it with relative ease. The result is that the potential energy of an ion in the crystal is simply

$$V = \frac{q_1 q_2}{4\pi\epsilon_v a_0} \alpha_M \tag{10-3}$$

The Madelung constant α_M has characteristic values for each crystal type and is always of order one. Other contributions to the cohesive energy are discussed in Chap. 14. The major contribution is that given by (10-3), however.

10-3 Covalent Crystals

The compounds of organic chemistry have bonds characterized by *electron sharing*. Methane, CH_4, for example, is bonded by the sharing of the 1s electrons of the four hydrogen atoms with the single carbon atom. In this way the molecule attempts to fill the 2p shell of C. Ammonia, NH_3, is another example in which sharing of the 1s electrons of three hydrogen atoms aids in filling the 2p shell of N. These bonds are termed *covalent*. This type of bonding is present for some solids too, forming an important class of materials.

We place the semiconductors within the covalent class on the basis of their covalent bonding. However, because of the distinctive electric character of the semiconductors, they are often assigned to a separate category. As we shall see, the primary distinction between semiconductors like Si and Ge and the insulators like diamond is only the width of the forbidden band gap, and it is natural to discuss the covalent semiconductors and covalent insulators together. The general properties of these crystals are described briefly here. In later chapters, the electrical properties, which are of greatest importance, are dealt with more completely.

Electrical conductivity

The conductivity of covalent crystals varies over a wide range. Some, like diamond, are excellent insulators. Others, like germanium, are

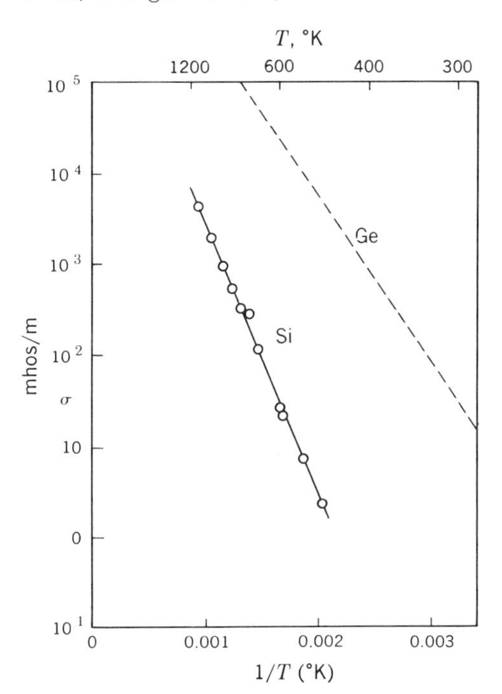

Figure 10-10 A plot of log σ versus $1/T$ for a pure semiconductor is a straight line with a slope determined by the band gap between a filled and an empty band. [*Data for* Si, *F. Morin and J. Maita, Phys. Rev.,* **96:** 28 (1954).]

fair conductors. Still others, like gray tin, have even higher conductivities, approaching those of the poorer metals. Since reliable data are available for the semiconductors, they are discussed here in considerable detail.

A pure semiconductor such as Si is about 10^{10} times a poorer conductor at room temperature than typical metals. Furthermore its conductivity σ increases with increasing temperature, whereas that of the metals decreases. Data for Si are presented in Fig. 10-10. They are plotted as log σ versus $1/T$. The fact that the data fit a straight line on such a plot shows that the temperature variation of σ is (as closely as these data allow)

$$\sigma = Be^{-H'/RT} \tag{10-4}$$

The constant H' is an energy; for Si it has the value of about 13,800 cal/mole. This expression is not strictly true, as theoretical considerations yield a temperature dependence of the constant B, but it is so slight relative to the exponential that the data of Fig. 10-10 are unable to show it. Explanation of H' in terms of a *band gap* between a *valence* and a *conduction* band is given in Chap. 12.

The electrical conductivity in covalent crystals is purely an electronic process. No matter transport, such as that which accompanies current flow in ionic crystals, is observed in good covalent crystals. The constant H' is therefore not related to diffusion in the crystals.

Impurities may affect the conductivity of semiconductors greatly. Some impurities have little or no effect, while others may increase the conductivity by many powers of 10. They do so, not primarily by impeding electron flow, as they do for metals, but by changing the number of charge carriers. The controlled addition of impurities to semiconductors, termed *doping*, is of immense technological importance. It forms the basis for the production of actual devices.

Optical properties

The optical properties of covalent crystals are much the same as those of ionic crystals. Since their electrical behavior is distinctly nonmetallic, the arrangement of full and empty levels must be such that low-frequency photons cannot excite electrons into the available empty levels. They are, therefore, transparent to radiation of long wavelengths. For example, pure diamonds are transparent to radiation longer than short ultraviolet wavelengths. Ge and Si are transparent for all wavelengths longer than the infrared. The covalent crystals have impurity absorption (e.g., colored diamonds). Exciton absorption peaks are also observed for the covalent crystals.

Band structure

All covalent crystals are basically insulators, and so their band structure has a similarity to that of ionic crystals. The deep-lying bands are unchanged when free atoms condense into solids, but the outermost levels assume the same full and empty band structure as ionic crystals. At first sight this appears strange, since pure covalent elements, say, diamond, silicon, and germanium, all have partly empty atomic p levels which might well be thought to be partly empty p bands for the solid. However, the s and p levels divide so that the total of *eight* $s + p$ states splits into two groups of four which are separated in energy by a gap with width characteristic of the given element.

The band structure just described is sketched in Fig. 10-11. At wide distances of atom separation the wave functions and energy levels

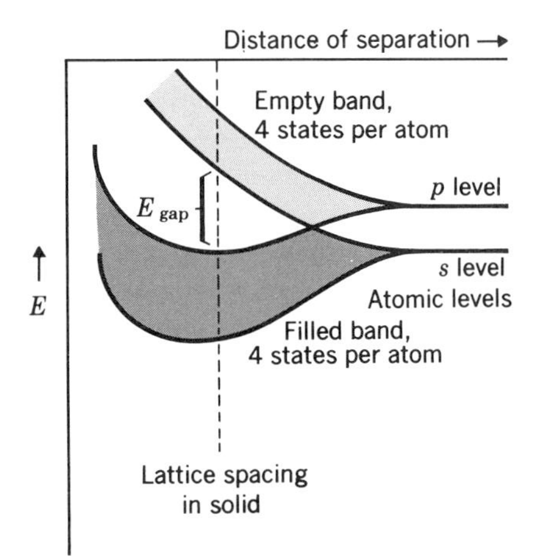

Figure 10-11 Energy levels of a semiconductor as a function of distance of atom separation.

are free-atom-like. At close distances, where the sharing of electrons occurs, the symmetry properties of the crystal give quite a different character to the wave function, and the bands form as is indicated. Since the covalent crystals have a total of four $s + p$ electrons in the outer shells, the lower band, again termed the *valence* band, is full and the upper band, termed the *conduction* band, is empty (except for electrons excited into it).

The effect of impurities on the electronic configuration is easily seen. If atoms with fewer than four $s + p$ electrons are added as impurities, say, Ga in Ge, then additional states appear in the forbidden gap near the upper edge of the valence band. If atoms with more than four $s + p$ electrons are added as impurities, say, As in Ge, then additional states appear in the forbidden gap near the bottom edge of the conduction band. Such impurities affect the conduction and optical properties to a great extent. Impurity effects form the subject of several later sections.

The sketch in Fig. 10-12 shows the density of states for the covalent crystal. The gap energy varies from near zero for gray tin to about 7 ev for diamond. A tabulation is given, for a number of important covalent crystals, in Table 10-3. Note that the last three entries are for materials which have four electrons per atom on the

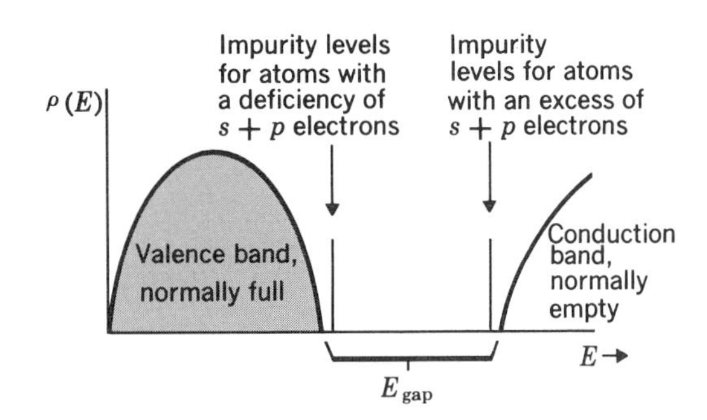

Figure 10-12 Density of states versus E for semiconductor.

average; each constituent element by itself cannot form a covalent crystal in the diamond structure.

Electrons in the conduction band of the covalent crystals have effective masses which are extremely dependent on direction in k space. Also, empty states at the top of the valence band, holes, have effective masses of wide variability, because the energy–wave-number relationship itself, that is, E vs. k, is much less simple than it is for the metals. These features are discussed in Chap. 12 on semiconductors.

The covalent bond is a strong one, and so these solids have high cohesive energy. Values of the order of 3 to 5 ev per atom are typical. Since the wave functions of the solid cannot be simply described, we make no attempt to justify these numbers.

Table 10-3

Material	E_g (at 0°K), ev
C (diamond)	7.0
Ge	0.75
Si	1.21
Sn (gray)	0.08
GaAs	1.45
GaSb	0.8
InAs	0.5

The crystal structure of the covalent solids is entirely controlled by the nature of the bond. Since each atom is surrounded by just four others with which it can share electrons, the nearest neighbors are arranged in a tetrahedral structure. The resulting crystal structure for the pure covalent crystals is the *diamond lattice* shown in Fig. 2-8*a*; for the compound covalent crystals it is the *zincblende* structure of Fig. 2-8*c*.

Covalent bonding has been discussed as though it exists only for elements or compounds whose atoms are bonded to exactly four others. Elements such as arsenic and antimony form solids in which electrons are shared by an atom with three nearest neighbors, atoms such as tellurium share electrons with two neighbors, and atoms such as iodine and bromine share electrons with only one neighbor. The crystal structures of the groups V, VI, and VII are not simple, however. The atoms of these crystals occur in sheets, chains, and molecules, respectively. The binding within a sheet, etc., is covalent; however, the binding between sheets, etc., is provided by the van der Waals forces, not related to the covalent bond.

10-4 Molecular Crystals

Molecular bonding represents the last possibility for the crystallization of atoms. It occurs for those elements or compounds with electron configuration such that little electron transfer occurs between atoms.

Consider, for example, argon. Since all atomic levels are completely occupied through the $3p$ shell, the only chance that the element could be metallic as a solid would be for the empty $3d$ or $4s$ levels to overlap the $3p$. This overlapping does not occur, since these levels are too far above the $3p$, which lies at about -20 ev. Argon cannot be ionic or covalent, and so some other force must cause it to become solid at low temperature. This force is a result of mutual polarization of argon atoms when they approach each other.

The polarization, which produces a bonding termed the *van der Waals force*, arises because of coordinated electron motion in the electron clouds of adjacent atoms. The electrons of each atom shift with respect to the nucleus in the presence of another atom; i.e., the atom becomes an electric dipole. This polarization of adjacent atoms is of such a sense as to produce a lowering of energy, and so the dipoles attract each other. Attraction must occur for spontaneous crystallization to take place. Of course, the electron shells cannot be very strongly distorted, because the energy gain arising from the interaction of adjacent dipoles must be greater than the energy necessary to distort the electron shells. The interaction is thus a weak one. Furthermore it is a short-range force, falling off rapidly as the atoms separate.

Table 10-4 Cohesive energy of some molecular crystals

Material	Cohesive energy kcal/g-mole	Melting point, °K
He (liquid)	0.025	
A	1.84	87
Kr	2.54	120
Xe	3.57	165
CH_4	2.3	112
H_2	0.24	20
Cl_2	6.41	239

The weak nature of the van der Waals attraction leads to a low cohesive energy, and values of a few thousand calories per mole are common. Some typical values are listed in Table 10-4.

In the table, the binding energy of He is so low that it does not even become a solid at the lowest attainable temperatures. In fact, there is good theoretical ground for the belief that He remains a liquid even at the absolute zero of temperature. The failure of He to solidify is a remarkable property, for one of the fundamental laws of thermodynamics states that, at absolute zero, the system falls into its absolute ground state. This ground state for a solid is very simple—it is just that state in which the solid is a perfect crystal lattice with no phonons present. The intuitive picture of a liquid is that the disorder present also implies that the system is in an excited state and that the crystalline ground state should lie lower. For He, however, the interaction between the atoms is so low that the uncertainty in position (from the Uncertainty Principle) for the atoms is larger than the distance between atoms, and the atoms thus lose sight of one another except in an average manner. If an external pressure is applied to the liquid, a phase transition occurs wherein the atoms are pushed sufficiently close to one another so that the forces become large enough to overcome the uncertainty in position and to allow the crystalline state to form.

Liquid He has a second anomaly which makes it famous. The He^4 isotope liquid shows the property of being a superfluid. Its meniscus is infinitely high, and it leaks from common containers. This property is not easy to understand from elementary arguments, but it is connected with the thermodynamics and statistics of a liquid made up of atoms with an even number of particles. He^3 isotope liquid does not show the superfluid properties, except possibly at very low temperatures, where they may exist owing to different causes.

The rare-gas solids are not the only cases of molecular bonding. Solids of saturated molecules like H_2, O_2, HCl, CH_4, etc., also form van der Waals solids because the molecule represents a completely saturated entity like the closed shell of an atom. As was pointed out previously, certain layerlike and chainlike arrangements of covalent-bonded atoms are bonded partly by van der Waals forces.

The crystals which are bonded almost completely by ionic or metallic forces also have van der Waals forces present. The latter usually account for only a small part of the cohesive energy of the lattice, from a few per cent up to 20 per cent in some instances.

The physical properties of crystals bonded with purely van der Waals forces are simple. They are insulators, of course, and they are usually transparent to electromagnetic radiation down to the very short ultraviolet. Their band structure is almost unknown, but the energy gap is large; in fact so little overlap of electron states occurs that the gap between valence and conduction levels is about the same as the difference in the corresponding atomic levels. The very fact that the atomic states in these crystals are so little different from those of the free atom makes these solids of considerable theoretical and experimental interest. These crystals are not, however, of much technological importance so far because they melt at low temperatures.

PROBLEMS

10-1 The cohesive energy of a solid can be expressed in any one of a number of energy units: calories per gram-mole or ev per atom (or molecule) are two of the units most commonly used. How many calories per gram-mole is equal to 1 ev per atom?

10-2 The ionization energy of a free atom of Na is 5.138 ev. The cohesive energy of Na is about 26,000 cal/g-mole at room temperature. Calculate E_{atomic} of Sec. 9-2.

10-3 (*a*) Pure alkali halides are transparent in the visible spectrum. The energy gaps of some of these crystals are listed below. Calculate the wavelength at which these crystals become opaque.

KCl	7.6 ev
KBr	6.5 ev
KI	5.6 ev

(*b*) Si has a band gap of about 1.2 ev and Ge of about 0.75 ev (at $T = 0°K$). At what wavelength do they become opaque?

10-4 The width of the highest filled band has been determined for some ionic crystals by using soft X-ray spectroscopy. Several measured values of bandwidth are tabulated below. Estimate the average density of states over the band for 1 g-mole of these crystals.

KF	1.5 ev
KBr	0.55 ev
Li_2O	12.8 ev

11

Transport properties and specific heat of metals

This chapter is devoted to a detailed discussion of the problems associated with the excitation of electrons in metals. The excitation may come from various sources, but the most common are temperature, electric fields, magnetic fields, or combinations of these. Not all excitations can be described with the same ease, but we can treat these three adequately enough to derive analytical expressions for the most important results for each type of excitation.

The simplest excitation is that caused by thermal fluctuations; this type of excitation leads to a contribution to the total heat capacity of a solid. While it is a small effect, it does have importance in elucidating the details of the band structure itself. More involved excitations are those associated with gradients in electron energy. The two most common are (1) a gradient in thermal energy of the atoms (measured by a gradient in temperature) and (2) a gradient in electrical potential. These gradients produce thermal conductivity and electrical conductivity. They are both complicated effects, since they involve a basic excitation process simultaneously with an interaction between the electrons and the lattice vibrations of the crystal. From a qualitative point of view these effects are simple enough to understand; however, great theoretical problems are associated with their complete quantitative description. Therefore the details of these transport properties in specific metals are only imperfectly understood. The same is true for the motion of conduction electrons in magnetic fields applied to the solid, though we are able to discuss one of the principal effects, the Hall effect, sufficiently well to show that it provides a useful tool for the study of some of the other excitations.

The plan of the chapter is to consider the simpler excitations first, then to go on to the more complicated ones and their combinations.

11-1 Heat Capacity of Metals

A simple metal should have a heat capacity of 6 cal/g-mole-degree from vibrations of the lattice, according to the classical theory. In addition to the lattice heat capacity, however, the valence electrons should also contribute to the heat capacity, since they can increase their translational energy with an increase in temperature. If each of the electrons can do this independently of all the rest, it should behave like an

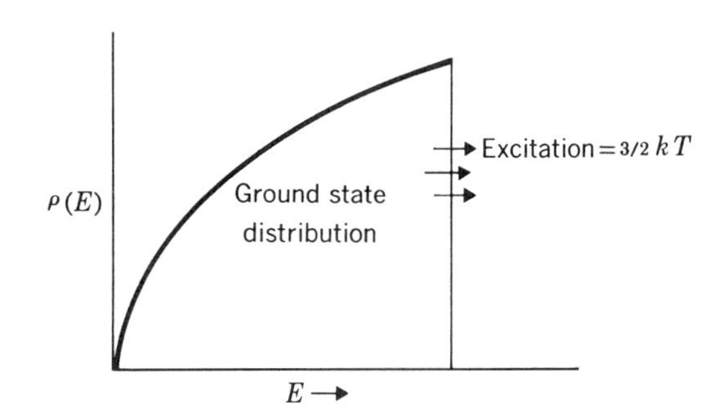

Figure 11-1 Only those electrons near the Fermi surface can be thermally excited into unfilled states at ordinary temperatures.

atom of a monatomic gas and therefore have thermal energy $\frac{3}{2}kT$. Its heat capacity should be

$$\text{Heat capacity/electron} = \frac{3k}{2} \tag{11-1}$$

The total electronic heat capacity for 1 g-mole of electrons should then be 3 cal/g-mole°K; so the actual heat capacity of a simple monovalent metal (lattice and electronic combined) should then be 9 cal/g-mole°K. Experiments prove this value to be too high: the measured heat capacity is never much more than 6 cal/g-mole °K.

To understand this difficulty, we must look more closely at thermal excitation of the valence electrons. Recall that the valence electrons are distributed in a band which is of order 5 ev in width. In the ground state all levels beneath the Fermi level are occupied, so any excitation of an electron has to be sufficient to excite it to one of the empty levels above the Fermi level. Thermal-energy fluctuations of size kT are only about $\frac{1}{40}$ ev at room temperature, so that electrons deep in the Fermi distribution cannot be excited. Therefore only those electrons contained within a thin skin of the Fermi surface can be excited by the thermal fluctuations. These features are pictured in Fig. 11-1.

A remarkably accurate expression for the heat capacity of the valence electrons can be obtained from two simple assumptions: (1) Assume that only those electrons within an energy kT of the Fermi level can be excited, all others being unable to absorb thermal energy. (2) Suppose that those electrons which *can* be excited behave like a simple gas, having thermal energy $\frac{3}{2}kT$. The total energy of N valence electrons at temperature T is therefore

$$\begin{aligned} E(T) &= E(0) + (\tfrac{3}{2}kT)\rho(E_f)kT \\ &= E(0) + \tfrac{3}{2}NkT\,\frac{3}{2}\,\frac{kT}{\mathscr{E}_f} \end{aligned} \tag{11-2}$$

where $\rho(E_f)$ is the density of states defined by Eq. (9-9), calculated at the Fermi level. The factor $\rho(E_f)(kT)$ is therefore the number of electrons which can be excited, and the factor $\frac{3}{2}kT$ is the thermal energy of each electron. The quantity $E(0)$ is the ground-state energy of the electrons in the solid and is independent of temperature.

From Eq. (11-2) the heat capacity can be calculated to be

$$C_v/\text{g-mole} = \frac{dE}{dT} = 3\rho(E_f)k^2T = \tfrac{3}{2}Nk\left(\frac{3kT}{\mathscr{E}_f}\right) \tag{11-3}$$

The units of C_v depend on the units of $\rho(E)$. Since it was derived in Eq. (9-9) for a cube of edge L containing N atoms, C_v can be described in calories per gram-mole per degree Kelvin for a given metal by calculating L for 1 g-mole of the metal. For the simple metals Eq. (11-3) gives $C_v \sim 10^{-4}T$ to $10^{-3}T$ cal/g-mole °K.

The form of Eq. (11-3) is important, for it shows that the electronic heat capacity is linear in temperature, contrary to the classical behavior of a gas. In the second of Eq. (11-3) the first factor is the classical result. However, the second factor, $3kT/\mathscr{E}_f$, shows that only a small fraction of the electrons contribute to the heat capacity. The magnitude of the electronic heat capacity relative to the lattice heat capacity depends strongly on temperature. At low temperatures (less than 5°K) the electronic contribution is larger than the lattice heat capacity, whereas at higher temperatures it is much less than that of the lattice vibrations. This behavior is shown in Fig. 11-2 for an alloy of V and Cr. At sufficiently high temperatures the electronic contribution could become quite significant again, since the lattice contribution levels off near 6 cal/g-mole-deg above the Debye temperature. Most metals melt before the electronic heat capacity becomes appreciable; however, the slow linear increase of the total heat capacity of the solid due to the linear electronic heat capacity can be detected at high temperatures where the lattice contribution levels off at 6 cal/g-mole.

The relative amount of thermal energy of the electrons compared with that of the lattice is important. The total thermal excitation of the electrons is always small at normal temperatures corresponding to solid crystalline metals. The condition that all the electrons be excited thermally is that the thermal energy kT be of the same order as the Fermi kinetic energy itself.

$$\mathscr{E}_f \approx kT_f \tag{11-4}$$

The temperature which satisfies this condition is called the *Fermi temperature.* Above this temperature the electrons behave like a classical gas, while far below this temperature the electrons approximate the ground state shown in Fig. 11-1, with small excitations at the Fermi level. For reasonable values of $\mathscr{E}_f$, the Fermi temperature is of order 10^4 °K; hence real metals fit the latter situation. The distribution in energy of the electrons is then approximately as sketched in Fig. 11-3 for an actual metal at reasonable temperatures.

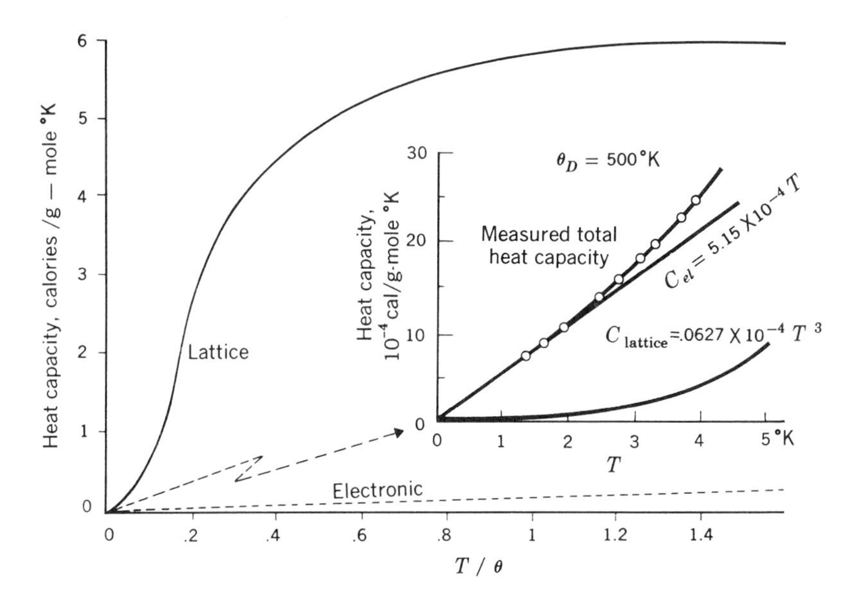

Figure 11-2 The electronic part of the heat capacity of a metal or an alloy varies linearly with temperature, while the lattice part varies according to the Debye equation. Only in the low-temperature region is the electronic part larger. The inset shows experimental data for the total heat capacity of an alloy 20 per cent V–80 per cent Cr below 5°K. [*Data from work of C. Cheng, C. Wei, and P. Beck, Phys. Rev.,* **120**: 426 (1960).]

A more rigorous theory of electronic heat capacity must take into account the fact that even the thermally excited electrons must obey the Pauli exclusion principle. Hence the excited electrons do not act completely like a classical gas. Furthermore, electrons deeper down in the Fermi distribution can be excited more than is assumed in Eq. (11-2). When these effects are treated properly, the electronic heat capacity is given by

$$C_v = \frac{\pi^2}{3}\rho(E_f)k^2T \tag{11-5}$$

This value is a little larger than that given by Eq. (11-3) and gives the same linear temperature dependence. The factor $\pi^2\rho(E_f)k^2/3$ is commonly designated γ.

The linear temperature dependence of C_v is found experimentally to be true. Such measurements serve for determination of the density of states at the Fermi level, and electronic heat capacity is one of the important direct measurements which can be made on the band structure of solids. Values of γ determined experimentally for some of the simpler metals are tabulated in Table 11-1. Calculations of values of $\rho(E_f)$ for metals show that it is of order 10^{41} states per joule gram-mole, or about 10^{46} states per joule per cubic meter (for the common metals).

Table 11-1 Electronic heat capacities of several metals†

Metal	γ, cal/g-mole °K × 10^4	Metal	γ, cal/g-mole °K × 10^4
Ag	1.5	Na	4.3
Al	3.4	Ni	17.4
Au	1.67	Pt	16.3
Co	12.0	Sn	4.3
Cr	3.75	Ta	13–14
Fe	12.0	U	26
La	16–21	W	1.8–5
Mn	33–43	Zn	1.25–1.50
Mo	5.0–5.25	Ar	6.92–7.25

† These representative values are taken from a more extensive table in the *AIP Handbook*, 2d ed., McGraw-Hill Book Company, Inc., New York, 1963.

11-2 The Fermi Function

The thermal excitation of electrons at the top of the distribution was treated in Sec. 11-1 in an approximate way by using the Boltzmann factor. Rigorously, the classical Boltzmann factor is not valid for a group of electrons which must obey the exclusion principle, because the Boltzmann law does not account for the fact that only one electron may occupy a state at a time. The probability function which describes the thermal behavior in quantum mechanics of a group of particles which obey the exclusion principle was first given by Fermi. It is written

$$w_{-}(E) = \frac{1}{e^{(E-E_f)/kT} + 1} \tag{11-6}$$

where $w_{-}(E)$ is the probability that a state having an energy E is occupied by an electron at temperature T. [*Note:* The number of electrons at energy E is the product of Eq. (11-6) and the density of states at E.] In Eq. (11-6), E_f is the Fermi energy.

The function $w_{-}(E)$ is plotted for $T = 0$ and for a finite temperature $T > 0$ in Fig. 11-3. At $T = 0$, the function is a step function, with all states below E_f filled and all states above E_f empty. At temperatures greater than zero, the function goes from unity to zero over a range in energy of about kT. The function has the value $\frac{1}{2}$ at $E = E_f$.

The expansion of Eq. (11-6) for $|E - E_f| \gg kT$ is

$$w_-(E) = e^{-(E-E_f)/kT} + \text{small terms for } E - E_f \gg kT \tag{11-7}$$

$$w_-(E) = 1 - e^{-(E-E_f)/kT} + \text{small terms for } E_f - E \gg kT$$

The two Eqs. (11-7) have the following physical significance: The first equation states that the probability of excitation of electrons into states with energy well above the Fermi energy is given by the Boltzmann factor with energy measured relative to E_f. The second equation

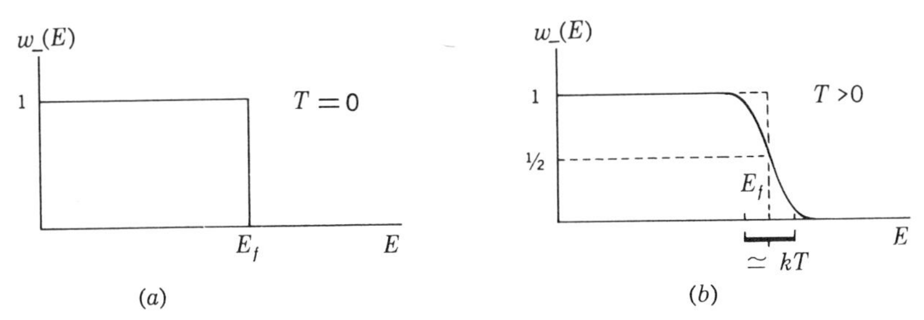

Figure 11-3 The Fermi function $w_-(E)$ for two temperatures. (*a*) At $T = 0$, the function is a step function centered at the Fermi energy E_f. (*b*) At $T > 0$, the steep step at E_f becomes rounded, with exponential "tails" extending approximately a distance kT above and below E_f. At $E = E_f$, $w_-(E)$ falls to $\frac{1}{2}$.

states that the probability of excitation of holes well below the Fermi energy is similarly given by a Boltzmann factor, since

$$w_+(E) = 1 - w_-(E) \tag{11-8}$$

where $w_+(E)$ is the probability of occupancy of a state by a hole.

Equation (11-7) gives the justification for the treatment of electron excitations used in the previous section. It states qualitatively that the electrons and holes behave much like a classical gas if the energy is measured from E_f. We might say figuratively that E_f is the energy from which electrons and holes can be thermally excited. The approximation becomes better as $|E - E_f|$ becomes progressively larger than kT. This approximation, useful in treating several properties of metals, is also employed to great advantage in the discussion of the Fermi level in semiconductors in Chap. 12.

11-3 Electrical Conductivity of Metals: Relaxation Time

Ohm's law of electrical conduction is probably the most widely known and practically useful law of physics, with the exception of Newton's laws of mechanics. This law, which expresses a linear relationship between the current density **J** and the electric-field vector **E**, applies both to the excellent metallic conductors and the poorer conductors, the insulators. The proportionality constant σ, which appears in this expression as it is usually written,

$$\mathbf{J} = \sigma \mathbf{E} \tag{11-9}$$

is, for most common materials, a scalar (called the *conductivity*). The difference between conductors and the insulators is simply in the magnitude of σ. It is of order 10^{+7} mhos/m for the metals and 10^{-10} mhos/m (or less) for good insulators.

Two things are now desirable: (1) a proof of Ohm's law from the atomic point of view and (2) calculation of the conductivity for the

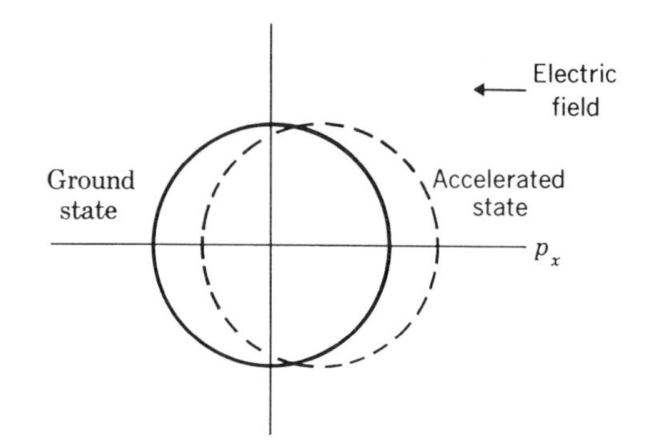

Figure 11-4 An external electric field produces a displacement of the center of the Fermi distribution. The average current of the electrons is then simply that corresponding to the center of the sphere.

different materials. The remainder of this chapter shows that a theoretical justification of Ohm's law is possible with certain starting assumptions and that a constant definable as the conductivity results from the analysis. In this chapter only conductivity in metals is considered, although the extension to semiconductors is easily made and is discussed in Chap. 13.

We start with the obvious fact that no steady net current flows in a material in the absence of an electric field. This basic fact follows from the property of the Fermi distribution that, for every electron with momentum in one direction in the crystal, another exists with momentum in the opposite direction. However, this momentum cancellation does not exist when a steady external field is applied to the crystal. Under such a field, each electron experiences a force in the *same* direction, and all electrons within the Fermi sphere accelerate as a unit (in a direction opposite to the field **E**) (Fig. 11-4). The coordinated motion of the valence electrons which is superimposed on their ground-state motion produces a net charge transfer in the solid, hence an electric current.

An expression can be written for the effect of the field by consideration of the force on each electron, after which summation is made over the entire group of electrons. If the field **E** is in the x direction, each valence electron experiences a force $q\mathbf{E}$ (where q for the electron is

-1.6×10^{-19} coulomb) and

$$m^*\ddot{\mathbf{x}} = q\mathbf{E} \tag{11-10}$$

In addition to the statement that the Fermi distribution corresponds to zero current where centered about $k = 0$, a second obvious fact is that Eq. (11-10) cannot by itself lead to Ohm's law. Integration of Eq. (11-10) would lead to a current which increases linearly with time and which, even at very weak fields, could cause arbitrarily large currents. Ohm's law states that, for a given external field, the current is limited to a fixed constant value with respect to time. Obviously, then, some additional force must be introduced to obtain the desired result. The situation is analogous to that of a ball on an inclined plane. When the plane is tilted, the ball accelerates forever; however, when proper frictional forces are introduced, the ball reaches a terminal velocity which is proportional to the net acceleration force (the angle of tilt). Hence, some additional frictional force on the electron must exist, a force which serves to limit the velocity to a finite value. A clue to the origin of this frictional force is provided by the additional experimental observation that the conductivity of metals decreases as the temperature rises. Since rising temperature produces increasing amplitude of thermal vibration, one possible origin of the frictional force is the interference by the thermal vibrations of the solid with the basic freedom of motion of the electrons within the crystal.

To understand this point completely, we must recall one of the most fundamental theorems concerning electrons in solids, namely, that the translational motion of a valence electron is not disturbed by its interaction with a perfect lattice of ions. This theorem is verified by the fact that the resistivity of a pure, annealed metal approaches zero near the absolute zero of temperature. However, if the crystal has some of its ions slightly out of place—say, because of vacancies and dislocations, as well as oscillation—the basic free translational property of the electrons is destroyed. The electrons are scattered by the imperfectly positioned atoms. The physical content of these statements is illustrated in Fig. 11-5. Of course, the exclusion principle

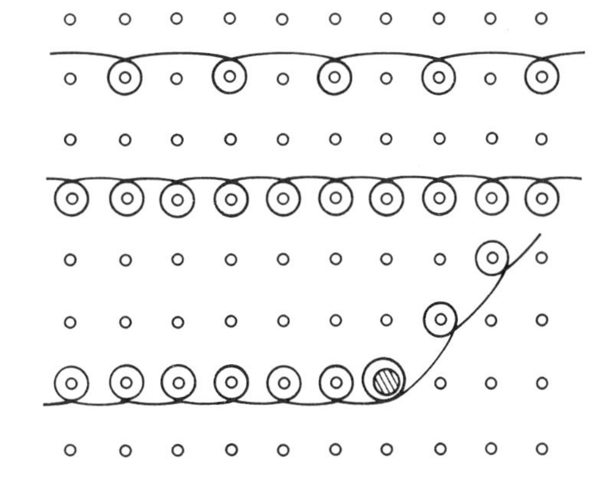

Figure 11-5 Electrons are not scattered in a perfect crystal, because the orbit of the electron exactly repeats itself through the perfect crystal. When an irregularity is present, the repetitive motion is broken.

restricts the scattering into states which are unoccupied. This has the result that only those electrons near the Fermi surface can be scattered, since the scattering collisions give relatively low energy transfer.

The frictional force which results from the scattering of individual electrons by the lattice has a pronounced effect on the entire Fermi distribution. Those electrons on the advancing nose of the Fermi sphere are scattered more frequently than those on the rear. They are therefore peeled off the advancing nose in a kind of ablative motion and are gradually returned to the lower energy states toward the rear. When the electron scattering can just keep up with the acceleration due to the field, a steady state is reached and the Fermi sphere remains

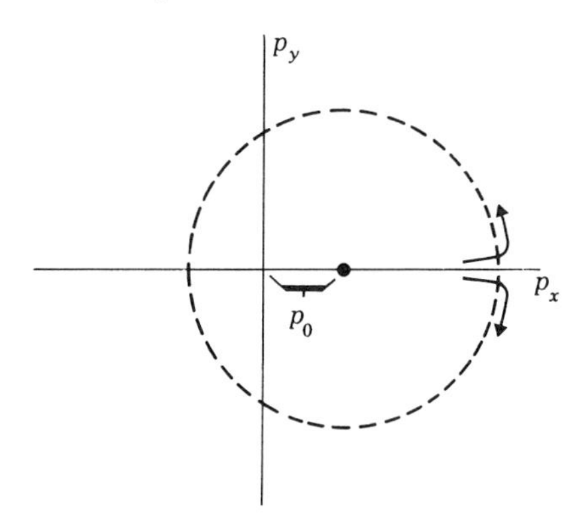

Figure 11-6 Scattering processes limit the displacement of the Fermi sphere to some value p_0.

stationary in its displaced position (Fig. 11-6). If $\mathbf{p}_0$ is the displacement of the center of the displaced sphere, then the net current is given by

$$\mathbf{J} = +\frac{Nq\mathbf{p}_0}{m^*} \qquad (11\text{-}11)$$

where N is the number of electrons in the Fermi distribution. The shift $\mathbf{p}_0$ of the Fermi sphere is estimated in the next section.

11-4 The Relaxation Time

The calculation of the shift of the Fermi distribution in Eq. (11-11) should be carried out completely within the framework of quantum mechanics. However, this difficult problem is entirely beyond the level of this book. Some physical insight into current flow can be gained by use of a mixture of Newtonian mechanics of accelerating charges and the concept of the Fermi distribution. This method stems from a very early, purely classical treatment of conductivity by Drude. We now sketch this classical theory and abstract that part of it which is still valid in quantum mechanics.

Newton's law for the acceleration of a charge q in a field E_x is

$$m^*\ddot{x} = qE_x$$

Integration gives for the velocity at time t

$$\dot{x} = v_{x0} + \frac{qE_x t}{m^*}$$

The first term is the initial velocity of the electron. We immediately discard this term, because the initial velocity in quantum mechanics corresponds to the velocity of the electrons in the initial Fermi distribution. We have already shown how, in the initial Fermi distribution, the average velocity of the electrons cancels out.

We now assume that after a certain time 2τ the electrons suffer a collision and that the result of the collision is to decrease to zero the

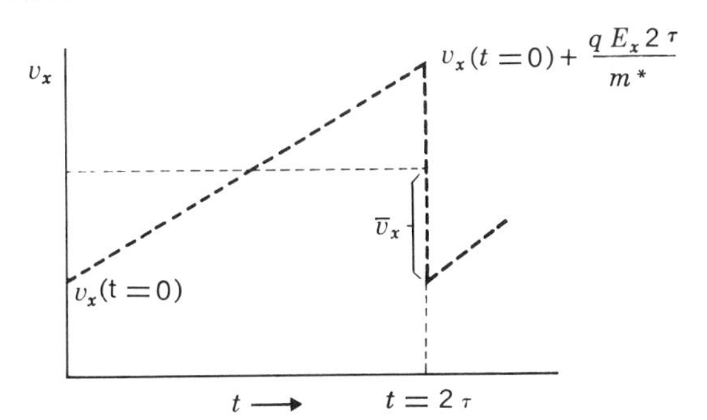

Figure 11-7 The acceleration of an electron occurs uniformly during the interval 2τ, when a collision causes the electron to return to its initial velocity.

excess velocity the electrons acquired from the field. After the collision, the electron accelerates for another period 2τ. The time τ is called the *relaxation time*. So far, it is defined only classically, but it can be given additional meaning. The average velocity *increment* $\bar{v}_x$ of the accelerating electron during its time of flight is simply half its final velocity increment,

$$\bar{v}_x = \frac{qE_x\tau}{m^*} \tag{11-12}$$

If all electrons of the solid participate in this saw-tooth motion (see Fig. 11-7), the charge flow for N electrons is (remember $\bar{v}_{x0} = 0$)

$$J_x = Nq\bar{v}_x \tag{11-13}$$

The current can be written

$$J_x = \frac{Nq^2\tau}{m^*} E_x \tag{11-14}$$

Since this equation states that the current is proportional to the field, it is a statement of Ohm's law. The conductivity is

$$\sigma = \frac{Nq^2\tau}{m^*} \tag{11-15}$$

Since the x direction is completely arbitrary in an isotropic substance, the last two equations hold for any arbitrary direction in the crystal.

This analysis must now be put into the framework of the quantum-mechanical description of a band. Rather than trace through the steps of the derivation and attempt to make a quantum analog, we simply interpret the result in different terms. The new quantity is the relaxation time, half the time between collisions. If we remember that, in the Fermi distribution, only the electrons near the Fermi energy can participate in collisions, the relaxation time is then the time between collisions for those electrons. In the terms of Eq. (11-11), $\mathbf{p}_0$ is given by

$$\mathbf{p}_0 = q\tau\mathbf{E}_x$$

The conductivity as expressed in Eq. (11-15) is a function of several parameters of the material. One of these is N, the total number of valence electrons per cubic meter. The conductivity is proportional to q^2; hence Ohm's law does not distinguish positive from negative charge carriers. The relaxation time enters explicitly. The mass, of course, must be the effective mass, since the inertia of the charge carriers to the external field is the quantity involved. The mass must always be taken to be a positive quantity, since the sign of the charge carriers is already taken into account in Eqs. (11-12) and (11-13).

A quantity related to the relaxation time is the mobility μ, defined as the average speed acquired in time 2τ in a field of unit strength.

$$\mu = \left|\frac{\bar{v}_x}{E_x}\right| = \left|\frac{q\tau}{m^*}\right| \qquad \text{(11-16)}$$

The mobility depends not only on the magnitude of the charge carrier but also on the relaxation time and effective mass. It can be measured directly and is therefore a more useful concept than τ itself. Expressed in terms of μ, the conductivity can be written

$$\sigma = N\,|q|\,\mu \qquad \text{(11-17)}$$

The relaxation time is also related to another parameter λ, the mean free path of the electrons capable of making collisions. If the velocity gained in the field is negligible compared with the Fermi velocity, then

$$\lambda = |v_f|\,(2\tau) \qquad \text{(11-18)}$$

where v_f is the speed of the electrons at the top of the Fermi distribution. When appropriate values are substituted in Eq. (11-18), values of λ for the common pure metals at ordinary temperatures are of order 100 A.

One further word is in order concerning the approximations which have been made. The entire derivation has been carried out on the supposition that both τ and m^* are independent of direction in the crystal. Experiments have shown that m^* is indeed a function of direction in crystals, so that σ is in general a function of direction. For cubic crystals, however, symmetry requirements of vector relations (in this instance σ) must be independent of direction, i.e., isotropic. The relaxation time itself then must assume a directional dependence such that the ratio τ/m^* is isotropic. For crystals of lower symmetry, say, hexagonal, this cancellation does not necessarily occur, and such materials (among them Zn and Cd) may have a directional dependence for σ.

11-5 Joule's Law

Heat is developed in a wire when a current passes through it. The heat energy is extracted from the source of the field and is transferred by the moving charges to the lattice itself. Extension of the ideas of the preceding section enables us to calculate the rate of heat transfer and hence to derive an expression for Joule's law.

One of the assumptions in Sec. 11-4 was that the electrons lose their added energy to the lattice by collisions with the lattice atoms. The external field gives extra energy to the electron during each trajectory, and we require that this energy be transferred to the lattice through the collisions. The kinetic energy gained by each electron between collisions is just

$$\tfrac{1}{2}m^*(2\bar{v}_x)^2 = \frac{2q^2E_x^{\,2}\tau^2}{m^*}$$

This amount of energy is transferred to the lattice once every 2τ sec by N electrons; so the total energy transferred per second is

$$\text{Heat/sec} = \frac{Nq^2\tau E_x^{\,2}}{m^*} = \sigma E_x^{\,2}$$

This is the familiar Joule's law.

11-6 The Additive Nature of the Resistivity: Mathiessen's Rule

The scattering of electrons may be caused by several types of lattice defects: by temperature vibrations (phonon scattering), by impurities (both interstitial and substitutional) and by mechanical deformation (which produces dislocations, vacancies, and interstitials). An experimental fact, known for a long time, is that the contributions to the resistivity ρ (defined as $1/\sigma$) from these several sources are additive. This rule, known as *Mathiessen's rule*, may be concisely written

$$\rho_{\text{total}} = \rho_{\text{temperature}} + \rho_{\text{impurities}} + \rho_{\text{deformation}} \tag{11-19}$$

The data in Fig. 11-8 further illustrate the rule.

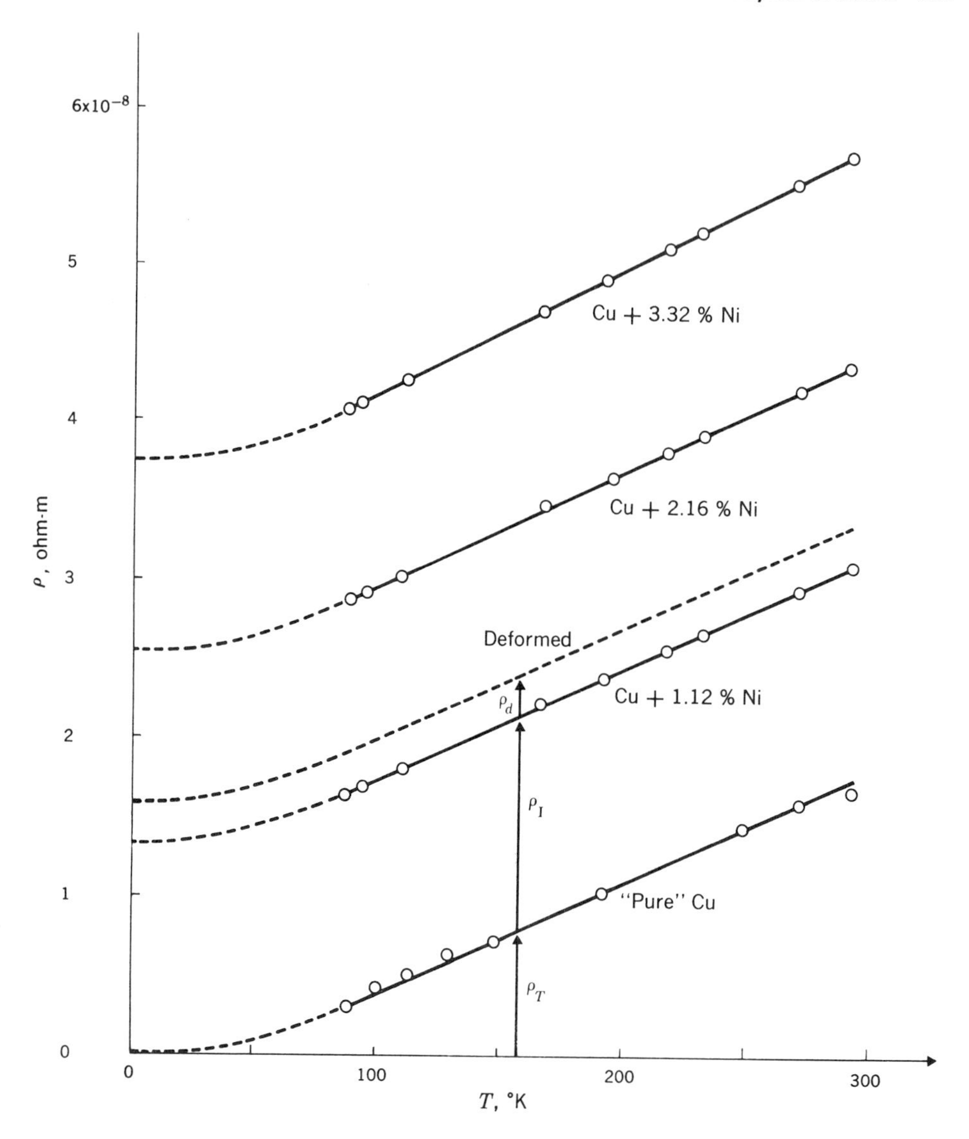

Figure 11-8 The additive nature of the resistivity. For these well-annealed alloys of Cu and Ni the resistive is made up of two parts: one part ρ_T caused by temperature vibrations and a second part ρ_I caused by impurity additions (indicated on the plot for Cu + 1.12 per cent Ni alloy). Deformation, either of the pure substance or of an alloy, would add a third part ρ_d (a hypothetical effect of deformation is indicated by the dashed line for the Cu + 1.12 per cent Ni alloy). Note that the last two parts, ρ_I and ρ_d, are independent of temperature (unless heat treatments alter the impurity or defect geometry). [*Data points for the undeformed alloys from J. Linde, Ann. Phys.*, **5:** 15 (1932).]

The basis for the behavior of metals described by Eq. (11-19) lies in the nature of the scattering process itself. To show this, we need to examine more closely the concept of mean free path. The constant λ, introduced in Eq. (11-18), is, in fact, not the distance between collisions for every trajectory: it is simply an average value. The scattering process can be more completely described by the following: For a group of electrons passing through the lattice in the x direction, the number $dN(x)$ that make a collision in some length of path dx at position x is given by

$$dN(x) = \frac{-N(x)\,dx}{\lambda} \tag{11-20}$$

where $N(x)$ is the number of electrons which have still to make a collision at position x.† In this equation, λ is simply a constant parameter to satisfy dimensional requirements. We shall show that it is identical to the quantity in Eq. (11-18). Equation (11-20) is based on the premise that an electron has a constant probability of making a collision in each interval dx. Equation (11-20) can be integrated to give the number of electrons remaining in the group after it has traversed a distance x,

$$N(x) = N_0 e^{-x/\lambda} \tag{11-21}$$

where N_0 is the number of electrons in the group at the beginning.

In Eq. (11-21), λ has the same meaning as "mean life" does in radioactive decay, since the number drops to $(1/e)$th of its original value when x takes the value λ. The parameter λ is the average distance between collisions. Since $N(x)$ is the distribution function for collisions, according to Eq. (9-10),

$$\bar{x} = \frac{N_0 \int_0^\infty x e^{-x/\lambda}\,dx}{N_0 \int_0^\infty e^{-x/\lambda}\,dx} = \lambda \tag{11-22}$$

Two independent scattering processes are easy to take into account. Suppose that they are temperature-scattering (i.e., from phonon collisions), with a mean free path λ_T, and impurity-scattering, with a mean free path λ_I. Then the number of collisions is given as a function of x by the expression

$$dN(x) = -\frac{N(x)\,dx}{\lambda_T} - \frac{N(x)\,dx}{\lambda_I} \tag{11-23}$$

The solution of this equation is

$$N(x) = N_0 e^{-(1/\lambda_T + 1/\lambda_I)x} \tag{11-24}$$

† The expression (11-20) is entirely analogous to the law of radioactive decay.

The total mean free path λ in this instance is given by

$$\frac{1}{\lambda} = \frac{1}{\lambda_T} + \frac{1}{\lambda_I} \tag{11-25}$$

The relaxation time $\tau = \lambda/(2|v_f|)$ can be written as

$$\frac{1}{\tau} = 2\,|v_f|\left(\frac{1}{\lambda_T} + \frac{1}{\lambda_I}\right) = \frac{1}{\tau_T} + \frac{1}{\tau_I}$$

where the τ's are defined through equations of type (11-17). Clearly, then, the resistivity, which is defined as $\rho = 1/\sigma$, can be written

$$\begin{aligned}\rho &= \frac{m^*}{Nq^2\tau} \\ &= \frac{m^*}{Nq^2\tau_T} + \frac{m^*}{Nq^2\tau_I}\end{aligned} \tag{11-26}$$

If we define the first term on the right as ρ_T and the second terms as ρ_I, then Mathiessen's rule results,

$$\rho = \rho_T + \rho_I$$

Clearly this derivation could be expanded to include a deformation-scattering term proportional to $1/\lambda_d$. Note that it is the *resistivities* which are additive, not the conductivities.

Numerous calculations about resistivities from the several sources can be made. One of these is described in Prob. 11-6.

11-7 Temperature Dependence of the Electrical Conductivity

Figure 11-8 shows an additional characteristic of metals. Except for the very lowest temperatures, the resistivity of pure metals is a linear function of temperature. For impure metals, the "thermal region" (i.e., temperatures above the point where $\rho_I \cong \rho_T$), the resistivity is also linear with temperature. At low temperatures, the temperature dependence of the resistivity is more complex.

11-8 The Hall Effect

Measurements of conductivity alone are not sufficient for determination of both the number of conducting charges, N, and their mobility μ. Neither do they permit determination of the sign of the predominant charge carrier. However, measurement of the *Hall effect* supplies the additional information required.

The Hall effect depends for its operation on the acceleration of a moving charge by a magnetic field. The force on the charge which produces this acceleration is given by

$$\mathbf{F} = q\mathbf{v} \times \mathbf{B} \tag{11-27}$$

where $\mathbf{v}$ is the velocity of the charge q, and $\mathbf{B}$ is the magnetic induction. The force depends on both the magnitude and the sign of the charge, and it is furthermore normal to both $\mathbf{v}$ and $\mathbf{B}$. In the present context the velocity $\mathbf{v}$ is assumed to be the average drift velocity due to the action of an electric field $\mathbf{E}$. Because of the symmetry of the Fermi sphere, the instantaneous velocity at zero field produces no average Hall effect. In general, the charge may be either + or −, so that $\mathbf{v}$ is either parallel or antiparallel to $\mathbf{E}$. This distinction is not important for the noble metals and the alkali metals, since the charge carriers are

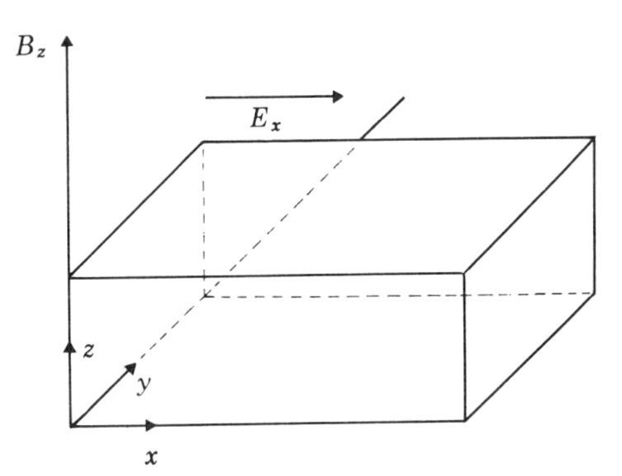

Figure 11-9 The geometry of the electric and magnetic fields for a simple Hall-effect calculation.

negative electrons, but it is particularly important for semiconductors, which can have either positive or negative charge carriers.

The force on a charge in the presence of this combination of electric and magnetic fields is not difficult to calculate. Let $\mathbf{E} = \mathbf{E}_x$ and $\mathbf{B} = \mathbf{B}_z$ (Fig. 11-9). The force on the charge from the electric field is $q\mathbf{E}_x$. Because of their drift velocity, the charges experience a force in the magnetic field, which changes their straight-line trajectories in the xy plane to curved paths.

The velocity of a charge moving in some arbitrary direction can be written as

$$\mathbf{v} = v_x\mathbf{i} + v_y\mathbf{j} + v_z\mathbf{k}$$

The force on the charge resulting from its motion in the magnetic field is

$$\mathbf{F} = qv_yB_z\mathbf{i} - qv_xB_z\mathbf{j} \tag{11-28}$$

No z component of force exists. Observe also that a component of force is changed in sign if the appropriate component of velocity is changed in sign.

In an electric field $\mathbf{E}_x$ the only component of velocity is $\mathbf{v}_x$. Hence the magnetic field produces a force in the y direction, drives electrons toward the near face of the solid (Fig. 11-9), and charges it negatively. The far face becomes positively charged, and a transverse electric field is set up in the $-y$ direction. This field $\mathbf{E}_y$ produces an additional force on the electrons of magnitude $q\mathbf{E}_y$, a force directed oppositely to that produced by the magnetic field. At equilibrium the net force must be zero.

$$q\mathbf{E}_y + q\mathbf{v}_x \times \mathbf{B}_z = 0 \tag{11-29}$$

Since $\mathbf{v}_x$ is the drift velocity

$$q\mathbf{E}_y = \frac{-q^2\tau}{m^*}\mathbf{E}_x \times \mathbf{B}_z \tag{11-30}$$

The scalar equation equivalent to this expression is

$$E_y = \frac{q\tau}{m^*}E_xB_z \tag{11-31}$$

Rewriting this expression with the fields collected as a common factor yields

$$\frac{E_y}{E_xB_z} = \frac{q\tau}{m^*} \tag{11-32}$$

The absolute value of this factor $|q|\,\tau/m^*$ as measured by $|E_y/E_xB_z|$ is called the *Hall mobility*. Replacing E_x in Eq. (11-31) by J_x/σ gives

$$E_y = \frac{1}{Nq}J_xB_z \tag{11-33}$$

Commonly the proportionality factor connecting the field E_y with the product of J_x and B_z is called the *Hall constant* R. Then

$$R = \frac{1}{Nq} \tag{11-34}$$

Conduction by electrons gives a negative value for E_y, *hence a negative value for* R. *Conduction by positive charges gives a positive value for* R.

Equations (11-32) and (11-33) are not independent. Since $|q|\,\tau/m^* = \mu$,

$$|R|\,\sigma = \mu \tag{11-35}$$

and these equations simply use the data in different ways. Table 11-2 lists the values of R and μ which have been obtained for a number of metals.

Table 11-2 Hall constants and Hall mobilities for some of the metals at room temperature

Metal	R, $\times 10^{10}$ volt m^3/amp weber	μ, m^2/volt sec
Ag	−0.84	0.0056
Al	−0.30	0.0012
Au	−0.72	0.0030
Cu	−0.55	0.0032
Li	−1.70	0.0018
Na	−2.50	0.0053
Zn	+0.3	0.0060
Cd	+0.6	0.0080

Measurement of the Hall effect gives three important quantities:

1. The sign of the current-carrying charges is determined. They are obviously electrons for the first metals listed in Table 11-2, down to Na.
2. The number of charge carriers per unit volume can be calculated from the magnitude of R.
3. The mobility is measured directly.

Not all metals have a negative Hall constant. Some metals (which almost surely have electronic conduction) are anomalous, with positive R. Zn and Cd, for example have positive R, indicating positive charge carriers. The interpretation is that the major charge carriers are vacant levels in the Fermi distribution, i.e., holes. In not all cases can one make the three statements above from the Hall effect. If both holes and electrons, for example, contribute to the conductivity then R can be either positive or negative, depending upon the relative densities and mobilities of the carriers.

11-9 Thermal Conductivity of Valence Electrons

All solids conduct heat. In general, the best heat conductors are the metals. Among the metals, the best electrical conductors (Ag, Cu, Al, Au) are also the best heat conductors (see Table 11-3). Since electrical conduction of metals has its origin in the behavior of the valence electrons, this similarity of behavior suggests that the conduction of heat in metals is in large part also a result of the behavior of valence electrons. In Table 11-3, the essential features of this behavior are displayed. (The column $K/\sigma T$ will be referred to at the end of this section.)

Table 11-3 Electrical and thermal conductivities at 293°K (room temperature)

Metal	σ, mhos/m, $\times 10^{-7}$	K, cal/m °K sec	$K/\sigma T$, $\times 10^{9}$
Ag	6.15	101	5.6
Cu	5.82	92	5.4
Au	4.09	71	5.9
Al	3.55	50	4.7
Na	2.10	32	5.2
W	1.80	48	9.1
Cd	1.30	24	6.3
Ni	1.28	14	3.7
Fe	1.00	16	5.5
Pb	0.45	8	6.1

Not all heat conduction is due to the electrons. Since heat energy resides primarily in the vibrations of the atoms about their lattice positions, energy can be transferred from one point to another by the collisions of the atoms with one another. We described in Chap. 3 the vibrations of the lattice in terms of sound waves called phonons. Since the phonons move through the crystal with the velocity of sound, one would imagine that heat conduction should also take place with the velocity of sound. However, the nonlinearity of the force law between atoms causes the phonons to scatter one another, and the mean free path of phonons at normal temperatures is much smaller than the mean free path of electrons in pure materials. Hence, when a solid has a conduction band with a significant number of electrons, the heat conduction by the electrons is usually much greater than the heat conduction by phonons. In general, if K_{ph} is the conductivity

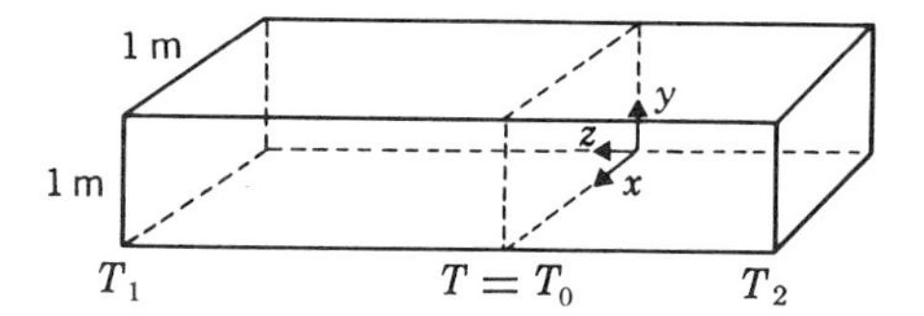

Figure 11-10 Geometry of bar for calculation of heat flow.

due to phonons and K_{el} is the conductivity due to electrons, the total conductivity K is

$$K = K_{ph} + K_{el}$$

For the good metallic heat conductors at room temperature, K is of the order of 10 to 100 cal/m °K sec, whereas for insulators, such as NaCl and quartz, K is of the order of 1 in the same units. Presumably for the metals, where K_{el} is dominant, K_{ph} is also about 1.

The physical model used here for the calculation of K_{el} is entirely consistent with that used in the development of Ohm's law. The electrons move through the lattice at high speeds, making collisions at intervals with displaced lattice ions. The scattering process occurs again after 2τ sec, and the common path length between collisions is therefore λ. No external electric field is presumed to exist; the only externally controlled variable is the temperature distribution in the solid. No transients are considered; so this is a steady-state calculation.

The macroscopic features of the problem are shown in Fig. 11-10. A temperature gradient is maintained along a rectangular bar of unit cross section. If $T_1 > T_2$, then heat flows down the bar in the negative z direction. The heat flux H is given by the simple law

$$H(\text{cal/m}^2\text{ sec}) = -K\frac{dT}{dz} \tag{11-36}$$

where K is the thermal conductivity. This expression should actually be a vector relation between H and the temperature gradient dT/dz, but for most materials of interest it can safely be written in this scalar form. The thermal conductivity is then a scalar, independent of direction in the crystal.

Calculation of heat conduction by the electrons rests upon the fundamental assumption that electrons come into thermal equilibrium with the lattice at the point where they make a collision. Electrons moving through a given element of area thus have different temperatures (i.e., velocities) if the solid is not homogeneous in temperature. The net heat flux through an element of area is then determined as the difference in thermal energy carried by electrons moving in opposite directions through the surface. This flux can be determined in terms of dT/dz and parameters of the valence band. The factor K is determined by use of Eq. (11-36). It is a function only of parameters of the metal.

The derivation will be made with the coordinates of Fig. 11-11. At the origin $T = T_0$. An area $dx\, dy$ in the xy plane is the cross

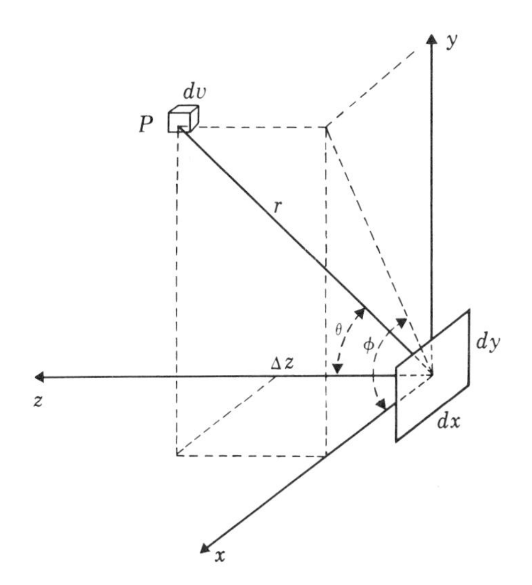

Figure 11-11 Coordinate system for calculation.

section through which the heat flux is calculated. Consider first all electrons which are on the $+z$ side of the xy plane at some instant of time. They all make a collision with a lattice ion at $t = 0$. According to our basic assumption, the electrons exchange energy with the lattice during this collision so that they have thermal kinetic energy characteristic of the temperature of the lattice at that point. After the collision, they fly off in all directions until they make another collision with a lattice ion. λ is the distance they travel between collisions. Some electrons pass through the area $dx\ dy$ in the interval between collisions and carry through that area an amount of thermal energy characteristic of the temperature at the point of the initial collision.

Consider all electrons which carry heat from the positive z side of $dx\, dy$. All these electrons must come from points lying inside a sphere of radius λ. For isotropic scattering, which is assumed, the probability that a given electron from some point P passes through $dx\, dy$

before it makes another collision is just $(dx\,dy\cos\theta)/4\pi r^2$. The number of electrons, N', which carry heat from the volume element dv about P through $dx\,dy$ is given by

$$N' = \frac{dx\,dy\cos\theta}{4\pi r^2}\,\frac{3}{2}\,\frac{NkT_p}{\mathscr{E}_f}\,dv \tag{11-37}$$

This expression uses the approximation of Eq. (11-2) that the fraction of electrons which can be thermally excited at temperature T_p is $3kT_p/2\mathscr{E}_f$. If the thermal energy carried by these electrons is $\frac{3}{2}kT_p$, then the heat which is carried through $dx\,dy$ by the N' electrons is

$$\text{Heat} = \frac{dx\,dy\cos\theta}{4\pi r^2}\,\frac{9Nk^2T_p{}^2}{4\mathscr{E}_f}\,dv \tag{11-38}$$

The total heat carried through $dx\,dy$ by all electrons from $+z$ is the sum over all volume elements within a radius λ of the origin. Before this summation can be carried out, however, account must be taken of the variation of temperature with z. If a linear relation between T and z exists,

$$T_p = T_0 + \frac{dT}{dz}\,\Delta z$$

which for small Δz and small temperature gradients permits the replacement of $T_p{}^2$ by $T_0{}^2 + 2T_0(dT/dz)\,\Delta z$.

The final expression for the heat H^+ carried by all electrons through $dx\,dy$ from the positive z side is then

$$H^+ = \int_0^\lambda\int_0^{2\pi}\int_0^{\pi/2}\frac{dx\,dy}{4\pi r^2}\,\frac{9Nk^2}{4\mathscr{E}_f}\left(T_0{}^2 + 2T_0\,\frac{dT}{dz}\,r\cos\theta\right) \times r^2\sin\theta\cos\theta\,dr\,d\theta\,d\phi \tag{11-39}$$

This integral can be evaluated to be

$$H^+ = H_0\,dx\,dy + \frac{3Nk^2T_0\lambda^2}{8\mathscr{E}_f}\,\frac{dT}{dz}\,dx\,dy \tag{11-40}$$

where H_0 is a constant independent of dT/dz. A similar expression for the heat, H^-, carried by the electrons from the negative z side of $dx\,dy$ is

$$H^- = H_0\,dx\,dy - \frac{3Nk^2T_0\lambda^2}{8\mathscr{E}_f}\,\frac{dT}{dz}\,dx\,dy \tag{11-41}$$

The net heat flux in the $+z$ direction, ΔH, is the difference between H^- and H^+.

$$\Delta H = -\,\frac{3Nk^2T_0\lambda^2}{4\mathscr{E}_f}\,\frac{dT}{dz}\,dx\,dy \tag{11-42}$$

The negative sign means that heat flows in the direction opposite to the temperature gradient. Recall that this heat is transferred in time

2τ sec. The rate per second is $\Delta H/2\tau$. Referring the heat flux to unit area by dividing ΔH by $dx\,dy$ yields

$$H \text{ (heat flux/sec unit area)} = -\frac{3Nk^2T_0\lambda^2}{8\mathscr{E}_f\tau}\frac{dT}{dz} \tag{11-43}$$

Since λ, $\mathscr{E}_f$, and τ are not independent of each other, finally

$$H = -\frac{3Nk^2T_0\tau}{m^*}\frac{dT}{dz} \tag{11-44}$$

Upon comparing this equation with Eq. (11-35), K can be written

$$K = \frac{3Nk^2T_0\tau}{m^*} \tag{11-45}$$

As an example of the use of this expression, a calculation of K is given for Na at $T_0 = 300°\text{K}$. Since $N = 2.5 \times 10^{28}$ per cubic meter and $\tau \approx 3 \times 10^{-14}$ sec (at 300°K), K becomes

$$K_{300°\text{K}}(\text{Na}) \approx 150 \text{ joules /sec m °K}$$
$$\approx 36 \text{ cal/sec m °K}$$

This calculated value is remarkably close to that measured for Na, 34 (see Table 11-3). Equation (11-45) is not actually that precise, since the approximations in its derivation are many, but it can be relied on in order of magnitude.

The conductivity K does not depend linearly on T as is suggested by Eq. (11-45), because τ also depends on temperature. In fact, τ varies about as $1/T$ for most metals; hence K is nearly temperature-independent. This is consistent with experimental fact—for Al, for example, K changes by less than 50 per cent between 120 and 700°K, while σ changes by 1,300 per cent over the same range. A plot of K and σ as a function of temperature for Ni is shown in Fig. 11-12.

More refined assumptions and approximations might be made in this derivation, as for the derivation of Ohm's law. These do not alter the result in principle; they only give the important factors $k^2NT_0\tau/m^*$ with slightly different numerical factors.

A further observation is significant. The ratio of K/σ yields

$$\frac{K}{\sigma} = \frac{3k^2T_0}{q^2}$$

or for electrons

$$\frac{K}{\sigma T_0} = \frac{3k^2}{e^2} = 5.5 \times 10^{-9} \text{ cal/sec mho deg}^2 \tag{11-46}$$

Hence $K/\sigma T$ should be a universal constant, true for all metals at all temperatures (if the great majority of heat is carried by the valence electrons). This law, known as the law of *Wiedemann-Franz*, is obeyed to about a factor of 2, as the data from Tables 11-3 and 11-4 show.

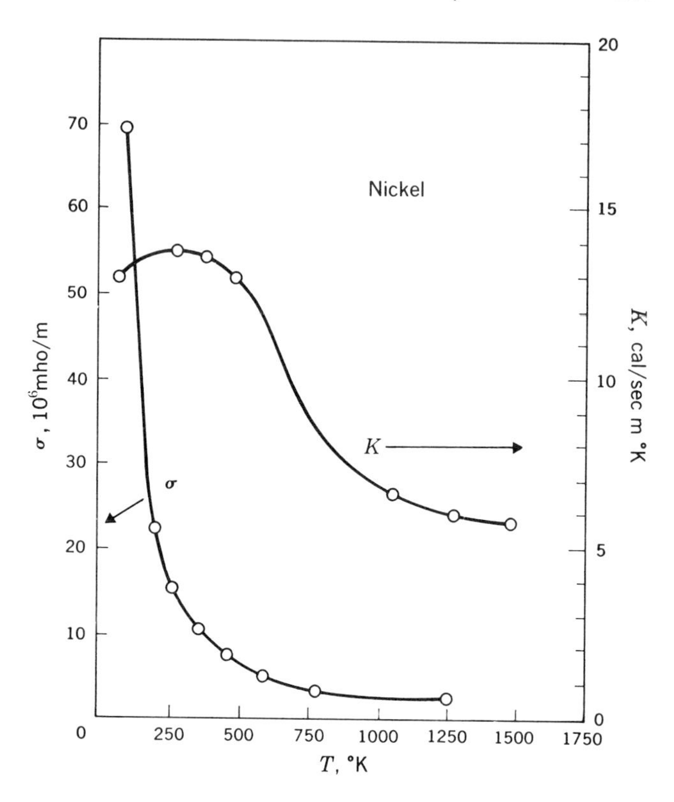

Figure 11-12 Thermal K and electrical σ conductivities of Ni as a function of temperature. Whereas σ changes by a factor 35 over the temperature range shown, K changes by a factor less than 3 over the same range.

The constant $3k^2/e^2$ is modified slightly, if a more precise theory is used in the development of Eq. (11-46); the value obtained by such theory is $\pi^2k^2/3e^2$. This latter number, termed the *Lorenz number*, is 5.8×10^{-9} cal/sec mho deg².

Table 11-4 Values of $K/\sigma T$ for several metals at several temperatures†

	T, °K	K	σ, $\times 10^{-8}$	$K/\sigma T$, $\times 10^9$
Ag	113	99.8	190.0	4.65
	291	100.6	62.0	5.55
	373	99.2	48.0	5.55
Fe	291	16.0	10.0	5.5
	373	15.0	6.0	6.7
	473	15.0	4.1	7.7
W	290	48.0	1.8	9.2
	1873	25.0	1.9	7.1
	3073	31.0	1.07	8.5
Na	280	32.0	23.0	5.0
	361	29.0	16.0	5.0

† Some of the values of K are interpolated from published data.

11-10 Superconductivity

The resistivity of many pure metals decreases with decreasing temperature in the manner shown for Cu in Fig. 11-8: the resistivity approaches zero smoothly as T approaches zero. The resistivity of some metals is strikingly different, however. For them, ρ does not go smoothly to zero at low temperature but drops precipitously from a finite value to zero at some temperature and remains at zero to $T = 0$. These latter metals (and alloys and compounds) are called *superconductors.*

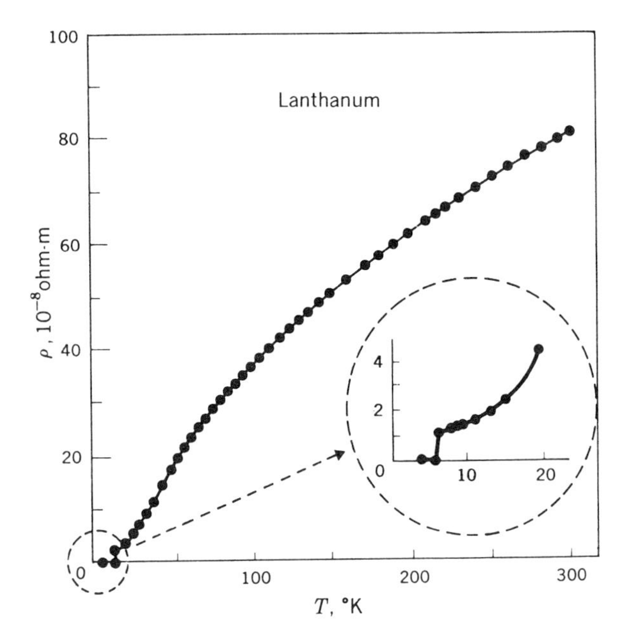

Figure 11-13 ρ versus T for La. The full curve shows the resistivity from room temperature to low temperatures, and the inset shows the region around the critical temperature of 5.8°K. [*After J. Alstad, R. Colvin, S. Legvold, and F. Spedding, Phys. Rev.*, **121**: 1637 (1960).]

The temperature at which the resistivity drops to zero is called the critical temperature T_c. It varies for different metals and alloys from less than 1°K to nearly 20°K (see Tables 11-5 and 11-6). The critical temperature is sharply defined for most superconductors. An example of the temperature variation of resistance is shown in Fig. 11-13 for a wire of La.

Many experiments have been carried out in the years since the discovery of superconductivity in Hg by K. Onnes in 1911. Numerous properties besides resistivity change as a result of the transition from the normal to the superconducting state. The most important property changes are:

1. The electrical resistivity goes to zero at the critical temperature. Experiments on the persistence of current induced in superconducting rings show that the resistivity change at T_c is at least a factor of 10^{17}.

2. The superconductor is a perfect diamagnet. As the metal is cooled below T_c, the magnetic flux is excluded from the material except

Table 11-5 Position of the superconducting elements in the periodic table. The numbers give T_c at zero field (top) and the approximate critical field $\mu_v H_o$ at $T = 0$ (bottom)

H																	He
Li	Be											B	C	N	O	F	Ne
Na	Mg											Al 1.17 0.01	Si	P	S	Cl	Ar
K	Ca	Sc	Ti 0.4 0.0056	V 5.3 0.1	Cr	Mn	Fe	Co	Ni	Cu	Zn 0.88 0.0052	Ga 1.1 0.0060	Ge	As	Se	Br	Kr
Rb	Sr	Y	Zr 0.75 0.0047	Nb 9.3 0.1985	Mo 0.9 0.01	Tc 8.2	Ru 0.47 0.0066	Rh 1.7	Pd	Ag	Cd 0.56 0.003	In 3.37 0.03	Sn 3.7 0.031	Sb	Te	I	Xe
Cs	Ba	Rare earths	Hf 0.37	Ta 4.48 0.083	W 0.01	Re 1.7 0.02	Os 0.71 0.0065	Ir 0.14 0.0019	Pt	Au	Hg 4.15 0.04	Tl 1.37 0.017	Pb 7.2 0.08	Bi	Po	At	Rn
Fr	Ra	Actinide series															

Rare earths	La 5.9 0.16	Ce	Pr	Nd	Pm	Sm	Eu	Gd	Tb	Dy	Ho	Er	Tm	Yb	Lu
Actinide series	Ac	Th 1.4 0.016	Pa	U 1.1	Np	Pu	Am	Cm	Bk	Cf	Es	Fm	Md	?	Lw

Table 11-6 T_c for some of the superconducting compounds†

Compound	T_c, °K
Mo_3Re	9.8
Nb_3Sn	18.2
Pb_2Au	7
NbN	14.7
MoC	~ 8
TaC	9.7
NbC	11.1
Cr_3Ru	3.3
Nb_3Zn	~10.5
MoTc	~14
$NbTc_3$	10.5
V_3Si	~17.0
V_3Ge	6.0

† These are not compounds in the ionic sense—the subscripts simply give the relative composition of the two elements.

for a thin region near the surface of the sample. This is called the *Meissner effect.*

3. Below T_c, superconductivity can be destroyed by application of a sufficiently strong magnetic field. At any temperature the minimum field (called the *critical field* H_c) necessary to destroy superconductivity varies with the material. Data for H_c as a function of temperature for Pb and Sn are shown in Fig. 11-14.

The first successful theory of superconductivity, that of Bardeen, Cooper, and Schrieffer (abbreviated BCS), is far too subtle for detailed description here. A qualitative description of it may be given, however. The theory is built around the fact that a gap of forbidden energy exists *just above the Fermi level* at absolute zero. The width of this gap, E_g, is about 3.5 kT_c at absolute zero. Such an energy gap was used previously in Chap. 10 in describing insulators, but a difference exists in the effect of the gap on the electrical properties of the two materials. For the superconductor, the Fermi distribution as a whole is translated by an electric field from $k = 0$ to a new position in the Brillouin zone. For an insulator, this translation is impossible, because the gap is at the zone boundary and all the states of the zone are filled. For the superconductor, the displacement of the Fermi distribution from $k = 0$ produces the supercurrent, and there is no tendency for it to return to the origin by thermal electron scattering as in a normal metal.

The existence of the gap has been demonstrated by several experiments, one of which is the electronic heat capacity of a superconductor. Experimentally the specific heat below T_c increases approximately

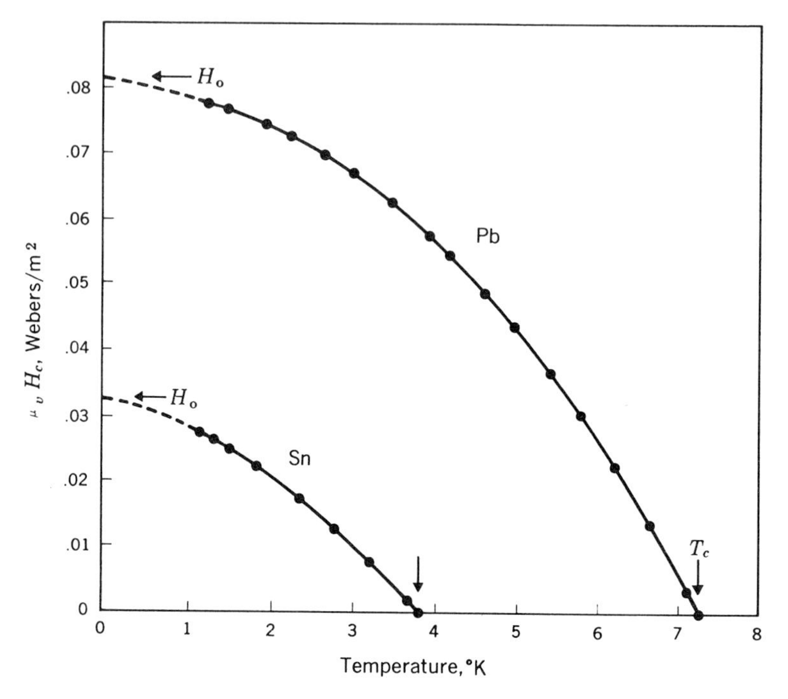

Figure 11-14 H_c versus temperature for Pb and Sn. [*Data from D. Decker, D. Mapother, and R. Shaw, Phys. Rev.*, **112:** 1888 (1958); *D. Shaw, D. Mapother, and D. Hopkins, Phys. Rev.*, **120:** 88 (1960).]

exponentially with a temperature increase. This behavior is interpreted in the following way: Because of the energy gap, the number of electrons excited across the gap is given roughly by a Boltzmann factor $e^{(-E_{gap}/kT)}$; hence the heat capacity varies essentially exponentially with temperature. Such exponential behavior of electronic heat capacity with temperature is in strong contrast to the linear behavior for a normal metal (Fig. 11-1). The energy-gap model is also consistent with the Meissner effect.

The superconducting state exists because of cooperative motion of the valence electrons. At $T = 0$, no electrons exist above the gap, and all electrons are superconducting (i.e., are not scattered by phonons in the same way as the electrons in a normal metal). As the temperature is raised above $T = 0$, more and more electrons become nonsuperconducting, less cooperative motion of the superconducting electrons exists, and the gap becomes narrower. Finally at $T = T_c$ the gap goes to zero, and the metal becomes normal at all higher temperatures. Because of detailed differences between metals, T_c varies from one to another. On a reduced temperature scale T/T_c, however, the superconductors are approximately similar to one another.

Cooperative phenomena of this type are not rare. Crystallographic ordering—the regular positioning of atoms of different types on crystal lattices—is an example. The aligning of electric dipoles in a ferroelectric material to produce a net electric moment for a solid is another example. A third is the alignment of magnetic dipoles to produce a net magnetic moment. The transition from the superconducting state to the nonsuperconducting state (as well as these other three transitions) in the absence of a magnetic field is called a *second-order transition,* since it is characterized by a discontinuity in specific heat at T_c but by no discontinuity in heat content at T_c.

Widespread practical use of superconductors is just beginning. Three prominent uses are rapidly being developed at present, each using one of the three attributes described at the beginning of this section.

The zero-resistance feature of superconductors is being exploited for high-current high-magnetic-field solenoids. Wire solenoids made of certain alloy superconductors have produced (at this writing) magnetic fields larger than 10 webers/m^2. They do so with no power dissipation, though conventional solenoids of Cu or Ag operating at room temperature develop an enormous amount of heat in producing a field of 10 webers/m^2.

The potential use of superconductors in electromagnets was obvious almost from the time of first observation of the phenomenon. Unfortunately, the disturbing influence of a magnetic field on the superconducting state was soon discovered, and the critical fields in which most superconductors become normal metals were found to be disappointly low—no more than 0.1 webers/m^2. About 1960, however, a group of intermetallic alloys based on Nb and V were discovered to have both high values of T_c (above 10°K) and high critical fields (up to 20 webers/m^2). Most development work has been carried out on alloys of composition Nb_3Sn and 3Nb-Zr. Coils of Nb_3Sn have produced fields of 10 webers/m^2, and coils of 3Nb-Zr, fields of at least 6 webers/m^2.

The most crucial feature of solenoid manufacture is production of the conductor itself. The alloy Nb-Zr is reasonably ductile, so wire production is not exceptionally difficult. The major difficulty comes in producing wire of sufficient length for winding a many-turn coil; the winding must obviously be a single continuous length. This difficulty has been overcome to such an extent that commercial production of solenoids is now common. The intermetallic compound Nb_3Sn is extremely brittle after the metal powders have been sintered near 1000°C to produce the alloy; it cannot be drawn into wire in this state. This difficulty was resolved in a laboratory demonstration in an ingenious way: a tube of Nb filled with a mixture of powdered Nb and Sn was drawn into a wire of proper diameter; this wire was wound into a coil, after which it was heat-treated to produce the superconducting Nb_3Sn.

For further information on this topic, the reader may consult the References.

The strong diamagnetism of superconductors offers the possibility of supporting loads by a magnetic field. Frictionless bearings for rotating equipment are thereby possible.

The destruction of superconductivity in a strong magnetic field makes possible a switching device called a *cryotron*. The resistance of a piece of superconducting wire which forms the core of a wire-wound coil can be changed tremendously by a small applied magnetic field. The cryotron can serve as a switch or as a computer memory element.

Enormous advances are being made in all important areas of superconductor knowledge and application. Basic experiments on the nature of the phenomenon coupled with theoretical advances in its explanation are yielding a better understanding of the phenomenon itself. The technological application of superconductivity devices is proceeding rapidly; perhaps the most serious problem to widespread application is that of cooling the devices to temperatures near absolute zero.

REFERENCES

1. M. G. Benz, "Superconducting Materials," in *Handbook of Magnetic Materials*, Reinhold Publishing Corporation, New York, 1968.
2. D. DeKlesk, *The Construction of High Field Electromagnets*, Newport Instruments, Limited, Newport Pagnell, Buckinghamshire, England.

PROBLEMS

11-1 If the lattice heat capacity is constant at 6 cal/g-mole °K for temperatures $\gg \theta_D$, the electronic heat capacity is equal to the lattice heat capacity at two temperatures (in addition to $T = 0$).

(*a*) From the data of Fig. 11-2, calculate what these temperatures are for the 20 per cent V–80 per cent Cr alloy.

(*b*) About what fraction of the total heat capacity at room temperature is contributed by the valence electrons?

11-2 The data given below are the measured molar heat capacities of an alloy 10%V–90%Cr near absolute zero. By plotting C/T as a function of T^2, show that the data fit the equation

$$C = \gamma T + \beta T^3$$

over the temperature range 1.4 to 4°K. Find γ and β. (Data of Cheng, Wei, and Beck.)

T, °K	C, 10^{-4} cal/g-mole °K	T, °K	C, 10^{-4} cal/g-mole °K
1.407	7.29	2.994	16.5
1.450	7.57	3.236	18.0
1.500	7.94	3.489	19.6
1.618	8.57	3.637	20.6
1.824	9.67	3.848	22.0
2.106	11.2	4.073	23.6
2.523	13.6		

11-3 Show that the fraction of electrons, F, within kT of the Fermi level is given by

$$F = \frac{3}{2}\frac{kT}{E_f}$$

if $\rho(E) \sim \mathscr{E}^{\frac{1}{2}}$.

11-4 For Na, $\sigma_{300°K} \approx 2.17 \times 10^7$ mhos/m and $m^*/m \approx 1.2$. Calculate:

(*a*) The relaxation time τ at 300°K.

(*b*) The mean free path between collisions at 300°K.

(*c*) The drift velocity $\bar{v}_x$ in a field of 100 volts/m. About how far does an electron move in a lamp filament 1 m long in a half cycle of 110 volts, 60 cycles alternating current?

11-5 The Hall mobility of Na is $+0.0053$ m^2/volt sec, and the Hall constant R is -2.50×10^{-10} volt m^3/amp weber, both at room temperature. Are these values consistent with the electrical conductivity of Na at room temperature?

11-6 The resistivity of pure Cu is increased by about 1×10^{-8} ohm m with the addition of 1 atomic per cent of a monovalent impurity in solid solution. Assuming random distribution of the impurities and considering the impurities to be spherical scattering centers of effective diameter several angstrom units, estimate λ_I from these data, and find the approximate scattering cross section of these impurities.

11-7 Crystal-lattice vacancies cause an increase in the electrical resistivity of a metal like Cu of about 10^{-8} ohm-m per per cent of vacancies. (*a*) Make a plot of the resistivity increase in Cu due to vacancies as a function of temperature for the equilibrium number of vacancies at temperature. (*b*) Do vacancies in thermal equilibrium add appreciably to the thermal resistivity at any temperature? (See Fig. 11-8 for ρ_T for Cu.) (*c*) Suppose that Cu is quenched from near its melting point (1083°C) to -190°C (liquid-air temperature) so rapidly that all vacancies in equilibrium at 1083°C are retained. Do these *excess* vacancies contribute significantly to the resistivity at -190°C? (Note that E_v for Cu $\approx$ 1 ev.)

11-8 The Hall constant for Al is about -0.3×10^{-10} volt m^3/amp weber. How many electrons per atom of Al take part in conduction of electricity? What does this information imply about the band structure of Al?

11-9 (*a*) Show that the integration of Eq. (11-39) leads to Eq. (11-40).

(*b*) Show also that Eq. (11-43) becomes Eq. (11-44) if the relationships between λ, E_f, and τ are inserted into Eq. (11-43).

11-10 Look up the original data from which Fig. 11-14 was plotted. Show that the critical-field data for Sn and Ta plot on the same curve if they are plotted as H_c/H_0 versus T/T_c.

12

The physical properties of semiconductors

12-1 Introduction

Semiconductors have had an interesting historical development, a development which shows in a classic manner the fundamental way in which the level of general scientific understanding depends on the sophistication of available experimental techniques. At the same time, semiconductor device technology depends in a crucial way on both the understanding of basic physical principles and the careful control of both the chemistry of the materials and the geometry of the devices. Of particular importance is the technique of zone refining for producing high-purity crystals.

Although the use of semiconductors in rectifiers goes back at least as far as the 1920s with the use of both cat-whisker rectifiers and copper oxide power rectifiers, their rectifying characteristics could not be well understood at that time. The operation of these simple devices depended in some unknown way upon the manner in which the junction was made. Later, realization came that the properties of semiconductors could better be understood in the elements germanium and silicon, where crystal purity could be more precisely controlled.

Two key events loosed a flood of discoveries. The basic transistor action was discovered in a rudimentary way by Brattain, Bardeen, and Shockley in the late 1940s. Shortly thereafter, Pfann developed the zone-refining process for the growth of silicon and germanium crystals of predetermined purity, a process which makes possible the technological exploitation of the transistor action.

Our discussion of semiconductors is composed of three stages. In Chap. 10, the general features of these materials were described and compared with those of other types of solids. In this chapter a detailed description of their electronic structure is made. In Chap. 13, a description of a few of the important devices is presented.

A semiconductor was defined in Chap. 10 as an insulator at low temperatures but which possesses moderate conductivity at higher temperatures. Completely filled bands of electrons are separated by forbidden energy gaps from other bands completely empty of electrons (Fig. 12-1). However, the width of the forbidden gap between filled and empty bands for the semiconductor is small enough so that thermal fluctuations create a measurable excitation of electrons across the forbidden gap at reasonable temperatures. Table 10-4 gives the forbidden-gap width for some important crystals.

Many different materials can be classed as semiconductors, but two in particular stand out because of their simplicity, Ge and Si.

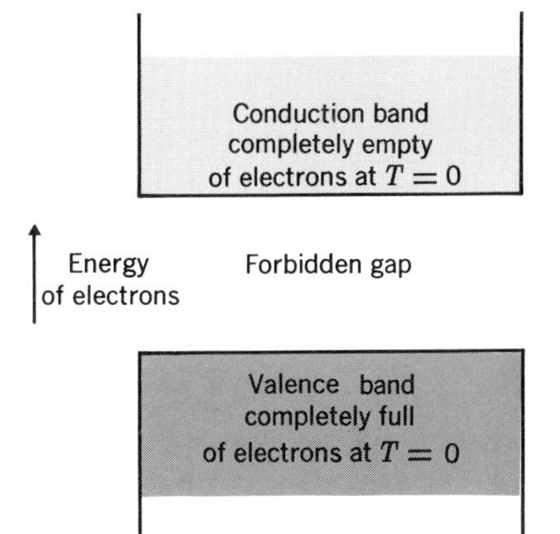

Figure 12-1 Band structure of a pure semiconductor. At absolute zero, the crystal has no current carriers, while, at elevated temperatures, a significant number of electrons can be excited across the forbidden gap.

Because they are technologically so important, they have received extensive study so that they are among the best-understood solids. The presentation of this chapter is in general terms for the most part, but Ge and Si are used throughout as specific examples. Part of the discussion pertains to the pure elements, termed *intrinsic* semiconductors, since their electrical properties are determined by the inherent nature of the elements themselves. Much of the discussion, however, pertains to impure semiconductors, termed *extrinsic*, or *impurity*, semiconductors, since their conductivity properties are controlled by the nature and amount of impurities added.

Some semiconductors are binary chemical compounds. For example, compounds made up of one element from the third and one from the fifth column of the periodic table are called III-V compound semiconductors. The concept of the covalent bond, which has been applied to Si and Ge, is also applicable here, because the average number of electrons per atom is also four. Figure 12-2 shows schematically the bond arrangement for both elemental Si and a typical

III-V compound, indium-antimonide. [This planar sketch of the bonding scheme should not be confused with the actual three-dimensional crystal. The actual crystal structure of Ge and Si is the diamond structure (Fig. 2-8*a*), and the crystal structure of the III-V compounds is the zincblende structure (Fig. 2-8*c*).]

The quasi-chemical picture of Fig. 12-2 is a concrete physical picture adequate for many qualitative features of semiconductor properties. Its simplicity does not permit it to be used for precise quantitative results, however. For example, appreciable forces are known to exist between next nearest neighbors of the elemental semiconductors, a fact not adequately depicted in this model. Furthermore, for the III-V compounds an extra complication occurs. Because the indium atoms of our example must borrow extra electrons

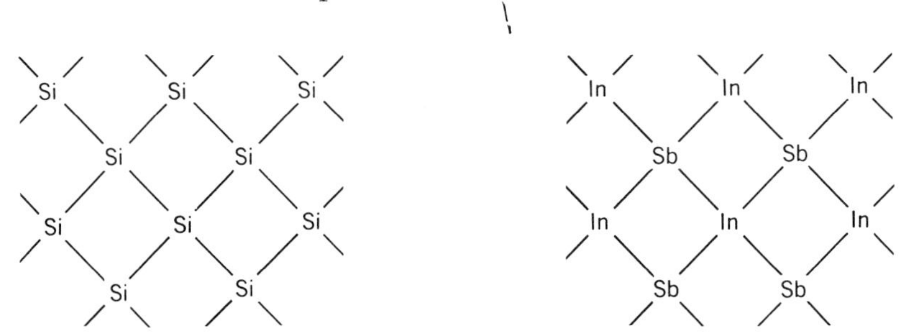

Figure 12-2 A two-dimensional representation of the bond arrangement in an elemental semiconductor and a compound semiconductor. (See Fig. 2-8 for a drawing of actual three-dimensional structures.) The bonds are the shared electron covalent bonds familiar in organic chemistry. Each atom is connected by bonds to its four nearest neighbors. In the compound semiconductor, a similar arrangement occurs. In this case, the In atoms have only three electrons to contribute to the bond arrangement, but the Sb have five, so the average still amounts to four, which just fills all the bonds.

from the antimony atoms, all atoms are ionized to some degree. Hence In-Sb and all III-V compounds should have in part the characteristics of an ionic crystal as well as the characteristics of a semiconductor.

The III-V compounds are not the only compound semiconductors. Others, like the II-VI compounds, have an average of four electrons per atom. Still others exist—copper oxide is an example—which do not follow this rule.

The remainder of this chapter has three divisions. First the major features of the band structure of semiconductors are presented, the discussion being limited to Ge and Si. Then the electronic states of impurity atoms are described; this can be done in a semiquantitative way. Finally, a quantitative description of the Fermi level of a semiconductor is given.

12-2 Band Structure of Si and Ge

A detailed qualitative description of the band structure of Si and Ge is presented in this section partly because of the great practical importance which these materials have. In addition the band structures of these two crystals are much more complicated than those already described in Chap. 9 for the simple metals, and the cases of Ge and Si thus also serve to demonstrate the general types of complexities to be expected in any intensive study of the electronic structure of solids. Although it is clearly beyond the scope of this treatment to go very far into this field, the special peculiarities of the bands of Ge and Si lead to many fascinating effects in the behavior of electrons and their interactions. Actually, a thorough reading of this section is not necessary for the later work of this and subsequent chapters, but it should give the reader a deeper understanding of semiconductors and electron structure of solids in general.

Most of the discussion of the band structure of solids in this book is based on the assumption that each band corresponds to one of the atomic levels and that these bands do not overlap each other. The kinetic energy for the metals has also been assumed to be a quadratic function of momentum

$$\text{KE} = \frac{h^2k^2}{8\pi^2m^*}$$

where m^* is an effective mass. For some metals, most notably Na, these assumptions are well satisfied, while, for numerous others, deviations are rather small. For Si and Ge, each of these assumptions breaks down in important ways, resulting in a band structure which is completely different from that for the simple metals.

The bands of Si and Ge have been stated in Chap. 10 to be a curious mixture of the valence levels of the free atoms. These levels split into two groups of states, forming two bands which do not overlap (Fig. 10-11). Each of these bands contains four electron states per atom. Because these elements (and also carbon) have the chemical valence 4, the lower band, called the *valence* band, is full and the upper band, called the *conduction* band, is empty. But these bands are by no means simple in the sense that the bands of Na are simple. Each of them is in reality an overlapping series of bands.

Detailed band-structure calculations make use of the fact that the crystal lattice often is highly symmetric. If the direction of motion of an electron in a crystal happens to coincide with one of these directions of high crystal symmetry, Schrödinger's equation is much simpler to solve than for a general direction of motion. Information about the band structure for nonsymmetric directions is very scanty and usually requires extensive approximations in solutions. Nevertheless, in any crystal with reasonably high symmetry, many directions exist for which solutions can be found, and it is reasonable to suppose that such a parameter as energy might vary in a reasonably smooth way for directions other than the directions of symmetry. Thus, knowledge obtained for the symmetric directions can be generalized to give qualitative information about the entire band structure of the crystal.

Symmetric families of the directions for the diamond lattice are the ⟨100⟩ and ⟨111⟩ directions, which between them give 14 directions in k space in which the energy might be calculated. Such calculations have been made for both Si and Ge for the [100] and [111] directions, thus giving information for all 14. These bands are plotted in Fig. 12-3 and show the energy of an electron as a function of the k vector in the two directions [111] and [100]. Each line in each diagram represents a separate band of energy levels showing the available states in the same way as Fig. 9-3 for the metals.

Consider first the bands for Si. The lowest line in the Si diagram has been labeled an s level. This line starts out from a minimum of

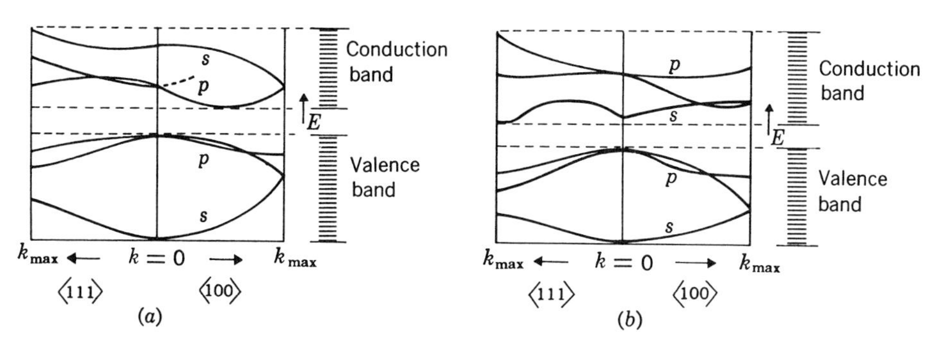

Figure 12-3 (*a*) The energy-band structure of silicon. (*b*) The energy-band structure of germanium. The function $E(k)$ is plotted in each of the two directions [100] and [111]. Curves in the [100] direction are plotted to the right of the vertical energy axis, while curves in the [111] direction are plotted to the left. (The left abscissa is thus not a negative k axis.) Each curve represents a band of energy states and is a plot of $E(k)$ for the direction shown. There are several bands which all belong to the $s + p$ valence levels of the atom. A completely forbidden region of energy is shown between the lower group of bands and the upper. Energy bands in other directions in k space are known only in an approximate sense by interpolation from the above curves.

energy at $k = 0$ and increases for either increasing [100] k vector to the right or increasing [111] k vector to the left. This band has in both sets of directions the qualitative shape postulated in Chap. 9. It has an approximate quadratic behavior near the minimum of energy and a maximum value at the top of the band. The same is also true for Ge. This s band holds *one* electron per atom.

The other levels in the valence band are those labeled p levels. These represent two overlapping bands of p-type levels. In the [100] direction, one of them is approximately an inverted parabola which curves smoothly down to meet the s band at the maximum k value. In the [100] direction the other p band starts off more steeply parabolic but levels off sooner and approaches the maximum k value at a higher energy level than does the other. In the [111] direction both bands approximate an inverted parabola to the maximum k value, though the curvature changes sign between $k = 0$ and k_{max}. Two facts are important about these p bands: (1) The p bands have their maximum energy values at $k = 0$. (2) One of these bands holds two electrons; the other, one.

Consider now the sequence of events as these bands are progressively filled with electrons. First electrons fill the lower s-like band in a conventional way. The band in the [111] direction fills out to its maximum value before the band is filled in the [100] direction. When the s band is entirely filled, electrons can immediately enter the p band, which bends down to meet the s band in the [100] direction. Thus no forbidden gap exists between this series of levels. Though gaps exist in some directions, e.g., in the [111], in other directions the levels overlap or at least meet. The band is therefore overall continuous. This kind of overlapping structure for energy bands is common and here gives rise to one continuous valence band of levels of mixed s and p type. The lower s band holds one electron per atom, the p bands three between them. The entire group holds a total of four. Since these elements have four valence electrons, the bands are just full.

Above the valence band in Si lies a second set of energy bands. Directly above the valence band on the energy axis is a set of p bands. In the [111] direction, there are two bands which cross each other before reaching k_{max}. One is a normal band with positive slope to k_{max}. The second goes through two inflexions and ends at k_{max} in a local minimum. In the [100] direction one of the p bands has a peculiar form. It goes through a minimum between $k = 0$ and k_{max} in the [100] direction. The other p band is shown dotted. Above the p bands, an s band has the shape shown. It lies entirely above the p bands in [111] but meets the p band at k_{max} in [100]. Again, this upper band, called the *conduction band*, is continuous. As before, the s band holds one electron per atom, and the p bands together hold three. Hence the conduction band holds four electrons per atom. These four plus the four from the valence band complete the band of levels corresponding to the original eight s and p levels of the free atom, as was pointed out in Chap. 10.

A new and startling feature of the conduction band of Si is the fact, shown in Fig. 12-3, that the minimum of energy does not lie at $k = 0$. It does not even lie at k_{max} but is at an intermediate point. Since this level lies in a [100] direction, there is not just a single minimum of energy: rather, six equal minima exist along the equivalent $\langle 100 \rangle$ directions. Of course, in neutral pure Si the valence band is filled, and the conduction band is empty. If, through some procedure, additional electrons are added to the conduction band of a Si crystal, they are distributed equally among these six minima.

The levels for Ge (Fig. 12-3*b*) are qualitatively the same as those for Si. The s and p levels form a continuous valence band containing four states per atom. An upper conduction band, also formed of mixed s and p states, is separated from the valence band by a forbidden

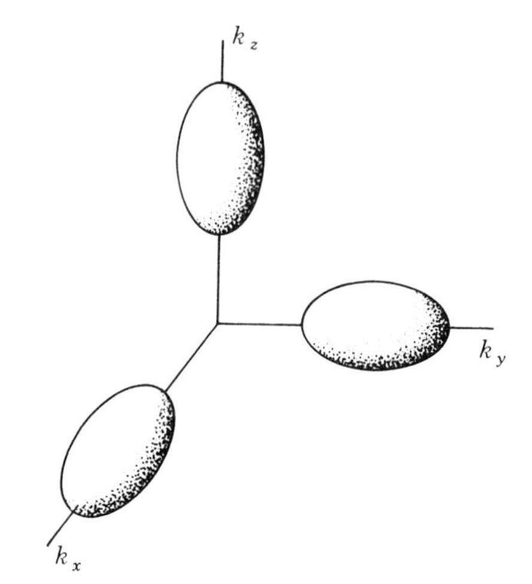

Figure 12-4 Surfaces of equal energy or energy contours in conduction band of Si. In the vicinity of the energy minima, the energy surfaces are ellipsoids of revolution. The electron effective mass is thus different in one direction from another as seen from the minima.

gap. For Ge the minimum in the conduction band occurs for an s band at $k_{\max}$ in the [111] direction. If additional electrons are added to neutral Ge, these electrons are distributed among the equivalent minima at $k_{\max}$ in the $\langle 111 \rangle$ directions.

An additional important feature of the band structure of Ge and Si concerns the effective mass of electrons in the conduction band and of holes in the valence band. Consider first the electrons. In a simple parabolic band (where $E = h^2k^2/8\pi^2m^*$) the energy surfaces in k space are spheres (see Fig. 9-6). The effective mass is simply the constant m^*. For Si and Ge, however, the effective mass is not a simple constant but depends on the direction of motion in the crystal. This effect results from the nonspherical, ellipsoidal shape of the constant-energy surfaces around the energy minima. The energy contours for Si and Ge are sketched in Figs. 12-4 and 12-5, respectively. The equations of an ellipsoid near a minimum can be written in the

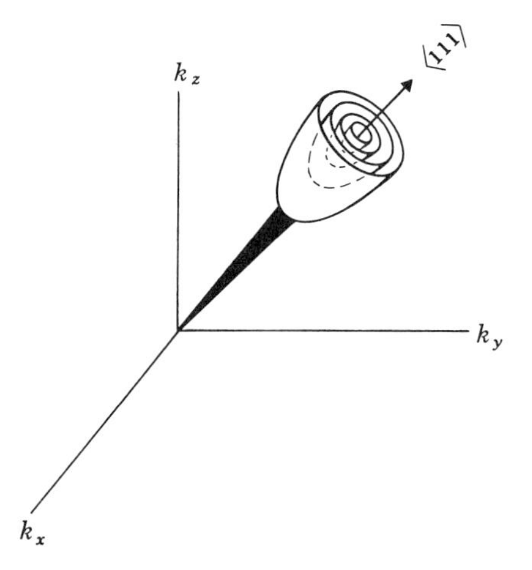

Figure 12-5 Energy contours in the conduction band of Ge. The minima are at $k_{\max}$ in [111] directions. (Only one is shown.)

form

$$E(k) = Ak_1^2 + Bk_2^2 + Ck_3^2 \tag{12-1}$$

where k_1, k_2, and k_3 are measured from the energy minimum in the directions of the principal axis of the ellipsoid. Here A, B, and C are constants. Analogously to the way in which spherical energy surfaces are written in terms of an effective mass, Eq. (12-1) takes the form

$$E(k) = \frac{h^2}{8\pi^2}\left(\frac{k_1^2}{m_1^*} + \frac{k_2^2}{m_2^*} + \frac{k_3^2}{m_3^*}\right) \tag{12-2}$$

Each effective mass corresponds to the inertia of the electron against acceleration away from the minimum in k space in the corresponding direction.

The precise shapes of the energy surfaces near the minima in the conduction bands of Si and Ge have been determined experimentally

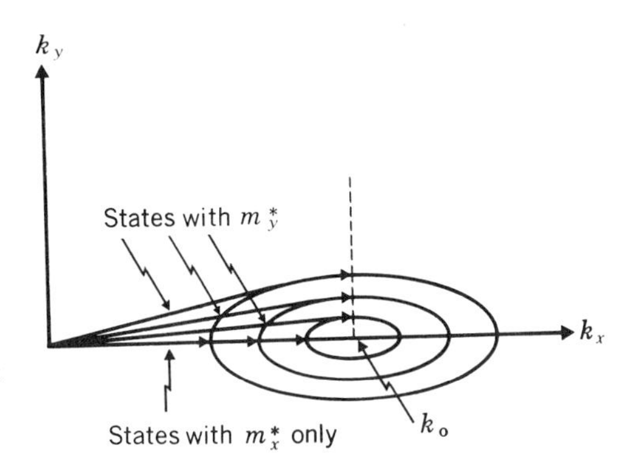

Figure 12-6 Cut in the k_x, k_y plane of energy surfaces for Si centered at a point away from the origin in k space. Electrons accelerated in the principal directions have inertial masses corresponding to the principal curvatures of the energy surfaces in those directions. [See Eq. (12-2).]

$$\frac{1}{m_i} = \frac{4\pi^2}{h^2}\left(\frac{\partial^2 E}{\partial k_i^2}\right)_{k_0}$$

For states not on the principal axes of the ellipsoid, the mass is not a simple number, but the acceleration of the electron is in a different direction from the force acting on it.

by cyclotron resonance (see Sec. 9-10). They have been found to be ellipsoids of revolution for both Si and Ge so that two of the masses in Eq. (12-2) are equal. The two remaining independent masses are: (1) The longitudinal mass, the mass measured with respect to accelerations along the direction from the origin to the energy minimum itself. (2) The transverse mass. The mass measured for accelerations in these directions are described for Si in Fig. 12-6; for Ge they are *along* [111] and *normal* to [111], respectively. The values of the masses are tabulated in Table 12-1.

The top of the valence band in Si and Ge is likewise complex; this complexity leads to directional effective mass for the holes. The maximum of the band occurs at $k = 0$, shown by Fig. 12-3. Resonance experiments show that the two separate p bands are not precisely degenerate at their maximum but that a slight separation in energy exists, with that p state containing one electron having a slightly lower energy. In addition, the upper p band is split into two states, the two being tangent at $k = 0$, but diverging slightly for other k values. Thus, in the vicinity of $k = 0$, three bands exist and are sketched in Fig. 12-7. The differing curvatures of the upper two

Table 12-1 Parameters of the band structure of Ge and Si

	Ge	Si
Electron transverse mass.......	0.082 m	0.19 m
Electron longitudinal mass.....	1.6 m	0.98 m
Mass of holes.................	0.042 m	0.16 m
	0.34 m	0.52 m
Band gap (thermal)†...........	0.75 ev ($T = 0$)	1.21 ev ($T = 0$)
	0.65 ev ($T = 300°K$)	1.08 ev ($T = 300°K$)
Band gap (vertical)†...........	0.803 ev	2.5 ev

† The thermal band gap is the distance between the lowest point of the conduction band and the highest point of the valence band. The vertical gap is the distance from the top of the valence band to the bottom of the conduction band at $k = 0$.

result in two effective masses for holes near $k = 0$. These two measured values are also listed in Table 12-1.

The comments in the preceding paragraphs show that, for detailed work on the nature of electrons and holes in Si and Ge, a single effective mass cannot be assumed consistent with the simple ideas originally suggested in Chap. 9. However, for many purposes, the

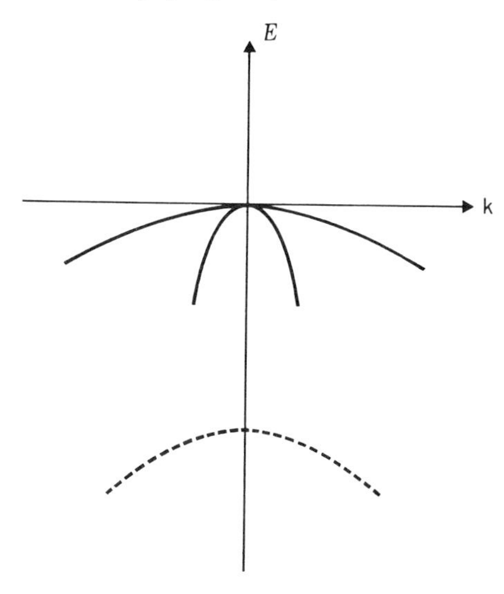

Figure 12-7 Details of the valence band near $k = 0$. The two bands shown in Fig. 12-3 split, leaving the one with two electrons per atom at the top. This band in turn splits into two bands as k increases, giving two different effective masses at the very top of the valence band. To show this effect clearly, the energy scale of this figure is much expanded over that of Fig. 12-3. Again, the mass in the [100] direction is different from that in the [111] direction, illustrated by the fact that the curvature of $E(k)$ is different in the two directions.

actual complex behavior is adequately approximated by a greatly simplified treatment. Fortunately, the two areas of greatest concern, the statistics of electrons, to be considered later in this chapter, and transport of charge, to be discussed in the next chapter, can be treated by using a simplified band picture and a single average effective mass. If more detailed calculations are to be made, however, the full generality of the actual band structure must be taken into account.

The energy gap between the conduction and valence bands of Si and Ge is an important quantity. Two values are important: one is the distance between the maximum of the valence band and the minimum of the conduction band, and the other is the vertical distance between the two bands at $k = 0$. The first is called the *indirect*, or thermal, gap; the second, the *direct*, or vertical, gap. Measured values of these quantities are listed in Table 12-1 also.

The band structure of other semiconductors and covalent crystals having the diamond lattice or the related zincblende lattice show complexities analogous to those of Si and Ge.

12-3 Electronic Impurity States in Semiconductors

In the Introduction, impurities in a semiconductor were stated to be important in determining the electrical conductivity. Their importance is due to the fact that when certain impurities are added to an otherwise perfect semiconductor they add either electrons to the conduction band or holes to the valence band of the crystal. The way in which the number of charge carriers changes both with impurity level and temperature is shown in Fig. 12-8 for Ge doped with B. At the

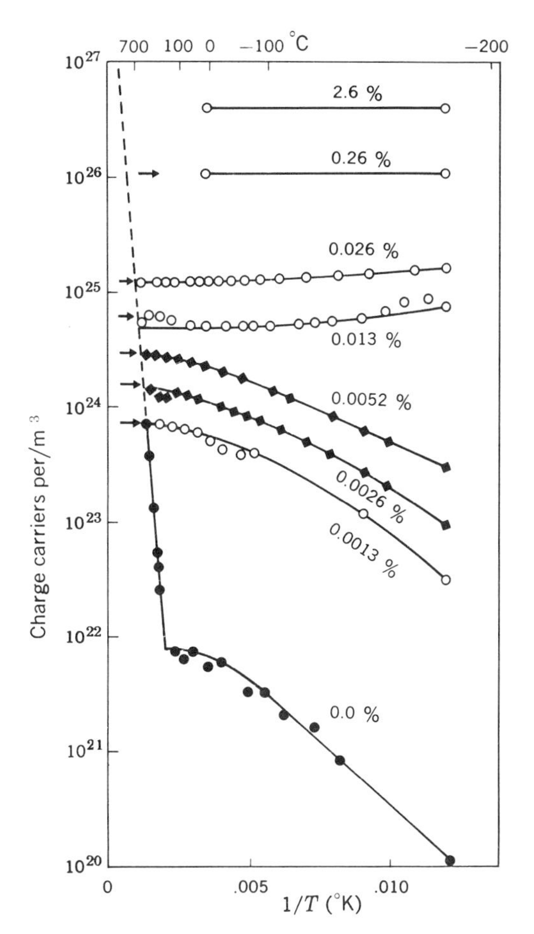

Figure 12-8 The number of charge carriers per cubic meter for B-doped Si. The numbers beside the curves give the B content in atomic per cent. In the low-temperature region the change in carrier density with temperature is a result of the ionization of B, which requires about 0.05 ev. The arrows beside the curves show the carrier density to be expected if all the B is ionized. (This Si has a small concentration of impurity even in the pure state, as the line for 0.0 per cent B shows.) The steep line above 300°C corresponds to intrinsic excitation of electrons. The gap energy measured from this line is about 1.1 ev. [*G. Pearson and J. Bardeen, Phys. Rev.*, **75: 865 (1949).**]

low temperatures impurities are the controlling factor, but at high temperatures all samples have nearly the same carrier density (for these low-impurity contents). At low temperatures the carrier density is controlled by the ionization energy. The impurities are ionized with the electrons going into the conduction band. At some higher temperature the impurities are virtually all ionized. Above this temperature additional electrons are excited across the entire forbidden gap, increasing the carrier density still more, independently of the impurity density.

The most interesting impurities so far as the electrical properties of germanium or silicon are concerned are those which come from

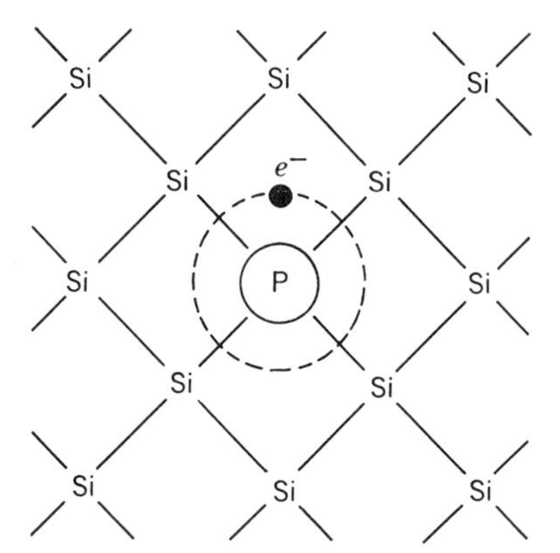

Figure 12-9 Representation of impurity atom for group V in a Si crystal. Four of the electrons enter bonds with the Si neighbors. The fifth enters the conduction band as a free carrier or, alternatively, at low temperatures becomes trapped at the impurity site.

either the third or the fifth column of the periodic table. These impurities have, respectively, one fewer valence electron or one more valence electron than an atom of the parent crystal. Group III or V impurities are found to enter the parent lattice substitutionally when present in small concentration. In Fig. 12-9 a phosphorus atom is depicted in a substitutional site in a silicon crystal. Because phosphorus is an atom from the fifth column of the periodic chart, it has five electrons in its $s + p$ valence shell. Four of these electrons are used in the regular bonding arrangement of the crystal, and one electron is left over, there being one more electron than the bond arrangement can absorb. A phosphorus impurity atom in a silicon crystal attempts to conform as much as possible to the bonding arrangements of the parent structure. The additional electron has no place to go except into the conduction band of the silicon crystal. Thus, it can be understood how a small addition of the right kind of impurity to a semiconductor crystal changes the conductivity of the crystal by a large amount, because additional impurity atoms contribute free carriers to a crystal which is an insulator at the lowest temperatures.

Adding impurities from the third group of the periodic table has a different effect. In this case, the impurity atom has only three electrons in the valence shell. Again, the impurities dissolve substitutionally in the parent lattice. For these group III impurities, however, the bonding structure around the impurity atom cannot be filled out completely for the lack of one electron. Then, as it attempts to conform to the parent structure, the impurity atom might borrow a bonding electron from a neighboring silicon atom. From the point of view of the band picture, the impurity atom contributes one less than the normal number of electrons to the valence band, the missing electron acting as a hole (Fig. 12-10). As many holes exist in the top of the valence band as there are impurities in the lattice. Furthermore, holes in the top of the valence band are efficient for carrying electrical current, just as are electrons in the conduction band.

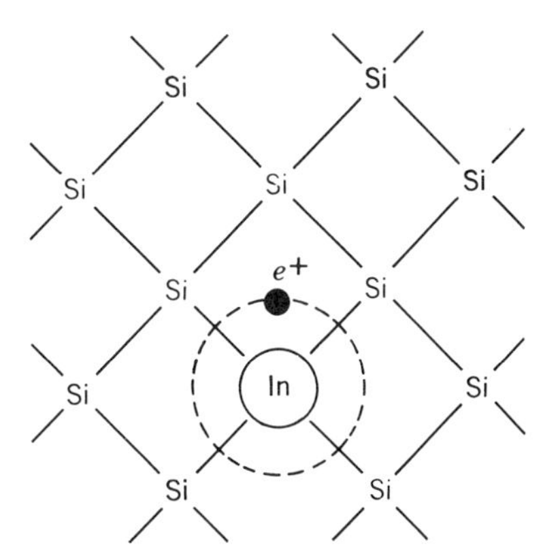

Figure 12-10 Indium atom from group III, substituting for a silicon atom in the silicon lattice.

Actually, a free carrier (electron or hole) does not result from each impurity atom at the lowest temperatures. As is shown in Fig. 12-8, at sufficiently low temperatures slightly impure crystals are seen to have very low charge-carrier densities. If the electrons from the impurity atoms were really free carriers in the crystal, a semiconductor should, of course, behave similarly to a metal, where the number of carriers is independent of temperature. This behavior is not observed for impurity semiconductors; hence the carriers induced by impurities cannot be free at all temperatures.

Impurity action is most easily understood by considering the charge neutrality in the vicinity of an impurity atom in a semiconductor crystal. When phosphorus or another group V atom donates an electron to the conduction band, the electron leaves the vicinity of the impurity atom and travels through the crystal. Thus, the impurity atom becomes singly ionized and possesses a charge of $+e$. When the electron is several atom spacings away from the impurity, it experiences a Coulomb force directed toward the position of the impurity atom. The arrangement of charges is analogous to that in the hydrogen atom. The only difference, of course, is that the electron sees the impurity ion through a medium composed of silicon atoms. The semiconductor crystal provides a dielectric medium for the interaction between the electron and the impurity atom. Therefore, a tempting approximation is to describe the impurity atom with its electron in a semiconductor lattice as a hydrogenlike atom in a dielectric, the capacitivity being that appropriate to the particular semiconductor crystal.

The conduction electron is bound to the impurity site by the potential

$$V = \frac{-e^2}{4\pi\epsilon r} \tag{12-3}$$

The factor ϵ is the effective dielectric capacitivity of the medium. In Prob. 8-7, the energy levels of a hydrogen atom in a dielectric medium and the Bohr radius of the lowest energy level were shown to be given, respectively, by

$$E = \frac{e^4 m^*}{8\epsilon^2 h^2 n^2} = \frac{R_\infty}{n^2}\frac{m^*}{\epsilon_r{}^2 m}$$

$$r_0^* = \frac{h^2\epsilon}{m^* e^2} = a_0 \frac{\epsilon_r}{m^*/m} \tag{12-4}$$

where R_∞ = Rydberg constant
n = principal quantum number
r_0^* = Bohr radius in the medium
a_0 = vacuum Bohr radius
ϵ_r = relative capacitivity.

The values of E and r_0^* calculated using Eq. (12-4) depend on the values of ϵ_r and m^*. Because m^* is not a scalar, Eq. (12-4) represents a simplification for a single mass value. Numerical calculations of the ground state have been performed which take account of the anisotropic mass for both Si and Ge and which yield

$$E_{\text{Ge}} = -0.0092 \text{ ev} \qquad \epsilon_r = 16$$

$$E_{\text{Si}} = -0.029 \text{ ev} \qquad \epsilon_r = 12$$

These values of binding energy correspond to an average scalar effective mass with which one obtains correct results in Eq. (12-4), which is

$$m_{\text{Ge}}^*/m = 0.17$$

$$m_{\text{Si}}^*/m = 0.31$$

Note that these values are some weighted average of the longitudinal and transverse masses given in Table 12-1. With these values of effective mass and the appropriate values ϵ_r, one obtains for the effective radius of the wave function in Eq. (12-4)

$$r_{\text{Ge}}^* = 54 \text{ A}$$

$$r_{\text{Si}}^* = 23 \text{ A}$$

These values are a reasonable average to the size of the actual ellipsoidal-shaped wave functions one obtains from a more accurate calculation using the anisotropic effective masses.

These results predict that the properties of the bound state in the impurity atom are not sensitive to the properties of the impurity atom except for its valence. (We have assumed that the impurity is from group V.) The observed impurity states do show a dependence on the particular impurity, as Table 12-2 shows. The average value

Table 12-2 Parameters of impurity semiconductors

Material	Impurities		Experimental binding energy, ev
Si	Li	donors group V	0.033
$\epsilon_r = 12$	P		0.044
	As		0.049
	Sb		0.039
	Bi		0.069
	B	acceptors group III	0.045
	Al		0.057
	Ga		0.065
	In		0.16
Ge	P	donors group V	0.0120
$\epsilon_r = 16$	As		0.0127
	Sb		0.0096
	B	acceptors group III	0.0104
	Al		0.0102
	Ga		0.0108
	In		0.0112

of the group V bound state in the table compares favorably with the calculated values above for Ge, but not as well for Si. These deviations show the extent to which the continuum approximation is valid. They show, for example, that one cannot neglect the nature of a particular impurity ion core, especially in Si, when seeking quantitative information. On the other hand, the continuum approach is seen to be quite good, for the reason that the orbit of electron about the impurity includes about 10^3 parent-lattice atoms. In other words, the de Broglie wavelength of the electron encompasses many lattice atoms, and so the discrete nature of the lattice is of small importance in determining the motion of the electron.

Holes on impurities from group III are treated in an analogous manner, except that there are two types of holes at the top of the band, and an accurate treatment is more complex. On the simplified model if one uses the same value for ϵ_r and takes an average value for the binding energy from the table, he finds similar effective masses and radii of wave function for the holes. This situation corresponds to a hydrogen atom in reverse, with a central negative charge and a positively charged "electron."

The physical model for a hole bound to a group III impurity is illustrated in Fig. 12-10 for an In atom (an acceptor) in a silicon lattice. The In atom has initially only three bonds to connect to its neighboring silicon atoms, but the empty bond has a tendency to migrate away from the In atom. Thus, the In atom fills out its bonding structure with an extra electron and becomes charged negatively. The extra electron on the In atom is borrowed from one of the silicon atoms of the crystal, and that silicon atom has one too few electrons, corresponding to a hole in the valence band at that point.

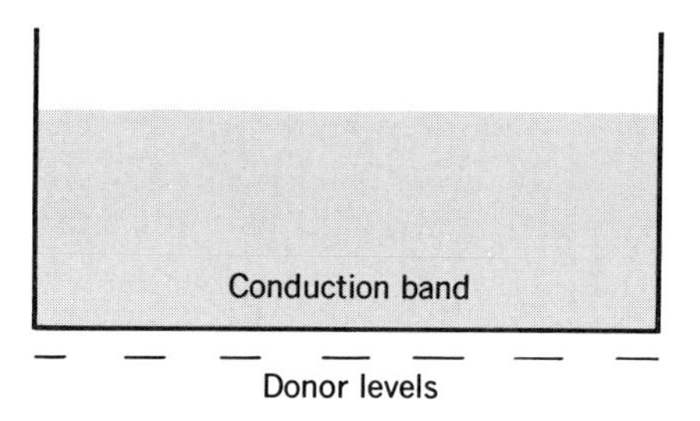

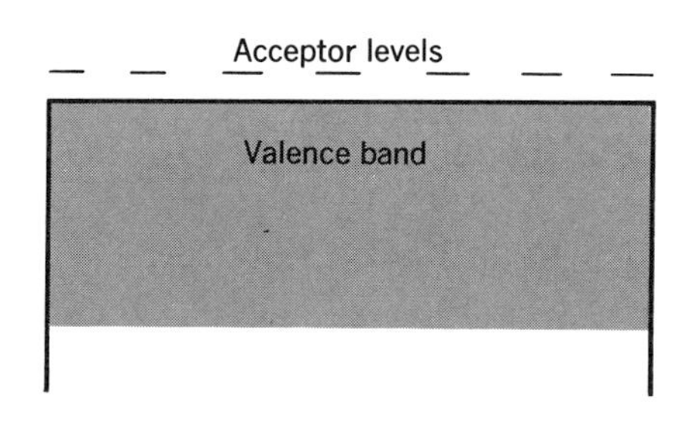

Figure 12-11 The forbidden band gaps of a semiconductor containing both *n*- and *p*-type centers. Excitation of electrons is upward in the figure, while excitation of holes is downward. Just as in the hydrogen-atom analogy, there are excited states for electrons and holes on the impurities, but for simplicity these are not shown.

The main feature of the temperature behavior of the conductivity of impurity semiconductors is now easy to understand. At 0°K, every impurity center captures either a hole or an electron in a bound state so that its charge becomes neutralized. Because these holes and electrons are not free to migrate through the valence or conduction bands, they cannot contribute to the conductivity of the crystal. As the temperature is raised, however, the impurity centers are easily ionized at temperatures well below room temperature. The carrier density has the behavior shown in Fig. 12-8.

The ground-state energy levels of the impurity semiconductors are summarized in Fig. 12-11. This figure shows the band gap of the intrinsic material with localized bound states within the forbidden-gap region corresponding to the bound holes or electrons. An electron bound to a group V impurity corresponds to a localized level in the forbidden energy region for the reason that, when the electron is freed from its impurity center, the binding energy must be supplied to free the electron to the bottom of the conduction band. Group V impurity centers are called *donors*, or *n-type impurities,* for the reason that they supply negative current carriers to the conduction band. Group III impurities are shown as localized states just above the valence band. For these impurities, electrons can be excited by a small excitation from the full valence band into the bound states, leaving behind a hole in the valence band. Alternatively, one can say that the energy of excitation of the hole is downward, opposite to the energy of excitation of the electron. Group III centers are called *acceptor centers,* or *p-type impurities,* because they take electrons from the valence band and leave positive current carriers.

One last point which must be emphasized concerns the relative size of the energies of Fig. 12-11, The forbidden gap is of the order of 1 ev, while the impurity centers have a binding energy of about 0.01 ev, which is about 1 per cent of the intrinsic band gap.

12-4 The Fermi Energy in Semiconductors

It can be inferred from the discussion so far and should certainly become clear from the next chapter, dealing with devices, that an important practical property of the semiconductor is its electrical conductivity. The electrical conductivity is primarily controlled by the number of carriers which are excited into either a valence or a conduction band by thermal fluctuations. Of course, just as for metals, the conductivity of semiconductors depends upon the mobility of the carriers as well as upon their density. Indeed, in the intermediate range (for example, $1/T \approx 0.002$ in Fig. 12-8), where the impurities are all ionized and fluctuations are not sufficiently energetic to create many electron-hole pairs, the temperature variation of the conductivity reflects the temperature variation of the mobility. At temperatures both higher and lower in the figure, the temperature variation of carrier concentration is so great that the temperature variation of mobility is unimportant. Carrier mobility in semiconductors is reconsidered at a later stage, but at this point the more important carrier-density problem is of primary concern.

The simplest case is that of the intrinsic semiconductor with no impurities present. For such material the only way in which carriers can be present in either the conduction or the valence band is to have a thermal fluctuation of sufficient size to bring an electron completely across the forbidden energy gap. For germanium and silicon, of the order 1 ev is necessary. It is important to remember in this regard that, for each electron excited in this way into the conduction band, a mobile hole is left in the valence band. Thus, electron excitation across the band gap creates a pair of current carriers. The problem is to compute the actual average densities of electrons and holes which the thermal fluctuations produce at a given temperature.

Computation of the carrier density requires use of the Fermi function, which was defined in Sec. 11-2. We use the form given in Eq. (11-7),

$$w_{-}(E) = e^{-(E-E_f)/kT} \tag{12-5}$$

We again adopt the view, explained in Chap. 11, that the Boltzmann factor can be used for computing electron excitation if energy is measured relative to the Fermi level, whose value we do not yet know for semiconductors. The factor $(E - E_f)$ must be present in Eq. (12-5), because if the energy E appeared only by itself in the formula, a simple change in a base line from which the energy is measured, which is always possible in a mechanical system, would have the apparent effect of changing the physical probability of occupancy of the state. Of course, simply changing the zero from which energy is measured cannot change such a physical quantity.

The density of electrons in the conduction band may be computed with the use of Eq. (12-5). Because $w_{-}(E)$ is the probability of occupancy by an electron of a given state with energy E, the total number of electrons in a given energy interval dE is the product of the probability function $w_{-}(E)$ and the number of states, $\rho(E)\,dE$, in the energy interval dE. If $n_{-}(E)\,dE$ is the number of electrons per cubic meter in the energy interval dE, then

$$n_{-}(E)\,dE = \rho(E)w_{-}(E)\,dE = \frac{4\pi}{h^3}(2m_e^*)^{\frac{3}{2}}(E - E_g)^{\frac{1}{2}}e^{-(E-E_f)/kT}\,dE \tag{12-6}$$

The factor $\rho(E)$ is assumed in this section to be the same as the simple density of states defined for metals in Eq.(9-30), and the factor $w_{-}(E)$ is given by Eq. (12-5). In Eq. (12-6) the origin of the energy scale is taken at the top of the valence band, and the bottom of the conduction band is at the point E_g (Fig. 12-12).

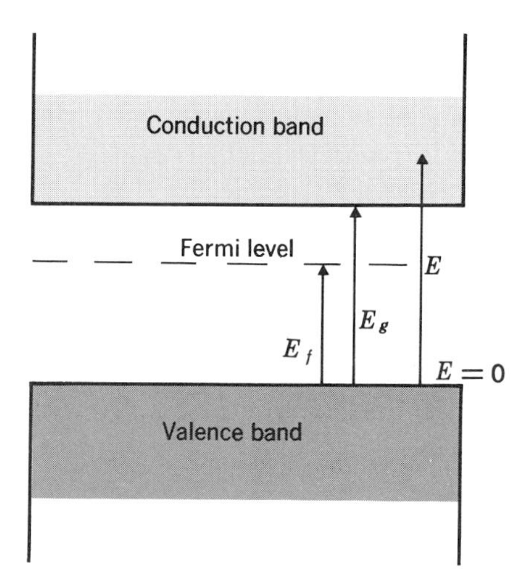

Figure 12-12 The energies of Eq. (12-7) are measured with respect to the top of the valence band as shown. (The Fermi level E_f is not necessarily at the donor energy.)

The total number of electrons in the conduction band, n, is obtained by integrating Eq. (12-6) over the entire band. The limits of integration are thus from E_g to E_{max}, which can be replaced by ∞ because of the exponential nature of the integrand.

$$n = \int_{E_g}^{\infty} \rho'(E) w_-(E)\, dE = \frac{2(2\pi m_e^* kT)^{\frac{3}{2}}}{h^3} e^{-(E_g - E_f)/kT} \tag{12-7}$$

Hence

$$n = 4.82 \times 10^{21} \left(\frac{m_e^*}{m}\right)^{\frac{3}{2}} T^{\frac{3}{2}} e^{-(E_g - E_f)/kT} \tag{12-7a}$$

In this equation [and in Eqs. (12-6) and (12-7)] the effective mass of the electron is designated m_e^* to distinguish it from the effective mass of the hole, m_h^*.

The functions $\rho'(E)$ and $w_-(E)$ are sketched in Fig. 12-13*a* and *b*. Their product $n_-(E)$ is sketched in 12-13*c*; the shaded area under the curve gives the value of n.

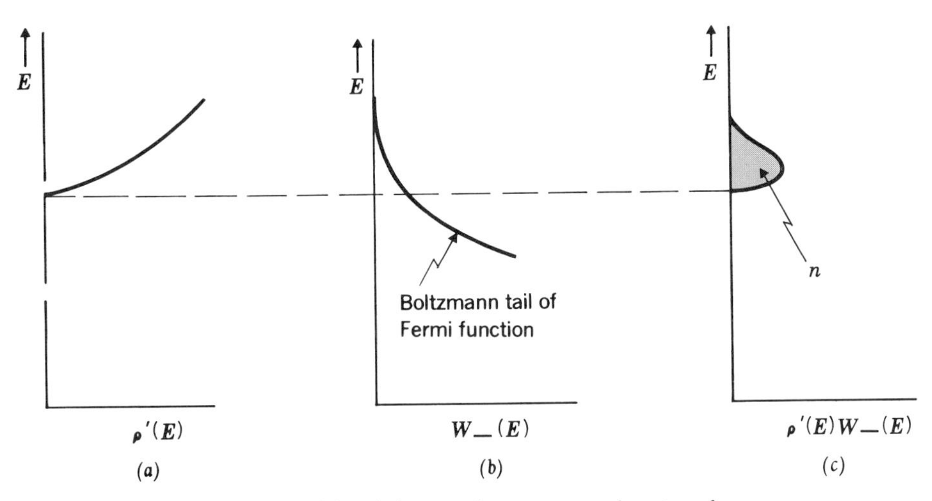

Figure 12-13 The functions $\rho'(E)$, $w_-(E)$ and their product curve as a function of energy at temperature T. The total number of electrons in the conduction band n is given by the shaded area.

The number of holes in the valence band for intrinsic material must, of course, be exactly equal to the number of electrons in the conduction band, for the reason that, whenever an electron is created in the conduction band by a thermal-excitation process, a hole must be left behind in the valence band.

The number of holes in the valence band is also given by an exponential factor, where the excitation energy is measured *down* from the Fermi level (Fig. 12-14). For holes

$$w_+(E) = e^{-(E_f - E)/kT} \tag{12-8}$$

[see Eq. (11-8)]. This equation assumes that E for holes is still measured from the top of the valence band, and it is therefore negative. We inject a word of warning at this point: care must be taken with the

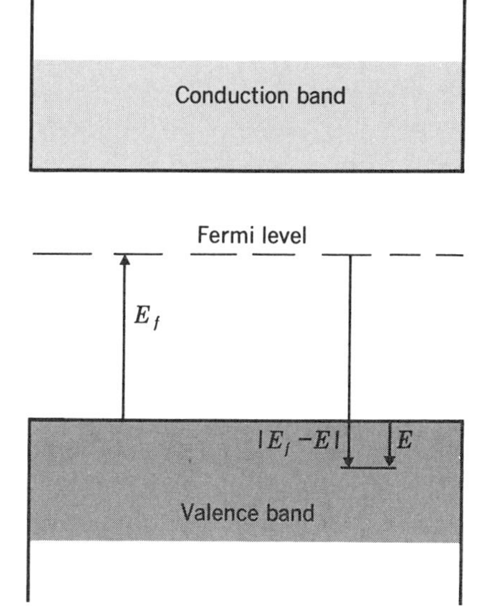

Figure 12-14 Excitation energies for holes.

sign of energy to get the correct type of Boltzmann factor for the creation of a hole. E_f has the same meaning as before: it is the reference level from which excitations of electrons and holes occur. We assume that E_f for holes is the same as for electrons.

The equation for holes analogous to Eq. (12-6) now becomes

$$n_+(E)\,dE = \frac{4\pi}{h^3}\,(2m_h^*)^{\frac{3}{2}}\,|E|^{\frac{1}{2}}\,e^{-(E_f-E)/kT}\,d\,|E| \tag{12-9}$$

Integration of this equation follows as before, and the result is that the total number of holes in the valence band, n_p, is

$$p = 4.82 \times 10^{21}\left(\frac{m_h^*}{m}\,T\right)^{\frac{3}{2}} e^{-E_f/kT} \tag{12-10}$$

Because the total number of excited holes in the valence band must be equal to the number of electrons in the conduction band, Eq. (12-10) must equal Eq. (12-8). For this to be true,

$$E_f = \frac{E_g}{2} \tag{12-11}$$

In writing Eq. (12-11), a term $3kT \ln (m_h^*/m_e^*)/4$ has been neglected. This term is always much smaller than $E_g/2$ for normal temperatures.

Equation (12-11) has a very fundamental meaning for semiconductors. It defines the generalization for semiconductors of the Fermi energy, which has already been defined for metals. The Fermi energy is that energy in a Boltzmann distribution from which electron and hole excitation seem to occur. According to Eq. (12-11), the Fermi energy in a pure semiconductor is at the midpoint of the gap of forbidden energies between the valence and conduction bands.

From a physical point of view, one can readily see why the Fermi level should be where it is for an intrinsic semiconductor. Suppose that an electron is taken from the top of the valence band and placed in the bottom of the conduction band. The energy E_g is required. In an intrinsic material a hole is created in the valence band when an electron is created in the conduction band and the energy is divided equally between the two. Since half the gap energy goes into the creation of the electron and half to the hole, the reference energy for each should be the midpoint of the gap.

12-5 The Fermi Level in Impurity Semiconductors

The point of view developed in the last section is now applied to the case of a semiconductor which contains impurity states within the forbidden energy gap. The main idea is that, when a group of donor states exists at a small distance below the edge of the conduction band, these donor levels can contribute electrons to the conduction band with an energy of excitation which is much less than the energy E_g. (The same statement also holds with respect to holes for acceptor states just above the valence band.) Because the Fermi level represents that energy from which electrons are excited on the average into the conduction band, one would expect the Fermi level to be near the donor levels in a crystal which contains an appreciable number of donor states. The word *average* has been used advisedly, because fluctuations can excite electrons completely across the energy band gap as well as electrons out of the donor impurity levels. This competition is not necessarily entirely in favor of the impurity states, because the fraction of the total from each depends upon the temperature. Near absolute zero, fluctuations capable of exciting electrons completely across the band gap are rare relative to those capable of exciting electrons out of the donor states. Hence, at low temperatures the Fermi level is nearly coincident with the donor states. As the temperature is raised, more and more electrons from the donor states become ionized, but, in addition, more fluctuations are large enough to excite electrons across the forbidden gap from the valence band. Because the number of donor impurities in a semiconductor crystal is small compared with the total number of atoms in the crystal, the small probability of a fluctuation across the forbidden gap multiplied by the number of parent-lattice atoms can become appreciable compared with the probability of the fluctuation at an impurity atom adequate to ionize the impurity times the number of impurity atoms. Thus, at higher temperatures, the Fermi level tends to decrease from the donor levels toward a limit at the middle of the band characteristic of the intrinsic material. The Fermi level is therefore a function of both the temperature and the density of impurity states in an impurity semiconductor.

To make these ideas quantitative, assume that the energies conform to the scheme shown in Fig. 12-15. As before, the probability with which a given quantum state in the conduction band is occupied is given by the Boltzmann formula

$$w_{-}(E) = e^{-(E-E_f)/kT} \tag{12-12}$$

The density of electrons in the conduction band is again given by the product of the probability of occupation of a state multiplied by the density of states and integrated over the entire conduction band. The result is the same as Eq. (12-7).

$$n = N_c e^{-(E_g - E_f)/kT} \tag{12-13}$$

where

$$N_c = 4.82 \times 10^{21} \left(\frac{m_e^* T}{m}\right)^{\frac{3}{2}}$$

Of course, in this expression, the Fermi energy is not yet determined. The number of holes is given also by expression (12-10),

$$p = N_v e^{-E_f/kT} \tag{12-14}$$

where

$$N_v = 4.82 \times 10^{21} \left(\frac{m_h^* T}{m}\right)^{\frac{3}{2}}$$

In addition to these two expressions, however, the solid has additional states for electrons in the donor levels. The density of electrons in the donor levels, n_d (measured per cubic meter), is the probability function [Eq. (12-14)], multiplied by the total number of donor states, which is taken to be N_d per cubic meter.

$$n_d = N_d e^{-(E_d - E_f)/kT} \tag{12-15}$$

Equations (12-13) to (12-15) still do not give a relationship from which the Fermi energy can be obtained. The necessary relation is obtained, however, from the requirement that the crystal be electrically neutral, i.e., that the total number of electrons in the conduction band must be equal to the total number of holes in the valence band *plus the total number of ionized impurity centers*.

$$n = p + (N_d - n_d) \tag{12-16}$$

Substitution of Eqs. (12-13) and (12-15) into Eq. (12-16) yields

$$N_c e^{-(E_g - E_f)/kT} - N_v e^{-E_f/kT} = N_d(1 - e^{-(E_d - E_f)/kT}) \tag{12-17}$$

This equation is a rather complicated transcendental equation for the Fermi energy E_f. By substitution, one can see that, at absolute zero temperature, the Fermi level lies at the donor level, provided that the donor level is near the conduction band. As the temperature increases to very large values, then N_c and N_v are larger than N_d and the term

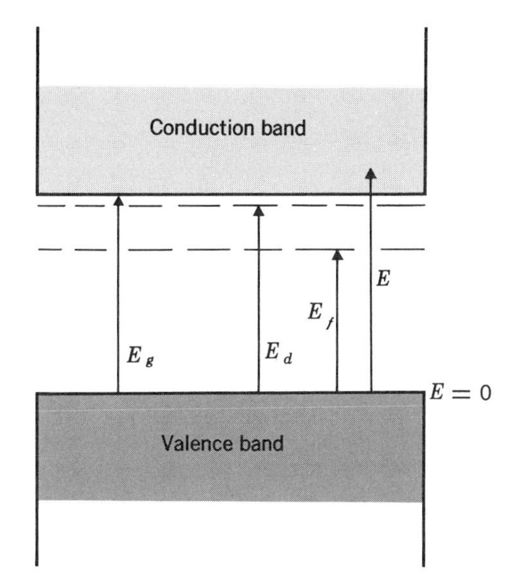

Figure 12-15 The energy scheme for computing the Fermi level in an impurity semiconductor. It is assumed that the semiconductor has impurity levels of only donor type.

on the right-hand side of the equation becomes small compared with the difference of terms on the left-hand side. Under these conditions the Fermi energy becomes the same as that of the intrinsic material. Note, in these formulas, that N_c and N_v are an effective number of states at the edge of the conduction and valence bands for excitation. E_f versus T is sketched in Fig. 12-16.

The variation of the Fermi level when acceptor levels are present (at energy E_a) is shown in Fig. 12-17. The behavior is just the opposite to that for donors.

Notice must be taken of the inaccuracies of these calculations at low temperatures. There the approximations of probability of occupancy of a state by a Boltzmann factor are not strictly true, and Eqs. (12-13) through (12-17) are in error. For n-type material E_f does not

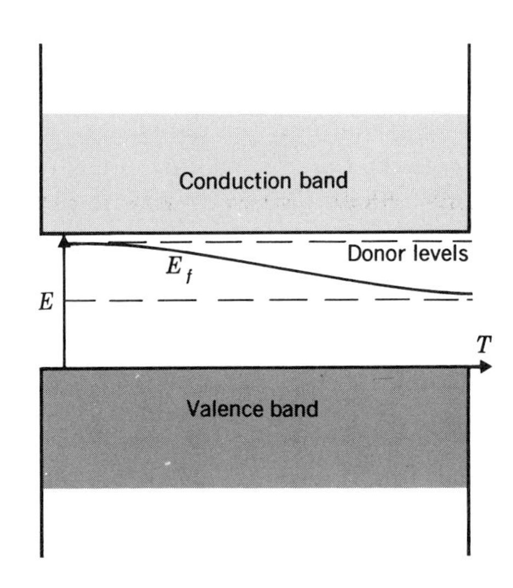

Figure 12-16 Variation of Fermi energy with temperature. As the temperature increases, the Fermi decreases from the donor band to the midpoint of the energy gap.

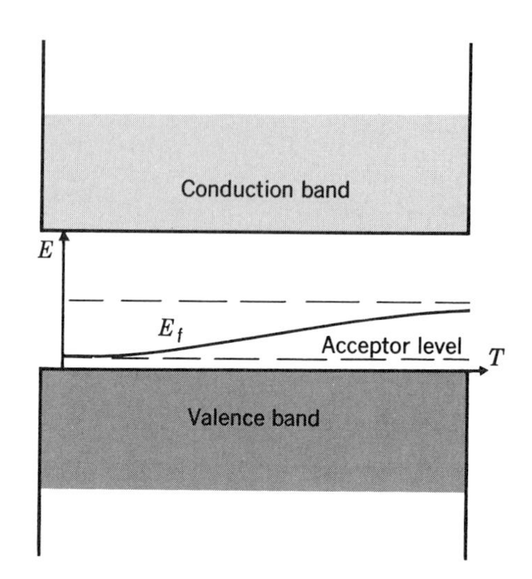

Figure 12-17 The variation of the Fermi level with temperature for acceptors. As the temperature increases, the Fermi level increases from the acceptor level to the midpoint of the gap.

approach E_d as T approaches zero: it actually approaches $(E_g + E_d)/2$. For p-type material it actually approaches $E_a/2$ as T approaches zero. For all moderately doped materials at temperatures above 100°K, however, the error is so slight that it is not worth worrying about, as Fig. 12-18 shows. This error is introduced because the approximation $|E - E_f| \gg kT$ is not valid (see Sec. 11-2).

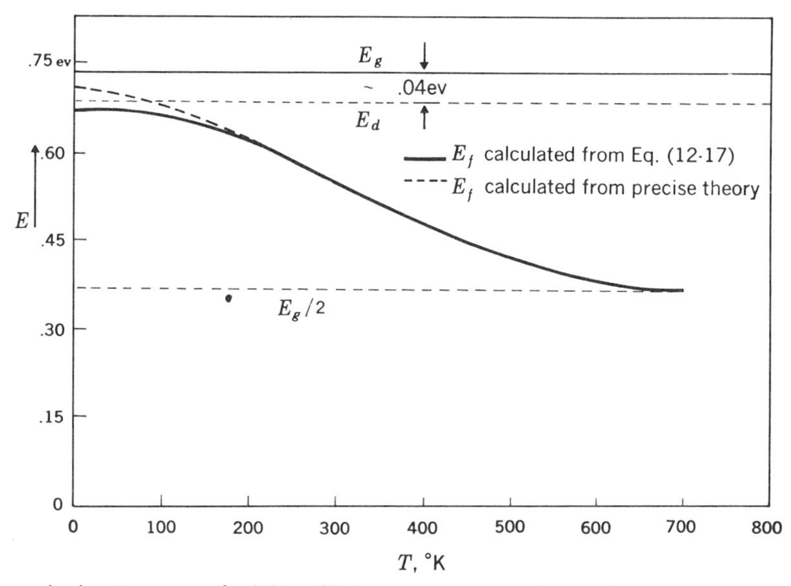

Figure 12-18 Example showing range of validity of Boltzmann approximation to the Fermi function. This is for Ge, assumed to maintain a constant thermal gap of 0.74 ev, doped with a donor (with assumed ionization energy of 0.04 ev) to a level of 10^{22} atoms per cubic meter. E_f calculated by using Eq. (12-17) deviates from E_f calculated by using the precise expressions appreciably only below 100°K.

12-6 Contact-potential Difference

In this final section we consider the problem of what happens to the Fermi level when two solids having initially different Fermi levels are brought together. The solution of this problem is important for the discussion of semiconductor devices. It is, however, a rather general problem, because it also arises whenever one touches two metals of a different kind together, and is especially important when two or more contacts are made at varying temperatures. Consider first two solids with different band structures completely isolated from one another. These solids are assumed for the moment to be two metals; however, exactly the same treatment is valid for semiconductors or for a metal with a semiconductor, etc. A diagram showing the relative Fermi levels of the two solids is given in Fig. 12-19.

Let the two solids be brought into contact with one another. Upon contact, electrons can flow freely from one solid to the other, and electron flow does actually occur until the Fermi levels of the two solids become equal. The Fermi levels must equalize, because some of the electrons near the Fermi energy in solid II are at higher energy states than some states in solid I which are above the Fermi energy of that substance and which are unoccupied. Therefore, electron flow occurs from solid II into solid I. When the Fermi levels of the two solids equalize, no further electron flow occurs. (From the point of view of thermodynamics, the following statements apply: The Fermi level represents the chemical potential of an electron in a solid. When the two solids come into contact, the chemical potential of the mobile electrons, and hence the Fermi level, of the two solids must be the same.)

Figure 12-20 shows the new situation after the two solids have equalized their Fermi levels. The surface of solid I is now at a different electrical potential from the surface of solid II so that the whole band structure has been shifted until the Fermi level is equalized. The potential difference between the two solids which is caused by the electron flow is never very large, usually being of the order of 1 volt or less. This difference in potential is termed the *contact*-potential difference, which we shall call V_c.

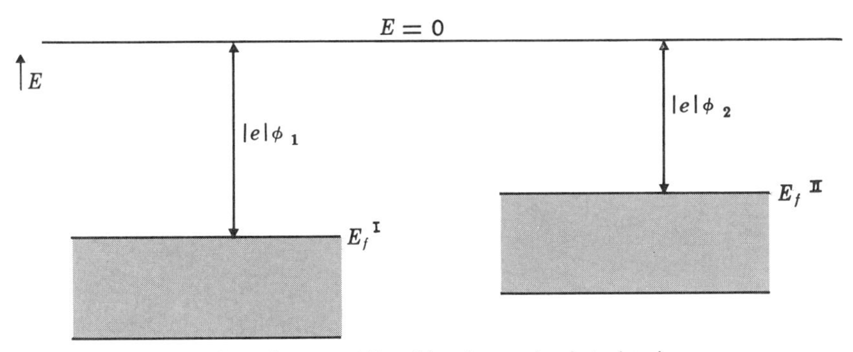

Figure 12-19 Two dissimilar metals with unequal Fermi levels completely isolated from one another.

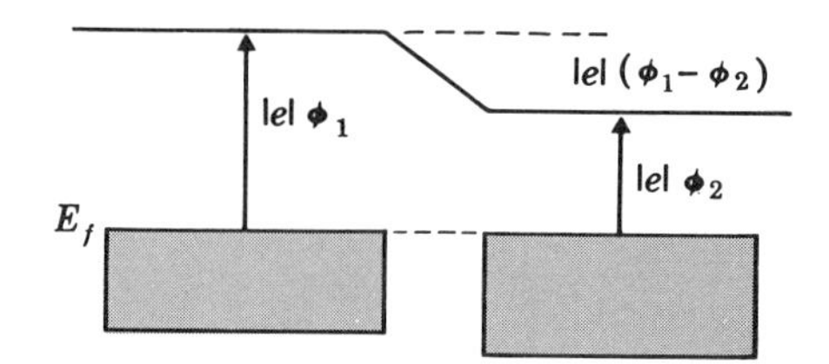

Figure 12-20 Origin of the contact-potential difference.

The actual number of electrons which flow when two solids are put into contact is rather small; hence the position of the Fermi level within the particular solid relative to the bottom of the band is not appreciably changed. Thus, so far as the local conditions in a solid are concerned, the electron flow can be neglected. In one of the Problems, the actual number of electrons which flow is shown to be a tiny fraction of the whole.

A property of a solid useful in describing the contact-potential difference is the *work function* of a solid. The work function is the energy required to remove an electron from the Fermi level of a solid out of the material to ∞. It is, for solid I, the energy $|e|\,\phi_1$ of Fig. 12-19. Different solids have different work functions; solid II of Fig. 12-19 is shown with a smaller work function $|e|\,\phi_2$. When the two solids are brought into contact so that the Fermi levels coincide, the contact potential difference becomes

$$V_c = \phi_1 - \phi_2 \tag{12-18}$$

PROBLEMS

12-1 Derive an equation for p-type semiconductor material similar to Eq. (12-17).

12-2 Calculate the position of the Fermi level of Ge at 300°K with the following concentration of impurities:

(*a*) 10^{23} In atoms per cubic meter
(*b*) 10^{22} Sb atoms per cubic meter
(*c*) 10^{23} In atoms + 10^{22} Sb atoms per cubic meter.

12-3 For each alloy of Prob. 12-2, calculate the majority and minority carrier densities at room temperature.

12-4 Suppose that τ for electrons in a particular piece of n-type Ge is 10^{-12} sec at 300°K.

(*a*) What increase in kinetic energy of an electron occurs in each time of flight for an electric field of 10^5 volts/m?

(*b*) How large an electric field is required to give an electron enough kinetic energy during each time of flight to cause ejection of an electron from a donor into the conduction band upon collision?

(*c*) From the valence band into the conduction band? (Assume for simplicity that $m_e^* = m$.)

12-5 Assume that two metals are brought into contact (i.e., separated by one lattice spacing across the surface of contact). The lattice spacing of each is $2A$, $V_c = 1$ volt. What is the charge flow per square meter from one metal to the next? If the two solids are two cubes of metal 1 m on an edge, find the fractional change in electron density of one of the cubes when the charge transfer occurs, assuming the metals to be monovalent. Assume the two cubes are placed in contact across one of the 1-m^2 faces.

13

Semiconductor devices

13-1 Introduction

The continuing development of the technology of semiconductor materials produces a seemingly endless succession of devices. To describe the operation of an appreciable number of these and to discuss their application in useful circuits is a long task, one inappropriate to this book. Nevertheless, discussion of several of the simpler devices can show a number of characteristics of basic importance. Therefore, a limited discussion is given of the principles of operation of the *p-n* junction and the transistor; this discussion is based upon differences in potential energy of charge carriers in different parts of the devices. We will see that a description of carrier motion in the devices can be made accurate enough even with simple models to permit reliable conclusions to be drawn concerning their operation.

The plan of the chapter is to consider first in Secs. 13-2 and 13-3 the basic electronic properties of homogeneous semiconductors, both intrinsic and extrinsic. Then follows (in Sec. 13-4) the physical description of the distribution of impurities in a *p-n* junction and the resulting electronic band structure of the junction. The rectifying characteristics of the biased junction and of the tunnel diode follow (Secs. 13-5 and 13-6).

The metallurgical and electronic characteristics of the transistor are discussed in Sec. 13-7, along with a description of the transistor's operation in the simplest circuit. The arrangement of impurities in the field-effect transistor and in some of the more elementary integrated circuits follows.

Final sections deal with purification of semiconductor materials, preparation of alloys containing particular impurity distribution, and descriptions of manufacture of some of the simpler devices.

13-2 Carrier Motion in Semiconductors

The conductivity of metals can be expressed in terms of the number of electrons in the Fermi distribution and the mean time between collisions, $\sigma = Nq^2\tau/m_e^*$ [Eq. (11-15)]. Only one type of carrier is present—electrons—and their number per unit volume shows so little variation between metals that differences in σ between them is mainly a result of differences in mean-free-time between collisions. Metals obey Ohm's law well, and the conductivity at a given temperature for an annealed metal of given purity is a well-defined constant.

Semiconductors, too, obey Ohm's law. For them, however, both positive and negative charge carriers are always present, and the conductivity must be expressed in terms of both.

$$\sigma = \frac{nq^2\tau_n}{m_n^*} + \frac{pq^2\tau_p}{m_p^*} \tag{13-1}$$

or alternately, in terms of mobilities,

$$\sigma = n\,|q_-|\,\mu_n + pq_+\mu_p \tag{13-2}$$

where n and p refer to properties of electrons and holes respectively. To be consistent with sign conventions used previously, we set q_- to be a negative charge and q_+ to be a positive charge. Observe that the sign of the conductivity, always a positive number, is independent of the sign of the charge carrier.

The Hall constant of semiconductors must also be expressed as a function of both types of carriers. Derivation of the proper equations in a manner similar to that carried out for metals shows that

$$R = \frac{1}{|q|}\frac{p\mu_p{}^2 - n\mu_n{}^2}{(p\mu_p + n\mu_n)^2} \tag{13-3}$$

The denominator is always a positive quantity; therefore, R may be positive, negative, or zero, depending on the relative magnitude of the two terms in the numerator. Since R is such a complicated function of the density and mobility of both carriers, Hall measurements alone are not sufficient to determine the mobility of the charge carriers, as was so for simple metals.

For intrinsic semiconductors, simplification of these equations can be made. Since $n = p$ in this case,

$$\sigma_{\text{intrinsic}} = n\,|q|\,(\mu_p + \mu_n) \tag{13-4}$$

The temperature dependence of σ for an intrinsic semiconductor thus depends both on the way in which the number of conduction electrons varies with temperature and on the way in which the mobilities vary with temperature. As was stated in Eq. (12-7a),

$$n = 4.82 \times 10^{21}\left(\frac{m_n^*}{m}\right)T^{\frac{3}{2}}e^{-E_g/2kT}$$

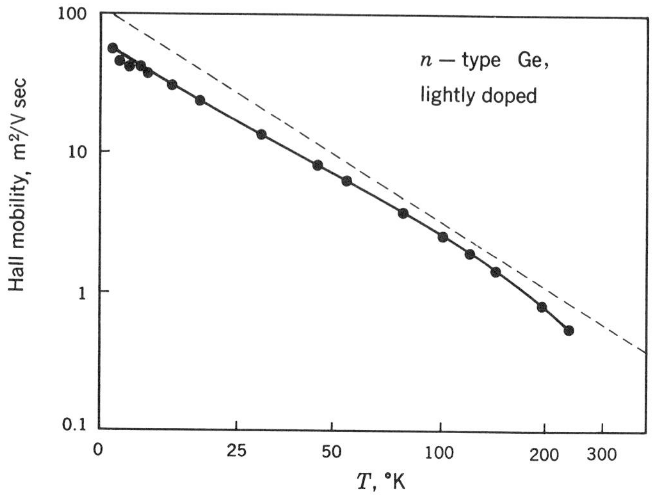

Figure 13-1 The mobility of charge carriers as a function of temperature, as determined from Hall measurements on nearly pure (but *n*-type) Ge. If $\mu_n = \text{constant} \times T^{-\frac{3}{2}}$, then $(d \ln \mu_n)/(d \ln T) = -\frac{3}{2}$. The circles are data points; the dashed line is drawn with a slope of $-\frac{3}{2}$. [*From E. Conwell and P. Debye, Phys. Rev.*, **93**:693 (1954).]

The mobilities, which are determined almost completely by the scattering of the charge carriers by lattice ions, are found to depend on temperature about as $T^{-\frac{3}{2}}$ (see Fig. 13-1). Thus

$$\sigma_{\text{intrinsic}} \cong \text{const} \times e^{-E_g/2kT} \tag{13-5a}$$

Experimental verification of this expression is excellent: $\ln \sigma$ versus $1/T$ for pure Ge and Si obey this relationship well (see Fig. 10-10).

In actual fact, the temperature dependence of the mobilities shown in Fig. 13-1 is not found to obey a strict $T^{-\frac{3}{2}}$ expression. For Ge, some observers have found that $\mu_p \sim T^{-2.3}$ and $\mu_n \sim T^{-1.7}$; for Si, variations such as $\mu_p \sim T^{-2.3}$ and $\mu_n \sim T^{-2.3}$ have been observed. These functions would imply a temperature variation of σ of the form

$$\sigma_{\text{intrinsic}} \cong \text{const} \times T^{\chi} \times e^{-E_g/2kT} \tag{13-5b}$$

where χ has a value between zero and -2. For the values of E_g (0.5 ev and higher) found in commonly used semiconductors, a temperature variation such as T^{-2} is insignificant compared to the variation in the exponential term. Hence, a temperature variation such as that shown in Fig. 10-10 still should be expected in any real case.

The Hall constant for intrinsic semiconductors is a little simpler than Eq. (13-3) indicates. Since $n = p$,

$$R = \frac{1}{n\,|q|}\,\frac{\mu_p - \mu_n}{\mu_p + \mu_n} \tag{13-6}$$

From this expression the equation for the mobilities is found to be

$$\left|\frac{E_y}{E_x B_z}\right| = \mu_p - \mu_n \qquad (13\text{-}7)$$

If the mobilities are equal, E_y and R are both zero. They are commonly not equal, however, so that real Hall constants do exist. Generally R is negative, since μ_n is usually greater than u_p.

Several constants are required to characterize the conductivity of intrinsic semiconductors. These constants are tabulated for Ge and Si in Table 13-1. Also listed for comparison are constants for two typical

Table 13-1 Constants for intrinsic Ge and Si together with approximate values for two typical intrinsic compound semiconductors

	Ge	Si	PbS	InSb
σ (300°K)	2.3 mho/m	1.6×10^{-3} mho/m	30 mhos/m	2×10^4 mhos/m
μ_n (300°K)	0.38 m^2/v sec	0.17 m^2/v sec	0.06 m^2/v sec	7.7 m^2/v sec
μ_p (300°K)	0.18 m^2/v sec	0.035 m^2/v sec	0.07 m^2/v sec	0.08 m^2/v sec
E_g (0°K)	0.78 ev	1.21 ev	0.37 ev	0.24 ev
n (300°K)	$2.5 \times 10^{19}/m^3$	$1.5 \times 10^{16}/m^3$	$2 \times 10^{21}/m^3$	$1.6 \times 10^{22}/m^3$

compound semiconductors, PbS (galena) and InSb. (For these two, the entries may not be entirely consistent, since growth of precisely stoichiometric crystals of these and many other compounds is extremely difficult.)

For extrinsic semiconductors the relations (13-2) and (13-3) cannot be simplified because they contain four unknown quantities. Therefore, other measurements are required to permit a complete description to be made. One of the most important of these measurements is that of the diffusion of carriers (and hence of mobilities) by means of photoexcitation experiments. In the extreme case where the material is strongly n- or p-type, all of the equations reduce to those of a single charge carrier.

The conductivity of impurity semiconductors has a characteristic temperature dependence (Fig. 13-2a) which can be deduced from consideration of Secs. 12-3 and 12-4. At low temperatures (large $1/T$), $\ln \sigma$ varies linearly with $1/T$. This is the region in which impurity ionization is important. The slope of the line is $E_i/2k$,—E_i is the binding energy of the electron or hole to the impurity.† At high

† Recall that the approximate equation for the position of the Fermi level in an impurity semiconductor, Eq. (12-17), is somewhat in error at low temperatures when the Fermi level approaches the conduction or valence bond. The approximation of the Boltzman function yields a slope of E_i/k for the line. The correct formulation of an expression for E_f using the Fermi function itself [Eq. (11-6)] yields the slope $E_i/2k$. This latter correct value is used in the text and in Fig. 13-2a.

temperatures where all of the impurities are ionized and the material is virtually intrinsic, $\ln \sigma$ versus $1/T$ is again a straight line with slope $E_g/2k$.

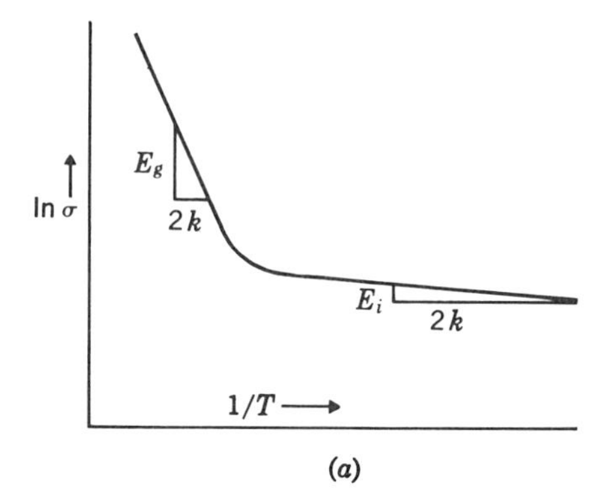

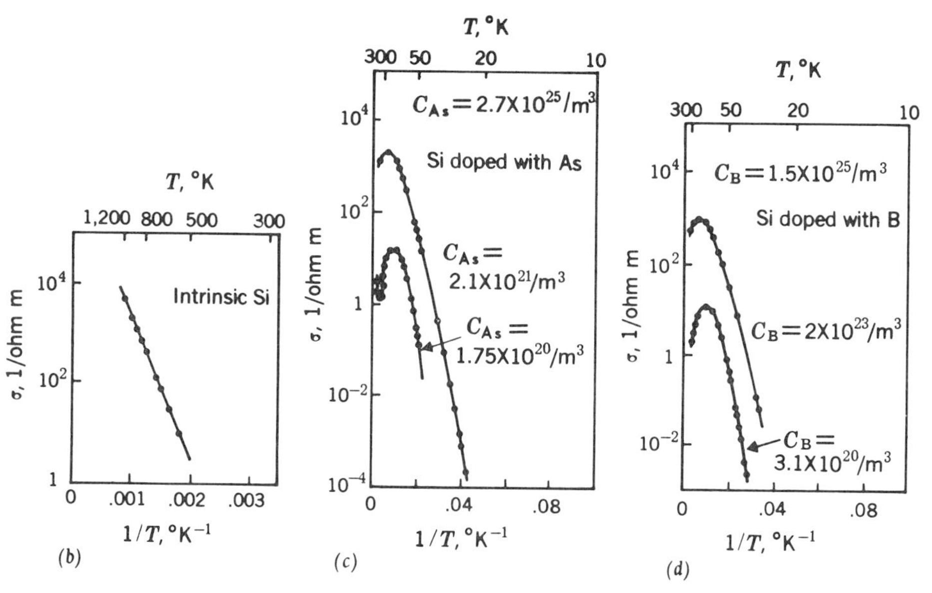

Figure 13-2 (*a*) Sketch of ln σ versus $1/T$ for an impurity semiconductor. (*b*) Plot of data for ln σ versus $1/T$ for intrinsic Si. The value of E_g from these data is 1.20 ev. (*c*) Plot of data for ln σ versus $1/T$ for As-doped Si. At the lowest level of impurity the intrinsic contribution is barely visible at the highest temperatures; for the more impure samples the intrinsic part is completely hidden. The slope of the lines around 40°K gives a value of $E_i \approx 0.048$ ev. (*d*) Plot of data for ln σ versus $1/T$ for B-doped Si. The slope of the lines below 50°K gives $E_i \approx 0.043$ ev. For both (*c*) and (*d*) the lines for highest doping levels gives semimetallic behavior. [*Data for* (*b*), (*c*), *and* (*d*) *redrawn from the paper of Morin and Maita, Phys. Rev.* **96**:28 (1954).]

Measurements on Si of several degrees of purity show the behavior sketched in Fig. 13-2*a*. For intrinsic Si the conductivity has the straight line behavior shown in Fig. 13-2*b*; for this line $E_g \approx 1.20$ ev. The extrinsic behavior of As-doped Si (n type) and B-doped Si (p type) is demonstrated in Fig. 13-2*c* and *d*. For both of these materials E_i has a value near 0.045 ev. The reader should note that ionization of the impurities is virtually complete at 100°K. Since interband excitation does not equal the 10^{20} number of extrinsic carriers in even the purest material until the temperature is well above 300°K, the total number of charge carriers in these samples shown in Fig. 13-2*c* and *d* does not change much in the temperature range 100°C to about 400°C. In this range, the mobility has a temperature dependence of about $T^{-2.4}$; the conductivity actually decreases with rising temperature in that range because of this fact.

For impurity semiconductors n is never equal to p (unless exact compensation occurs). The carrier most prevalent is called the majority carrier, that least prevalent, the minority carrier. Observe that the concentration of majority carriers in a doped semiconductor is larger than that for the intrinsic material; furthermore, the concentration of minority carriers in the doped material is less than that present in intrinsic material. In fact the number of electrons and holes present at any level of doping obeys the relation [derived by multiplying (12-13) and (12-14)]

$$np = N_c N_v e^{-E_g/kT} = n_i^2 \tag{13-8}$$

Here the constant n_i^2 is a function of temperature but not of the impurity concentration.

13-3 Generation and Recombination of Charge Carriers

The equilibrium concentration of charge carriers given by Eqs. (12-13) and (12-14) need not always exist in semiconductors. Thermal fluctuations may cause local variations with time. In addition, external disturbing influences such as light radiation of proper wavelength, rapid temperature change, electron or ion bombardment, and carrier injection may cause departure from equilibrium; the characteristic time required for equilibrium to be reestablished once the disturbing influence has been removed is called the *carrier lifetime*.

Let the equilibrium concentration of electrons and holes in a homogeneous semiconductor at a given temperature be n_0 and p_0. Suppose that light of wavelength sufficiently short to excite additional electron-hole pairs is incident on the sample. Suppose further that these pairs are excited uniformly throughout the sample and that their concentration per unit volume is $\Delta n(0)$ and $\Delta p(0)$. (Recall that these two quantities are equal.) Then, if the light is turned off, the excess number of electrons and holes disappears with time according to the following law:

$$\begin{aligned} \Delta n(t) &= \Delta n(0) e^{-t/\tau} \\ \Delta p(t) &= \Delta p(0) e^{-t/\tau} \end{aligned} \tag{13-9}$$

Here $\Delta n(t)$ and $\Delta p(t)$ are the excess concentrations at any time t, and τ is the carrier lifetime. These functions are plotted in Fig. 13-3. This behavior of excess carrier concentrations has been confirmed by experiment many times.

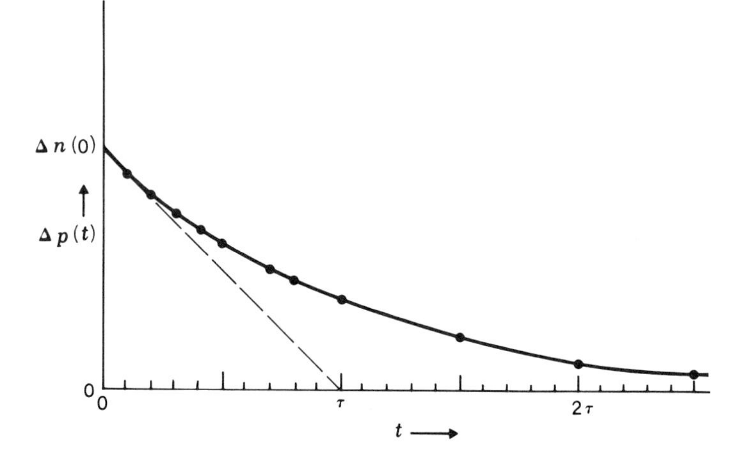

Figure 13-3 Decay of excess charge carriers as a function of time. $\Delta n(t)$ has fallen to $1/e$ of its value in time equal to τ. Note that in time 2τ it is down to nearly $\frac{1}{10}$ of its initial value. The slope of the line at $t = 0$ has the value $\Delta n(0)/\tau$.

One should note that τ is a constant of the material. It depends on the purity and perfection of the crystal. It can be as long as milliseconds in highly perfect, pure crystals or shorter than a nanosecond. If the charges undergo random walk during their lifetime, then they may diffuse an rms distance L_n given by the expression $\sqrt{Dt}$. For electrons in Si with a lifetime of a millisecond, L_n is about 3 mm at room temperature; for Si heavily doped with Au so that the lifetime is a nanosecond, L_n is about 0.003 mm. Diffusion lengths L_p for holes are comparable in magnitude.

Generation of electron-hole pairs may take place by thermal fluctuations of electron energy at least as large as the gap energy. This being so, the rate of generation, $G(T)$ (i.e., the number of n-p pairs produced per second per unit volume) is proportional to the Boltzmann factor, and $G(T) \propto e^{-E_g/kT}$. However, because the beginning and final states are not discrete but are in bands, this temperature dependence must be modified slightly by the addition of a term linear in T to some power. For parabolic bands this becomes T^3. Thus

$$G(T) = gT^3e^{-E_g/kT} \tag{13-10}$$

where g is a constant. For Si and Ge, the value of E_g is so large that the exponential term overwhelms the term T^3, and the generation rate is governed principally by the Boltzmann factor.

Recombination of electrons and holes takes place predominantly at special sites in the crystal called, appropriately, *recombination centers*. They may be imperfections (vacancies, dislocations, impurities) throughout the bulk of the material, or they may be special sites on internal or external surfaces. Since the recombination rate R depends on the concentration of both electrons and holes, we can write

$$R = rnp \tag{13-11}$$

where r is some constant. For intrinsic material, $n = p$ so that

$$R = rn_i^2 \tag{13-12}$$

At equilibrium, the generation and recombination rates are equal. Equating (13-10) and (13-12) yields the result for n_i

$$n_i = \frac{\sqrt{g}}{r} T^{\frac{3}{2}} e^{-E_g/2kT} \tag{13-13}$$

This expression is in agreement with the results of Sec. 12-4 for an intrinsic semiconductor.

One of the most important ways of controlling the recombination rate in Si (and hence the carrier lifetime) is provided by the control of the gold content. Gold is an efficient recombination center because it is interstitial and hence exhibits a large capture cross section for charge carriers. Furthermore, its diffusion coefficient is much higher at a given temperature than any other dopant commonly used in Si. Thus, after all other doping agents are added, gold can be dispersed through a device by diffusion without appreciably disturbing the geometry. The diffusion coefficients (as a function of temperature) for common doping agents in Si are sketched in Fig. 13-4; D for Au is at least 10^5 larger at 1000°C than dopant alloying elements used in Si, such as B, P, As, Sb.

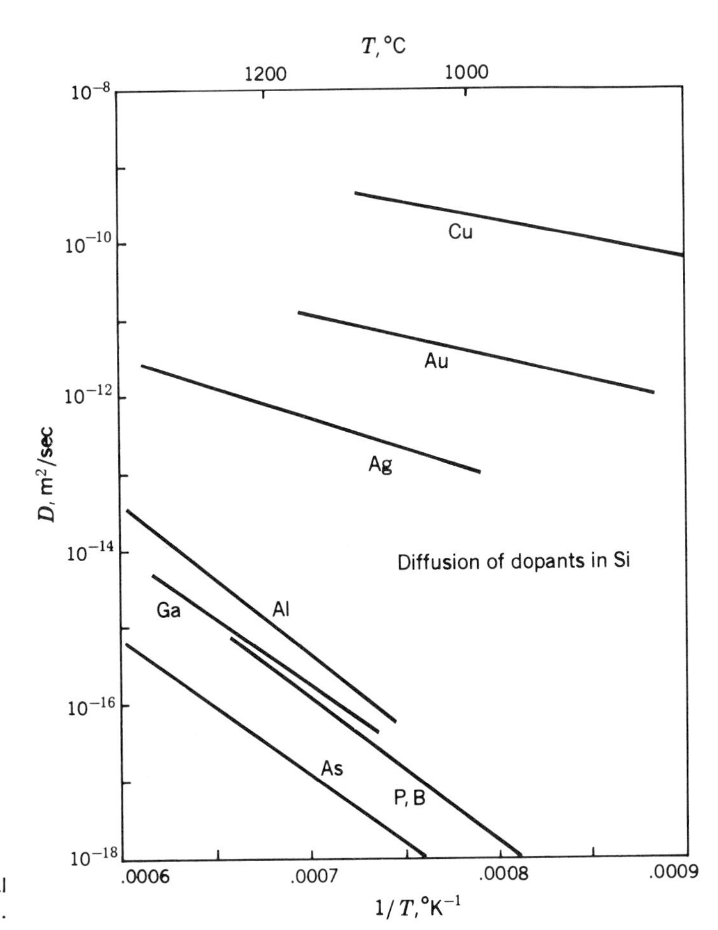

Figure 13-4 Diffusion of several doping agents in Si. They divide into two groups, the slow-diffusing substitutional impurities, such as As and B; and the fast-diffusing interstitial impurities Ag, Au, and Cu.

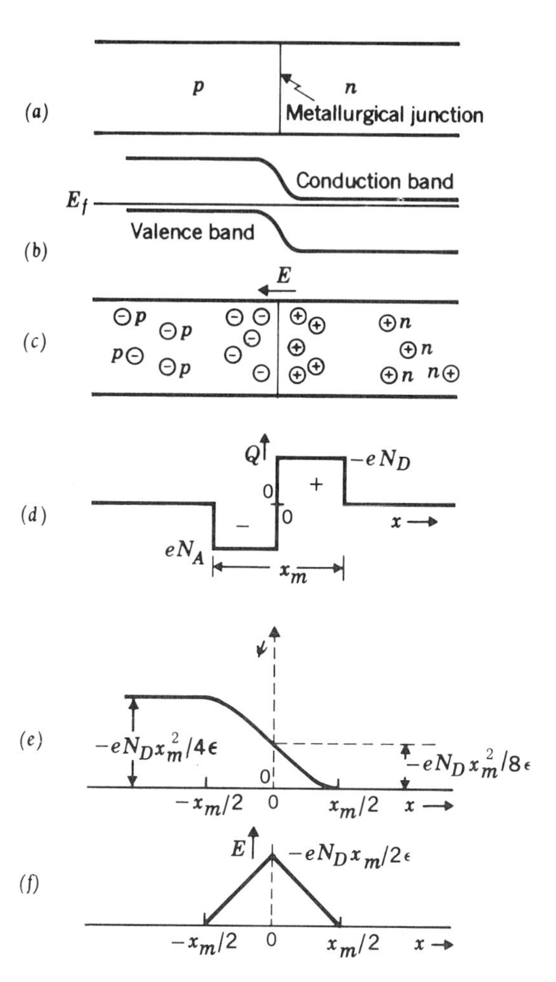

Figure 13-5 Charge and potential variations across a *p-n* junction. (*a*) The metallurgical junction of an abrupt junction. (*b*) The Fermi level is constant through an unbiased junction. (*c*) A local field $\mathscr{E}$ is set up in the vicinity of the junction by the formation of a double layer of uncompensated ions. Far away from the junction the ions are compensated by free charges over the volume. (*d*) An approximation of the uncompensated charge in the vicinity of the junction. (*e*) The electric potential in the vicinity of the junction. (*f*) The electric field $\mathscr{E}$ in the vicinity of an abrupt junction.

Recombination at surfaces is an important phenomenon. Interior boundaries such as grain boundaries are highly undesirable, so devices are always made of single-crystal material. Control of recombination at external surfaces is dependent on development of treatments which will give stable and desirable conditions at the surface in both transistor and integrated circuit production.

13-4 The *p-n* Junction

The homogeneous semiconductor has some usefulness as a resistor which may have interesting optical, particle radiation, temperature, or other properties. Semiconductors are most useful, however, when both *n*- and *p*-type materials are combined in the same device. In this section, we consider the properties of the unbiased *p-n* junction. For simplicity, we consider a planar symmetrical junction, one in which the equivalent net doping levels of the two halves are equal, though of opposite sign (see Fig. 13-5*a*). We also suppose that our junction is a perfect atomic bond and that the change from n- to p-type occurs in one atom spacing.

Thermodynamic equilibrium in the unbiased junction exists when the Fermi levels in the two halves of the device are equal. Since the position of the edges of the band relative to the Fermi level for a homogeneous semiconductor (or for material sufficiently far from the junction) is a function only of temperature, the band edges in a p-n junction must vary across the junction. This variation must have the form sketched in Fig. 13-5b; however, the details of the change in band edge in the boundary region need to be determined.

The variation in bond edge can be deduced by considering the nature of the carrier redistribution which would attend the sudden formation of a junction. Because of large gradients in charge carrier concentration at the instant of joining, some electrons would diffuse from the n side to the p side, and some holes would diffuse the other way. This event would upset local charge neutrality on either side of the junction; a region (called the *space-charge* region) of negatively charged acceptor and positively charged donor immobile ions would be left in the p- and n-regions (Fig. 13-5c). An electric field E would then be set up which would oppose further diffusional drift, and a dynamic equilibrium would result. Determination of this field and from it the electric potential ψ at all points in the junction permits complete description of the band structure through the junction.

Solution of this problem is difficult for the general case, but a simplifying assumption concerning the distribution of charges in the space-charge region permits the calculation to be made easily. Assume the immobile ions on either side of the junction to be uniform in density and to extend a distance $x_m/2$ into both the p and n regions. This distribution of charge would have the slope sketched in Fig. 13-5d. The total charge is assumed to be $-|e|\,N_D$ and $|e|\,N_A$ on the two halves. The electric potential ψ in the junction is then found by solving the following differential equation for the p and n regions:

$$\begin{aligned} \frac{d^2\psi}{dx^2} &= \frac{-eN_D}{\epsilon} \qquad 0 < x < \frac{x_m}{2} \\ \frac{d^2\psi}{dx^2} &= \frac{eN_A}{\epsilon} \qquad \frac{-x_m}{\epsilon} < x < 0 \\ \frac{d^2\psi}{dx^2} &= 0 \qquad x > \frac{x_m}{2} \qquad x < \frac{-x_m}{2} \end{aligned} \tag{13-14}$$

Solution of the first of these equations yields the result for ψ that

$$\psi = \frac{-eN_D}{\epsilon}\left(\frac{x_2}{2} - \frac{x_m}{2}\,x + \frac{{x_m}^2}{8}\right) \tag{13-15}$$

where ψ and $d\psi/dx$ have been set equal to zero at $x = x_m/2$. This expression is plotted in the right half of Fig. 13-5e. A similar equation is found for negative values of x; the two are matched at $x = 0$. (Recall the assumption that $N_A = N_D$.)

The electric field across the junction is, of course, found at an intermediate step in the derivation of Eq. (13-15). This field is plotted in Fig. 13-5f.

The total variation of ψ across the junction is $eN_Dx_m{}^2/4\epsilon$. Thus, an amount of work $e^2N_Dx_m{}^2/4\epsilon$ must be done to transport an electron in the conduction band of the n side to the conduction band of the p side of the junction. A similar amount of work must be done to transport holes across the junction.

13-5 The Junction Rectifier

The p-n junction delivers no current in spite of the difference in potential energy of charge carriers in different parts of the device. For it to do so would be a violation of the second law of thermodynamics. However, it does pass current in either direction under appropriate external voltages. The junction is not an ohmic device, though; its resistance in one direction is much larger than that in the other. In the discussion which follows, we shall both see the reason for this and develop analytic expressions for the resistance in the two directions.

The fact that no net current flows in the unbiased p-n junction may be shown from consideration of the energy-level scheme shown in Fig. 13-6. The number of electrons in the conduction band of the p-type

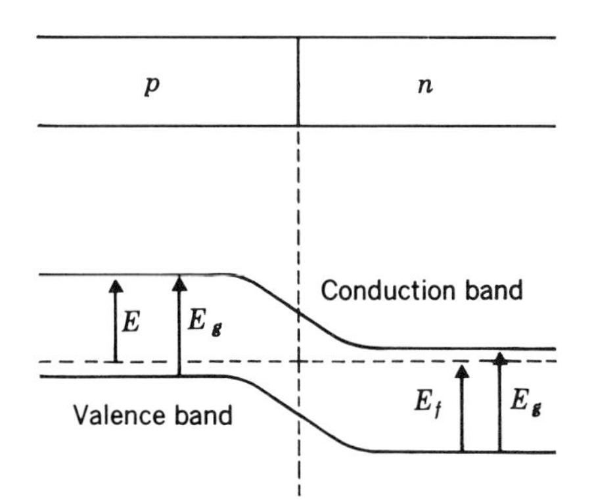

Figure 13-6 The energy levels of electrons in the conduction and valence bands across an unbiased p-n junction.

material is proportional to $e^{-E/kT}$ where E is the excitation energy required on the p side. The number of electrons in the conduction band of the n-type material is proportional to $e^{-(E_g-E_f)/kT}$. One sees that the ratio of the number of electrons in the p-type material compared with the number in the n-type material is a Boltzmann factor with an energy factor equal to the difference between the bottom of the conduction band in the two materials.

The currents may now be calculated. The electron flux to the right, call this $I_e(\rightarrow)$, is proportional to $e^{-E/kT}$, since any electron which diffuses to the edge of the barrier moves down it to the n side. The electron current to the left, call this $I_e(\leftarrow)$, is proportional to a product of two Boltzmann factors, one the number of electrons on the n side, the other the probability that each such electron has enough energy to climb the barrier. The first factor is proportional to $e^{-(E_g-E_f)/kT}$, the second is equal to $e^{-(E-E_g+E_f)/kT}$. Thus:

$$\begin{aligned} I_e(\rightarrow) &= Ce^{-E/kT} \\ I_e(\leftarrow) &= Ce^{-(E_g-E_f)/kT}e^{-(E-E_g+E_f)/kT} \\ &= Ce^{-E/kT} \end{aligned} \tag{13-16}$$

One sees that $I_e(\rightarrow) = I_e(\leftarrow)$, and no net current flows due to hole motion. The constant C, which must have the dimensions of current density, is given approximately by $qD_n n_p/L_n$, where D_n is the diffusion coefficient of the electrons, n_p is the number of electrons on the p side, and L_n is the diffusion length given by $\sqrt{D_n \tau_n}$.

A voltage V applied to the junction with the n side positive with respect to the p side alters the energies as shown in Fig. 13-7. The

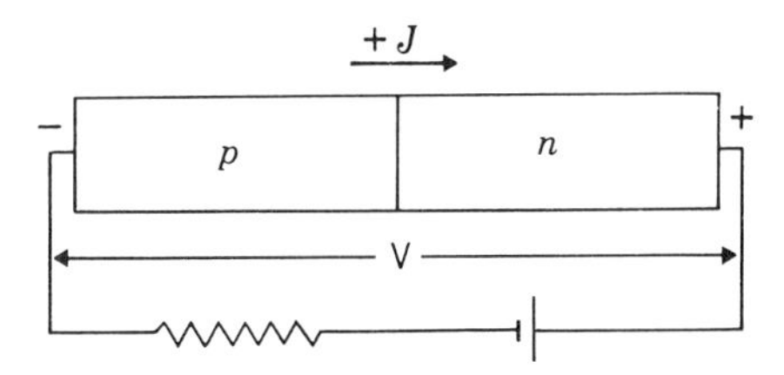

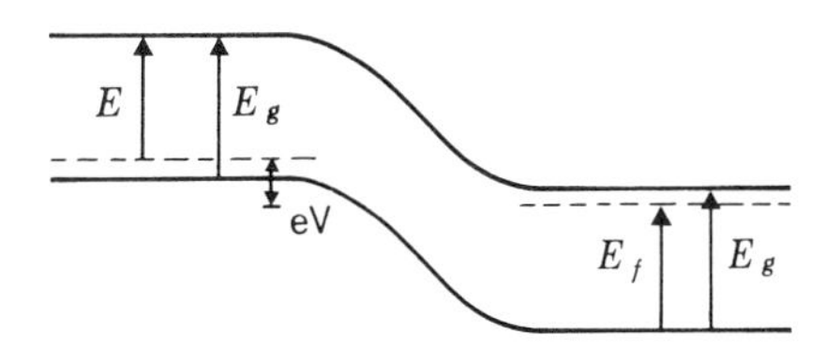

Figure 13-7 Electron energies in a reverse-biased p-n junction. For this battery connection V is negative since the direction of current flow through the device is opposite to that labeled positive.

Fermi level is depressed at the positive terminal relative to the negative terminal by an amount eV joules. If the solid were a homogeneous conductor, i.e., either completely n-type or p-type, this energy increment would be distributed along its length uniformly. However, most of the energy eV (and hence change in Fermi level) occurs in the region of the junction itself. This important fact results in a depression of the energy of an electron in the n side relative to the p side even further than it was in the unbiased junction. Note that the sign convention is established that V is negative in this instance.

Consider now the electron flow in the two directions. The electron current to the right is not changed as compared to the previous case since this current depends only on the electron density in the conduction band of the p-type material. The electron current to the left decreases, however, since an increase occurs in the potential barrier up which the many electrons in the conduction band of the n-type material must climb. These two currents are

$$I_e(\rightarrow) = Ce^{-E/kT}$$
$$I_e(\leftarrow) = Ce^{-(E_g-E_f)/kT}e^{-(E-|e|V+E_f-E_g)/kT} = Ce^{-(E-|e|V)/kt} \qquad (13\text{-}17)$$

the net charge flow in an electron current to the right (since V is negative). In terms of conventional (i.e., positive) current flow to the right

$$J = Ce^{-E/kT}(e^{|e|V/kT} - 1) \qquad (13\text{-}18)$$

J is a small negative current whose magnitude approaches a constant as $|e|\,V$ increases beyond several kT. This battery connection is termed reverse bias.

A voltage V applied with the n side negative alters the potential across the device in a significant way. The Fermi level of the n-type material is raised by an amount $|e|\,V$ relatíve to that in the n-type (Fig. 13-8). This change in potential produces no change in the electron flow to the right, hence $I_e(\rightarrow)$ is the same as before. $I_e(\leftarrow)$ is altered significantly, however, since the potential hill which the electrons in the n-type material encounter is decreased.

$$I_e(\rightarrow) = Ce^{-E/kT} \qquad (13\text{-}19)$$
$$I_e(\leftarrow) = Ce^{-(E-|e|V)/kT}$$

The difference current, again given in terms of conventional current flow, is then

$$J = Ce^{-E/kT}(e^{|e|V/kT} - 1) \qquad (13\text{-}20)$$

For this battery connection, J increases rapidly with increase in V; this is called the forward-bias connection.

Equations (13-18) and (13-20) are identical if the direction of current flow and of application of potential are considered consistently.

$$J = Ce^{-E/kT}(e^{-|e|V/kT} - 1) \qquad (13\text{-}21)$$

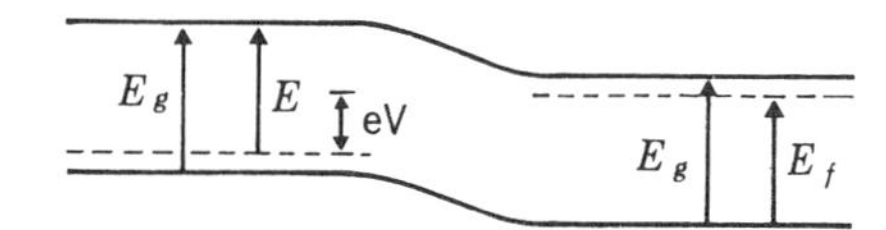

Figure 13-8 Electron energies in a forward-biased p-n junction.

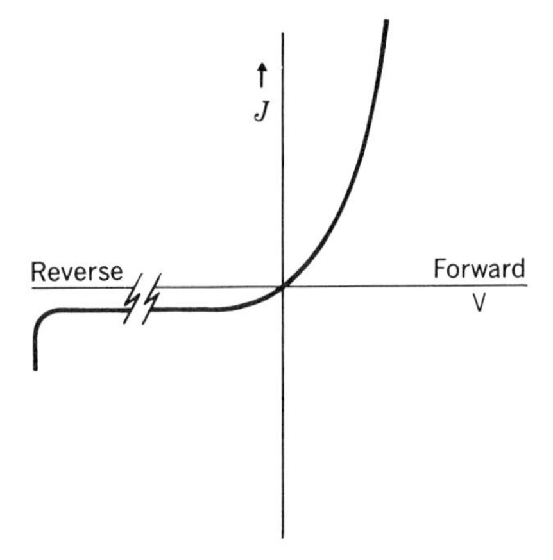

Figure 13-9 Current-voltage characteristics of a biased *p-n* junction at some temperature *T*.

Here current flow in the forward direction and forward-biasing energy $|e|\ V$ are termed positive, while current flow in the reverse direction and reverse-biasing energy $|e|\ V$ are termed negative. This equation is plotted in Fig. 13-9. The maximum reverse current is of order a few milliamperes per square centimeter of junction area; it remains steady at this value until the voltage reaches sufficient size to cause an avalanche of electrons to pour through in the reverse direction. This voltage varies from a few volts to some hundreds of volts, depending upon the doping geometry of the junction. The maximum permissible forward current is many amperes per square centimeter and occurs for a forward voltage of only a few volts.

13-6 The Tunnel Diode

The operation of the p-n junction depends in a critical way on the states at the edge of the band. The position of the Fermi level is determined by the ratio of intrinsic carriers to the extrinsic carriers introduced by doping agents which have states near the band edge. Although our examples of Sec. 12-3 described only the ground states of isolated impurities immersed in the semiconductor host, a series of excited states also exists between these ground states and the band edge. When the impurities are present in sufficient quantity, indeed, a band of levels replaces the localized ground state levels where overlap of wave functions of electrons on adjacent impurity atoms occurs. At sufficiently high impurity concentrations, in fact, this impurity band merges with the adjacent conduction or valence band.

The quantity of doping agent thus critically determines the character of the band edge. For material used in ordinary p-n junctions, the impurity concentration is of order 10^{20} to $10^{22}/\text{m}^3$; at this concentration the states are localized. When the concentration reaches $10^{24}/\text{m}^3$ appreciable overlap occurs, and when it reaches $10^{26}/\text{m}^3$, the impurity band merges with the adjacent intrinsic band (see Prob. 13-7). At all temperatures this material is highly extrinsic. The Fermi level

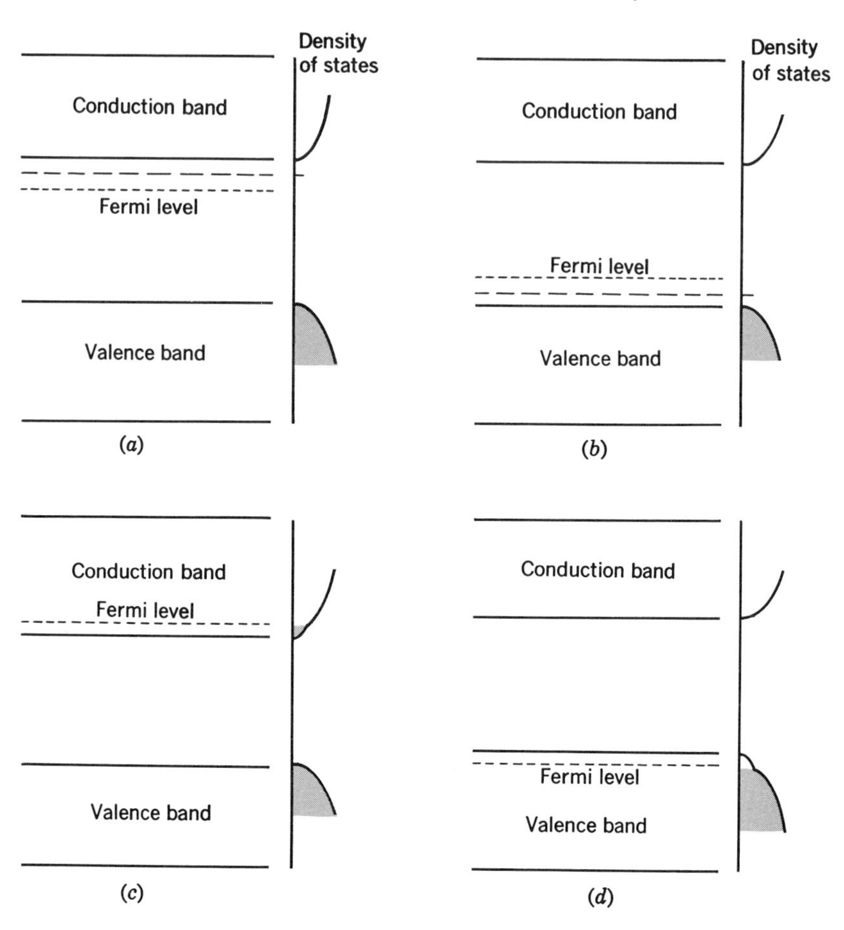

Figure 13-10 Band structure of doped semiconductors. (*a*) Lightly doped *n*-type semiconductor; (*b*) lightly doped *p*-type semiconductor; (*c*) heavily doped *n*-type semiconductor; (*d*) heavily doped *p*-type semiconductor.

is no longer in the forbidden gap; it moves into this hybrid impurity-intrinsic band. The exact shape of the density of states curves near the band edges is unknown, but the bands are assumed to have the character of the sketches in Fig. 13-10*c* and *d*. This heavily doped material is thus seen to have something of the character of a metal; empty states lie immediately adjacent to filled states, although the density of these states is much lower than that for a normal metal.

The practical significance of material with these characteristics is seen only if one examines the energy-level diagrams of a *p-n* junction made from it. In the region of the junction, the bottom of the conduction band and the top of the valence band cross the Fermi level (Fig. 13-11*a*). No net current can flow, of course, with no voltage applied. If the *n*-type side is biased positively with respect to the *p*-type side, the energy levels shift in the usual way (Fig. 13-11*b*). This arrangement of levels produces no startling effects for the typical wide (i.e.,

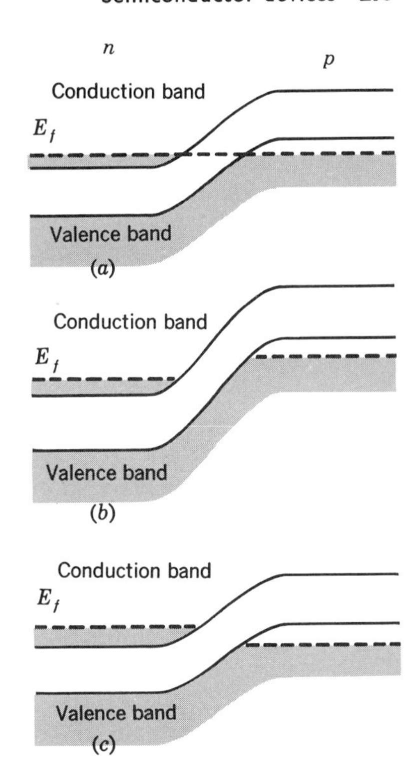

Figure 13-11 Electron energy of heavily doped *p-n* junctions in several conditions of bias. (*a*) No bias; (*b*) reverse bias; (*c*) forward bias.

some 0.001 in.) junction, even though the electrons at the Fermi surface in the *p*-type material have a higher potential energy than some empty states in the *n*-type material. The electrons cannot cross the wide junction. The forward-biased material has energy levels also arranged in the usual way and would behave about like a normal forward-biased junction.

An unusual event occurs if the width of the junction is made extremely narrow, 150 A or less. Electrons can then tunnel through the high-energy barriers at the junction, emerging in empty states at the far side. The characteristics of such a diode are interesting: not only does a large current flow for the reverse-biased arrangement, but also a peculiar (and useful) current-voltage characteristic is found for the forward-biased arrangement. The overall current-voltage characteristic is sketched in Fig. 13-12 for a typical diode.

Tunneling also occurs for a forward-biased junction, for here too the filled and empty levels overlap (Fig. 13-11*c*). The current increases toward *a* in Fig. 13-12. It does not increase indefinitely, however, since the filled electron levels are raised ultimately to the point where they are opposite the forbidden region in the *p*-type material. As this occurs, the tunneling current decreases from its maximum, finally dropping to some point *b*, near zero. At higher forward bias, the current-voltage characteristic is nearly that of a conventional diode.

Tunneling currents in common diodes are typically small; the current at point *a* might be only a few milliamperes. Similarly the

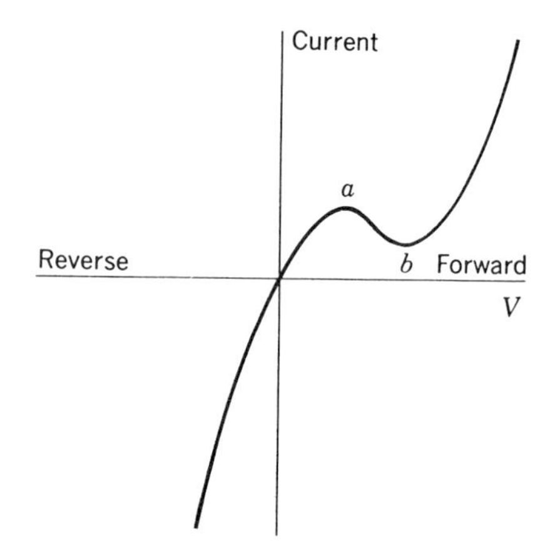

Figure 13-12 Current-voltage characteristics of a tunnel diode.

critical voltage at b might be only a few tenths of a volt. Nevertheless, from a to b the device does have a negative-resistance characteristic and is extremely useful because of it.

The crucial part of the operation of this device is the tunneling action of electrons; they pass from one side of the barrier to the other simply because the wave function exists on both sides. Quantitative calculation of the amount of the current is *not* simple, however, both because the tailing of the band is difficult to describe quantitatively and because the tunneling probability itself is difficult to calculate. In particular, the magnitude of the current at b for a long time seemed to be larger than it should be; this "excess current" seems likely to be explained by taking proper account of the way the bands tail off into the forbidden gap.

The circuit characteristics of this device are valuable in several respects. The high-frequency behavior of the negative-resistance region is excellent; the negative resistance is nearly independent of frequency from near zero to well above the thousands of megacycles region. The response time for the device to a signal can be as rapid as 10^{-9} sec. One design characteristic needing careful control is the relatively large capacitance (of order 1 $\mu f/cm^2$ of junction area) of the junction. The capacitance of a particular device can be reduced, of course, by using junctions of small area, but then corresponding reduction in current-handling capability occurs. In addition, the device becomes increasingly fragile as the cross-sectional area is decreased.

The manufacture of tunnel diodes with small capacitance and high tunneling efficiency requires application of several interesting metallurgical principles. The doping agents must be soluble enough in the semiconductor host to give impurity concentrations at least as high as 10^{25} atoms per cubic meter. Yet the diffusion coefficients of the doping agents in the host at the manufacturing temperature must be small enough so that an abrupt and not a diffuse junction results.

Consider how these requirements are met in practice. A tunnel diode can be prepared by alloying a small bead of an appropriate

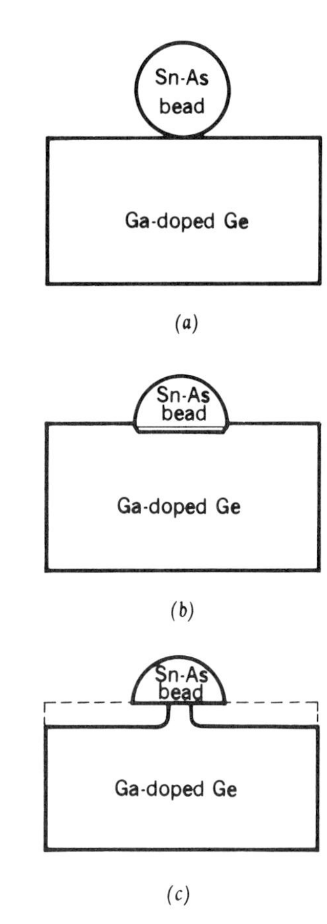

Figure 13-13 Steps in manufacture of tunnel diode.
(*a*) Bead of As-doped Sn is placed on *p*-type wafer.
(*b*) Sn bead is melted and solidified to produce *n*-*p* junction.
(*c*) Ge is etched away to reduce size of junction to proper area.

metallic alloy into the surface of a Ga-doped germanium wafer (Fig. 13-13*a*). The bead is melted by heating the pair to an appropriate temperature, and the junction is formed during solidification of the molten material. Arsenic is an appropriate doping agent for the n side of the diode since it diffuses slowly in Ge at a convenient alloying temperature (say 450°C) and has a relatively high solubility at that temperature (about 10^{26} atoms per m^3). It will not solder to lead wires, however, and it sublimes rather than melts as its temperature is raised. It can be applied as an alloy with tin, which is a neutral doping agent, solders well, and has a satisfactory melting temperature (231°C).

Formation of the junction follows well established thermodynamic and kinetic principles. The molten bead dissolves a small amount of Ge which crystallizes back onto the solid Ge during cooling, carrying with it a small amount of the bead material. The p-n tunnel junction then forms at some interface between the Ga-doped wafer and the resolidified Sn-As bead (Fig. 13-13*b*). A chemical etchant which attacks Ge but not Sn may be used to produce a junction of proper area (Fig. 13-13*c*). A typical diode produced in this way might have a peak current of a few milliamperes and a junction capacitance of 50 pf.

13-7 The Transistor

Semiconductor rectifiers were used extensively early in this century. The three-element device analogous to the vacuum tube triode, the transistor, was developed much later (in the 1940s). This development required that the physics underlying the properties of extrinsic semiconductors be well understood and the metallurgical and chemical art of doping semiconductors in prescribed ways be adequate to the task. The alloy junction transistor, useful as it is as an electrical circuit component, is itself in part a stage in the development of the miniature integrated circuit.

A junction transistor is formed by putting two p-n junctions back to back, say in n-p-n configuration, Fig. 13-14. If the p-type region is thick enough so that the elements of the junction function completely independently of one another, it is not an interesting device. If, however, the p-type region is thin compared to the diffusion length L_n, the junctions do not function independently and the phenomenon known as transistor action occurs.

The operation of a transistor depends on the way in which the energy of the electrons (and holes) varies from one end of the device to the other and on the way in which biasing voltages affect this energy.

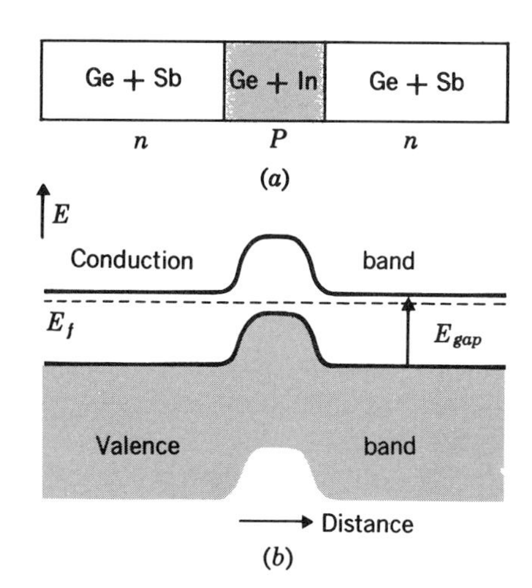

Figure 13-14 (*a*) An *n-p-n* transistor. (*b*) The energy levels of the unbiased transistor. This is drawn as a symmetric transistor. The *n*-type ends are assumed to have the same composition of Sb; the *p*-type common layer may or may not have the same amount of In.

The base device with no biasing voltages attached is sketched in Fig. 13-14*a*; the energy of the electrons in various parts is sketched in Fig. 13-14*b*. The central p-type region is thin; it may be no thicker than a few thousandths of an inch, for reasons stated below. The n-type pieces on the end may be much larger, perhaps as much as an eighth of an inch in size.

This device may be put into circuits in innumerable ways, but the flow of charge in the device may best be analyzed in the simple common-base connection, Fig. 13-15*a*. The left hand n-p junction is biased in the forward direction by a small battery of about 1.5 volts through a small resistor R_i of perhaps 500 ohms across which a signal can be

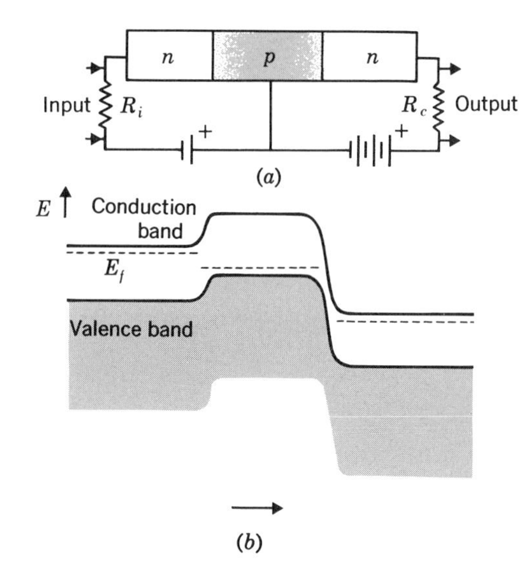

Figure 13-15 (*a*) A common-base *n-p-n* transistor circuit; (*b*) the electron energy levels of the biased transistor.

developed. The right-hand junction is biased in the reverse direction by a larger battery, perhaps 9 volts, through a larger resistor R_c, of some 10,000 ohms, across which the amplified signal can be developed. The left-hand side of the n-p junction is called the emitter; the right-hand side of the p-n junction is called the collector. The central region of p-type material, though thin in the junction itself, has come to be called the base.

Electron flow in the transistor, biased as the sketch shows, is not difficult to analyze. A current of electrons flows through both junctions because of the relative positions of the energy levels of the electrons in various parts of the device, as they are sketched in Fig. 13-15*b*. Electrons flow through the forward-biased emitter junction just as they do in a forward-biased junction diode. The behavior of the emitter current is then just as it was for the diode described in Sec. 13-5. If the base were thick, 0.1 inch or more, these electrons would flow about the emitter circuit, and the collector circuit, reverse-biased as it is, would not sense the existence of this current. However, if the base section is thin enough, an important event occurs there. The electrons emitted into the base find themselves in a region with small electric field; they diffuse about rather randomly and may ultimately find themselves in the strong field of the reverse-biased p-n junction. At this point they are swept through the collector junction and drawn around through the collector circuit, producing a voltage across the output resistor R_c. If the base is thin enough, a large fraction of the electrons (99 per cent or more) emitted by the forward-biased junction pass through to the collector, and the efficiency of the device is high.

The gain of the transistor may now be calculated. The forward resistance of the emitter is low compared with R_i, so that any input signal V_i developed across R_i produces a forward current $i = V_i/R_i$. The collector current i_c, passing through the resistor R_c, produces an output voltage $V_c = R_c i_c$. If the current efficiency of the transistor is unity, i.e., if $i_c = i_i$, then the voltage gain V_c/V_i is given by R_c/R_i. Observe that the high reverse resistance of the reverse-biased collector does not enter into the calculation. Note also that the power generated in the output resistor is derived from the collector-biasing battery.

Thus, the charge flow is not difficult to understand. The key features of its operation are (1) proper geometry of the junctions with the base thin enough to give high efficiency of transfer of electrons and (2) low recombination rate of the minority charge carriers in the base. The first of these requirements is met in practice by making the base about 0.2 of the diffusion length, the second by doping the p-type base relatively lightly.

Making transistors with high efficiency and high yield requires careful design and control of doping procedures during manufacture. Several processes have been developed, two of which are described.

Alloy transistors are made by alloying emitter and collector dopants into a semiconductor wafer of opposite conductivity; this is basically the same procedure as was described for the production of tunnel diodes. As an example, see the schematic of processes used to make an indium-doped p-n-p transistor in Fig. 13-16. Small beads of indium metal are

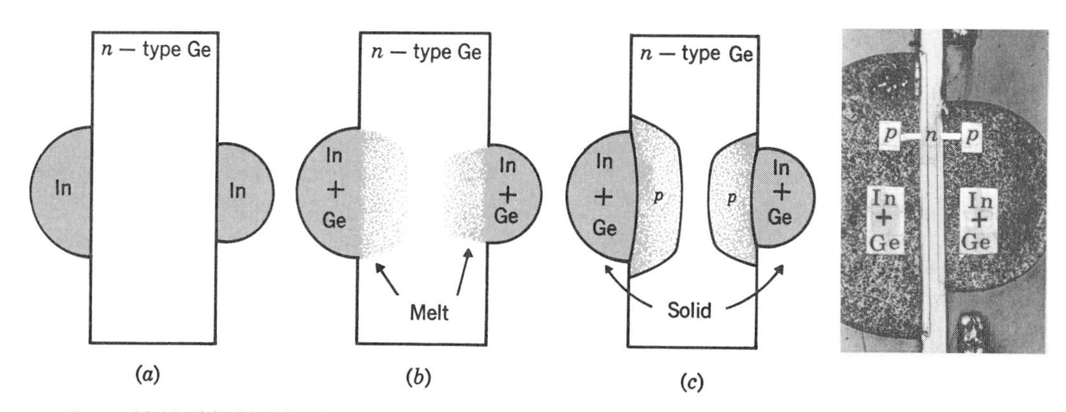

Figure 13-16 (*a*), (*b*), (*c*) Schematic of processes for production of a p-n-p transistor; (*d*) photograph of an actual p-n-p transistor. (*Photograph courtesy of J. C. van Vessem.*)

placed on opposite sides of a wafer of n-type Ge (doped with As, say). The assembly is heated well above the melting point of In (155°C). In the molten state, the In dissolves some of the Ge. In equilibrium it would dissolve a few per cent of Ge at a temperature of 400°C, say. Whether this much is indeed dissolved depends on the time at temperature (In has a very high boiling point, about 1450°C, so not much In will evaporate during this process). Then the assembly is cooled so that the n-type Ge conducts away the heat. The first solid to form is Ge, plus a small amount of In; with proper control of the initial n-type dopant, time, and temperature, this layer is p-type. When the temperature falls sufficiently low, the remaining liquid freezes into a two-phase eutectic solid with a final configuration about as is shown in Fig. 13-16*c*. Two junctions are formed, a *p-n* junction on the right and an *n-p* junction on the left. The In metal in the resolidified portion makes good metallic contact to the *p*-type regions and provides good solder contacts for external leads. A magnified picture of a *p-n-p* transistor made in this way is shown in Fig. 13-16*d*.

Diffused-base transistors are made by selective diffusion of dopants into solid wafers of semiconductor. This process is useful for producing not only independent transistors, but also transistors in integrated circuits. It is not a particularly new metallurgical technique, since it has been used for many decades to case-harden steel shafts and bearing surfaces.

To make this transistor, one starts with an *n*-type wafer such as As-doped Si. A *p*-type dopant such as Ga is diffused into the wafer from one of its faces. The initial concentration of Ga and the time and temperature of diffusion are controlled so that a surface layer is *p*-type while the interior remains *n*-type. Along some plane, the *p-n* junction, the Ga, and the As just compensate for each other electronically. A second *n*-type layer can be diffused in on top of the *p*-type layer, say by boron diffusion, yielding an *n-p* type junction. Of course, the profile of the Ga will change slightly during this second diffusion step, thus moving the first *p-n* junction slightly farther into the wafer. However, by proper control of time, temperature, and concentration of impurities, almost any desired planar configuration of junctions can be produced. After these two diffusion steps, contacts may be made to the three elements by alloying or evaporation with metals.

A process which combines solid-state diffusion and alloying provides good control of the junction geometry and lead wires. A wafer of *p*-type material (which will be the collector) has an *n*-type layer diffused into it, forming an *n-p* junction. Beads of *p*-type and *n*-type alloy (there might be an aluminum-tin alloy and an antimony-gold alloy, respectively) are alloyed into the surface of the *n*-type material. The first bead forms a *p-n* junction with an external ohmic contact, the second an ohmic contact to the *n*-type base. These processes are described in more detail in Reference 3.

The crux of the kinetics of these processes is the relative interdiffusion rate of the several impurities. In the molten-bead processes, solid-state diffusion is negligible compared with diffusion in the liquid alloys at the temperatures used in alloying. However, production of a flat junction requires careful control of manufacturing variables, including proper orientation of the cube axes of the unit cell relative to the face of the wafer.

Interdiffusion in the solid poses a different set of problems. The diffusing "interface" is not abrupt; rather, the profile of concentration depends on time, temperature, and initial boundary conditions. If a layer is formed by depositing a *fixed quantity* of dopant Q on the surface and permitting it to diffuse inward, then the concentration has the

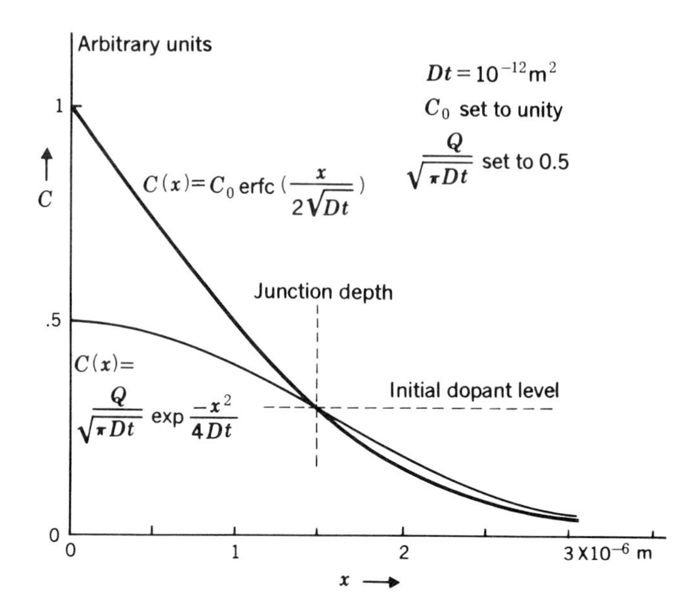

Figure 13-17 Diffusion of dopant into silicon wafer. The two lines are plotted according to Eqs. (13–22) and (13–23). The *p-n* junction forms at a depth where compensation occurs. By appropriate choice of constants in the two equations, the junction has been made to form at the same depth for both. This example would be appropriate for phosphorus diffusion into *n*-type Si for about three hours at 1100°C.

shape of a Gaussian distribution

$$C(x) = \frac{Q}{\sqrt{\pi Dt}} \exp\left(\frac{-x^2}{4Dt}\right) \tag{13-22}$$

That curve has the shape sketched in Fig. 13-17; an actual example is the plot of data in Fig. 4-10. If a *fixed concentration* of dopant C_o is maintained at the surface (as with a gas of large volume), the concentration has the distribution of the complementary error function

$$C(x) = C_o \operatorname{erfc} \frac{x}{2\sqrt{Dt}} \tag{13-23}$$

that function is also sketched in Fig. 13-17. In either case, the *p-n* junction forms at that point where compensation of the initial dopant and the diffusing dopant occurs, see dotted line in Fig. 13-17.

These are the elementary principles of transistor operation and manufacture. The circuit applications are seemingly endless, but unfortunately the details of charge motion and their significance in device operation become hazy with the more complicated circuits; for these authors that point comes where equivalent circuits must be drawn. Similarly, device manufacture is so tied up with proprietary "art" and so quickly gets into nuances of detailed procedure for specific devices that further discussion here seems unwarranted. References which deal with many of these topics are listed at the end of the chapter.

13-8 Field-effect Transistors

The geometrical distribution of selected impurities is highly important in each of the three devices considered so far: the junction diode, the tunnel diode, and the transistor. It would be surprising if additional geometrical distributions of impurity did not exist which would give rise to devices with additional properties, and indeed there are. One of these devices is the *field-effect transistor*.

The geometry of the devices considered up to this point have had planes of variation of dopant normal to the direction of charge flow. The field effect transistor has its *p-n* junctions *parallel* to the planes of current flow. When a potential is applied between the source and the drain of such a device (see Fig. 13-18), charge flows through a narrow *n*-type channel between the *p*-type regions. The charge carriers are solely electrons, the majority carrier in the *n*-type material. A variable current carrying capacity results from the fact that the width of the depletion layer of a *p-n* junction is a function of the bias on the junction; the width of the channel, and thus the conductivity of the channel, is therefore a function of biasing voltage.

A wide variety of voltage configurations is possible with the terminals shown in Fig. 13-18*a*. With the potential of the drain positive with respect to the source and both gates, the channel narrows as the depletion layer increases in width, as in Fig. 13-18*b*. This

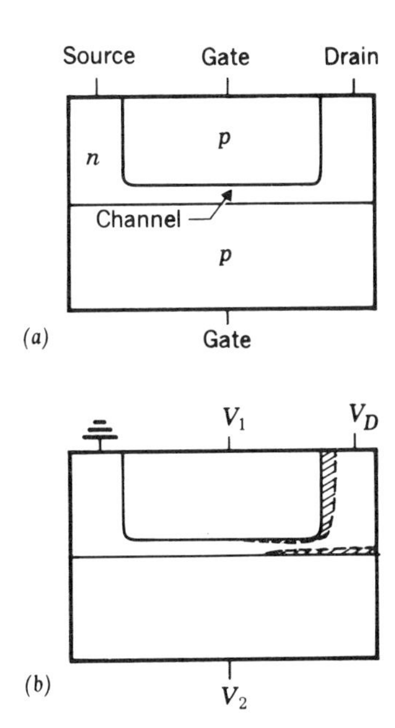

Figure 13-18 The field effect transistor. (*a*) Geometry of dopant. (*b*) Pinching-off of channel with proper biasing voltages.

narrowing is not uniform along the channel since a potential drop exists along its length causing the reverse bias across the junctions to vary along the length. With large enough reverse bias, the channel can pinch off completely. For larger relative values of V_o, the current increases only very slightly until breakdown occurs.

This sketch of the principles of operation of the field-effect transistor may be amplified by the interested reader by consideration of the vast literature which exists on their properties. The original paper of W. Shockley, an excellent review in Chap. 8 of Reference 3, and many original sources cited at the end of that chapter give excellent additional information.

This device required an exceptionally long period of development, more than 10 years. This long time occurred because the techniques of making very thin (about 1 micron) layers demands much greater control of impurity distribution than in conventional transistors. If the manufacturing process be by diffusion, extremely fine differences in diffusion distances of the two dopants must be carefully controlled.

13-9 Integrated Circuits

We have considered, up to this point, a group of individual active devices which can be combined in various ways with resistors, capacitors, inductors and voltage sources to perform useful functions. Techniques have been developed which combine several of these components on a single semiconductor chip, the unit being called an *integrated* circuit. In its ideal form, such a circuit has but three types of connections: input and output signals and biasing voltages. Actually, the development of integrated circuits was preceded by the use of a partially integrated circuit, a *hybrid*, which in part uses integrated components. We consider some of these circuits and their components, but this entire topic is so vast and the manufacturing techniques so intricate that attention can be paid here only to the underlying principles.

The chip of semiconductor, usually silicon, on which the components are formed is a single crystal doped either *n*- or *p*-type. The components are then formed by diffusing into this chip in selected areas a succession of doping agents in such geometry as to produce elements with the desired properties. For example, a *p*-type substrate might be altered by diffusing successively *n*-, *p* and *n*-type dopants into selected areas to produce two *n-p-n* transistors side by side, Fig. 13-19. Leads between the transistors and to external signals and voltages can be made by selective deposition of aluminum insulated, where necessary, from other parts of the circuit by layers of a dielectric material such as SiO_2. These two transistors are isolated from each other by the two *p-n* junctions between the *n*-type collector and the *p*-type substrate. This is not perfect isolation since these junctions do provide some capacitive coupling.

Another method of isolation may be provided by insulating the active devices by a passive dielectric layer. Such a geometry is given in Fig. 13-19*b*, where a passive layer of SiO_2 has been deposited (by a necessarily intricate several-step process) to provide isolation between the two devices. Again, external and internal leads can be provided by appropriate deposits of a metallic conductor such as Al.

Capacitors for integrated circuits may be either appropriately biased *p-n* junctions or thin-film parallel plate capacitors. Each type has its special advantages and disadvantages in circuit usage.

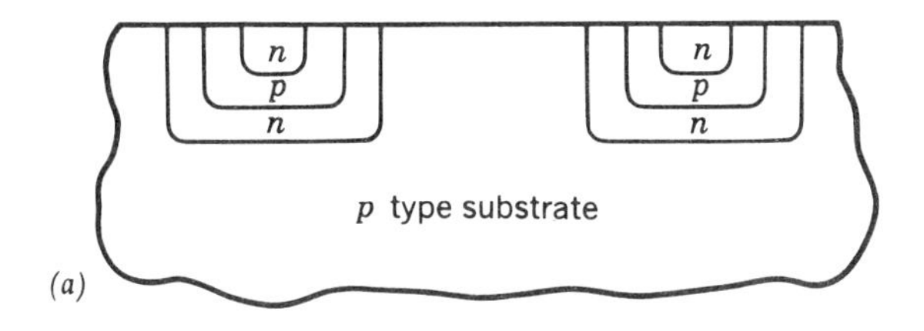

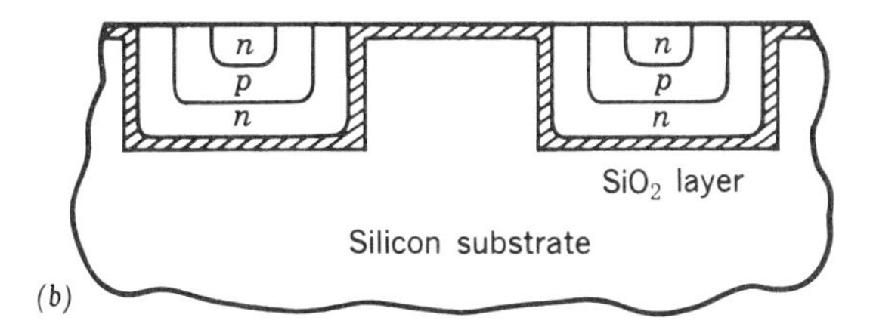

Figure 13-19 (*a*) Two *n-p-n* transistors produced in a single chip of Si by successive diffusion of properly selected dopants in selected areas. (*b*) Isolation of the transistors by a passive layer.

The capacitance of a *p-n* junction arises from the double layer of ionic charge. The boundaries of the depletion layer correspond to the "plates" of the capacitor; the distance between them is about typically 1 micron (i.e., about 10^{-6} m) for a doping level of about 10^{20} atoms per cubic meter. Since the relative dielectric constant of Si is 12, the capacitance of such a junction is about 10^{-4} f/m², or about 0.1 pf/mil². Areas on the integrated circuit up to size 1,000 mil² can be used readily, so capacitors of order several hundred pf can be attained.

The biased *p-n* junction has several disadvantages as a capacitor. It is usually formed by diffusion of several layers into the Si chip at the same time as other components are often determined by requirements on other components. Furthermore, the junction capacitor and the substrate together form a transistor; hence the junction must be maintained in reverse bias to avoid transistor action (see Fig. 13-20*a*). Finally, the depletion layer of the junction is a function of the bias across the junction, so the capacitance will be modulated by a variable voltage across it.

Thin film capacitors possessing discrete layers of true dielectric materials overcome some of these undesired characteristics. They are usually formed of two metallic (or semiconducting) plates separated by a thin deposited layer of a dielectric such as SiO, SiO_2, Ta_2O_5, Al_2O_3. (See Fig. 13-20*b*.) Since the layers of dielectric can be made thin (500 to 1000 A), the capacitance per unit area is about 10 times that of a junction capacitor of the same area. Thin film capacitors with breakdown voltage of 10 to 50 volts and with capacitance around 1 pf/mil² can be prepared.

Resistors for integrated circuits may be either a properly doped

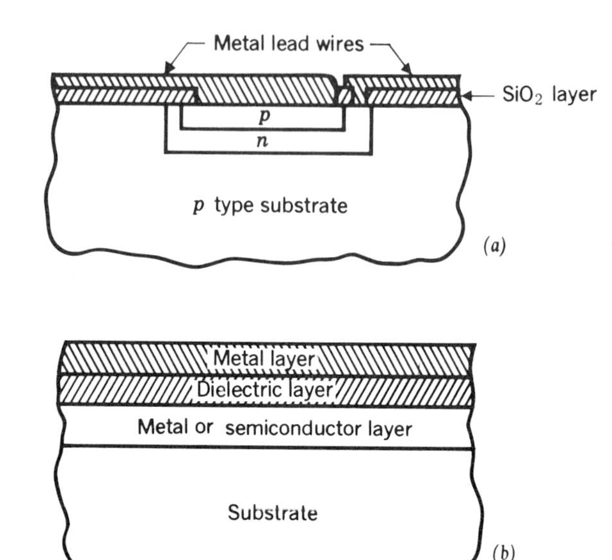

Figure 13-20 Capacitors for integrated circuits. (*a*) Typical *p-n* junction capacitor. (*b*) Thin film capacitator.

section of the semiconductor chip itself or a deposited strip of a metal or alloy. The first type, called a *diffused* resistor, may be prepared by diffusing a proper dopant to a specified depth over a carefully marked geometrical strip of the chip. Resistors in the range of a few tens of ohms to more than ten thousand ohms can be prepared in this way. Metallic connections of aluminum metal can readily be deposited and evaporated layers of insulator serve to isolate the resistor on the surface.

Such resistors have a significant defect. Since they are formed in the chip itself, they must be isolated by a properly biased *p-n* junction (see Fig. 13-21*a*). Consequently, they have certain constraints in use. Furthermore, the distributed capacitance of a *p-n* junction appears along their length; parasitic transistors are also present.

As a consequence of this, *thin-film* resistors of deposited metal or alloy are often produced in specified geometries on the surface of the chip. Such metals as Al, Ta, Cr, Ni and the alloy nichrome (80 per cent Ni and 20 per cent Cr) may be used. Resistance values from near zero to the range of 100,000 ohms are possible. Since the resistors are commonly insulated from the substrate by an insulating film, Fig. 13-21*b*, they have many fewer parasitic features. In particular, their distributed capacitance is lower, so they are useful to higher frequencies.

The reader can find in the literature many excellent discussions, micrographs and analyses of these and many other circuit components. References 2, 3, and 4 are excellent places to start. Perhaps the most striking feature of their fundamental character is this: truly monolithic circuits are a continuously ordered array of Si atoms with local regions doped *p*- or *n*-type in minute amounts. The doping levels commonly used, from 10^{-4} per cent to perhaps 10^{-1} per cent, are very small compared to the alloying levels used in almost any other material of

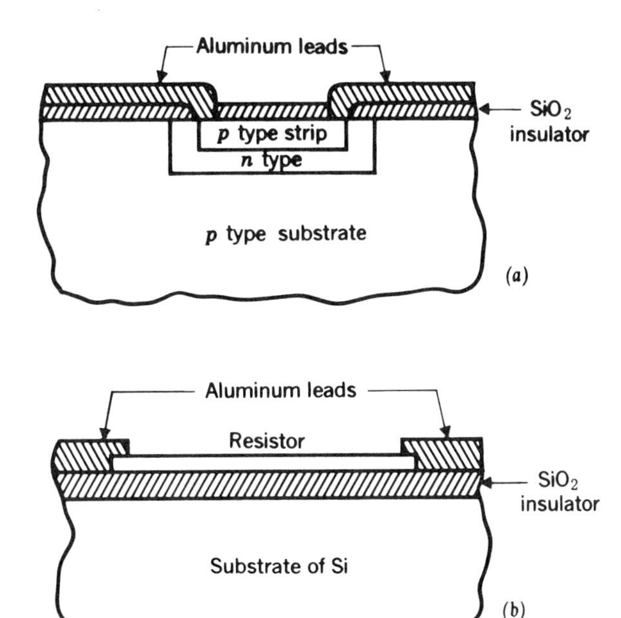

Figure 13-21 Resistor for integrated circuits. (*a*) Longitudinal section of a diffused resistor. The *p*-type strip is insulated from the rest of the circuit by a *p-n* junction. (*b*) Longitudinal section of a thin film resistor.

technological importance (most steels contain at least a few per cent of alloying addition).

13-10 Summary

This chapter contains a discussion of some of the elementary aspects of electron circuit devices with emphasis on the influence of materials on their properties. A myriad of details might have been supplied and many other optical, thermal, piezoelectric, and magnetic devices might have been considered. To attempt a topic so broad would have been a futile task. Probably no area of investigation and technology is expanding at a faster rate than that of devices based upon electronic and magnetic properties of solids.

Another area which must unfortunately be passed by is that of purification of semiconductors and of the subsequent alloying processes. The thermodynamics of the phase diagrams alluded to in Chap. 5 and the kinetics of interdiffusion of alloying elements are fascinating subjects. The successful technological development of device manufacture and the continued evolution of more intricate, minute, and reliable devices has resulted from a marvelous coalition of scientists and engineers, mainly physicists, chemists, metallurgists, ceramists, and electrical engineers.

Finally, the use of semiconductors and semimetals seems to be in its infancy as far as materials themselves are concerned. Though Ge and Si have been extensively used, their use has been predominant because purification and alloying techniques for them have been relatively simple. As material development in compound semiconductors proceeds, materials better tailored to specific uses will be found. The coming generation of scientists and engineers will enjoy the pleasures of taking part in this development.

REFERENCES

1. N. A. Lee, B. Easter and H. A. Bell, *Tunnel Diodes*, Chapman and Hall, Ltd., London, 1967.
2. K. J. Dean, *Integrated Electronics*, Chapman and Hall, Ltd., London, 1967.
3. R. M. Warner, Jr., and J. N. Fordemwalt, *Integrated Circuits, Design Features and Fabrication*, McGraw-Hill Book Company, New York, 1965.
4. D. K. Lynn, C. S. Meyer and D. J. Hamilton, *Analysis and Design of Integrated Circuits*, McGraw-Hill Book Company, 1967.
5. W. Shockley, "A Unipolar Field-effect Transistor," *Proc. IRE*, **40**: 1, 365 (1952).
6. L. B. Valdes, *The Physical Theory of Transistors*, McGraw-Hill Book Company, Inc., New York, 1961.
7. James F. Gibbons, *Semiconductor Electronics*, McGraw-Hill Book Company, Inc., New York, 1966.

PROBLEMS

13-1 (*a*) Calculate the drift velocity of electrons in a homogeneous bar of Si in a field of 1000 volts/m.

(*b*) How long is required for an electron to drift one centimeter under such conditions?

(*c*) How much heat is generated in the crystal during that period? Where did this heat come from?

(*d*) For an average effective mass of about 0.3 m_0, find the value of τ.

(*e*) If the electron had the average velocity of a monatomic gas, $(\frac{3}{2})kT$, find its thermal velocity. Determine then the mean free path between collisions, λ.

13-2 Calculate the ratio of total electron conduction current to total hole conduction current in (*a*) intrinsic Ge and (*b*) 0.05 ohm-m *p*-type Ge.

13-3 (*a*) Find the necessary density of Sb atoms in Si to produce material of resistivity 10^{-4} ohm-m. (Assume that μ_n has the value of about 0.10 m^2/volt-sec.)

(*b*) Find the thickness of a square film of such material required to produce a sheet resistance of 200 ohms per square.

(*c*) Find the length to width ratio of a strip of this material required to produce a resistor of 10,000 ohms.

13-4 Assume the following facts about the random diffusion of electrons (or holes) in a semiconductor:

1. That D for the charge carriers is given by $D = fd^2/6$
2. That $f = \frac{1}{2}\tau$
3. That $d = \lambda$
4. That the charge carriers have kinetic energy characteristic of a monatomic gas, i.e., that $KE = \frac{3}{2}kT$.

Show that a quantity μ, defined as $|q|\ \tau/m$ is related to D by the equation

$$\frac{D}{\mu} = \frac{kT}{|q|}$$

13-5 Describe the operation of a *p-n-p* transistor in the common base connection.

13-6 Show that τ in Eq. (13-9) is the average lifetime of a charge carrier. Show also that the half-life of excess charge carriers (i.e., the time at which $\Delta n(t) = \Delta n(0)/2$) is given by $\tau \ln 2$. Thus $\tau_{\frac{1}{2}}$ and τ are analogous to the half-life and mean-life in radioactive decay.

13-7 (*a*) The cube edge of Ge at room temperature is 5.66 A. Show that a crystal of Ge contains about 4.4×10^{24} atoms per cubic meter.

(*b*) For an impurity density of 10^{22} atoms per cubic meter, show that the impurity atoms are on the average, some 500 A apart. They are then some 10 Bohr radii apart (see Sec. 12-3) and interact only slightly.

(*c*) For an impurity density of 10^{24} atoms per cubic meter, show that the impurity atoms are about 60 A apart. They are, then, some 1 or 2 Bohr radii apart and would interact appreciably.

(*d*) At an impurity concentration of 10^{26} atoms per cubic meter, show that one impurity atom exists for a volume of about 50 unit cells. Hence they are only about 25 A apart. Overlap of wave functions is thus great and the exclusion principle gives rise to a band. According to Eq. (9-7*a*), the band would be filled to a level of about 0.08 ev if this were an *s*-state (make this calculation). Thus, the broadening is enough to cause overlap of this band with the valence or conduction band of the semiconductor (see Table 12-2), but it is not enough to cause overlap completely across the intrinsic gap.

13-8 Derive the general expression for the Hall constant in a two-carrier system. Show that it reduces to the simple expression derived in Chap. 11 for a one-carrier system.

14

Ionic crystals

14-1 Introduction

Ionic crystals are generally transparent insulators, whose electrical conductivity, though small, is due to the diffusive motion of the ions themselves. They are, of course, never pure elements, since the ionic bond requires electron transfer from one chemical constituent to another. The crystal structures of some ionic crystals are simple, as the structures of NaCl and CsCl show (Figs. 2-9 and 2-10), while others are extremely complex.

Many materials classed as ceramics are ionic crystals and have great practical usefulness. Some other ionic crystals, especially the simpler alkali halides (the salts), have an additional scientific usefulness, since they permit the study of phenomena in solids which are difficult to study in other crystal types. For example, the unique relationship between conductivity and ionic diffusion permits detailed study of atom motion in ionic crystals not possible for metals and semiconductors. Furthermore, crystallographic defects can be studied in these crystals in ways not possible for the good conductors (e.g., by resonance and optical methods).

Phenomena which might be described in a complete discussion of the electronic and crystallographic features of ionic crystals are too numerous for a single chapter. The selection made here is limited to those which can best be explained with simple models. The sections which follow concern:

1. The nature of the ionic bond
2. Imperfections and ionic conductivity
3. The electronic band structure
4. Color centers and luminescence

14-2 The Ionic Bond—Madelung Energy

One of the elements in a crystal composed of more than one element tends to capture electrons from the other (or others) because of the difference in electronic structure of the two. The structure which results is called an *ionic compound*. In some crystals containing more than one element the exchange of electrons from one element to the other is complete, and the crystal is entirely ionic in character; in other crystals only partial exchange of electrons occurs within the crystal and the crystal is not completely ionic.

In chemical compounds, some atoms have a tendency to gain electrons, forming negative ions, while other atoms lose electrons, forming positive ions. The energy given up when a neutral atom gains an electron and becomes a negative ion is called the *electron affinity*. The energy necessary to remove an electron from an atom creating a positive ion is, of course, simply the *ionization energy*. The electromotive series (Table 14-1) is a list of the elements in the order of ease of ionization in solution. At the top of the list are elements with the smallest ionization energy which form positive ions. At the bottom of the list are atoms with the maximum electron affinity which form negative ions. Atoms in the top part of the list therefore lose electrons to atoms toward the bottom in chemical compounds.

Table 14-1 Partial list of elements arranged in the electromotive series

Li
Rb
K
Ca
Na
Al
Zn
Cd
Sn
H
Cu
Hg
Br
Cl
F

A more quantitative description of the ionic bond can be made by considering the sequence of events which occurs when two atoms of different species are brought together to form a chemical compound. Sodium chloride may be used as an example. Since sodium and chlorine occur at opposite ends of the electromotive series (as do other examples of alkali metals and halogen gases), sodium chloride is one of the best examples of an ionic compound. First, consider the sodium and chlorine to be free atoms at an infinite distance of separation. If sodium atoms are ionized by the removal of the valence electrons, an energy per atom equal to the ionization energy, 5.14 ev, is required. If these electrons are transferred to the chlorine atoms to form chloride ions, an energy gain of 3.75 ev occurs. Thus positive work equal to 1.4 ev per "molecule" is necessary to form the ions at infinity. If finally the ions are allowed to come together to their equilibrium distance with the Coulomb force field acting between them, the solid is stable if the energy gained by the Coulomb force between oppositely charged ions is greater than the electron transfer energy of 1.4 ev.

The Coulomb energy per ion is not difficult to compute, because it is simply half the potential energy of an ion in the final structure. Surrounding any positive ion in the sodium chloride structure are six neighboring negative ions, 12 next-nearest-neighbor positive ions, and so forth. The Coulomb potential V at a positive ion is then a sum over all the ions in the solid. The final result is

$$V = \frac{-e^2}{4\pi\epsilon_v a_0}\left(6 - \frac{12}{\sqrt{2}} + \cdots\right) \tag{14-1}$$

where a_0 is the lattice spacing of the sodium chloride lattice. In sodium chloride, if the sum in the last equation is performed properly over an infinite crystal, the result is

$$V = \frac{-e^2}{4\pi\epsilon_v a_0}\alpha_M \tag{14-2}$$

where $\alpha_M = 1.748$. The constant α_M is called the *Madelung constant* after the man who first performed the sum. The constant is the same for all crystals of the sodium chloride type, because the sum in the parentheses in Eq. (14-1) is independent of the particular ions. For other crystal types, Madelung's constant has slightly different values, because the sum appearing in the parentheses is slightly different in each case. In Table 14-2, Madelung's constant for several crystal types is listed.

Table 14-2 Madelung constants

Crystal type	α_M
NaCl	1.748
CsCl	1.763
ZnS (zincblende)	1.638
ZnS (wurtzite)	1.641

The Coulomb energy and electron transfer energies are the primary contributions to the attractive energy of the ion. As for all crystals, however, a repulsive energy must be considered when the ions come close together. The repulsive energy is a result of the Pauli exclusion principle for a closed ion shell. The repulsive potential, however, decreases rapidly as the distance between ions increases, and it does not contribute appreciably to the overall energy of the crystal at the actual distance of ion separation.

Including all the energy terms, the cohesive energy of the crystal can be written as

$$E_c = \tfrac{1}{2}[A - I - R + V] \tag{14-3}$$

where A = electron affinity
I = ionization energy
R = repulsive energy
V = Coulomb energy of attraction given by Eq. (14-2).

The various terms in this equation are listed in Table 14-3 for sodium chloride. The slight discrepancy between the observed cohesive energy and the proper sum of the various components in the table is due to small neglected contributions, such as the van der Waals energy between the ions.

Table 14-3 Cohesive energy of sodium chloride (relative to free neutral atoms), $a_0 = 2.81$ A.

Electrostatic energy, V	8.86 ev
Repulsive energy, R	1.02 ev
Ionization energy of Na, I	5.14 ev
Electron affinity of Cl, A	3.75 ev
E_c [calculated by using Eq. (14-3)]	3.25 ev
Observed cohesive energy per ion (E_c)	3.3 ev

The primary feature in the calculation of the cohesive energy of sodium chloride in Table 14-3 is that the major contribution is the electrostatic energy. The electron transfer takes very little energy, since the ionization energy of the positive ion and the electron affinity of the negative ion very nearly cancel each other. The simple Coulomb-force law between the ions of the ionic crystal is seen in later sections to exert a major influence on other characteristics of the ionic crystals, in addition to that of the energy. A tabulation of binding energies of a number of compounds relative to the free ions is given in Table 14-4.

Exchange energy

One feature of the preceding calculation of the cohesive energy of ionic crystals has not been adequately explained. The discussion so far gives one no reason to expect an atom like fluorine or chlorine to accept an additional electron to form an ion with a release of energy. The additional factor needed to understand the peculiar stability of a closed electron shell in an atom is called the *exchange energy*, and we make a small digression to describe its origin. The electronic binding energies in atoms were explained in Chap. 8 in terms of the Coulomb energy of attraction between electrons and the nuclei, together with

Table 14-4 Measured lattice energies of some ionic crystals†

Crystal	Lattice constant, A	Lattice energy, kcal/g-mole	
		Theoretical	Experimental
LiCl	5.13	199.2	198.1
LiBr	5.49	188.3	189.3
LiI	6.00	174.1	181.1
NaCl	5.63	183.1	182.8
NaBr	5.96	174.6	173.3
KCl	6.28	165.4	164.4
KBr	5.59	159.3	156.2
KI	7.05	150.8	151.5
CsCl‡	4.11	152.2	155.1
CsBr‡	4.29	146.3	148.6
CsI‡	4.56	139.1	145.3
AgCl	5.55	203	205.7
AgBr	5.77	197	201.8
AgI§	6.47	190	199.2
CuCl§	5.41	216	221.9

Structures are similar to NaCl except when marked as follows

† After Mayer.

‡ CsCl structure.

§ zincblende structure.

the screening effect of the electrons. This is also sufficient to explain the ionization energy of the electron in lithium. When one writes the correct wave function for those electrons which overlap one another, as in the valence shells, and uses this wave function to compute the energy of the state, there is an additional contribution to the energy of the state called the exchange energy. While it is related to the screening energy, it cannot be described on the basis of classical ideas, since it is a quantum-mechanical effect due to the form which the wave functions take when many electrons with different wave functions occupy the same region of space. The exchange energy becomes appreciable whenever a given shell has more than one electron.

We have already seen that a neutral chlorine atom gains energy by gaining an additional electron in the valence shell to fill it. This energy gain occurs in spite of the fact that no Coulomb energy of attraction exists between the electron and the neutral atom. However, the exchange-energy force acts between the electron and the neutral atom when the valence shell is almost full, and it is a considerable effect.

14-3 Imperfections in Ionic Crystals: Ionic Conductivity

The general discussions of imperfections in the earlier chapters of this book must be extended for ionic crystals, for two reasons. First, special effects exist because at least two elements are present. Second, the presence of ions brings up questions of charge neutrality in the crystal. A review of the basic imperfections follows, with the modifications which are necessary for ionic crystals.

Effective charge of point defects

Since defects in ionic crystals represent rearrangements of the ions, they create long-range electric fields in the crystal. For example, the charge neutrality in the immediate vicinity of a positive-ion vacancy

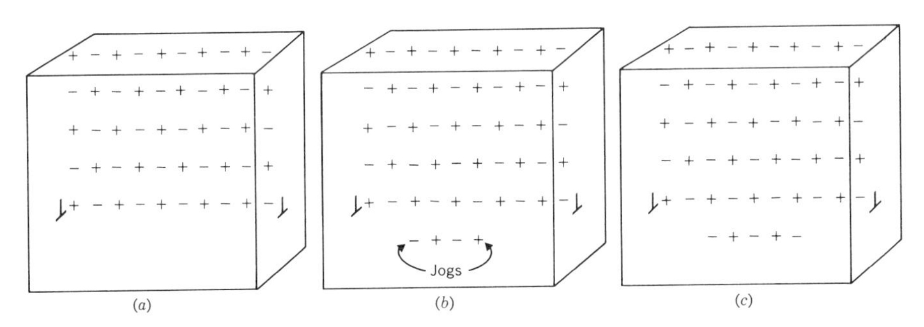

Figure 14-1 Charged jogs on edge dislocations. (*a*) This figure represents the extra half plane inserted in an ionic crystal. There are exactly the same number of positive ions and negative ions so that the simple edge dislocation represents no charge singularity in the crystal. The dislocation is along the bottom edge of the extra half plane, and the slip plane is perpendicular to the plane of the figure at the bottom edge of the half plane. (*b*) In this figure some ions have been added to the extra plane, leaving vacancies at lattice sites within the crystal. It is seen that now the dislocation is not a straight line, since the bottom edge of the extra half plane has had a partial row of atoms added to it. The slip plane of the dislocation therefore has a jog in it at the points where the partial line of ions ends. The two ends of the partial row of atoms are therefore called *jogs* in the dislocation. (*c*) The jogs on the edge dislocation possess an effective charge of $\pm e/2$. As successive atom ions are added to one of the jogs, the jog must change its sign by one whole unit of electric charge. This means that the jog ending in a negative ion differs by one electronic unit from the jog ending in a positive ion. Since the jog ending in a positive ion must be exactly symmetrical about zero with the jog ending in the negative ion, the charge on each jog must be a half unit of electronic charge.

is disturbed. The local region containing the vacancy has one fewer positive than negative ions, and from the point of view of the perfect crystal an effective charge is present at the point of vacancy. Since the positive-ion vacancy is caused by the removal of the positive ion, the effective charge at the vacancy is the negative of the charge on the ion. Thus electrons or other vacancies interact with the vacancy as though it is an effective negative point charge in the crystal.

A somewhat more complex point-charge effect is observed on dislocations. Although a pure edge dislocation is uncharged, a net charge exists at points on the dislocation called *jogs*. The geometry of the jogs and the reason for the existence of the charge are described in the caption of Fig. 14-1.

Vacancies

Figure 14-2 shows a (100) plane of a sodium chloride crystal with vacancies on the sodium sublattice. Sodium-ion vacancies are formed by transfer of ions from the interior lattice positions to new lattice points on the surface. Thus the interior of the crystal does not have a stoichiometric ratio of sodium and chloride ions, while the surface has an excess of sodium ions. The surface is therefore charged with

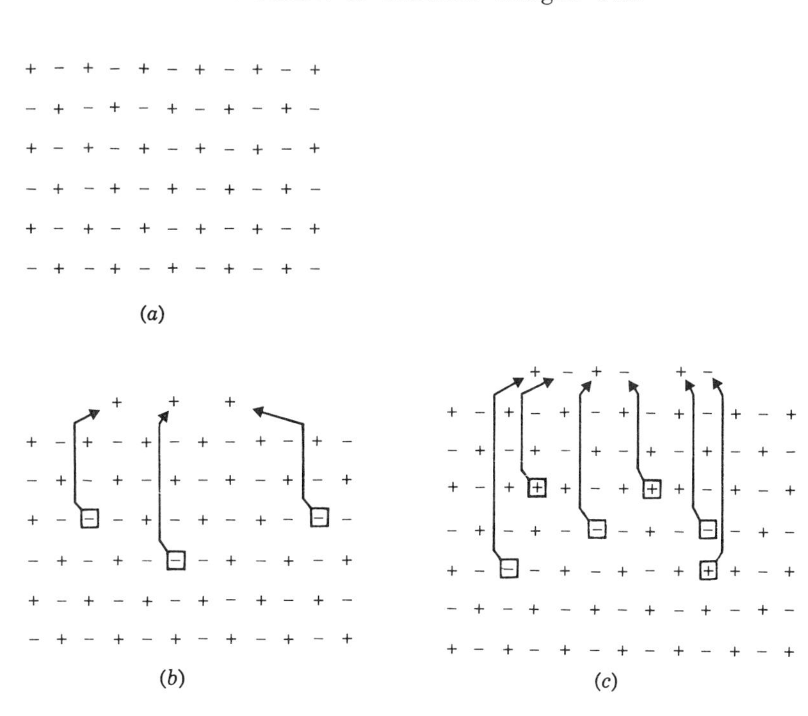

Figure 14-2 Schottky defects in one plane of a sodium chloride crystal. (*a*) The perfect lattice containing no vacancies. (*b*) Sodium ions are removed from some lattice sites in the interior of the crystal and placed upon the surface. When the process is completed, a large voltage is generated between the surface and the interior, which is not observed. Note that, in all figures, we have placed minus signs in the boxes representing the absence of Na (positive) ions. This minus sign is to remind the reader that the Na vacancy really represents a net negative charge in an otherwise perfect crystal. (*c*) The charge neutrality of the crystal can be restored by creating exactly the same number of vacancies in the chloride sublattice. The basic unit is therefore the *pair* of vacancies, one from each sublattice, called the Schottky defect.

respect to the interior. Obviously, this charge displacement causes the energy necessary to form additional vacancies to be larger so that the number of excess vacancies on one sublattice in the crystal is small. However, if the same number of vacancies are produced in each sublattice of the sodium chloride crystal, then the interior of the crystal is, on the average, neutral. One therefore expects that vacancy formation will occur in ionic crystals so as to preserve charge neutrality. Thus, the basic vacancy unit in an ionic crystal such as sodium chloride is composed of one positive-ion vacancy and one negative-ion vacancy, a defect called a *Schottky defect*. In a crystal like magnesium chloride, $MgCl_2$, where the magnesium ion is doubly charged, each vacancy on the magnesium sublattice is compensated by two vacancies on the chloride sublattice.

Some charge separation between the surface and the interior of an ionic crystal such as NaCl does actually occur. Sodium ions being

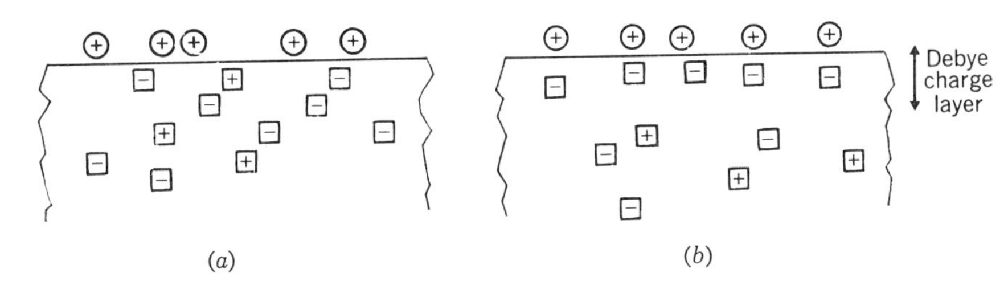

Figure 14-3 Formation of Debye layers in ionic crystals due to unequal vacancy-formation energies. Positive-ion vacancies are represented again by ⊟ because of their effective negative charge and negative ion vacancies by ⊞. (*a*) Uniform distribution of positive- and negative-ion vacancies, which are present in different amounts because of different energies of vacancy formation. A voltage is formed between the surface and the interior of the crystal. (*b*) The overall energy of the crystal is decreased when the excess positive-ion vacancies, ⊟, migrate toward the surface, forming a Debye space-charge layer. The interior of the crystal below the Debye layer now possesses equal numbers of positive-ion vacancies and negative-ion vacancies.

different from the chloride ions, the energy of formation of a single vacancy in the sodium sublattice is different from the energy of formation of a vacancy in the chloride sublattice. Thus, in spite of the obvious requirements of charge neutrality, the probability for a (positive-ion) vacancy, $p_v{}^+$, should be related to the Boltzmann factor which contains the energy of formation, E^+,

$$p_v{}^+ = e^{-E^+/kT} \tag{14-4}$$

Likewise, the probability of a negative-ion vacancy, $p_v{}^-$, is given in terms of the energy of formation E^- by

$$p_v{}^- = e^{-E^-/kT} \tag{14-5}$$

A possible arrangement of charge for $E^- > E^+$ is sketched in Fig. 14-3, where an excess of positive charge exists on the surface of the crystal and an excess negative charge inside. However, if the vacancies are mobile, then the excess vacancy component (in the figure, the positive-ion vacancies) moves toward the surface, forming a dipole layer with excess positive ions on the surface and neutralizing positive-ion vacancies lying a small distance beneath. The width of the dipole layer depends on the temperature and the density of the excess positive-ion vacancies. In general the layer is several hundred atom spacings in width. The interior of the crystal is neutral with this charge configuration. The argument has been stated here in terms of a surface-charge layer (called a Debye layer); however, any source of vacancies within the crystal possesses an enveloping layer or a charge cloud. Such Debye clouds are important on dislocations, which are postulated to be the major source of vacancies in real crystals.

Interstitials

Interstitials must also satisfy charge conservation in the crystal. When a positive ion is brought from the surface and inserted into an

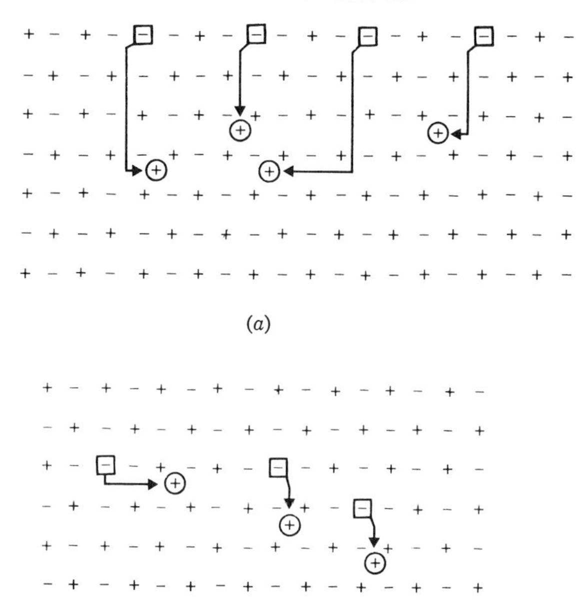

Figure 14-4 Frenkel defects. (*a*) Interstitial positive ions are inserted into the crystal from the surface, creating a distribution of positive-ion interstitials. The interior of the crystal is charged relative to its surface. (*b*) If equal numbers of positive-ion vacancies are also created, the interior of the crystal again becomes locally neutral. One can of course neglect the surface in the formation of a Frenkel pair. The pair may be formed by taking an ion out of its normal lattice site and forcing it into an interstitial site somewhere in the crystal.

interstitial position, the interior of the crystal contains one excess positive charge, while the surface has lost the same charge. A distribution of such interstitials also produces large electric fields within the crystal. Charge balance in the crystal is maintained, however, if the positive ions are not taken from the surface but are removed from lattice positions within the crystal. For this case, the basic lattice imperfection is the interstitial ion together with the vacant lattice site from which it was taken. This defect is called the *Frenkel defect* (Fig. 14-4).

The energy of formation of a Frenkel defect is the formation energy of positive-ion vacancy plus the formation energy of the positive-ion interstitials. A Frenkel defect as described may be created in one sublattice (say, the positive ions); however, they may exist in both. Since the energy of forming a Frenkel pair in one sublattice is different from the energy of formation in the other, one type of Frenkel defect predominates. In general, since the positive ion is the smaller of the two, producing less lattice strain, the positive-ion Frenkel defect is predominant.

Dislocations

Because edge dislocations in crystals are the actual sources and sinks for vacancies and interstitials, the above discussion suggests that, if

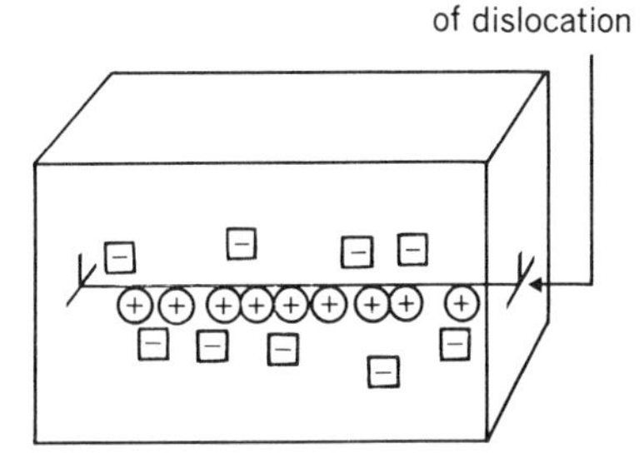

Figure 14-5 Charged dislocations. As sources of vacancies in the crystal, the dislocations develop charge clouds due to the difference in the formation energy between positive- and negative-ion vacancies. The dislocation is drawn showing the excess positive ions, ⊕, distributed along it, which cancel the charge due to the excess positive-ion vacancies, ⊟, in the surrounding Debye cloud.

the formation energy of one vacancy from one sublattice is different from the formation energy of the vacancy from the other, a Debye cloud should be associated with the charged dislocation (Fig. 14-5). The charge cloud around dislocations has been demonstrated experimentally. Since the dislocation is more mobile than its neutralizing charge cloud, the dislocations can be moved away from their neutralizing charge clouds by mechanical stress. The electric dipole thus created can be measured.

Ionic conduction

Vacancies and interstitials in an ionic crystal are charge centers. Therefore, the charge is transported with them when they move through the crystal. When an external electric field of magnitude E is placed across the crystal, positive-ion vacancies (representing negative charges in the crystal) diffuse toward the positive electrode, while negative-ion vacancies diffuse toward the negative electrode. (The reader should not in this section confuse E without subscript, which stands for the magnitude of the electric field, with the various activation energies E_m, E_v, which are always written with subscripts.) This drift motion of vacancies, a small effect superimposed on the random thermal motion, represents that part of the conductivity in the crystal which is associated with the diffusion of the vacancies. The quantitative discussion of ionic diffusion is based upon an inspection of Fig. 14-6. In the electric field, a vacancy with charge q in the valley at $x = 0$ has a lower activation energy for moving to the right than for moving to the left. The probability of jumping to the right is given by

$$f_r = \nu e^{-(E_m - qEa/2)/kT} \tag{14-6}$$

The probability of jumping to the left is given by

$$f_l = \nu e^{-(E_m + qEa/2)/kT} \tag{14-7}$$

In these equations ν is the vibrational frequency of a vacancy, and E_m is its motional energy.

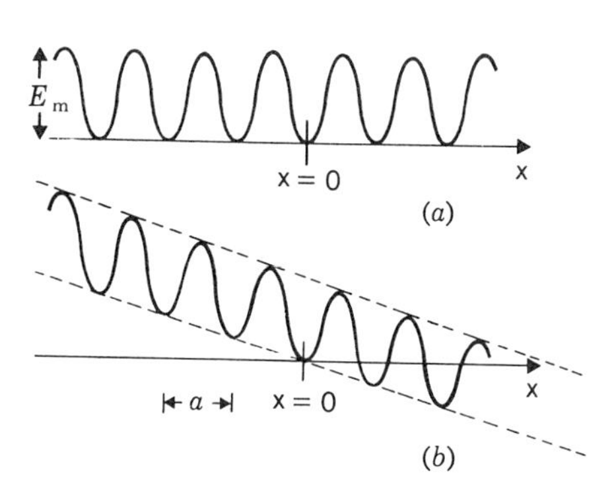

Figure 14-6 Energy of a vacancy as a function of position in the lattice. (*a*) Without external electric field. (*b*) With external electric field of magnitude *E*. The slope of the envelope shown by the dotted line is $-qE$. The lattice spacing is shown as *a*, and the effective charge on the vacancy is *q*.

The net rate of jumping of the vacancy down the external field gradient, f, is the difference between the jumps to the right and the jumps to the left.

$$f = f_r - f_l = 2\nu e^{-E_m/kT} \sinh \frac{qEa}{2kT} \tag{14-8}$$

A positive f means a net frequency of jumping to the right, a negative f to the left. Since the change in potential in the external field from one lattice site to another is small compared with thermal fluctuations, $\sinh (qEa/2kT)$ can be approximated by qEa/kT to give the expression

$$f = f_R - f_L = \frac{\nu qEa}{kT} e^{-E_m/kT} \tag{14-9}$$

The total net flux of vacancies in the field, of course, depends also upon the vacancy density; so the rate of jumping must be multiplied by the number of vacancies in a plane of width a to get the total vacancy flux $n_v af$.

$$n_v af = \frac{nqEa^2\nu}{kT} e^{-(E_v + E_m)/kT} \tag{14-10}$$

$$n_v af = \frac{nqE}{kT} D \tag{14-10a}$$

where $n_v a$ = density of vacancies within the plane of atoms at $x = 0$
n = density of positive ions in the crystal per cubic meter
D = self-diffusion coefficient of the ions
$= a^2 \nu e^{-(E_v + E_m)/kT}$

The mobility μ of a diffusing vacancy can be written in terms of the average velocity fa

$$\mu \equiv \frac{v_{\text{drift}}}{E} = \left|\frac{fa}{E}\right| = (\nu a^2 e^{-E_m/kT})\frac{|q|}{kT} = \frac{D_v\,|q|}{kT} \tag{14-11}$$

where D_v, the vacancy-diffusion coefficient, is equal to the factor in brackets. Note that the right-hand side does not contain E_v. Equation (14-11) is called the *Einstein relation* between the drift velocity of the vacancy in the external field and the diffusion coefficient of the vacancy. This formula has a universal validity when a particle diffuses in an external force field and has already been referred to in Eq. (10-1) and Prob. (13-4).

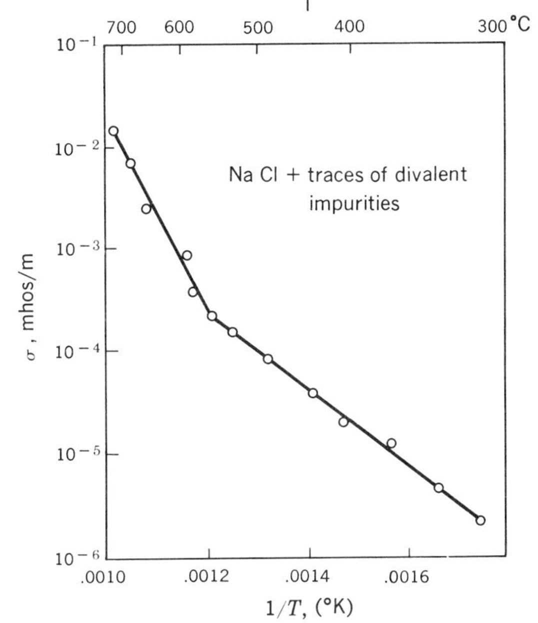

Figure 14-7 The ionic conductivity of NaCl as a function of temperature. The logarithm of the conductivity is plotted as a function of $1/T$, and the slope of the line is proportional to the activation energy. [*After H. Etzel and R. Maurer, J. Chem. Phys.*, **18:** 1003 (1950).]

Equation (14-10) can be written in a form that demonstrates Ohm's law for ionic crystals. If J is the electrical current flux density and σ the conductivity, then

$$J = n_v qaf = \sigma E$$

from which

$$\sigma = \frac{nq^2a^2\nu}{kT}\, e^{-(E_v+E_m)kT} \tag{14-12}$$

Equation (14-12) shows that the ionic conductivity follows a simple exponential law containing both the formation energy and the motion energy of the ions. The ionic conductivity in *real* crystals is, however, a more complicated function of temperature. Figure 14-7 shows a plot of the logarithm of the ionic conductivity of a sodium chloride crystal as a function of reciprocal temperature. The slope of the line is a measure of the activation energy for the process. The activation energy at high temperatures is that found for diffusion and satisfies Eq. (14-11). However, at lower temperatures, the activation energy decreases to a much lower value. The temperature at which the change occurs depends upon the divalent impurity content in the crystal.

To explain the effect of impurities on the conductivity, suppose that a perfectly pure crystal of sodium chloride has some magnesium ions substituted for sodium ions. Since the magnesium ion is doubly ionized and the sodium ion is only singly charged, the process of substitution charges the crystal electrically. However, if an additional sodium ion is removed from the crystal when the divalent positive ion is added, then charge neutrality of the crystal is preserved. Thus, when divalent-ion impurities are added to a monovalent salt, positive-ion vacancies in like concentrations are also added. At low temperatures, where the equilibrium density of vacancies is very low, a constant density of vacancies exists equal to the divalent-ion density. The vacancies contribute to the ionic conductivity of the crystal. The role of fluctuations now is only to move the existing vacancies, since it does not need to produce them also. The activation energy in the low-temperature "impurity range" is therefore just the motion energy for the vacancies. The break in the curve occurs at the point where the thermal or intrinsic vacancy density about equals the impurity density, and its position can be used as a measure of the purity of the crystal.

14-4 Electron Band Structure

Ionic crystals are insulators, and so their band structure must consist of a full valence band separated by a finite energy gap from an empty conduction band. No free Fermi surface exists, and considerable energy is required to excite electrons into the conduction band. A partial list of the widths of the band gap in some ionic crystals is given in Table 14-5. The alkali halides, which are among the most stable ionic crystals, have large band gaps.

The band structure of ionic crystals has not had the detailed study accorded to either semiconductors or metals, partly because ionic crystals do not have the practical electrical importance of either the metals or the semiconductors (except for their use as insulators, where detailed knowledge of the band structure is unimportant). Furthermore, free electrons do not participate in the normal properties of the ionic crystals. Electrons are produced in the crystal, however, by photons of sufficient energy, and the electronic structure manifests itself in the photoconductivity of the crystal and the general interaction of the crystal with the exciting light.

Table 14-5 Band gaps of several ionic crystals†

Crystal	Approximate band gap, ev
LiF	11
LiCl	9.5
NaF	11.5
NaCl	8.5
NaBr	7.5
KF	11
KCl	8.5
KBr	7.5
CsCl	8.4
AgCl	6
AgBr	6

† These values are only approximate, because of the difficulty of distinguishing the exciton absorption from the true fundamental absorption.

The "normal" band structure of an ionic crystal, such as NaCl, is sketched in Fig. 14-8. Here k is plotted in the [111] direction on the left and in the [100] direction on the right. The maximum of the valence band and the minimum of the conduction band occur at $k = 0$. This picture seems to be valid for the alkali halides, although, at this writing, sufficiently detailed calculations are not available to be sure that no surprises exist. The valence band corresponds to the filled p band of the halide and the conduction band to the s band of the alkali metal. The minimum energy necessary for photoexcitation is the excitation of the electrons for levels at $k = 0$ in the valence band to levels at $k = 0$ in the conduction band.

Photoexcitation in an ionic crystal such as AgCl shows that the band structure has quite a different shape from that of Fig. 14-8. For this crystal large excitation occurs at a wavelength below 2,000 A.

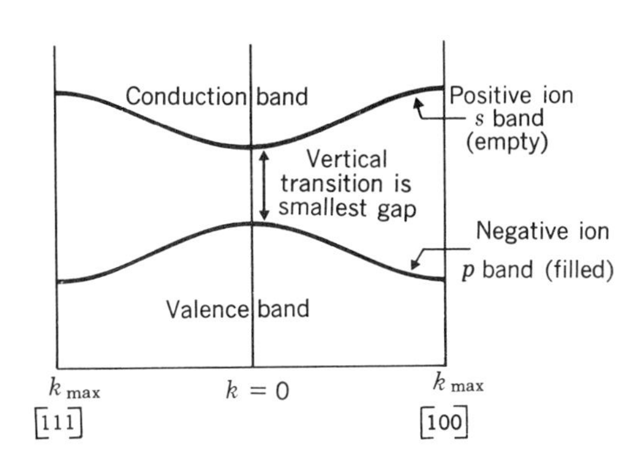

Figure 14-8 Schematic band structure for a "normal" ionic crystal. The maxima and minima of the bands occur at $k = 0$, and light absorption can occur in a "vertical" transition—i.e., direct excitation of electrons and holes at $k = 0$. Standard notation is used, with [100] plotted to right and [111] plotted to left.

Lesser excitation occurs for wavelengths longer than this in a so-called "tail" extending out to some 4,000 A. The band structure consistent with these observations is shown in Fig. 14-9. The maximum energy in the valence band occurs, not at $k = 0$, but rather at the Brillouin-zone boundary in the [111] direction. The main excitation of photo-electrons is across the gap at $k = 0$. Momentum is readily conserved in this process. Excitations from other values of k in the valence band to $k = 0$ in the conduction band are more difficult; i.e., they have smaller absorption coefficients, because the momentum change of the

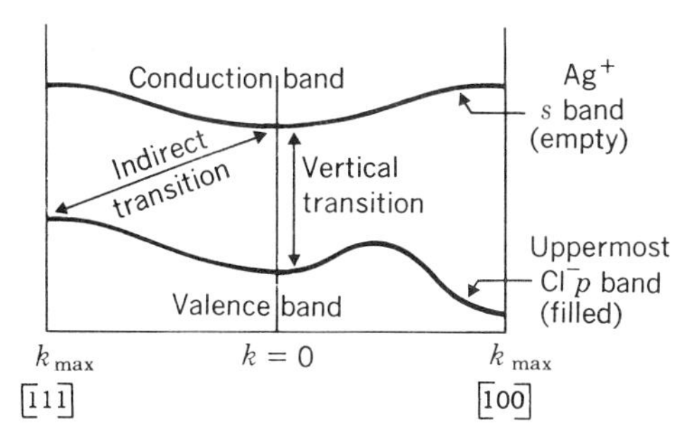

Figure 14-9 In AgCl, the maximum in the valence band occurs at k_{max} in [111], thus leading to indirect light absorption.

electron requires that phonons be produced in the crystal simultaneously. However, excitation from $k_{max}[111]$ in the valence band to $k = 0$ in the conduction band (called an indirect transition in Chap. 12) does occur, and since it requires less energy than the transition at $k = 0$, it produces the long-wavelength tail on the excitation curve.

Excitons

Some absorption of incident light occurs in ionic crystals for photon energies just less than that necessary to excite electrons completely across the energy gap. This absorption manifests itself in an *exciton peak* at energies just less than that of the absorption edge, Fig. 10-8. The characteristic of the exciton peak which makes it different from the fundamental absorption is that the crystal does not show extra electrical conductivity when the exciton light is being absorbed. This experimental fact is explained by assuming that the exciton absorption produces bound pairs of electrons and holes. Because electrons and holes are oppositely charged, they attract each other, producing bound-state levels similar to those of the hydrogen atom. Since the bound states have energies less than that of the separated (i.e., ionized) electron-hole pair, the exciton energy levels lie in the forbidden band gap below the fundamental absorption limit by an amount approximately equal to the hydrogenlike bound-state levels in the dielectric continuum. If the electron and hole are bound to one another, the pair is a neutral entity and does not contribute to electrical conductivity.

The hydrogenlike model for similar bound states in semiconductors is valid only when the orbit of the electron covers many electronic sites of the crystal. The static dielectric constant must be quite large, and this requirement is well satisfied for the semiconductors. For ionic crystals, dielectric constants are normally not sufficiently high for this approximation to be really satisfactory, however. For example, for Si, $\epsilon_r \simeq 12$, whereas, for NaCl, $\epsilon_r \simeq 2$ at high frequencies. Consequently the hydrogenlike model is only qualitatively applicable for ionic crystals. In the normal ionic crystal, the electron belonging to the cation is simply transferred to a neighboring anion during formation of the exciton, and the electron remains bound to the anions in the nearest-neighbor shell of the excited negative ion, Fig. 14-10.

An exciton is not bound to a particular pair of ions in the crystal, since the excitation can move from one set of ions to a neighboring pair because of the overlap of wave functions. From the alternative point

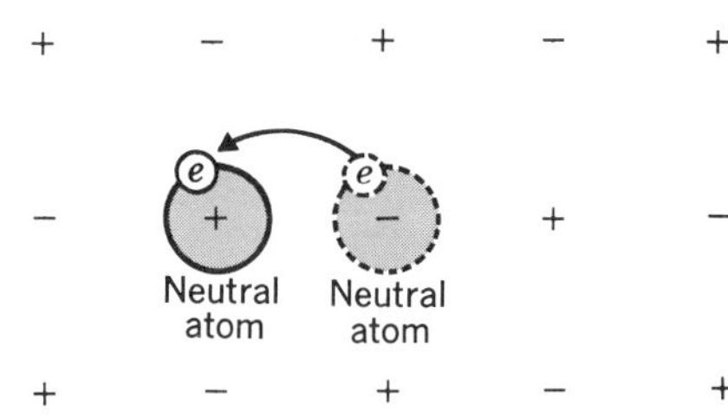

Figure 14-10 Localized exciton. The electron from a negative ion is transferred to a neighboring positive-ion site, creating a neutral pair of atoms in the crystal.

of view of the nonlocalized exciton, the hydrogenlike state may have translational energy in the crystal, and the excitation can move bodily through the crystal with a given velocity, corresponding to a hydrogen atom moving through a dielectric medium. Since the lifetime of an exciton is not usually very great, experimental evidence for exciton migration has been very difficult to obtain, however.

Polaron

Conduction electrons and holes in ionic crystals are much more complicated than those in semiconductors. In a semiconductor in an external field, the response of the electrons is described in terms of the effective mass of the electrons, which is in turn directly related to the electronic band structure. However, suppose that a free electron is created in an ionic crystal. Since the atoms surrounding it are ionized, the charge of the electron tends to polarize its immediate surroundings, i.e., to cause relative displacement of + and − ions. The electron tends to carry this polarized region with it when it moves. Since the polarization consists in ion displacement, the forced motion of an electron through the crystal causes some rearrangement of ion positions and the effective mass of the electron is thereby increased. For this reason, the conduction electron in an ionic crystal is often called a *polaron*. Theoretical estimates of the contribution to the effective mass of the moving charge from the polarization field suggest that about 30 per cent of the mass in the silver halides and more than half of the mass in the alkali halides are due to the polarization field. Thus the polaron effect is not small.

14-5 Color Centers and Luminescence

In Chap. 10 the representative ionic crystals were described as transparent in the pure state without imperfections. At wavelengths longer than the ultraviolet absorption, only in the infrared range is absorption of light observed. This absorption is due to the vibration of entire ions under an exciting electromagnetic wave. It occurs for a frequency of about 10^{13} cps. Thus, in general, pure salts tend to be completely transparent in the visible region of the spectrum. However, impure salts are often colored, because energy is absorbed in certain regions (called absorption bands) of the visible spectrum owing to the presence of various impurities and other point imperfections. A variety of absorption bands is found for even the simplest alkali halides, and a complete list of their properties is too large to incorporate into this book.

The more elementary absorption bands are caused by the presence of vacancies in the crystal. When the vacancies capture electrons or holes, they are capable of absorbing light and are called *color centers.* Sodium chloride is a good example of a crystal which can contain color centers. If an NaCl crystal is heated in an atmosphere of sodium vapor, it becomes slightly nonstoichiometric, with more sodium than chlorine ions. Excess sodium atoms absorbed from the vapor become ionized sodium ions in the crystal. The additional sodium ions require the formation of vacancies in the chloride lattice, and the extra electrons from the absorbed sodium atoms are injected into the conduction band. However, the excess electrons do not remain in the conduction band very long, since they are soon captured by the chloride-ion vacancies. Since the chloride-ion vacancy represents a net positive charge within the crystal, it attracts the electron to itself in exactly the same way as electron-deficient impurities attract electrons in the semiconductor crystal. Again the electron is not trapped in a large hydrogenlike orbit around the chloride-ion vacancy, but since ϵ_r for the alkali halides is only about 2 or 3, the orbit is not nearly so large as that of the semiconductors. Hence, the hydrogenlike orbit of such a bound electron in an ionic crystal encompasses only a few neighboring ions around the vacancy, and the approximation of a uniform dielectric medium is not accurate. In this case, a more accurate statement is that the electron is shared by all the neighboring positive ions which surround the vacant site (Fig. 14-11).

The electron captured by a negative ion vacancy in an alkali halide is termed an *F center* (from the German *farbzentre*). NaCl is colored a deep yellowish brown and KCl is colored blue by *F* centers. The absorption peak of the crystal containing color centers occurs over a band of frequencies which may be about 200 A wide. The energy of the peak of the F-center band is listed in Table 14-6 for some crystals.

The chlorine-ion analog to an F center also exists. If some of the alkali halides are heated in a halogen vapor, a different kind of absorption peak is observed. This absorption peak corresponds to positive-ion vacancies with captured holes, and the centers are called *V centers*. They are not as stable, however, as F centers and disappear at room

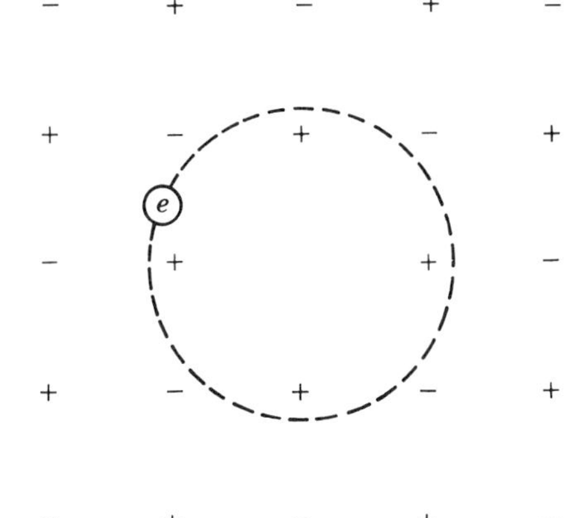

Figure 14-11 A schematic model of an F center in a sodium chloride type of lattice. The negative-ion vacancy has captured an excess electron. Since the dielectric constant is not large, the hydrogenic orbit extends not much farther than the neighboring shell of atoms around the vacancy.

temperature. This disappearance points up the fact that the thermal properties of the color centers are sometimes quite complex. As another example of thermal effects at higher temperatures, diffusion can take place and vacancies coagulate to form more complex centers with different absorption frequencies.

Color centers have been described as being formed by the nonstoichiometric addition of one of the components of the crystal.

Table 14-6 F-center absorption frequencies (peak value) for alkali halides

Salt	Energy, ev
LiCl	3.3 ev
NaCl	2.7 ev
KCl	2.3 ev
KBr	2.0 ev
BbCl	2.1 ev
CsCl	2.1 ev

However, they can also be generated in crystals by irradiation either by X rays or by energetic nuclear particles. The coloration induced in a crystal by the addition of a stoichiometric excess of one component yields a crystal with uniform color if the crystal is well annealed. Normally, however, X rays are absorbed very close to the surface, and crystals colored in this manner usually have a very densely colored layer near the surface, with the bulk of the crystal either uncolored or only very lightly colored. Energetic gamma rays, on the other hand, are absorbed throughout the crystal and give a uniform coloration.

Impurities in the salts also produce absorption bands in the crystal and often lead to a different type of behavior. For example, potassium chloride containing thallium impurities absorbs energy at a characteristic frequency, but reemits it at a different frequency. This reemission is called *luminescence* and occurs in "activated" potassium chloride when the concentration of thallium ions is small. Luminescence occurs by the following process: When the thallium ion is excited by a photon, its apparent size in the crystal changes and the neighboring ions change their positions. With the neighboring ions in a different position, the excited state itself shifts its energy to a lower energy state in an adiabatic way and reemits a light photon in a short time. However, the excited state at the time when the photon is emitted has a lower energy than it had when the original photon was absorbed, and so the emitted photon has a lower frequency than the absorbed photon. Thus, a crystal absorbing ultraviolet light can reemit visible light.

The length of time required for the excited impurity to reemit a photon varies considerably with impurity. If the reemission occurs immediately, i.e., in the normal time for atomic light emission—about 10^{-8} to 10^{-7} sec—the process is called *fluorescence*. However, the emission often occurs after a delay of seconds or even hours; then it is called *phosphorescence*.

Again, the properties of luminescence with various types of impurities in various salts is a broad field and is inappropriate for further description in this book.

PROBLEMS

14-1 Compute the Madelung's constant for two-dimensional NaCl. The series is not absolutely convergent, and so the sum must be carried out over successive shells of atoms, each shell enclosing a neutral group of atoms. Carry the summation out to about $4a_0$.

14-2 Consider the existence of positive and negative vacancies in an ionic crystal of the NaCl structure. Suppose that the number of positive-ion vacancies n_v^+ obeys the expression $n_v^+/n = e^{-E_v^+/kT}$ and the number of negative-ion vacancies $n_v^-/n = e^{-E_v^-/kT}$, where n is the number of positive (or negative) ions in the crystal. Suppose that $E_v^+ = 0.75$ ev and $E_v^- = 1.25$ ev. (*a*) If the crystal is a sphere 1 cm in radius, compute the surface charge on the sphere at 1,000°K. Assume that the vacancies are uniformly distributed in the interior but that the neutralizing charge is placed exactly on the surface. (*b*) With the uniformly distributed vacancies assumed in (*a*) calculate the energy necessary to form a new vacancy of each type at the center of the sphere, taking account of the electrostatic field generated through the sphere by the unequal number of vacancies.

14-3 Suppose that the surface of an ionic crystal is a plane (the xy plane). Let the formation energy of a positive-ion vacancy be E_v^+ and of a negative-ion vacancy E_v^-. Find the actual distribution of vacancies as a function of the distance z from the surface. Assume that the density of vacancies is given by the formulas

$$n_v^+(z) = ne^{-(E_v^+ - |e|\Phi)/kT}$$

$$n_v^-(z) = ne^{-(E_v^- + |e|\Phi)/kT}$$

where E_v^+ and E_v^- are the formation energies for single positive-ion and negative-ion vacancies, respectively. Φ, which is a function of z, is the electrostatic potential at the point z produced by the unequal density of negative- and positive-ion vacancies. (Recall that a positive-ion vacancy has an effective negative charge in the crystal.) The net charge density at the point z is thus $\rho(z) = e(n_v^- - n_v^+)$, and the potential function is the solution of Poisson's equation $\nabla^2\Phi = -4\pi\rho$. Hence

$$\frac{d^2\Phi}{dz^2} = 4\pi\,|e|\,(n_v^+ - n_v^-)$$

Assume in the calculation that (high T)

$$\frac{n_v^+}{n} = e^{-(E_v^+ - |e|\Phi)/kT} = 1 - \frac{E_v^+ - |e|\Phi}{kT}$$

and similarly for n_v^-. The final solution should be constructed so that, as $z \to \infty$, the crystal is locally neutral,

$$n_v^+ = n_v^- = ne^{-(E_v^+ + E_v^-)/2kT}$$

14-4 In the preceding problem, assume that $E_v^+ = 0.75$ ev and that $E_v^- = 1.25$ ev. Compute the width of the Debye charge layer near the surface at 1,000°K and 300°K for a crystal with nearest-neighbor ion distance of 2.0 A.

14-5 Consider atoms in a biased random motion in one dimension. The atoms can jump forward or back a distance d in each jump. The probability of any jump being to the right is p, that of its being to the left is $1 - p$.

(*a*) Show that the simple average of displacement, X, after n jumps is, not zero, but

$$X = nd(2p - 1)$$

(*b*) Show that the mean square displacement $\overline{X^2}$ is

$$\overline{X^2} = nd^2 + n(n-1)d^2(2p-1)^2$$

(*c*) If $\overline{X^2}$ has the value given by (*b*), show that, for $p = \frac{1}{2}$, it reduces to the value of $\overline{X^2}$ for true random motion $\overline{X^2} = nd^2$.

(*d*) If $\overline{X^2}$ has the value given by (*b*), show that, for $p = 0$ or 1, $\overline{X^2}$ reduces to the value $\overline{X^2} = n^2d^2$.

14-6 Compute the volume density of divalent ions in a crystal of the NaCl type (with lattice spacing 2.0 A) if the "knee" in the conductivity curve comes at 700°K and the activation energies for vacancy formation and motion are $E_m = E_v = 1.0$ ev.

14-7 An ionic crystal can contain both Schottky and Frenkel defects. If the lattice frequency ν, which appears in the formula for conductivity, is the same for both these defects and if the Frenkel mechanism is dominant at low temperatures, show that it is dominant at all temperatures.

15

Electronic Structure of Alloys

15-1 Introduction

In several chapters, we have considered some of the changes which occur in the electronic structure of a perfect crystal when various types of impurities and other imperfections are introduced. In this chapter, we deal more systematically with this problem, with emphasis on the metals. In particular, we are concerned with the disturbance created in a metal lattice by the substitution of atoms of different valence and with the changes in the electron wave functions and in the density of states caused by that disturbance. As in the earlier discussion of impurities in semiconductors, we are also concerned with the possibility of the formation of bound states in metals, which represent localized charge distributions in the vicinity of impurity sites.

Before becoming involved in the detailed discussion, we begin by stating the central fact about imperfect crystals: that bands of energy levels persist. This is true even in such a highly disordered array of atoms as exists in a liquid. Even though Bloch's theorem is invalid in highly disordered crystals, a valence electron does not remain on a single atom. The very fact that phenomena assumed to involve electrons hopping from atom to atom are observed in liquid metals (for example, electrical conductivity) implies that the electrons possess translational kinetic energy. Hence, just as in perfect crystals, the spread in translational kinetic energy available to the electrons leads to spreading of the energy levels into the familiar band of the solid.

The same conclusion is reached if one considers the disordered crystal as being made up of N atoms brought together from infinite separation. As the atomic wave functions overlap, the N-fold degeneracy of the atomic level (for the system of N atoms) is lifted by the perturbations of the atoms on one another, and a spread of energy levels arises. (See the argument associated with Fig. 9-1.)

Even though it is thus demonstrated that bands exist for all types of condensed matter, whether perfect crystal or disordered array, it does not follow that the bands of the imperfect crystal are exactly like the bands of the perfect crystal, and differences do exist. In fact, a major difference occurs in the nature of the forbidden band gaps. The rule absolutely forbidding electrons from the band gaps in perfect crystals becomes somewhat relaxed as the crystal becomes disordered. For the case of a slightly disordered crystal, such as a lightly doped semiconductor, we have seen that the forbidden energy gap is partially violated because donor and acceptor states do exist in the "forbidden" gap. For a more disordered crystal, say a heavily doped semiconductor, the band edges lose their definition, and the density of states tails off into the previously forbidden gap region in an exponential fashion, as is demonstrated in Fig. 15-1.

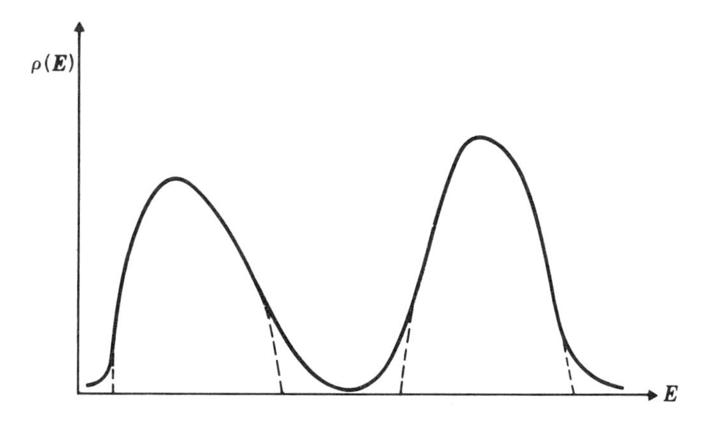

Figure 15-1 Density of states as a function of energy for a perfect crystal (dotted line) and for a highly disordered crystal (solid line). The main features of the bands are preserved, but the sharp band limits disappear with states falling off into the forbidden gap.

As a practical matter, this spreading of states into the forbidden continuum as disorder increases may or may not be important, depending upon the circumstances. For example, liquid nonconductors generally do not become conductors merely by randomizing their atoms, since the exponential tails do not extend very far. On the other hand, semiconductors do become partially metallic when they are highly doped, because electron transfer from donor to donor can take place easily when overlap occurs between donor atoms.

We now turn to the case of small densities of impurities in metals, the major topic of this chapter. We begin with an early success of alloy theory, first expounded by Hume-Rothery as an empirical set of rules, and then explained on the basis of the rigid-band model in one of the earliest major triumphs of the quantum theory of solids. Succeeding sections probe more fully into the types of disturbances caused by imperfections in metals, and take up the tendency of the conduction electrons to screen out these disturbances. The screening effects of the conduction electrons in impure metals are found to be fundamentally different from the effects of impurities on the electronic structure of insulators. The final section takes up an application of screening effects to the problem of localized states in metals.

Since the field of alloy physics is under intense development at the time of writing, only the barest sketch of the applications is possible here. In particular, the especially important case of overlapping bands is barely touched. Nevertheless, the general principles guiding this field are described in broad outline.

15-2 Rigid-band Approximation

The rigid-band concept is based on the simplest possible assumptions regarding the change which occurs when impurities are added to metals. Suppose we consider substitutional addition of zinc (valence 2) to copper (valence 1). The most elementary reasoning would suggest that the main effect would simply be a greater filling of the copper conduction band by the additional electrons from the zinc atoms. Such a model was constructed long ago on inadequate quantum-mechanical grounds, but it has served to classify the findings of a whole generation of metallurgists working in the field of alloys, as we shall discuss in the next section. We now demonstrate a more complete quantum-mechanical basis for this model.

The average energy of a quantum-mechanical system is given by

$$E = \int \psi^* H \psi \, dx \tag{15-1}$$

For the solid, the wave function is labelled by its k vector, and $E(k)$ may be written

$$E(k) = \int \psi_k^* H \psi_k \, dx$$

$$H = \frac{-\hbar^2 \nabla^2}{2m} + V(x) \tag{15-2}$$

The integration is over the entire solid. H is the total "hamiltonian operator," which for a particle in state k contains both the kinetic energy operator and the potential energy of the electron as a function of its position in the solid.

When an impurity is substituted for one of the metal atoms, the potential energy of the electrons in the vicinity of the impurity changes as compared to the situation in the pure metal. In the metal, the changes in potential due to the impurity are rather complex for reasons which occupy us in the latter half of this chapter. In this section, we suppose merely that the potential changes are weak and localized, momentarily begging all questions regarding the conditions under which these assumptions are valid. Let

$$H = H_0 + \delta H \qquad \psi = \phi_k + \delta\psi \tag{15-3}$$

where δH is the change in the potential due to the impurity, and $\delta\psi$ is the change thereby generated in the wave function. Then H_0 and ϕ_0 are quantities corresponding to the perfect crystal and satisfy the perfect crystal Schrödinger equation,

$$H_0\phi_k = E_k\phi_k \qquad H_0 = -\frac{\hbar^2}{2m}\nabla^2 + V_0(x) \tag{15-4}$$

Then, substituting the expression for H and ψ into (15-1), one obtains

$$E = \int (\phi_k^* + \delta\psi^*)(H + \delta H)(\phi_k + \delta\psi)\,d\tau \tag{15-5}$$

$$\delta E = \int \phi_k^*\delta H\phi_k\,d\tau + \int \phi_k^* H_0 \delta\psi\,d\tau + \int \delta\psi^* H_0\phi_k\,d\tau$$

The second and third terms of the latter equation are zero to first order in δV for reasons of "orthogonality" we will not discuss. Remembering that δH is just the change in local potential δV,

$$\delta E = \int \phi_k^* \delta V(x)\phi_k\,d\tau \tag{15-6}$$

This equation is only valid to first order in the parameter $\delta V/V_0$, and is thus valid only for small changes in the potential. If the crystal is one of the simple metals, then we know the wave function is nearly a plane wave, and hence $\phi_k = e^{\mathbf{ikx}}$. Then

$$\delta E_k = \int \delta V(x)\,d\tau = \langle \delta V\rangle_{\text{av}} \tag{15-7}$$

In this approximation, the shift in energy throughout the band is uniform and constant and is just the integrated average of the potential change.

If the density of impurity atoms is dilute so that the impurities are well screened from one another by the conduction electrons, the regions between the impurities (being regions of pure crystal) must have a band structure characteristic of the pure metal, and hence a Fermi level equal to that of the pure metal. Since the entire solid must be in thermal equilibrium, the impure solid, with regions containing impurities adjacent to regions containing none, must have the same Fermi level as the pure solid.

The new density-of-states function for the total solid can now be drawn. Since each energy level in the band is displaced by the same amount, the entire band is displaced rigidly, and the result is shown in Fig. 15-2. The Fermi level remains unchanged, but the band shifts,

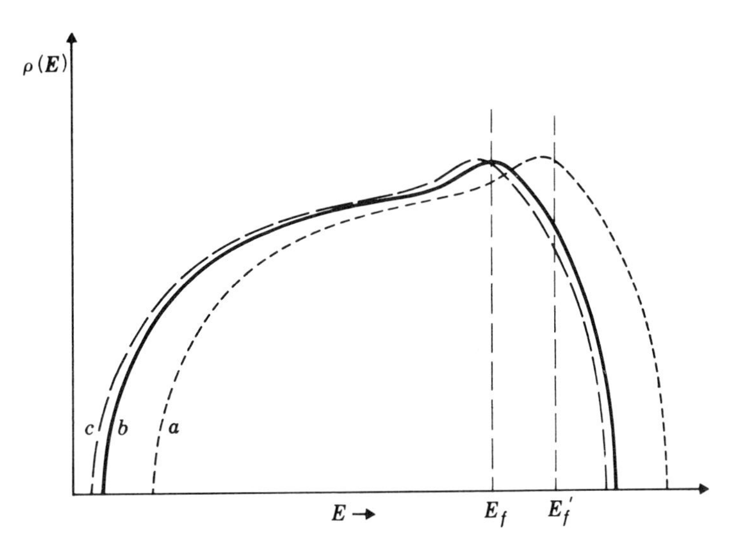

Figure 15-2 The density of states of a dilute alloy (*b*) compared to that of the pure metal (*a*). Each state is shifted by the amount $\langle V \rangle_{av}$ so the entire density of states function is shifted rigidly by the same amount. In the dilute alloy, E_f is the same as in the pure metal. The third curve (*c*) shows the result in a heavily doped metal, where the Fermi level increases to E_f' after significant overlap of the wave functions of impurity atoms occurs.

and more electrons are contained in the shifted band beneath the Fermi level than in the pure metal.

At this point, the reader may note that the problem seems overdetermined. The Fermi level must remain the same, but the band is shifted by $\langle \delta V \rangle_{av}$, which is presumably calculable. Actually, $\langle \delta V \rangle_{av}$ is not calculable, a priori, but must be "self-consistent." This is a point to which we shall return in a later section, and it is sufficient here to say only that the scattering of the conduction-band electrons makes the value of $\langle \delta V \rangle_{av}$ come out to be just that value which makes the Fermi level of the impure metal equal to that of the pure metal. In the present context this statement will seem somewhat obscure; in the later section the basis for it should become clear.

Thus the value of $\langle \delta V \rangle_{av}$ is adjusted until the states of the band below the Fermi level contain all the electrons of the metal, including those additional ones contributed by the impurity atoms. Note that so far as the *relative position within the band* is concerned, the effect on the impurities of adding electrons is the same as if the electrons from all atoms, including the impurities, were simply dumped into the band, as though it were a bucket, thus filling the band to a new relative level. The *Fermi kinetic energy* in the impure metal is now higher than before.

In this derivation of the rigid-band model, the essential assumptions were (1) small δV, (2) nearly free conduction-band electrons, and (3) dilute impurities. The last assumption leads to a constant Fermi level but is not strictly necessary for the "rigidity" of the band. In fact, as the impurity content increases beyond a few per cent, significant overlap must occur, and then the curve (*c*) of Fig. 15-2 results.

15-3 The Hume-Rothery Rules

The rigid-band model was the basis of the early successes in the theory of alloy formation. The classic problem of the metallurgy of alloys amounts to rationalization of the vast amount of data pertaining to the phase diagrams of alloys in terms of some kind of theory. The specific problem is the search for the features which control changes in the crystal structure of metals upon addition of various alloy elements. One of the major triumphs of that early work was the promulgation by the English metallurgist, Hume-Rothery, of some empirical rules governing alloy formation; these rules were named after him. The rules, which are stated briefly on page 99, may be stated more completely as follows:

Size Factor

Hume-Rothery's size-factor rule, though not directly related to the material of this chapter, is stated here for completeness. If one element is to be highly soluble in another to form a substitutional alloy, the "size factor" must be favorable: For large solubility, the atomic radii of the two elements must not differ by more than 15 per cent. (For interstitial alloys, the interstitial atom must be small in order to fit into the lattice interstices. For an example, see the data in Sec. 5-10 and the examples given there.)

The size factor implies that atoms cannot be accommodated in an alloy unless the local strain about them is much less than unity. One must not look too closely at this rule, because the "atomic size" of an atom is not always a unique choice, and the reasons that a lattice can accommodate local deviations in atom spacing of up to 15 per cent and nor more are rather vague. Also, the rule has an empirical and qualitative sense in that any impurity atom has greater than an infinitesimal solubility in any lattice. The rule above is meant to imply that under favorable conditions alloys of *significant* solubility are formed. On the other hand, in cases where we wish to apply it, the solubility need not extend to complete solubility across the entire composition range.

Chemical Factor

The chemical factor is the interesting effect with regard to this chapter because it is based on the rigid-band model. It states that characteristic phases or crystal structures appear in the alloy composition diagrams over and over again, and that these phases are related to the average number of valence electrons per atom in the alloy.

Perhaps the best known of these phases (usually referred to as electron compounds) occurs when one starts with an fcc lattice of a monovalent atom like copper and adds a divalent impurity like zinc. At an average electron concentration of approximately 1.5 valence electrons per atom, the alloy changes from an fcc lattice to a bcc lattice. Apparently as the fcc Fermi surface fills up the Brillouin zone, the bcc lattice becomes more stable than the fcc lattice, and the so called $\frac{3}{2}$ electron compound is formed.

The $\frac{3}{2}$ electron compound has been explained in terms of overlapping of the first Brillouin zone boundary by the Fermi surface. The effect is illustrated in Fig. 15-3 as the Fermi surface nears the top of the zone.

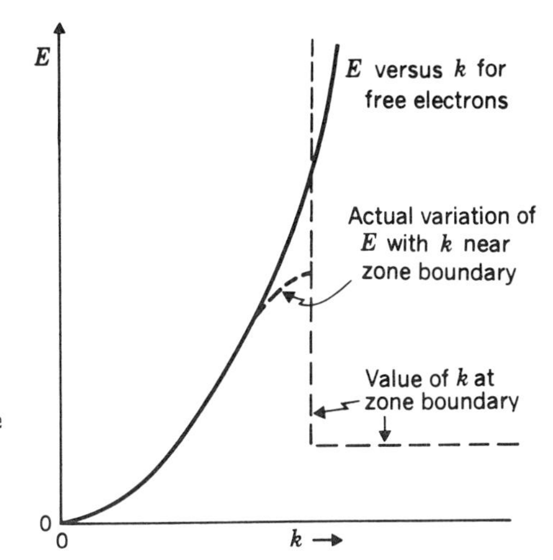

Figure 15-3 In the free-electron approximation, the energy is a parabolic function of k, but near t top of the zone it flattens out. Thus, a system with electrons near the top of the band will have an energy lower than the free-electron value The lowering of energy as the Fermi surface nears the Brillouin zone boundary can be thought of as an attractive force between the Fermi surface and the zone boundary.

Using the idea of a rigid band, we see that the $E(k)$ function bends down as the electrons fill the zone. Comparison of the perfect parabola with the actual curve of $E(k)$ shows that the energy is lowered by the approach to the zone boundary, and one may say an attractive force exists between the Brillouin zone boundary and the Fermi surface. It is this attractive force, for example, which creates the bumps on the otherwise spherical copper Fermi surface in the [111] directions illustrated in Fig. 15-4.

The reader will note that although we have assumed that the band in Fig. 15-3 becomes rigid, on the addition of electrons, we do not use the other result from the previous section, and assume that the Fermi

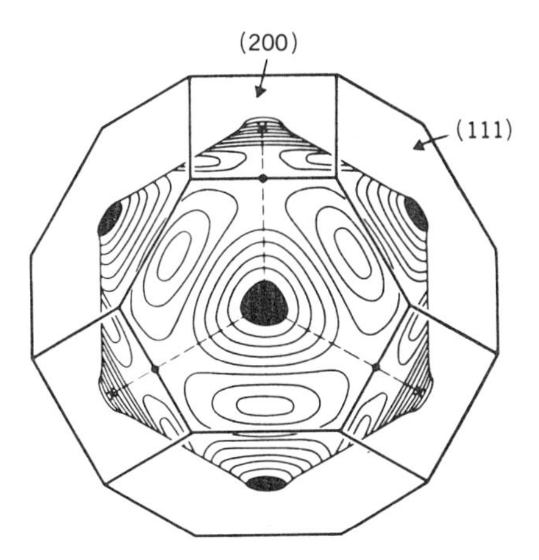

Figure 15-4 Fermi surface of copper showing humps in the [111] directions which overlap the Brillouin zone boundaries in these directions.

level is constant. Instead, we compare the two distributions with $E(k)$ starting at the same $E(0)$. The reason for this is that here we deal with large impurity densities, and the constant Fermi level theorem no longer applies. Of course, we probably cannot justify pegging the two bands being compared at the same $k = 0$ level either, but in the rest of this section we shall consistently do so for simplicity, remembering that the arguments made thereby are to be understood in a qualitative sense. Even for the case of heavy alloying however, δV must be small if the rigid band model is to remain valid, hence electrons of the conduction band must have a state function which is relatively like a plane wave.

In Prob. 9-9 we saw that as the Fermi sphere (considered to be a perfect sphere) grows in size into the fcc and bcc zones, it touches the fcc Brillouin zone boundary first at $e/a \simeq 1.4$ along the 8 [111] directions, hence stabilizing the fcc lattice. The 12 sides of the bcc Brillouin zones touch at $e/a \simeq 1.5$, become overlapped for $e/a > 1.5$, and, presumably because of their larger number, they now stabilize the bcc phase. Reference is made to Fig. 15-5, showing again the Brillouin zones of the two structures, and to Figs. 15-6 and 15-7 where the densities of states are compared for the two. In the vicinity of the larger peak, comparison of the average energy of a given total number of electrons in bcc with the same number of electrons distributed over the fcc density of states shows that the bcc distribution has the lower energy.

The picture presented here was surmised by Jones early in the 1930s, but even today sufficient understanding is not available to adequately predict phase changes in alloys. For example, even for pure copper (Fig. 15-4), the Fermi surface already overlaps the zone boundary extensively in the [111] directions, but how this overlapping

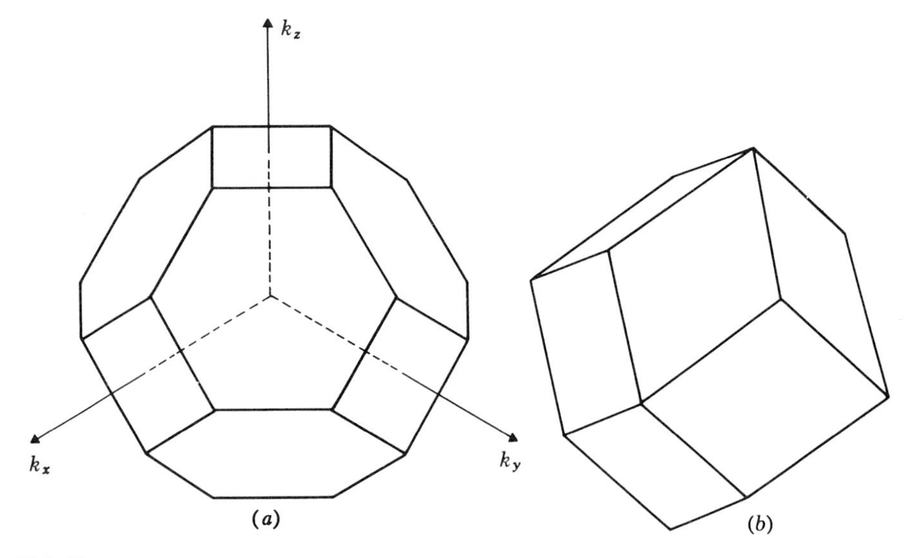

Figure 15-5 First zone boundaries for (*a*) fcc and (*b*) bcc.

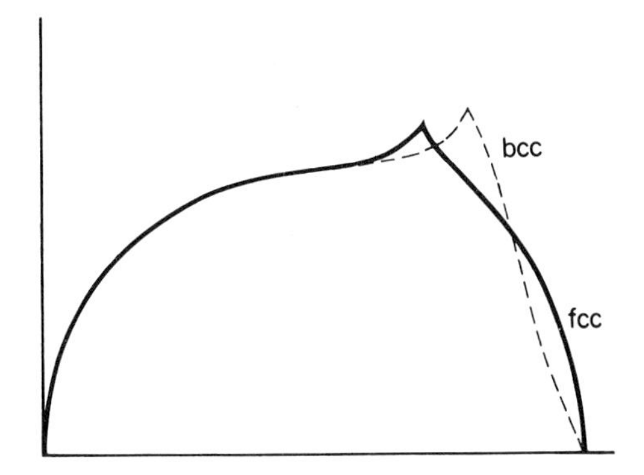

Figure 15-6 Schematic comparison of density of states for fcc and bcc. The peaks on the parabolas are caused by the effects described in Fig. 15-3. The fcc show a small peak early, while the bcc show a larger peak later.

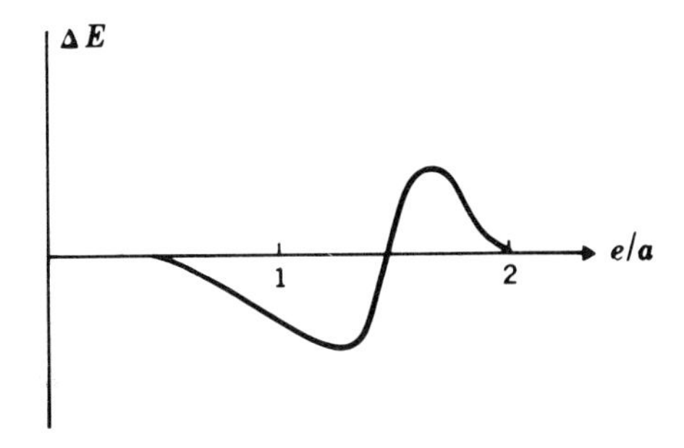

Figure 15-7 Presumed energy difference ΔE between fcc and bcc as a function of e/a.

increases with e/a is unknown. In particular, detailed information about the Fermi surface of the bcc alloy is unavailable. The extensive formation of $\frac{3}{2}$ electron compounds (see Table 15-1), and the appearance

Table 15-1 A few of the many electron compounds identified in binary phases and their approximate e/a ratios.†

e/a ratio	1.5			1.6	1.75
Crystal type	bcc	hcp	β-Mn	γ-brass	hcp
	CuZn	CuGa	CuSi	CuZn	CuZn
	AgZn	AuCd	AgHg	AgZn	AgZn
	CuSn	AgHg	AgAl	CuSn	CuSn
	FeAl	AgSn	AuAl	CuGa	AuSn
	NiIn	AuIn	CoZn	MnZn	AuAl
	AuCd	AgSn		FeZn	AgCd
	AuZn	AgSb		PtCd	CuSi

† Much more extensive information about these phases and others can be found in Chap. 10 of Barrett, C. S. and T. B. Massalski, *Structure of Metals*, 3d edition, McGraw-Hill Book Company, 1966.

of characteristic electron compounds at other electron concentrations for the other systems, show that this picture must have considerable validity. Of course, many metal alloys having the prescribed electron concentrations do not take on the predicted structures, and one must conclude in these cases that other factors in the solid predominate, or that the Fermi surface in these cases is not as simple as it is given here, or that the rigid-band assumption is invalid. Deviations from the rigid-band model might imply the growth of additional humps on the density-of-states curves or shifts of overlapping bands relative to one another, etc.

15-4 Screening by Conduction Electrons in Metals

The reader will remember that in Chap. 12 a charged impurity in an insulating crystal was seen to produce a Coulomb field in the crystal, with the crystal acting as a dielectric. The impurity states were then derived from the model of a hydrogen atom in a dielectric. We now consider in detail the analogous situation of an ionized impurity added to a metallic crystal (with the electron added to the conduction band to provide electric neutrality for the crystal in toto). This case is fundamentally different, because a good metallic conductor should support no direct-current electric field, classically. When a Coulomb field is established in the metal crystal by an ionized impurity, currents flow within the crystal until the field is damped out. One must be careful in applying classical physics to very short distances, however, because the electrons cannot damp out the Coulomb field of the impurity ion over distances shorter than the de Broglie wavelength of the conduction electrons.

The valence electrons of an impurity atom immersed in a metal join the conduction band, leaving behind a charged ion, which is represented in the following by a point positive charge at the lattice site. The conduction electrons then redistribute themselves about the ion in some manner, which leads to a deviation of charge $\Delta\rho_-$, where $\Delta\rho_-$ is measured relative to the electron distribution of the pure crystal. We shall make a calculation of $\Delta\rho_-$ in one-dimensional geometry by means of Poisson's equation,

$$\nabla^2\Phi(x) = -\frac{1}{\epsilon}\Delta\rho_-(x) \tag{15-8}$$

The potential function $\Phi(x)$ is the additional electric potential generated within the crystal by the charge deviation, $\Delta\rho_-$. The potential energy of an electron in this potential field is $-|e|\,\Phi(x)$; the complete energy diagram of an electron about an impurity ion is that sketched in Fig. 15-8*a*. The electron distribution in the entire solid must be in thermal equilibrium, hence the (total) Fermi energy must be constant everywhere within the solid. The Fermi kinetic energy $E_f(x)$ must then be

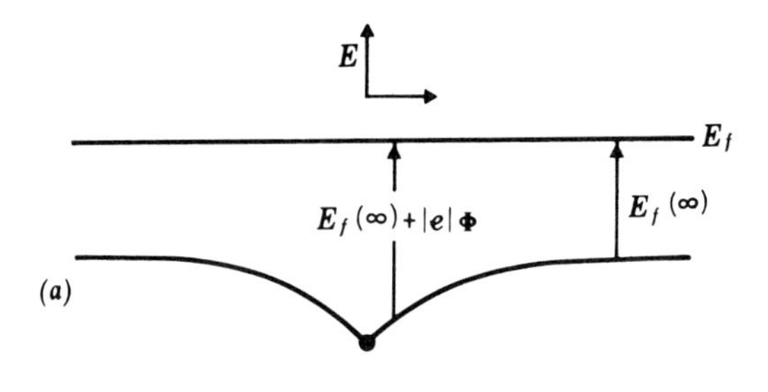

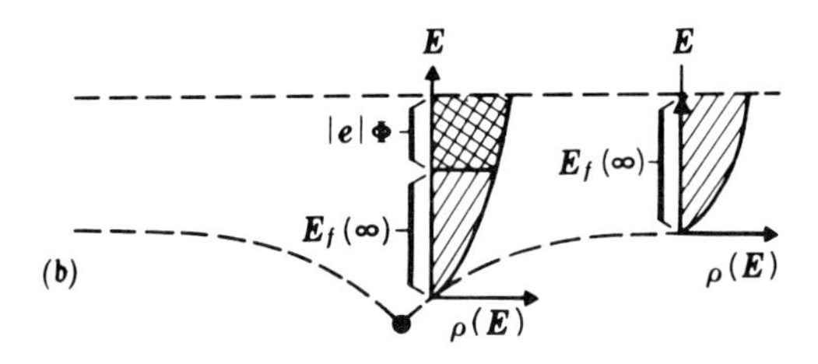

Figure 15-8 (*a*) Energy band of a metal with an impurity. The Fermi energy is a constant, but $|e|\Phi(x)$ is added to the bottom of the band.
(*b*) Density of states of electrons near the impurity and far away. The cross-hatched area corresponds to the additional electron states added below the Fermi energy near the impurity.

related to $\Phi(x)$ by the expression,

$$E_f(x) = E_f(\infty) + |e|\,\Phi(x) \tag{15-9}$$

This equation is graphed in Fig. 15-8, where the positive ion impurity is seen to cause a local lowering of the bottom of the band. Hence, as one goes near the impurity, the band is filled higher because the Fermi level must be everywhere a constant. The additional filling of the band near the impurity is of course just the excess electron density, $\Delta\rho_-$. At a point near the impurity where the band is shifted by the amount $|e|\,\Phi = \delta E_f$ (see Fig. 15-8*b*), the additional electron density is

$$\Delta\rho_- = |e|\,(\delta E_f)\rho(E_f)\Phi(x) \tag{15-10}$$

Here $\rho(E_f)$ is the ordinary density of states given in Eq. (9-9). This equation is known as the Thomas-Fermi equation. Combining Eqs. (15-8) and (15-10), we can write (in one dimension)

$$\frac{d^2\Phi(x)}{dx^2} - \frac{e^2}{\epsilon}\,\rho(E_f)\Phi(x) = 0 \tag{15-11}$$

The physically reasonable solution of this equation is

$$\Phi(x) = e^{-x/\lambda} \qquad \text{for } x > 0$$

where

$$\lambda^2 = \frac{\epsilon}{e^2\rho(E_f)} \tag{15-12}$$

The important features of this solution are the exponential distribution of electrons around the disturbance at the origin and the dependence of the screening length λ on the density of states. The solution to Prob. 15-2 shows that λ is of order 1 to 2 A, so the effective volume occupied by the excess electrons is about a unit cell.

The more appropriate physical case is that for a point charge in three dimensions. The radially symmetric solution of Poisson's equation corresponding to a charge at the origin is

$$\Phi(r) = \frac{e^2}{r} e^{-r/\lambda} \tag{15-13}$$

In this expression λ has the same value given in Eq. (15-12). Physically, Eq. (15-13) means that the normal Coulomb field of a point charge is screened by an electron distribution which dies away about it exponentially with a certain screening length, λ. The form of the potential (15-13) is called the Yukawa potential.

15-5 Charge Oscillations

The previous discussion of screening was semiclassical in the sense that the result was not obtained by solving the complete Schrödinger equation. Instead, the charge distribution was found using classical electrical laws, but with the Fermi distribution welded to it, ad hoc. When the problem is investigated rigorously, a startling new phenomenon occurs. Rather than decaying with the Yukawa potential given in Eq. (15-13), the screening charge *oscillates* as it dies away in the vicinity of the impurity and has a range greater than that given in (15-13).

The reader will not be surprised that we shall not enter into a rigorous treatment of the full quantum-mechanical problem; fortunately, it is possible to solve a much simpler problem which indicates the character of the true solution. We start from the observation that for densities of electrons in metals the screening length of Eq. (15-13) is of atomic dimensions, and the impurity thus causes a sudden or sharp change in the potential of the electrons within the metal. We then ask: What is the charge distribution of conduction electrons in the vicinity of a sharp change in the potential?

The physics of the problem is, in fact, not completely lost if the potential change is assumed to be an infinitely sharp discontinuity. We are thus in the presence of an old friend, the one-dimensional box problem, and we calculate the charge distribution in the vicinity of the edge of the box.

Normalized solutions of the Schrödinger equation within the one-dimensional box of length L are

$$\psi_k = \sqrt{\frac{2}{L}} \sin kx \qquad k = \frac{n\pi}{L} \tag{15-14}$$

The charge density ρ of N electrons within the box is given by the sum over the individual charge densities of the electrons.

$$\rho(x) = \sum_k e[\psi_k(x)]^2 = \frac{2e}{L} \sum_{n=1}^{N} \sin^2 \frac{n\pi x}{L}$$

We approximate the discrete sum by a continuous integral and obtain

$$\begin{aligned} \rho(x) &\simeq \frac{2e}{L} \int_0^N \sin^2 \frac{n\pi x}{L}\, dn \\ &\simeq \left(\frac{N}{L} - \frac{\sin(2\pi N x/L)}{2\pi x} \right) e \end{aligned} \tag{15-15}$$

Writing $eN/L = \rho_0$, this becomes

$$\frac{\rho(x) - \rho_0}{\rho_0} = - \frac{\sin 2k_F x}{2k_F x} \tag{15-16}$$

Here $k_F = \pi N/L$, the k vector at the Fermi level.

Thus the charge distribution near a planar potential discontinuity is by no means uniform, as it (nearly) is in the center of the box. Instead, it oscillates with a wavelength corresponding to electrons at the top of the Fermi distribution, and the oscillations die out with the distance from the edge as $1/x$. This function is schematically shown in Fig. 15-9. Rigorous solution of the problem of a point charge in three

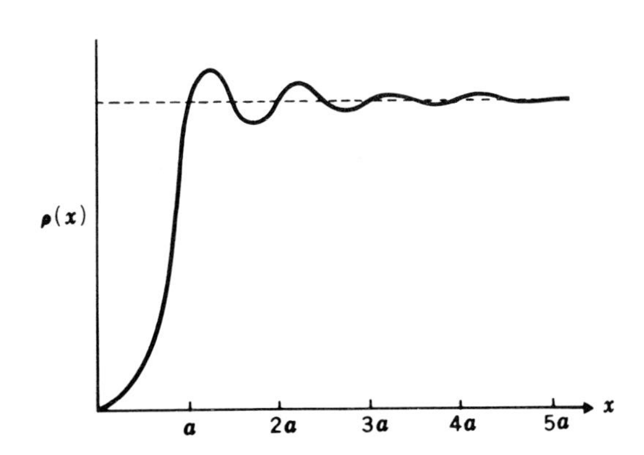

Figure 15-9 A plot of the charge density near the end of the one-dimensional box. The wavelength of the oscillations is the wavelength of electrons at the Fermi energy, where in this simple one-dimensional problem the wavelength is also equal to the distance a between atoms.

dimensions shows that the most important change is the rate of decay, in this case $\sim 1/r^3$ instead of $\sim 1/r$. More precisely, around a spherical hole (i.e., a spherical potential discontinuity caused by an impurity), the charge distribution is

$$\begin{aligned} \rho(r) &\propto \frac{\alpha \cos \alpha - \sin \alpha}{\alpha^3} \\ \alpha &= 2k_F r \end{aligned} \tag{15-17}$$

where r is the distance from the edge of the hole. Equation (15-17), which is only valid far from the impurity (more than a de Broglie wavelength of the most energetic electron), retains the oscillations with the Fermi wavelength. This function thus has the same qualitative shape as that in Fig. 15-9.

The reader may note at this point that he now has two descriptions of the region around an impurity in a metal. Equation (15-13) states that the *potential function* is basically exponential, while (15-17) states that the *charge density* is oscillatory, dying off as $1/r^3$. Of course, charge density and potential function are related to one another through Poisson's equation, (15-8). In fact, the charge density of screening electrons consistent with the Yukawa potential (15-13) has also the same functional form, i.e., $\rho_- \propto (e^{-r/\lambda})/r$. Thus one set of arguments leading through Poisson's equation and the Fermi density gives the Yukawa potential, and the arguments of this section lead to the oscillatory potential. The reader will note, however, that the argument was a two-step process. The fact that the crystal is a metal subject to screening results in a sharply decaying potential rather than a slow Coulomb potential. Then, if the screening length is short, the electrons cannot respond ideally because of their finite wavelength even at the top of the Fermi distribution. For normal metal crystals, the screening length in Eq. (15-6) is sufficiently short (a few angstroms) so that the oscillations are already well pronounced even for nearest neighbors. The main feature of the potential around an impurity in a metal, as it has emerged in the past two sections, is thus a rather deep well in the cell of the impurity ion itself with oscillatory fluctuations outside the central cell which die off as $1/r^3$.

The charge distribution has highly interesting consequences, and it is one of the triumphs of modern solid-state physics that the oscillations have been observed directly in magnetic resonance experiments in solids (see Chap. 23). In such experiments, the electron density is measured at the lattice sites of atoms surrounding an impurity in a solid, and the results give a dramatic confirmation of Eq. (15-17). The oscillations in charge also have been used to explain interactions between defects in a solid. Thus, if the charge trough of one impurity should overlap the charge maximum of a second, Coulomb attraction will result for that configuration, while if a charge maximum overlaps a trough, repulsion results. The details of such interactions are only just being worked out experimentally at this writing, but they do serve as the beginning of a chemical basis for precipitation reactions in solids.

15-6 Bound States on Impurities in Metals

To the extent that the screened Coulomb field with charge oscillations (just derived for a single impurity in a metal) can be called a weak field, we have shown that the effect is simply to shift the band to lower energies in the rigid-band approximation. From the success of the Hume-Rothery rules, the rigid band must be an excellent approximation in many systems; however, we should also investigate the extent to which a metal can have localized impurity states like those found in semiconductors and, if possible, gain some understanding of the varying conditions under which one should expect either a rigid band or bound states.

The entirely proper procedure would be to solve the Schrödinger equation with the potential function given by Eq. (15-17). For simplicity, however, the screened Coulomb potential is approximated by a square-well potential, Fig. (15-10). The zero of potential is the bottom of the band far from the impurity. The width of the well is 2λ, where λ is the screening length; its depth remains arbitrary for the moment.

It is instructive to consider this problem in two steps, first in one dimension, then in three dimensions. For the first case, the state functions are found by solving the one-dimensional Schrödinger equation.

$$\frac{d^2\psi}{dx^2} + \frac{2m(E - V)\psi}{\hbar^2} = 0 \tag{15-18}$$

For region I outside the well, $V = 0$, and two types of solutions are possible. The first are the free-running wave solutions

$$\psi_{\text{I}} = e^{ikx}$$
$$E_{\text{I}} = \hbar^2 k^2/2m \tag{15-19}$$

These are the solutions first discussed in Sec. 7-4. The energy of the states is positive, and the state functions are those of the normal conduction electrons in the metal.

The second set of solutions is of a different form;

$$\psi_{\text{I}} = Ae^{-kx} \qquad \text{for } x > \lambda$$

and

$$\psi_{\text{I}} = A'e^{kx} \qquad \text{for } x < -\lambda$$
$$E_{\text{I}} = \frac{-\hbar^2 k^2}{2m} \tag{15-20}$$

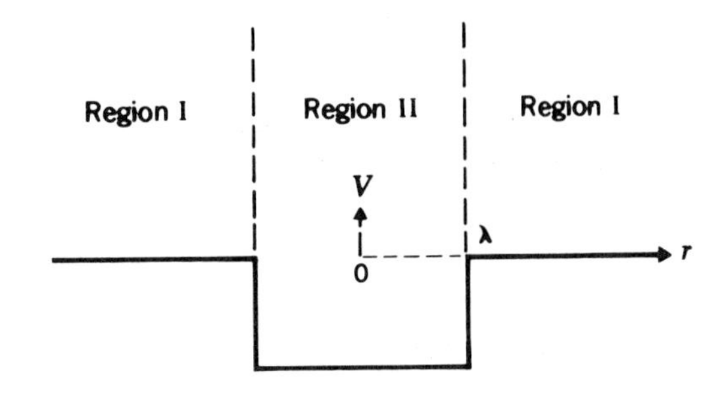

Figure 15-10 Approximate potential well for electrons in the vicinity of an impurity ion in a metal. The radius of the well is the screening length λ. Two regions are distinguished: I, outside the well, and II, inside the well.

The two forms of ψ_{I} are used for the different regions to keep the solutions finite. Note that the energy is negative; consequently, these states do not normally represent allowed state functions in the absence of the well.

The state functions for region II are similar to those of Eq. (15-19). For simplicity we write them in two forms, one set is an even function about $x = 0$, the other set as an odd function.

$$\psi_{\mathrm{II}} = B \cos \kappa x$$

$$E_{\mathrm{II}} = \frac{\hbar^2 \kappa^2}{2m} + V_0 \tag{15-21a}$$

and

$$\psi_{\mathrm{II}} = c \sin \kappa x$$

$$E_{\mathrm{II}} = \frac{\hbar^2 \kappa^2}{2m} + V_0 \tag{15-21b}$$

Note that V_0 is a negative quantity so that E_{II} is negative for sufficiently small values of κ.

Matching of the state functions in regions I and II puts restrictions on the state functions at $x = \pm\lambda$. Both the state functions and the first derivatives are required to be continuous at that boundary; i.e.,

$$\psi_{\mathrm{I}}(\lambda) = \psi_{\mathrm{II}}(\lambda) \qquad \psi_{\mathrm{I}}'(\lambda) = \psi_{\mathrm{II}}'(\lambda) \tag{15-22}$$

Hence, for the even functions, Eq. (15-21a), the conditional equation is

$$k = \kappa \tan \kappa\lambda \tag{15-23a}$$

and for the odd functions, Eq. (15-21b), the conditional equation is

$$k = -\kappa \cot \kappa\lambda \tag{15-23b}$$

These transcendental equations cannot be solved analytically, so numerical solutions must be used. Several methods of carrying out such solutions exist; a simple one is the following.

The equality of the energy level of the state in the two regions demands that a third equation be valid, namely that

$$\kappa^2 + k^2 = |V_0| \frac{2m}{\hbar^2} \tag{15-24}$$

With the substitutions $r = \kappa\lambda$ and $s = k\lambda$, Eqs. (15-23) and (15-24) become, for the even functions,

$$s = r \tan r$$

and

$$s^2 + r^2 = \lambda^2 |V_0| \frac{2m}{\hbar^2} \tag{15-25a}$$

the same substitutions transform Eqs. (15-23b) and (15-24) into

$$s = -r \cot r$$

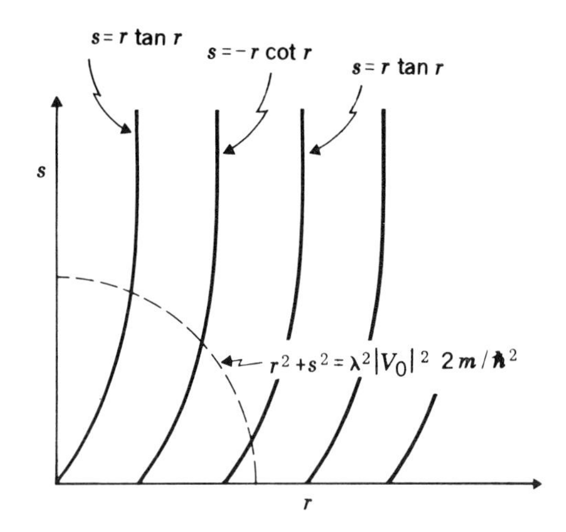

Figure 15-11 Plots of the functions $r^2 + s^2 = \lambda^2 |V_0| 2m/n^2$, $s = r \tan r$ and $s = r \cot r$. Points of intersection of these functions give appropriate values of r and s from which the state functions and energy levels can be determined.

and

$$s^2 + r^2 = \lambda^2 |V_0| \frac{2m}{\hbar^2} \tag{15-25b}$$

The solution of these two sets of equations permits both the state functions and energy levels of the states to be determined. One can see how the solutions for s and r can most easily be found by plotting the tangent and cotangent functions in the rs plane along with the circle of radius $\lambda\sqrt{|V_0|\, 2m/\hbar^2}$, Fig. 15-11. Points of intersection of the circles with the trigonometric plots give values of r and s from which the state functions and energy levels can be calculated.

As an example of the use of this technique, consider the following example:

Let the potential well be of size $\lambda = 2 \times 10^{-10}$ m and of depth $V_0 = -15$ ev. Find the state function and energy level of the lowest bound state. By trial, one finds r to be 0.54 and s to be 0.32. Thus

$$\begin{aligned} \psi_{\mathrm{I}} &= Ae^{-0.16\times 10^{10}x} && x > \lambda \\ \psi_{\mathrm{II}} &= B\cos 0.27 \times 10^{10}x && -\lambda < x < \lambda \\ \psi_{\mathrm{I}} &= Ae^{0.16\times 10^{10}x} && x < -\lambda \\ E_{\mathrm{I}} &= E_{\mathrm{II}} = -3.9 \text{ ev} \end{aligned} \tag{15-26}$$

The state function and energy level for this example are sketched in Fig. 15-12. This state is a bound state, i.e., an electron remains bound to this well. Other possible state functions for higher states are sketched in Fig. 15-13. Problem 15-4 requests the reader to show that no higher bound state exists for this particular well. (Inspection of Fig. 15-11 shows that one bound state always exists.)

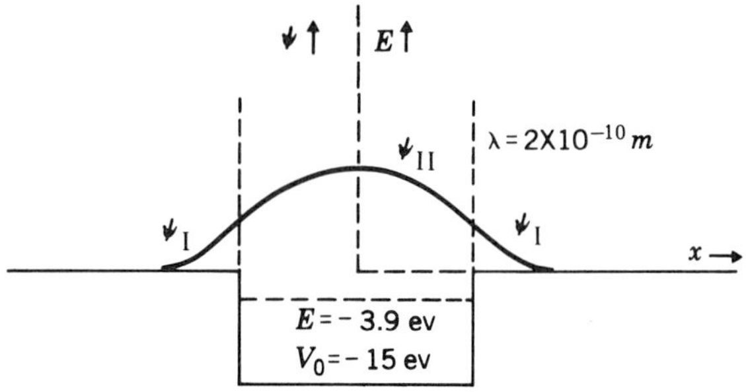

Figure 15-12 State function and energy level for the lowest bound state of an electron in a well of depth −15 ev and total width 4 angstrom units.

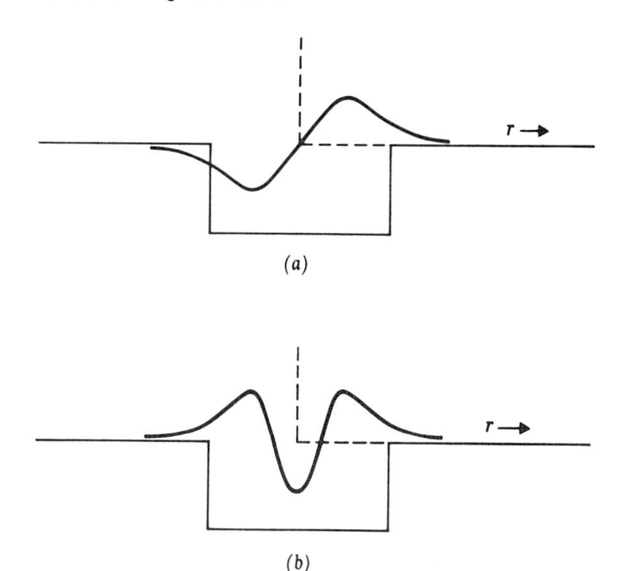

Figure 15-13 State functions for excited states of an electron bound to a potential well.
(*a*) First excited state.
(*b*) Second excited state.

A well in a three-dimensional crystal containing but a single impurity atom might naturally be assumed to have spherical symmetry. For this case, one must use the Schrödinger equation in spherical polar coordinates; it can be written for zero angular momentum as

$$\frac{d^2}{dr^2}(r\psi) + \frac{2m}{\hbar^2}[E - V(r)]r\psi = 0 \tag{15-27}$$

Note the additional complexity of the radial derivative. Outside the well, in the region II where $V(r) = 0$, this equation is

$$\frac{d^2}{dr^2}(r\psi) + \frac{2mE}{\hbar^2}(r\psi) = 0 \tag{15-28}$$

From the form of this equation, it is clear that the solution is similar to that for Eq. (15-18), i.e.,

$$r\psi_{\text{I}} = Ae^{-kr} \qquad E = \frac{-\hbar^2k^2}{2m} \tag{15-29}$$

The solution within the well (region II of Fig. 15-12) is given by

$$r\psi_{\text{II}} = B \sin \kappa r \qquad E = \frac{\hbar^2\kappa^2}{2m} + V_0 \tag{15-30}$$

Note that one cannot choose $r\psi_{\text{II}} = \cos \kappa r$; such a choice would require that ψ_{II} be divergent at the origin and hence nonphysical. The lowest state, then, must be analogous to Eq. 15-21*b* The conditions on the continuity of the function and its first derivative at the radius of the well lead, in just the same way as before, to

$$k = -\kappa \cot \kappa\lambda \tag{15-31}$$

(The reader is asked to verify this expression in Prob. 15-6.)

This equation is curiously similar to Eq. (15-23), yet the replacement of the tan $\kappa\lambda$ by the $-\cot \kappa\lambda$ has the important consequence just mentioned that the lowest bound state has one node. An additional consequence is that no bound state may exist if the well is not sufficiently wide or deep (these factors enter in the combination $\lambda^2 \, |V_0|$, as Eq. (15-24) shows).

The depth of the potential function has so far been left as a variable parameter, but further consideration shows that it is not the same kind of parameter as we have used for impurities before. Suppose that the normal valence of an impurity atom is one greater than the host atoms of the metal. Suppose further that the well in three dimensions was sufficiently wide and deep that two bound states were present. The results just derived show that two extra electrons, in addition to the normal valence conduction electron, will appear in the immediate vicinity of the impurity ion. Then an effective positive charge of one electron unit is present on the impurity atom, but two electrons are bound there. The result is that in the vicinity of the impurity atom a net negative charge exists. Since the conduction electrons act to screen out additional charge, they rush away from the vicinity of the impurity ion until the region is neutral and only one electron is bound there. Thus the net potential created by an impurity ion is not determined until the screening character of the electrons themselves is taken into account.

The effect just described is a phenomenon encountered in our discussion (here and in the last section) for the first time. In all problems dealt with previously, the potential function for the electrons was presumed to be known before the electrons were put into the system. In this case, however, the electrons help to create their own potential through correlation effects between the electrons. The solution to the problem of correlation in this instance must be that exactly one bound state must be taken from the bottom of the conduction band to provide for the additional electron of the impurity ion.

15-7 Band Structure of Impure Metals

At this point, the chapter has yielded two somewhat contradictory descriptions of the band structure when impurities are added to metals. According to the rigid-band model, any extra electrons are simply added to or subtracted from the band; the band shape remains the same. On the other hand, the quantum mechanics of an isolated potential well corresponding to the screened potential surrounding a single impurity show that bound states can appear if the well is sufficiently deep. Obviously, a bound state is a major change in the shape of the band.

One can reconcile the two pictures by remembering that the distortion in the potential must be small for the rigid-band model to be valid, although the well must not be too shallow if a bound state is to appear. These are apparently extreme special cases of strong and of weak potentials. But the story is not finished, because no mention has been made of the changes which the impurity causes in the band when the bound state appears. Does it, for example, remain unchanged, or do lumps appear in it? And what kind of shifts must one expect?

The reader may be relieved to learn that a proper treatment of the band states requires mathematics well beyond the level of this book. However, the results can be made quite plausible by looking again at what one would expect for a particle with a potential such as a one-dimensional box.

Keeping the radius of the well fixed, and the k vector fixed, we look at two extreme cases. In the first, the well is very shallow, and the running wave is sketched in the vicinity of the impurity in Fig. 15-14. The kinetic energy of the electron is large compared to the additional potential energy of the shallow well, and the wave is only very slightly

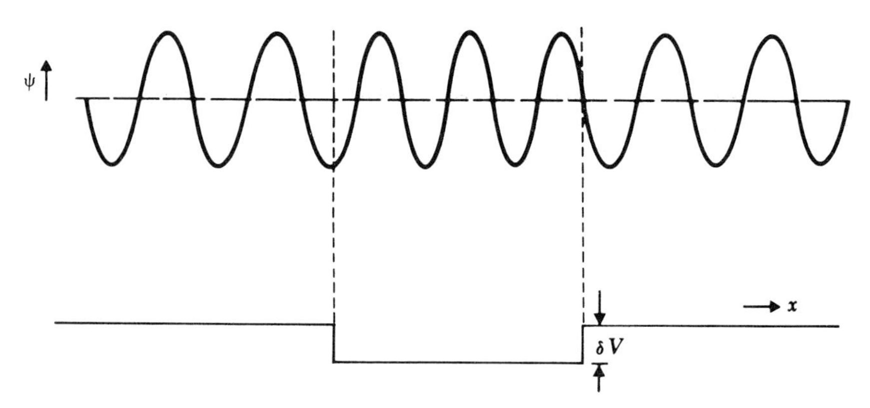

Figure 15-14 Wave function for a well where the kinetic energy of the particle is much larger than the potential energy of the well.

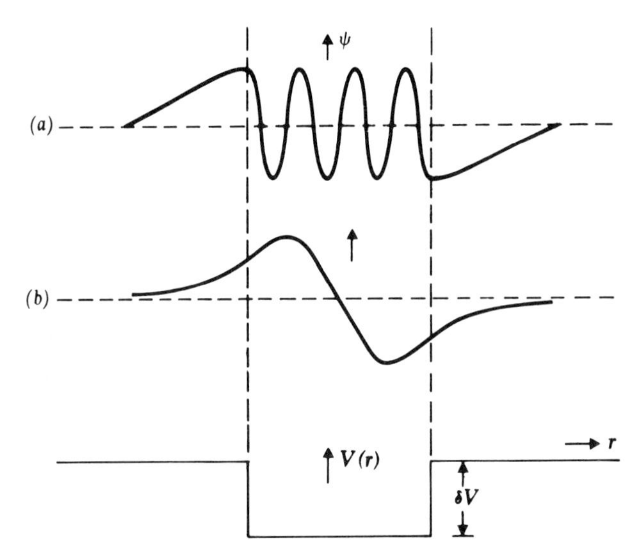

Figure 15-15 Wave function for a well where the kinetic energy of the particle is not large compared to the potential energy of the well. (*a*) Kinetic energy is greater than the well depth; a band state. (*b*) Kinetic energy is less than the well depth; a bound state.

perturbed by the well. The total energy of the electron is changed only by the δV of the well averaged over the "box." This case corresponds to the rigid-band model.

However, consider what occurs when there is a bound state in three dimensions. (Remember this corresponds to the first excited state in one dimension.) The wave function now looks like the sketch in Fig. 15-15. The wave is now strongly perturbed in both wavelength and amplitude in the vicinity of the impurity. Obviously, the energy change of the electron is more complicated and may even be a function of the wavelength of the electron.

A more rigorous treatment confirms this crude analysis, and the density of states for increasingly deep impurity wells is sketched in Fig. 15-16. In the case of the small well, the band is uniformly shifted. As the well is deepened, a lump appears at the bottom of the band which breaks off to become a bound state for a deep well. Also, other lumps can appear in the band itself.

In Fig. 15-16, the Fermi level is shown fixed for all cases, and in this way we remind the reader of the self-consistency which is necessary in the problem. As long as no overlap exists between impurity wells, the Fermi level of the impure solid must be the same as that of the pure metal.

As an example of the kind of argument which one runs into, consider the case of one mercury atom in a copper crystal. Mercury has valence one greater than copper. If the potential is deep, as in Fig. 15-16*d*, one bound state exists and the band is filled up to E_f. The entire crystal contains one more electron than a pure copper crystal. Using the

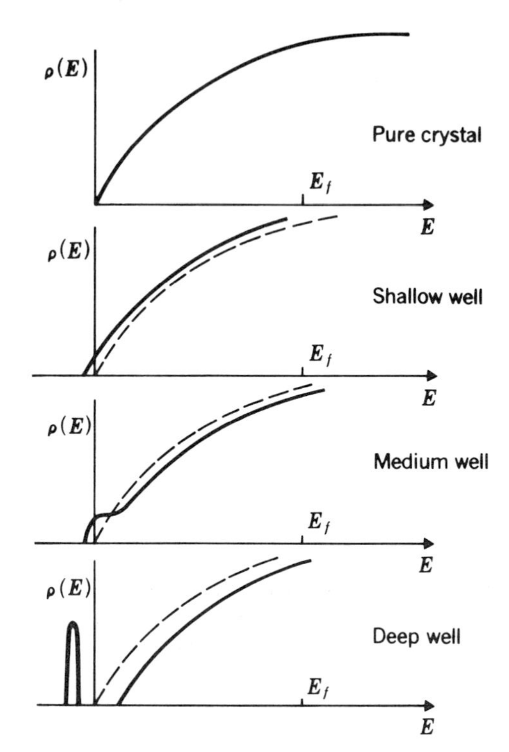

Figure 15-16 Density of states as a function of energy for metals containing potential wells of increasing depth.

argument first sketched in Sec. 15-2, one sees that the copper lattice far from the mercury atom is unaware of the presence of the mercury, so the Fermi level there is the same as in pure copper. On the other hand, the copper far from the impurity is in thermodynamic equilibrium with that near the impurity, and E_f must be the same for the two regions. Hence the depth of the well is adjusted so that the bound states which develop, together with whatever lumps appear in the band itself, must add up to just the right number of electrons at the Fermi level. Since the conduction electrons themselves provide the screening of the Coulomb potential of the mercury atom, the amount of screening is such as to make the overall situation self-consistent.

If the well is not deep enough to cause a bound state, the argument still applies for the rigid-band shift. The shift must be just sufficient to cause the Fermi level to come out the same as in the pure metal.

One final comment is in order. The rigid-band model might most likely apply to atoms that are neighbors in the periodic table. This case would correspond to the smallest distortion of the potential. Examples might be Cu with Zn, or Ag with Cd. On the other hand, transition metal impurities in nontransition metal crystals might be expected to produce localized bound states.

15-8 Localized States in Metals

Probably the best way to violate the conditions for rigid-band behavior is to pick an alloy in which the host electrons have s character and the impurity has d or f electrons. As a case in point, the resistivity caused by transition metal impurities in copper is found to be a strong function of the atomic number of the impurity. Two strong peaks in resistivity as a function of e/a are observed (see Fig. 15-17). These two peaks are due to the d shell electrons on the impurity. The normal d shell is split by its interaction with the lattice, and as one moves from Ni to K, the localized d level rises through the Fermi surface. Remember that the normal copper lattice has the d band well submerged below the Fermi level.

The level on the transition metal ion is not narrow, but has a breadth caused by its mixing with the s electrons of the copper conduction band. This spreading of the bound level by the overlapping s band is an effect not predicted by the earlier sections because of the restriction there to a single band case. The mixing of s and d bands and levels in this case means that the sharp "resonance" corresponding to the bound state is broadened because of the overlap with the s conduction band, and the state becomes a mixture of s- and d-like states. Another way of stating it is to say that the d state is not completely a bound state but rather has a finite lifetime as a bound state until it leaks out into the s conduction band. The lifetime is connected with the width of the state by the uncertainty relation, $\Delta E \, \Delta\tau \cong h$.

A number of other interesting effects are also observed in these alloys. In addition to the large scattering when the peak of the local level goes through the Fermi surface, local magnetic moments are

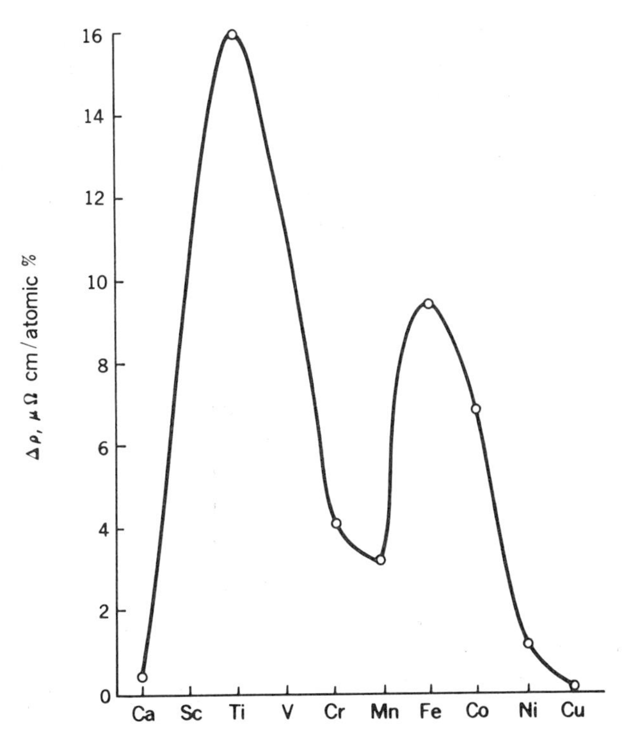

Figure 15-17 The increase in resistivity of copper per atom % of impurity from the 3-*d* series.

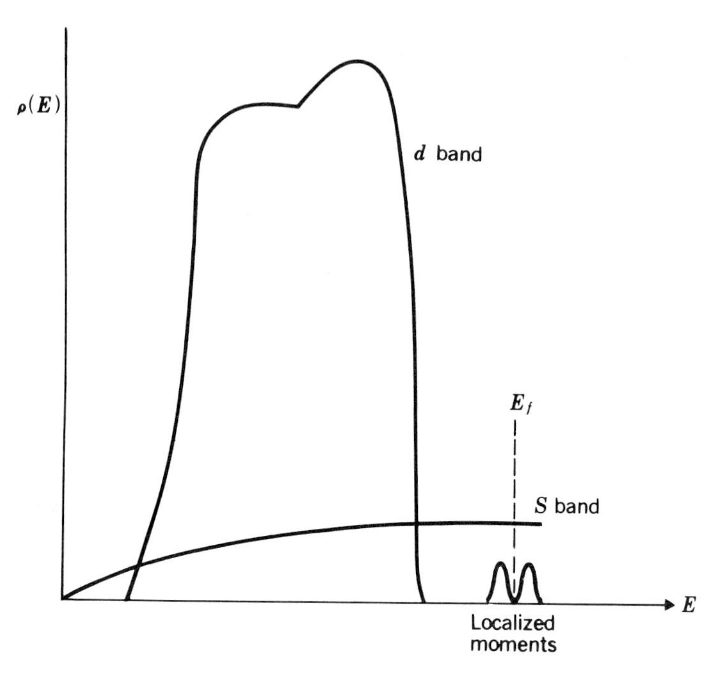

Figure 15-18 A schematic drawing of the density of states for a copper crystal with a few transition metal impurities. The normal *d* shell of copper is shown well below the Fermi energy, with two levels splitting off the top of that band and localized at the impurities. In the figure, one level is above, the other is below the Fermi level, which corresponds approximately to Mn.

observed on the impurity. In Fig. 15-17, those impurities which have local moments in copper are shown in boldface type. The splitting of the *d* levels is into spin-up and spin-down states, analogous to the situation in free atoms, which leads to Hund's rule. Those impurities near the two ends of the transition series which form no permanent magnetic moments presumably do not have split bound states in the manner shown in Fig. 15-18.

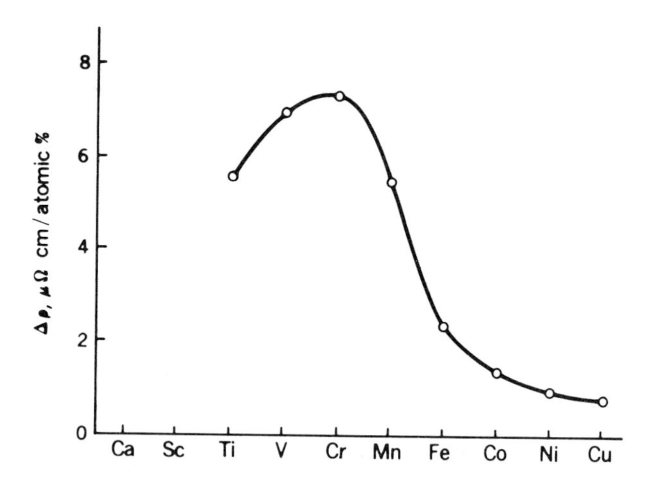

Figure 15-19 Similar resistivity data are plotted for aluminum as were plotted for copper in Fig. 15-17.

Indeed, the story connecting the localized moments on transition impurities to the splitting of the local level can be carried farther. The resistivity of transition metal impurities in aluminum metal is shown in Fig. 15-19. There is only one rather broad resistivity peak, suggesting that the localized d levels in this case are not split. A check on whether these impurities possess local moments shows that they do not.

PROBLEMS

15-1 Show that Eq. (15-13) satisfies the radially symmetric Poisson's equation

$$\frac{1}{r^2}\frac{d}{dr}\left[r^2\frac{d\Phi}{dr}\right] - \lambda^2\Phi = 0$$

15-2 Assuming that the dielectric constant of a metal is about ϵ_v, find λ for a metal such as sodium for a dioalent impurity.

15-3 From Eq. (15-22), show that for $\Psi_{\mathrm{I}} = Ae^{-kx}$, $x \geq \lambda$, and $\Psi_{\mathrm{II}} = B \sin kx$, $x \leq \lambda$, the condition (15-23a) follows. Prove that (15-23b) also follows from the use of the cosine function in region II.

15-4 Show that only one bound state exists for the well of Fig. 15-8.

15-5 Problem 8-3 stated that the Schrödinger equation in spherical coordinates for $l = 0$ was

$$\frac{1}{r^2}\frac{d}{dr}\left[r^2\frac{d\psi}{dr}\right] + \frac{2m}{\hbar^2}(E - V)\,\psi = 0$$

Show that Eq. (15-27) follows.

15-6 Verify that Eq. (15-31) follows from the continuity conditions at $r = \lambda$ for the sine function within the spherical well.

16

Surfaces

Surfaces were listed earlier in the book as one of the imperfections always present in a real solid. Since, in a solid of macroscopic dimensions, the fraction of atoms of the total at or near a surface is of the order 10^{-8}, the contribution of these atoms to normal bulk properties of the solid is negligible. However, as with all imperfections, the atoms of the surface can be crucial in special circumstances, such as when the surface atoms are conducting and the interior atoms are insulating, etc. Beyond this, however, special importance is attached to the properties of the surface because it is through the surface that all charge must flow from one conductor to another; it is through the surface that charge must flow in thermionic emission; it is on the surface that all chemical reactions with gaseous environments must occur; and it is the surface atoms, rather than the interior atoms of the solid, with which we are in actual contact when we interact mechanically with the solid. Thus, the surface and its properties have an importance far greater than the ratio of surface atoms to interior atoms. Clearly, however, we shall not be able to cover intensively all the important aspects of surfaces.

One of the early scientific successes in surface studies, and presumably one of the most important technological applications of the properties of surface, is the emission of electrons from the surface of metals at high temperature. This effect, called thermionic emission, is crucial for all devices requiring electron guns, e.g., the familiar radio tube, tv tubes, electron microscopes, betatrons. We shall thus begin the chapter with an exploration of this topic.

In other chapters of this book we have seen that when defects are introduced into a perfectly periodic (infinite) crystal, changes occur in the band structure, and in particular, bound states often appear. Thus, an important question is what effect the presence of a real surface has on the band structure. The quantum mechanics of a surface has intrigued theorists since the late 1930s. Unfortunately, only recently have the experimental tools been developed sufficiently to enable the theories to be tested, and recent experiments have shown that surface structure is often far more complicated than the simple theories assumed. Hence, theory has fallen far behind experiment in this field. However, because of the fundamental nature of the question, we discuss here the quantum mechanics of a simple model of a surface in order to see whether localized bound states might be a possibility for surfaces as they are for other defects.

Finally, we review the current experimental situation regarding surfaces. The fact that atomic forces are short-ranged means that the only atoms affected by the surface are those in the first few atom planes. Hence there is a considerable experimental problem in getting detailed information about surface structure because the technique must be sensitive to atomic dimensions, and any tendency for the formation of even monolayers of oxides, etc., must be controlled. For these two reasons, until only a few years ago, surface science was highly inexact. However, during the 1950s several new tools were developed which have transformed the entire field.

First, techniques were developed for completely cleaning a surface of foreign atoms in a sufficient vacuum to keep it clean. Sometimes vacuums of 10^{-10} torr are required. Also, tools such as field-emission microscopes and low-energy electron diffraction techniques were developed which respond directly to the surface layer itself. A discussion of these exciting developments forms the latter portion of the chapter.

16-1 Thermionic Emission

The fact that electrons are emitted from a hot filament was discovered shortly after the invention of the electric lamp in the late nineteenth century. Quantitative study of the phenomenon eventually led to the development of the Richardson-Dushman equation, which adequately describes the effect,

$$j = A_0 T^2 e^{-e\phi/kT} \tag{16-1}$$

where j is the current density leaving the surface of the metal,

T is the temperature, e is the electronic charge, and

ϕ is the work function

The fact that a metal should emit electrons is easily understood when one remembers that only a finite energy is required to lift the electron from the Fermi level over the work function barrier to the exterior of the crystal, and that this energy can be supplied by thermal excitations (see Fig. 16-1). The minimum energy required for an electron to leave the crystal is $e\phi$. In order to calculate the total current, we shall calculate the net average current normal to the surface

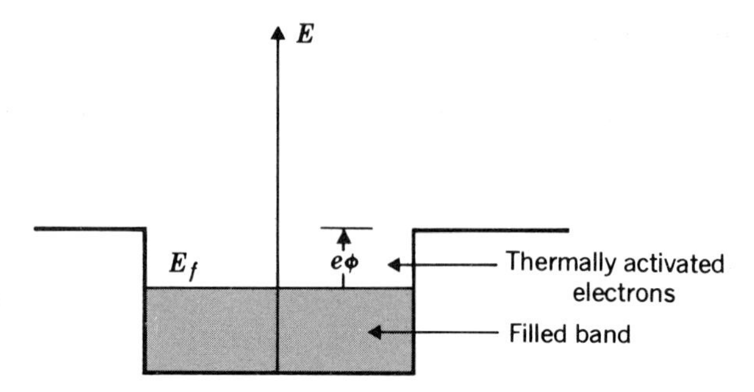

Figure 16-1 Energy level diagram of electrons in a metal showing the energy necessary to remove an electron from a crystal. ϕ is the work function.

for those electrons possessing sufficient momentum to overcome the surface barrier. If the surface normal is along the Z axis, the Z component of momentum p_z must be given by

$$p_z \geq \sqrt{2m(E_f + e\phi)} \tag{16-2}$$

(We count the zero of energy at the bottom of the band.)

The probability of an electron possessing the required momentum is given by the Boltzmann factor,

$$\begin{aligned} W(p) &= e^{-[E(p)-E_f]/kT} \\ &= e^{E_f/kT}e^{-(p_x^2+p_y^2+p_z^2)/2mkT} \end{aligned} \tag{16-3}$$

because for energies far above the Fermi level, classical statistics are valid. The current density normal to the surface for electrons having momentum p is thus given by

$$j_z(p) = \frac{e}{m}\,p_z W(p) = \frac{e}{m}\,p_z e^{[E_f-(p_x^2+p_y^2+p_z^2)]/2mkT} \tag{16-4}$$

Multiplying (16-4) by the density of momentum states at p, g/h^3, and integrating, we obtain for the total z component of current j

$$j_z = \int_{-\infty}^{\infty} dp_x \int_{-\infty}^{\infty} dp_y \int_{\sqrt{2m(E_f+e\phi)}}^{\infty} dp_z\,\frac{g}{h^3}\,p_z\,\frac{e}{m}\,e^{[E_f-(p_x^2+p_y^2+p_z^2)/2mkT]} \tag{16-5}$$

The factor g is the degeneracy factor explained in Chap. 9. For s electrons, it is 2, corresponding to the two values of spin. The three integrals are readily evaluated, resulting in the expression

$$\begin{aligned} j_z &= A_0T^2e^{-e\phi/kT} \\ A_0 &= \frac{2\pi gemk^2}{h^3} \end{aligned} \tag{16-6}$$

in agreement with the Richardson-Dushman equation, (1).

This equation, as we have derived it, is not sensitive to the properties of the surface, but does contain parameters such as m and g that are sensitive to the band structure. However, we have omitted the fact that when an electron impinges on the potential barrier at the surface

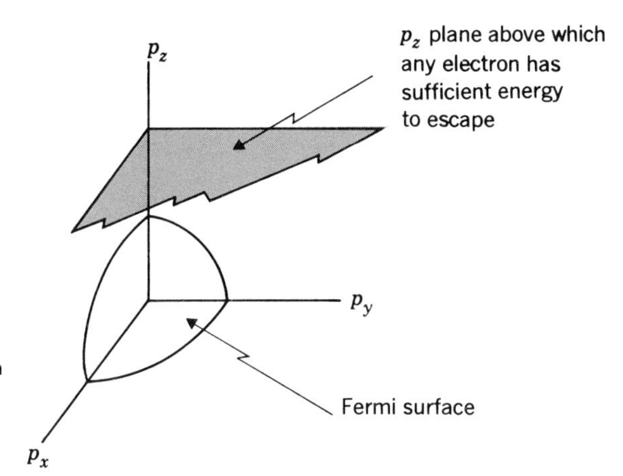

Figure 16-2 Momentum space diagram showing region containing electrons with p_z greater than the minimum required to escape from the surface.

shown in Fig. 16-1, even if the electron has enough energy to transcend the barrier, part of the electron wave will be reflected, just as in optics a light wave is partially reflected when it impinges upon a medium of a different refractive index. Hence for this reason, A_0 is too large even for a clean surface. Also, in practice, A_0 will change drastically if the surface layer becomes contaminated with oxide or other chemical reactants. In addition, ϕ is actually a function of crystal facet.

A further complexity of importance when an external field is present is the effect of the image charge in the metal as an electron evaporates. Since the metal is a conductor, as the electron leaves the surface it induces electrostatic redistribution of charge on the surface of the metal because the surface of the (grounded) conductor must always have zero electrostatic potential. This is tantamount to postulating an *image* charge of equal and opposite charge equidistant from the surface in the interior of the metal as shown in Fig. 16-3. The image potential of the evaporating electron thus modifies the potential box as shown in Fig. 16-1, and the actual potential is more nearly like that shown in Fig. 16-4*a*. When an external electric field $\mathscr{E}$ is switched on, as in Fig. 16-4*b*, the evaporating electron sees an additional potential of amount $-e\mathscr{E}z$, and the maximum height of the barrier to be surmounted is decreased. Thus, the current increases when there is an external field; this phenomenon is called the Schottky effect and is treated further in the problems.

One final topic which deserves comment is the nonthermal emission of electrons from the surface. When an external field is switched on, as in Fig. 16-4*b*, the surface presents a barrier of both finite height and width to the electrons, and the height is, in addition, lowered by the external field, as explained. Thus, electron tunneling through the barrier is theoretically possible, and is observed for very strong fields. In fact, even the thermionic emission is increased for fairly strong fields by tunneling. The so-called cold emission forms the basis for the action of the field-emission microscope, to be discussed in a later section. Tunneling is very strongly affected by the height and width of the

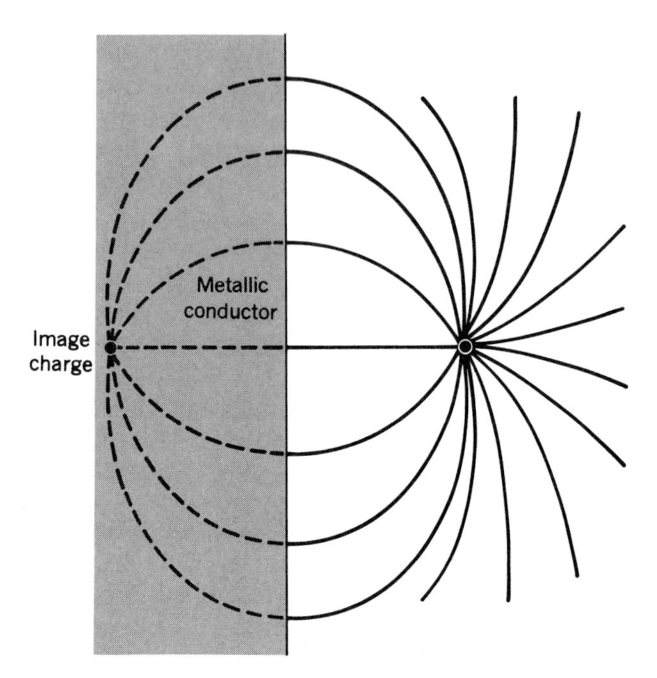

Figure 16-3 The image charge of an evaporating electron. Since the metal is electrostatically at constant potential, electric field lines must meet the surface normally. The potential distribution is thus equivalent to having an image charge of opposite sign embedded within the solid.

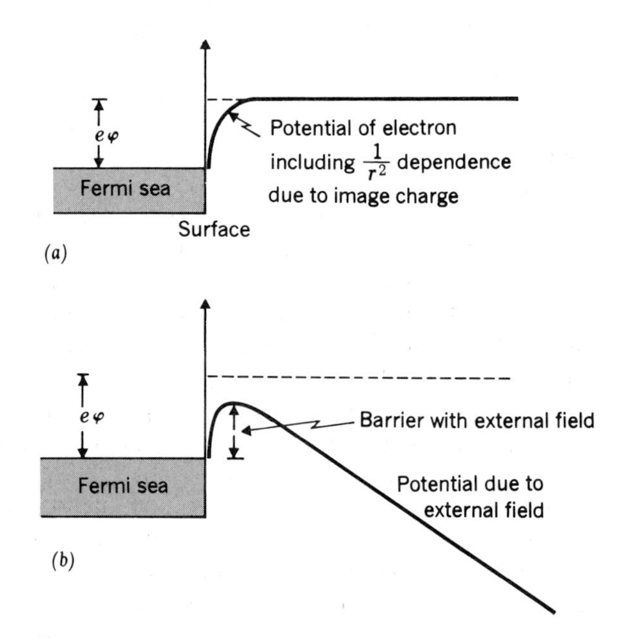

Figure 16-4 (*a*) The potential of an evaporating electron as a function of distance from the surface, showing the effect of the image charge.
(*b*) The same potential when an external electric field, $\mathscr{E}$, is imposed.

barrier, so that strong fields are necessary to bring these quantities into the physically interesting range. Further discussion is left for the problems.

16-2 The Quantum Mechanics of the Surface

Although the treatment of thermionic emission could almost neglect the actual properties of the surface, for other purposes, such as studying surface traps in semiconductors, one must know more about how the presence of the surface affects the atomic structure of the solid. In particular, one needs to know what to expect about how the surface alters the energy level and band scheme of the solid, and in this section we give an elementary account of this topic.

The nature of the surface state

The wave function of a particle in a one-dimensional box is shown in Fig. 16-5a for the case where the walls of the box are infinitely high and

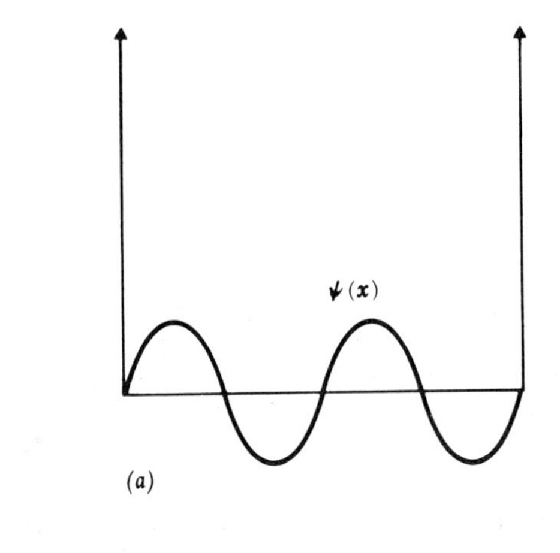

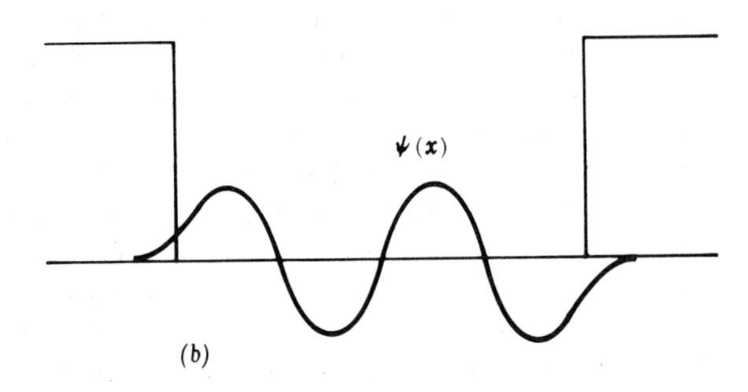

Figure 16-5 (a) The wave function in a box with infinite potential sides. At each end of the box the wave function must have a node.
(b) When the potential outside the box is finite, the wave function there is a decreasing exponential, so that the wave function at the surface is intermediate between a node and a maximum.

in Fig. 16-5*b* for the case where the walls are only finite. In the first case, the values of k are given by the condition that the wave function must go to zero at the two ends of the box, but we have never discussed how to get the values of k in the case of the finite box. Clearly, the values of k for the second case lie between the values of the first case, because we must connect the sinusoidal functions inside the box with the real exponentials outside. The wave function inside the box must then meet the surface with a slope intermediate between zero (corresponding to putting the wave function maximum at the surface) and its maximum value (corresponding to putting the wave function node at the surface, as in the case of the infinite wall). There will be just as many k values as before, but the lattice of allowed values in k space will have been shifted very slightly off the original points, Fig. 16-6.

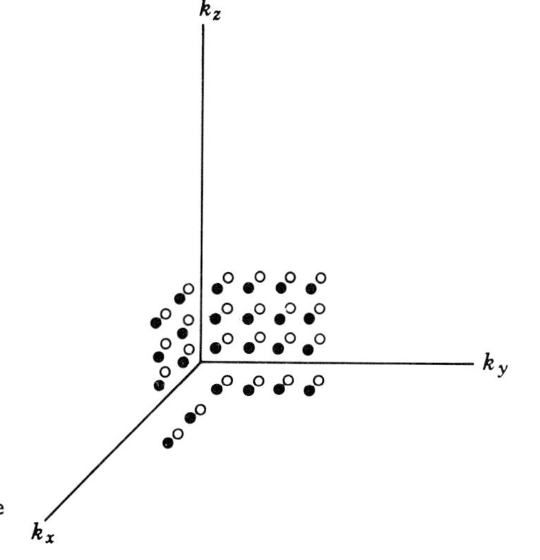

Figure 16-6 Allowed k values. The solid dots represent the values of k for a box with infinite sides. Corresponding to Fig. 16-5*b*, each allowed k value is slightly displaced, and the new values are indicated by open dots.

The crystal with real surfaces hence has a band of states only infinitesimally different from the crystal with infinite walls. However, we have been used to finding that when an imperfection occurs in a crystal which breaks the complete periodicity of the lattice, a new type of state often occurs which is localized in the vicinity of the imperfection. Such localized states also occur under certain circumstances for surfaces, and were theoretically predicted by Tamm in 1932. Such a localized state would have to look something like the drawing of Fig. 16-7 and is fundamentally different from any of the solutions derived for the free particle in a box. Since the particle in a box problem does not admit of surface-localized solutions, it is clear that it must be necessary to put real atoms in the model in order to find localized states. In order to do so, we must first discuss the quantum mechanics of the simplest crystal possible containing atoms; this is called the Kronig-Penney model.

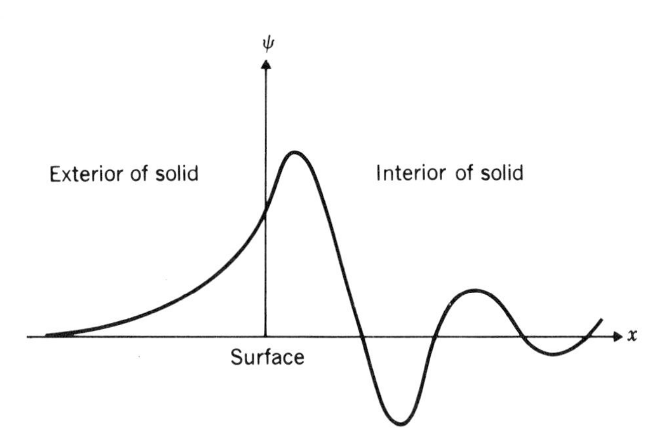

Figure 16-7 A schematic diagram of a surface state. Outside the box, a simple decreasing exponential is obtained, while inside the box the wave function envelope decreases exponentially from the surface.

The Kronig-Penney model

The Kronig-Penney model begins with a potential function in one dimension like that shown in Fig. 16-8. The crystal consists of a series of rectangular spikes, with the height and width of each potential spike given by V and δ respectively. We shall now assume that the spikes become very sharp, and take the limit where the height of the spikes times their width is finite. Thus $\lim_{\substack{\delta \to 0 \\ V \to \infty}} V\delta = \mathscr{V}$. The limiting spike function is a well-known artifact in mathematics called the *delta function.*

The delta function has the interesting property that when it is multiplied by any other well-behaved function and integrated over the region of the spike, it gives the value of the function at the point of the spike multiplied by the strength of the spike. Thus if $f(x)$ is any

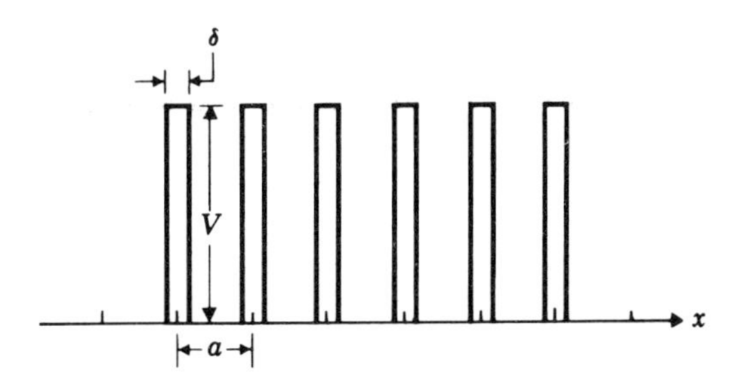

Figure 16-8 A potential for the Kronig-Penney model of a crystal. The potential spike function is allowed to go to the limit of infinite height and infinitesimal width.

function,

$$\int_{\substack{\text{region of}\\ \text{spike}}} f(x)V(x)\,dx = \lim_{\substack{\epsilon\to 0\\ V\to\infty}} \int f(x)\epsilon V\,dx = f(x_0)\int_{-\delta/2}^{\delta/2} \epsilon V\,dx = \mathscr{V} f(0) \tag{16-7}$$

Equation (7) follows from the fact that as we approach the limit where $\delta \to 0$, $V \to \infty$, the function $f(x)$ becomes approximately flat in the interval δ, and its value in the interval can be taken out of the integral sign, with the integral now giving the area $\mathscr{V}$ of the spike.

Coming back to the quantum mechanical problem, the wave function between the spikes must be a sinusoid because the potential in that region is zero.

$$\psi(x) = \cos \lambda x + A \sin \lambda x$$

$$\lambda = \frac{1}{\hbar}\sqrt{2mE} \tag{16-8}$$

For simplicity, we do not use the completely general function, $C \sin(\lambda x) + D\lambda \cos(x)$, because this function differs from (16-2) by a trivial normalization constant, which we suppress. The question now must be answered how to continue the wave function through the region of the spike into the next cell of the "crystal," where another sinusoid must be found. We find the appropriate conditions by integrating Schrödinger's equation through the region of a spike.

$$\psi'' = \frac{2m}{\hbar^2}(V(x) - E)\psi(x) \tag{16-9}$$

Integrating Eq. (16-9) over just the region of one spike, which we take at $x = 0$, then

$$\int_{-\delta/2}^{\delta/2} \frac{d^2\psi}{dx^2}\,dx = \frac{2m}{\hbar^2}\int_{-\delta/2}^{\delta/2} (V - E)\psi\,dx \tag{16-10}$$

If ψ is assumed to be continuous (though ψ'' cannot be!), with the use of (16-7), we obtain

$$\left[\frac{d\psi}{dx}\right]_{-\delta/2}^{\delta/2} = \frac{2m}{\hbar^2}\mathscr{V}\psi(0) \tag{16-11}$$

This equation thus shows that there is a discontinuous change in slope in ψ' at the spike, depicted by Fig. 16-9.

In addition, of course, since the lattice is perfectly periodic, Bloch's theorem must be valid,

$$\psi(x) = \psi(x + a)e^{ika} \tag{16-12}$$

This equation will be used to write the function to the right of a spike in terms of the function to the left of the same spike. Thus if we label the cell to the left as I, and that to the right as II, then

$$\psi_{\mathrm{I}}(0) = \psi_{\mathrm{II}}(a)e^{ika} \tag{16-13}$$

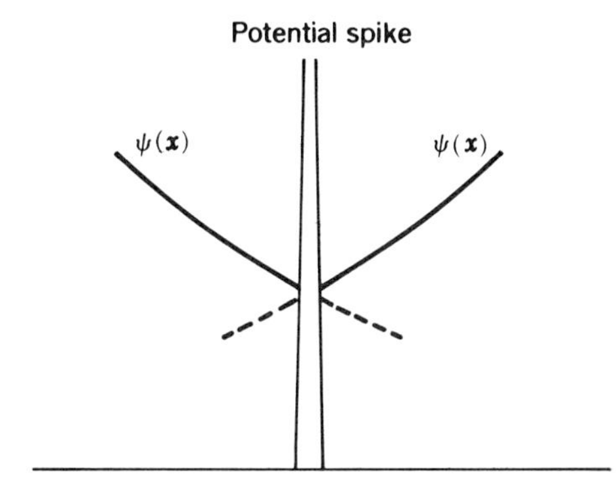

Figure 16-9 Continuation of a wave function through a spike. The wave function is continuous, but its first derivative has a discontinuity proportional to the strength of the spike.

Returning to (11), can write the left hand side as

$$[\psi']_{\delta/2}^{\delta/2} = \psi_{\mathrm{II}}'(0) - \psi_{\mathrm{I}}'(0) = \psi_{\mathrm{II}}'(0) - \psi_{\mathrm{II}}'(a)e^{ika} \tag{16-14}$$

Then (16-11) becomes

$$\psi_{\mathrm{II}}'(0) - \psi_{\mathrm{II}}'(a)e^{ika} = \frac{2m}{\hbar^2}\mathscr{V}\psi_{\mathrm{II}}(0) \tag{16-15}$$

Using a function of the form (16-8) for ψ_{II}, this equation finally becomes

$$A\lambda - e^{ika}[-\lambda \sin \lambda a + A\lambda \cos \lambda a] = \frac{2m}{\hbar^2}\mathscr{V} \tag{16-16}$$

The equation for A finally becomes

$$A = \frac{2\alpha - e^{ika} \sin \lambda a}{1 - e^{ika} \cos \lambda a}; \qquad \alpha = \frac{m\mathscr{V}}{\lambda\hbar^2} \tag{16-17}$$

In addition, $\psi_{\mathrm{I}}(0) = \psi_{\mathrm{II}}(0)$ at the spike, writing (16-13) for $\psi_{\mathrm{I}}(0)$, and using (16-8) we get another relation for A,

$$A = \frac{e^{-ika} - \cos \lambda a}{\sin \lambda a} \tag{16-18}$$

Eliminating A from (16-18) and (16-17), we get the relation between E and k (remembering that E is contained in λ),

$$\begin{aligned} \cos ka &= \cos \lambda a + \alpha \sin \lambda a \\ \alpha &= \frac{m\mathscr{V}}{\lambda\hbar^2} \\ \lambda &= \sqrt{\frac{2mE}{\hbar^2}} \end{aligned} \tag{16-19}$$

Equation (16-19) is the solution to the problem, since it gives the allowed values of E in terms of k. We note that when $\mathscr{V}$ is small, then $\cos ka \simeq \cos \lambda a$. In this approximation, $k \simeq \lambda$, which is the free-electron approximation. We note further that for large values of E, the extra term falls as $1/\sqrt{E}$, and hence for large E, the free-electron picture becomes more nearly valid. The allowed values of E are found by numerical solution of the equation, but we see immediately from the equation that $\cos ka$ must be between -1 and $+1$, so that whenever E has a value which takes the right side outside this range, that value of E is forbidden. For purposes of visualization, we plot the right side of Eq. (16-19) in Fig. 16-10. We note that the curve takes excursions out

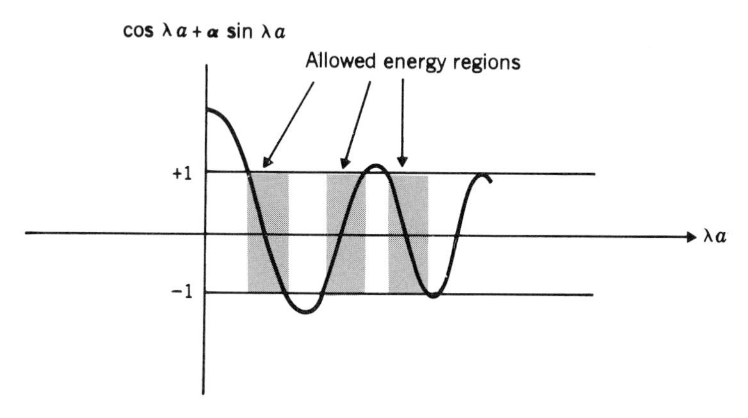

Figure 16-10 A schematic drawing showing the numerical solution of Eq. (16-19). Allowed values of k are found in the region where the solid curve is between + and -1 on the vertical axis.

of the range between -1 and $+1$ at just those k values where $\cos ka = \pm 1$ or $ka = n\pi$, which is the condition for the diffraction of the electrons at the Brillouin zone boundary. Thus, the model demonstrates all the features we have been led to expect in Chap. 9.

The state at the surface

In the previous paragraphs, Block's theorem was used, and this use implies the validity of a perfectly periodic solid, presumably with periodic boundary conditions. The question remaining is, what happens when the series of spikes is terminated? We take the spike distribution shown in Fig. 16-11. For reasons which will become

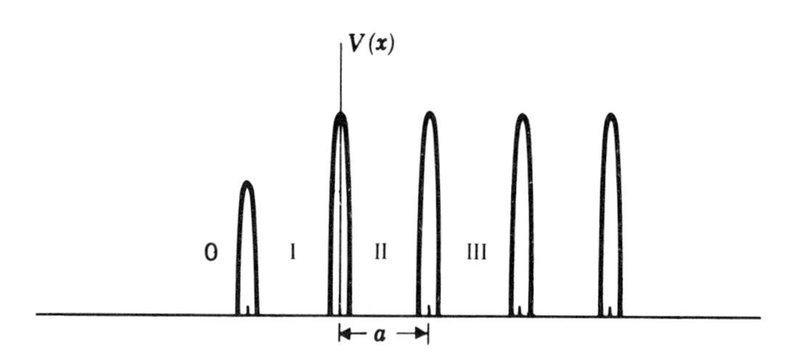

Figure 16-11 Modification Kronig-Penney potential for a one-dimensional crystal. The end atom has a different potential from all the others. The spacing between all atoms is a.

apparent, we assume the surface spike at $x = -a$ is different from the others, and denote its strength $\mathscr{V}_s$.

When we consider the band states represented by the wave function Eq. (16-8), the boundary condition at the surface yields the same results already noted in the discussion of Fig. 16-6. Outside the crystal, a wave function of the form

$$\psi_0(x) = Be^{|\lambda|x} \tag{16-20}$$

is matched to the band states inside the crystal. Within the crystal, Block's theorem is still assumed valid, but the allowed values of k are very slightly shifted from those given by periodic boundary conditions as shown in Fig. 16-6. Thus the changes in the band states imposed by the presence of the surface are relatively trivial.

The major result discovered by Tamm, however, is that a new type of solution exists for the case of the surface with the potential of Fig. 16-11, which is not a part of the band states, and which lies below these states in energy. We investigate the properties of these postulated states by asking if states of negative energy can exist. For these negative energy states, λ is imaginary, and Schrödinger's equation yields an exponential wave function between the spikes, which we take in the form

$$\begin{aligned} \psi_{II}(x) &= \cosh \gamma x + A' \sinh \gamma x \\ E &= -\frac{\hbar^2\gamma^2}{2m} \qquad i\gamma = \lambda \end{aligned} \tag{16-21}$$

(This function is labelled ψ_{II} for the reason that we wish to make use of the Kronig-Penny results, and the analysis lead to Equation (19) concerns the wave function to the right of the reference spike, which in Fig. (16-11) would be the wave function in region II.) The wave function in region I will be given by Bloch's theorem, which we still assume to be valid, subject to proof that the overall wave function exists.

$$\psi_I(x) = (\cosh \gamma(x + a) + A' \sinh \gamma(x + a))e^{ika} \tag{16-22}$$

The wave function outside the crystal in the region labelled 0 in Fig. (16-11) is given again by Eq. (16-20).

The results of Eqs. (16-12) through (16-19) are still valid if we remember that the energy is now negative and λ in those equations is imaginary. Rewriting Eq. (16-19) for an imaginary argument, we have

$$\begin{aligned} \cos ka &= \cosh \gamma a + \epsilon \sinh \gamma a \\ \gamma &= \sqrt{\frac{2m\,|E|}{\hbar^2}}; \qquad \epsilon = \frac{m\mathscr{V}}{\gamma\hbar^2} \end{aligned} \tag{16-23}$$

The right-hand side is now plotted in Fig. (16-12) and it is quickly seen that the right-hand side never enters the region between +1 and −1, where $\cos ka$ has real values. Thus, there are no solutions for real k. However, if k is purely imaginary, there is only a small region near $ik = \kappa = 0$ where $\cosh \kappa a$ does not overlap the right side of 17, and

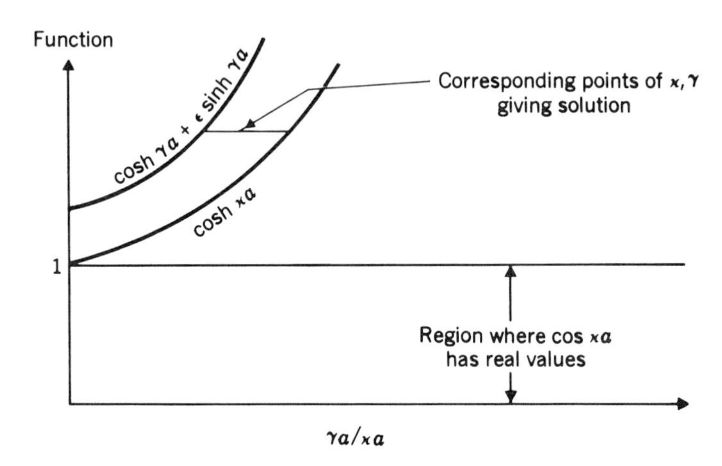

Figure 16-12 Diagram showing numerical solution of Eq. (16-23). When k is real, the right-hand side and the left-hand side do not overlap in value. However, when k is imaginary, the overlap is complete.

solutions are forbidden.

The fact that no solutions for real k are possible is interesting, because Bloch's form of the wave function *with a real exponential instead of the sinusoidal form* is precisely what is required for a localized state. In this case the wave function is exponentially damped from cell to cell as one progresses toward the interior from the surface, giving rise to a wave function effectively localized in the vicinity of the surface.

We now take up the problem of matching wave functions at the surface spike. The matching conditions at the spike for the wave function and its first derivative from Eq. (16-11) become

$$\begin{aligned} \psi_0(-a) &= \psi_{\mathrm{I}}(-a) \\ \psi_{\mathrm{I}}'(-a) - \psi_0'(-a) &= \frac{2m\mathscr{V}_s}{\hbar^2}\,\psi_{\mathrm{I}}(-a) \end{aligned} \tag{16-24}$$

with the substitution of (22) and (20), the conditions become

$$\begin{aligned} B &= e^{(\kappa+\gamma)a} \\ A' &= \epsilon_s + 1 \\ \epsilon_s &= \frac{2m\mathscr{V}_s}{\hbar^2\gamma}\,; \qquad \kappa = ik \end{aligned} \tag{16-25}$$

Rewriting Eqs. (17) and (18) for an imaginary λ, and remembering that since $\sin ix = i \sinh x$, $A' = iA$, we have, respectively

$$A' = \frac{\epsilon e^{-\kappa a} + \sinh \gamma a}{e^{\kappa a} - \cosh \gamma a} \tag{16-26}$$

$$e^{-\kappa a} = \cosh \gamma a + A' \sinh \gamma a$$

Substitution of the second of these equations into the first gives

$$\coth \gamma a = \frac{(A')^2 - 1 - \epsilon A'}{\epsilon} \tag{16-27}$$

Finally, substitution of Eq. (16-25) for A' gives

$$\coth \gamma a = \frac{\epsilon_s}{\epsilon}[\epsilon_s - \epsilon + 2] - 1 \tag{16-28}$$

Coth x has the form shown in Fig. 16-13. ϵ_s and ϵ are proportional

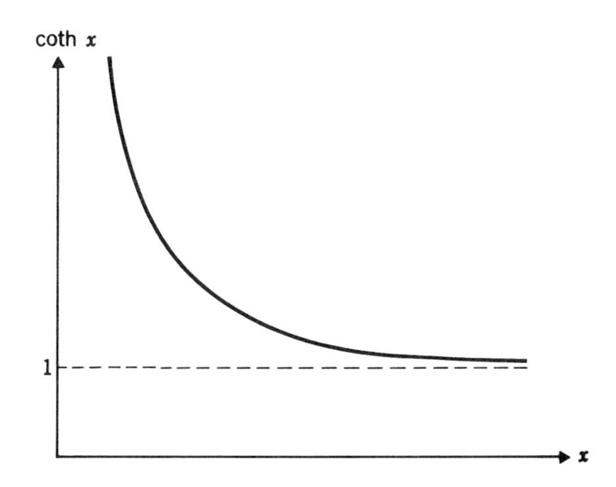

Figure 16-13 Schematic diagram showing numerical solution of Eq. (16-28). The right-hand side is always greater than 1 and thus overlaps values of the left-hand side whenever ϵ_s is greater than ϵ.

to the heights of the surface and interior spikes, and when $\epsilon_s = \epsilon$, no solution exists except at $\gamma = \infty$. However for all values of $\epsilon_s > \epsilon$, a solution does exist.

The solution obtained here relates to a bound state below the lowest band of the Kronig-Penney model. Other bound states appear in the forbidden regions between the first and second bands, etc., and are obtained by going through the same process as here, using $\cos \gamma x + A \sin \gamma x$ instead of Eq. (16-21), but requiring imaginary values of k.

Summing up, we have found that when the surface spike is different from those in the interior, a localized state has appeared that is reminiscent of the state of an impurity in a semiconductor. The state we have derived is called a Tamm state after its discoverer. The result found here may be generalized for real crystals. No Tamm states are localized on surfaces when the atomic potential there is completely symmetrical. However, Shockley has shown that for crystals possessing more than one atom per unit cell, localized states can again be found due to band overlaps. For real crystals, the situation is thus quite complicated because the existence of surface states depends subtly upon the kinds of wave functions in the solid (i.e., the possibility of overlapping bands) and upon relaxations at the surface. Often one would expect both, since atomic relaxation when it occurs is difficult to include quantitatively. Thus, fully realistic calculations are not yet available. The situation is further complicated by the influence of impurities when they are present, because they contribute additional states of their own. Overall, then, the quantum mechanics of surfaces is in a highly ambiguous state as far as theory is concerned, although our simple models provide considerable quantitative insight.

16-3 Experimental Tools

The present advanced state of progress in surface science relies upon three major experimental developments. The first and most fundamental advance was in vacuum technology. Even in a vacuum chamber, any exposed surface attempts to come into chemical equilibrium with any ambient atmosphere. In order to make meaningful studies of the surface, it is obviously necessary that the complete chemical composition of the surface be under the control of the experimenter. If the surface tends to form an oxide, then the oxide partial pressure in the vacuum chamber must be sufficiently low over the period of the experiment so that significantly less than a monolayer of oxide forms. Problem 16-4 shows that at 1000°K if the sticking probability for impinging gas atoms is 1, then the pressure must be $<10^{-9}$ torr in order that less than 1 per cent of a monolayer of oxide be formed in 10^3 sec for an atom of mass 10^{-22} g. A vacuum of this magnitude was not attainable on a regular basis until the middle of the 1950s. A significant part of the achievement of such vacuums was the development of pressure gauges which work in this range.

High-field microscopes

The second major experimental achievement was the development of high-field microscopes. The field-emission microscope was first developed in 1936 by Müller and was followed in 1951 by the field-ion microscope, a major improvement again developed by Müller. The field-emission microscope operates on the cold electron emission principle, and requires fields at the metal surface of order 10^9 to 10^{10} volts/m to give practical electron currents. These fields are generated by charging a system containing a very sharp point of the specimen material, see Fig. 16-14. Points of radius of order 100 to 1000 A are typical and are obtained by etching procedures. The highest resolution obtained with this instrument is in the range of several tens of angstroms.

The electron emission is strongly dependent upon work function. The apparatus is in fact used to measure work functions of different facets of the same crystal and the effect of impurity layers on the work function. Surface diffusion as a function of crystal facet can be measured somewhat crudely by observing the blunting or growth in radius of the tip caused by atom migration at the higher temperatures. Chemical reactions are also observable when new phases form, or monolayers of gas reactants build up.

The field-ion microscope is a far more powerful tool than the electron-emission microscope for studying surface features because it has higher resolution, with single atoms easily observed in favorable

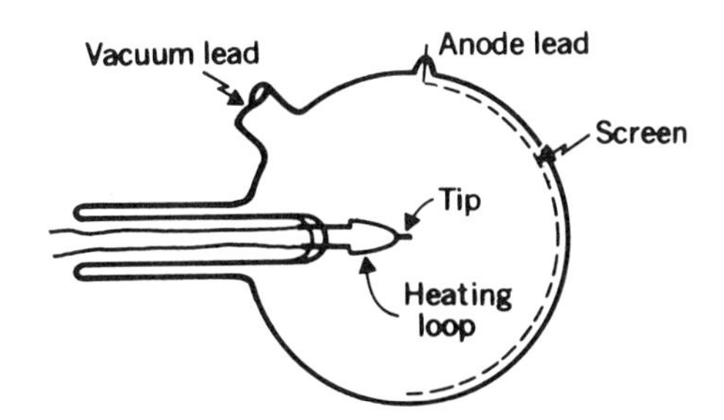

Figure 16-14 Field-emission microscope. The specimen in the form of a sharp tip of 100A to 1000A in radius is placed at high negative potential relative to the screen. The electrons are emitted normally to their surface, and impinge on the fluorescent face of the tube after a straight-line flight. The physical arrangement of the ion scope is the same. However, the polarity of point and screen are reversed, and a small amount of gas is injected into the system (about 10^{-3} torr).

cases. The physical arrangement is similar to that of the emission microscope; however, the polarity of the tip is reversed. The tip is now *positive* with respect to the surroundings, and a small amount of gas is injected into the apparatus. Usually the gas is an inert gas such as He, or A, but H_2 or other molecules can be, and sometimes are, used. In the vicinity of the tip in the region of large field strength, reverse tunneling occurs in which valence electrons tunnel across their ionization barrier into the positively charged metal. An appropriate sketch of the potential energy situation for tunneling electrons is given in Fig. (16-15).

A comparison of the resolution achievable with the two techniques leads into a complex argument. One would suppose that the resolution

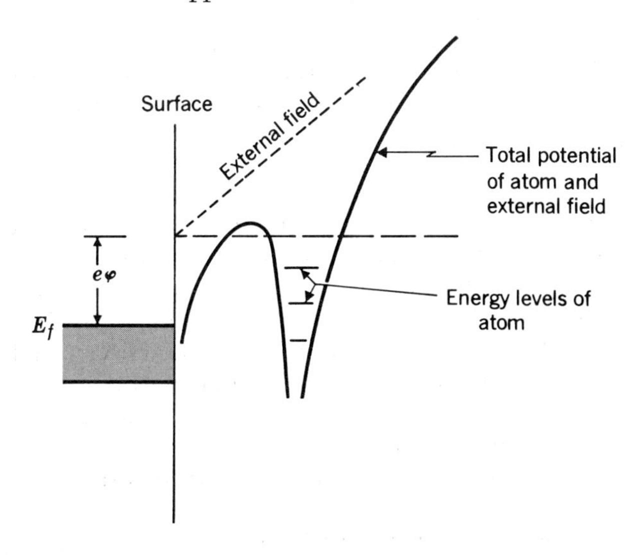

Figure 16-15 Schematic drawing of the potential energy of a valence electron on a gas atom in the vicinity of the positive tip. As the atom approaches the tip, both the height of the barrier and its width decrease. The rounded region shown at the metal surface is due to the image force discussed in Sec. 16-3.

in the emission case is limited by the uncertainty principle, but it turns out that velocities tangent to the surface in the original metallic electron velocity distribution are more important in spreading the image. In the case of the ion-emission instrument, the factors governing resolution are even more complex but depend upon the tangential velocities of the ions and the degree to which the actual potential distribution along the surface between ions of the surface is smooth or lumpy.

The path of a gas molecule in the tip region is not as simple as one might expect. The strong fields at the tip polarize the atoms, and in the inhomogeneous field they are attracted to the tip until they rebound after elastically colliding with it. Eventually, on one of their flights near the surface, tunneling occurs, and the resulting ion is ejected toward the screen (see Fig. 16-16.) Depending upon conditions,

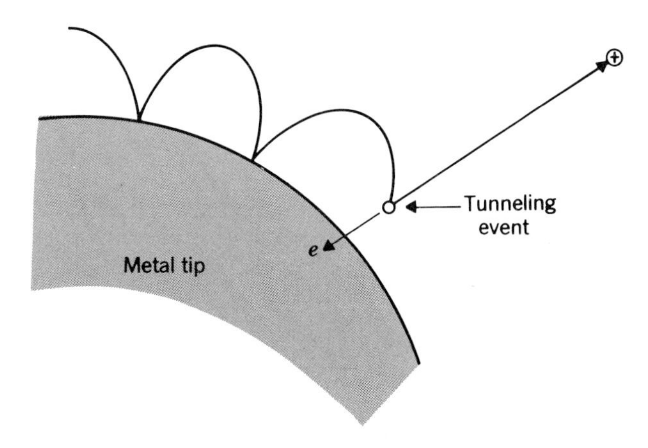

Figure 16-16 Hopping motion of gas atom in vicinity of the tip. The field polarizes the atom, and since a dipole experiences a net force in an inhomogeneous field, the atom is drawn to the tip. After several collisions with the surface, tunnelling occurs, and the charged ion is ejected toward the screen. Tunnelling obviously occurs most easily in regions of maximum electric field, which occur at corners and edges of the surface.

hundreds of passes into the immediate region of the tip can occur before the final tunneling ionization process takes place.

The precise features of the patterns observed in the instrument have only recently come under close scrutiny. The light and dark regions of a pattern obviously depict atomic force fields and correspond to the individual surface atoms. The pictures thus show individual atom placement, but a detailed study of the intensity variations also should eventually yield wave function details of the surface atoms also. An excellent example showing the power of this instrument is reproduced in Fig. 1-3.

One of the major drawbacks to the ion microscope is that low temperatures ($\simeq 20°K$ with He) are required for high resolution. Hence one only sees frozen "still" pictures, even though flashing to high temperatures can be done between "takes."

The ion microscope has been used to study defects in surface layers such as vacancy concentrations, atomic arrangements around an interstitial below the surface, dislocations, positions of adsorbed atoms, etc.

Low energy electron diffraction (LEED)

Electron diffraction by crystals has been known since the Davisson-Germer experiments originally confirmed the wave nature of electrons in 1927. These experiments were made with energies in the range up to 370 volts. In order to be certain that only the first layer of surface atoms contributes to the major part of the pattern, the electron energy must be kept low, usually less than 100 ev. This requirement makes it very difficult to detect the electrons because such efficient methods as letting the electrons impinge on the fluorescent screen require kilovolt electrons. Thus the main emphasis for work in this field dates from the development of an apparatus by Germer and his co-workers in 1960, which makes use of a post-acceleration technique, Fig. 16-17.

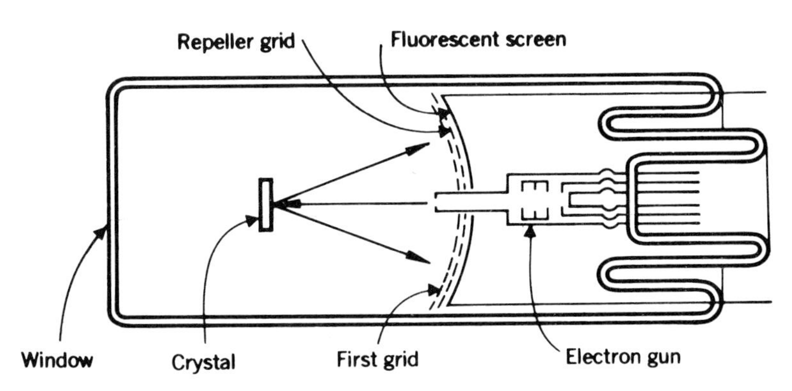

Figure 16-17 LEED apparatus. The monochromated and collimated electrons are directed at the specimen surface at voltages of a few to several hundred volts. Scattered electrons are then allowed to go through a series of screens near the face of the tube which sort out the electrons of the beam and accelerate them to the kilovolt range before allowing them to strike the fluorescent surface. The first grid is at the same potential as the sample, giving a field-free region in the space between it and the sample. The second grid is at a small *negative* potential relative to the first grid to repel any inelastically scattered electrons. The fluorescent screen then is placed at a large positive voltage and accelerates the electrons into the kilovolt range before they strike the screen.

The unique feature of this apparatus is that electrons are scattered at the surface at controllably low voltages, but after the scattering occurs they are then accelerated through a grid system to several thousand electron volts for display on the face of the cathode ray tube.

The field-ion and field-emission techniques are primarily used for the study of metal surfaces, but LEED is applicable to any surface. The high-field techniques, of course, have the advantage of high resolution, but there remain questions concerning distortions, etc., induced by the extremely high fields necessary. By contrast, in the diffraction technique, electrons are sprayed over a macroscopic region of the surface, and the resultant pattern averages over large numbers of atoms. LEED thus infers information about the structure of surfaces in the same way as X-ray or neutron scattering infers information about the structure of normal three-dimensional crystals.

The display as seen in a LEED experiment, then, has much similarity to that of an ordinary X-ray picture. If the electrons are diffracted only by atoms on the surface (an assumption which will be discussed in due course), the diffraction is that of a two-dimensional lattice.

A typical picture of a LEED pattern is shown in Fig. 16-18. The

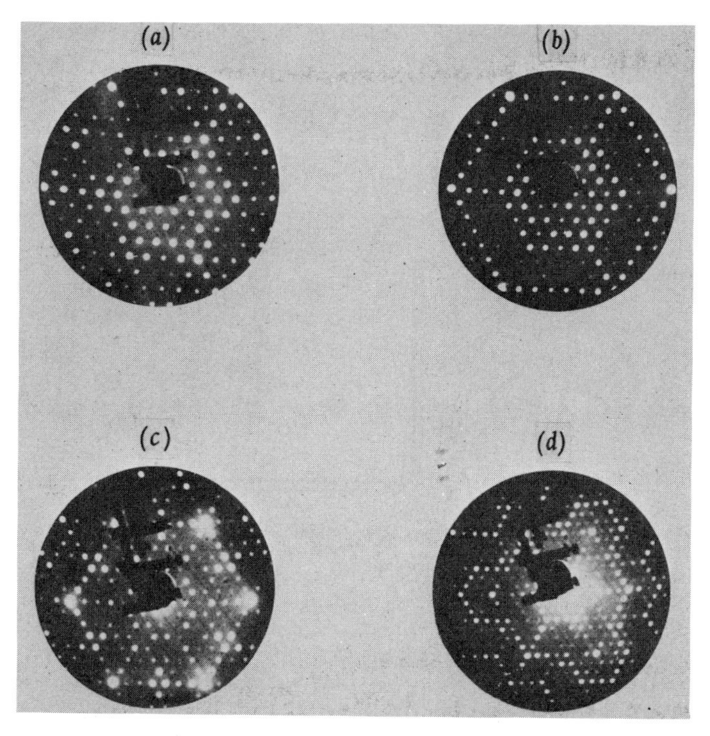

Figure 16-18 Diffraction pictures of the Si (111) surface taken with (*a*) 30, (*b*) 37, (*c*) 52, (*d*) 86 volts, respectively. The basic hexagonal nature of the surface is demonstrated by the strong reflections evident especially in (*b*) and (*c*). The weaker spots [especially the seven weak reflections between the corners of the hexagon in (*b*)] indicate that additional structure is present. (Photographs courtesy of F. Jona.)

diffraction conditions are a special case of the Bragg law for three-dimensional diffraction of X rays, but we will derive them here for the two-dimensional case anew.

The condition for constructive interference between wavelets scattered from different atoms on a plane is given by the requirement that the path difference between waves scattered by neighboring atoms be an integral number of wavelengths. Thus, when the incident wave is normal to the scattering plane (Fig. 16-19),

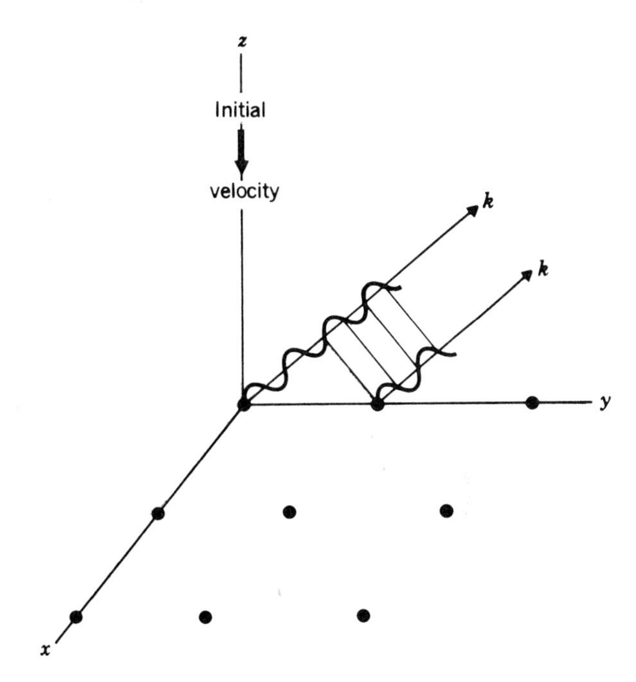

Figure 16-19 Geometry for scattering from a surface; shows the path difference equal to an integral number of wavelengths.

$$\hat{\mathbf{k}} \cdot \mathbf{a} = n\lambda \tag{16-29}$$

where **a** is a lattice vector on the plane, n is an integer, and λ is the wavelength. $\hat{\mathbf{k}}$ is a unit vector in the scattered-wave direction **k**. Rewriting the equation, we obtain

$$\mathbf{k} \cdot \mathbf{a} = 2\pi n \qquad n = 1,2, \ldots \tag{16-30}$$

Provided k is larger than the minimum required to obtain diffraction for $n = 1$, Eq. (16-29) predicts that diffraction will always occur for some direction k. This direction is depicted by Fig. 16-20, where two cones are drawn defined by the two component equations,

$$\begin{aligned} \mathbf{k} \cdot \mathbf{a}_x &= 2\pi n_x \\ \mathbf{k} \cdot \mathbf{a}_y &= 2\pi n_y \\ n_x + n_y &= n \end{aligned} \tag{16-31}$$

If diffraction occurs for a given k_x,k_y, then increasing either k_x or k_y for a given value of n simply opens up the angle of the cone, and inspection of Fig. 16-20 shows that intersection always exists.

Making observations with LEED is more complicated than using X rays. One reason is that for low energy (i.e., less than a few hundred

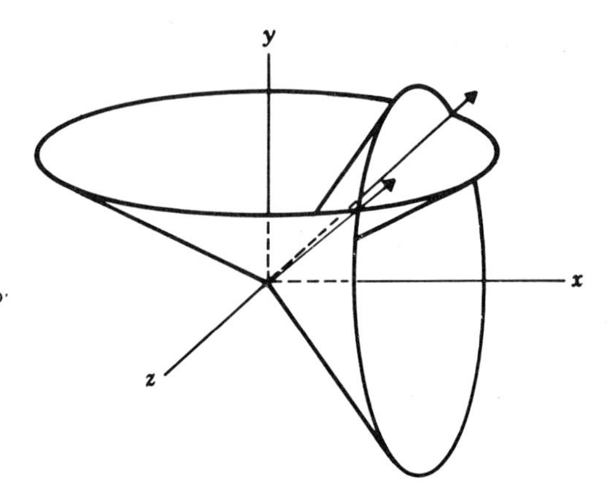

Figure 16-20 Eqs. (16–31) are depicted as two cones lying in the x and y directions respectively. The intersection of the cones gives the diffracted direction k. Increasing k simply opens out the apex angle of the cones, causing k to approach the z axis as a limit.

ev), the scattering probability is high, multiple scattering of the electrons by the atoms can be important, and the theory is more complicated than that given by Eq. (16-30). In addition, above about 100 ev, the electrons penetrate deeper into the lattice beyond the first layer of atoms and are scattered by atoms below the first surface layer. In this case, the scattering is partially three-dimensional, and the two-dimensional equations must be modified. The problem here is that, if the atoms on the surface are removed from their normal bulk lattice positions, the third dimension does not yield a true diffraction condition but leads to diffuse scattering of a more complicated character. For all these reasons, the theory of electron-scattering from surfaces has been slow to develop. Figure 16-18 shows a picture of a surface of Si, and although the bright dots correspond to the basic structure of the Si lattice, other weaker spots also are present. The basic hexagonal symmetry of the surface is demonstrated by the hexagonal bright spots. The additional weaker spots forming on a line between the bright spots are additional features that demonstrate that the surface has relaxed in some important way from its normal lattice position. Such additional features are common in LEED work, and the symmetry of the new diffraction spots and their dependence on electron voltage form the take off point for theoretical interpretation.

16-4 Surface Structure

Clean surfaces

One of the striking findings of recent surface work has been the discovery, alluded to earlier, that even the atoms of an apparently completely clean surface often take on a different lattice structure from that of the underlying crystal. Further, the restructured surface may exist over only a restricted temperature interval. Such surface rearrangements occur for metals, semiconductors, and insulators, although, because LEED interpretation is often difficult, some of the reported cases are still controversial. In addition, in some cases where the surface atoms are not rearranged, the atoms of the surface are relaxed outward from the bulk of the crystal. Another complexity often seen is for a high-energy surface to decompose into a pitted surface composed of projecting cones with low-energy sides.

The most highly studied case of a rearranged surface is that of the semiconductors, Ge and Si. The [111] surface of the diamond lattice should have a series of dangling bonds normal to the surface (Fig. 16-21),

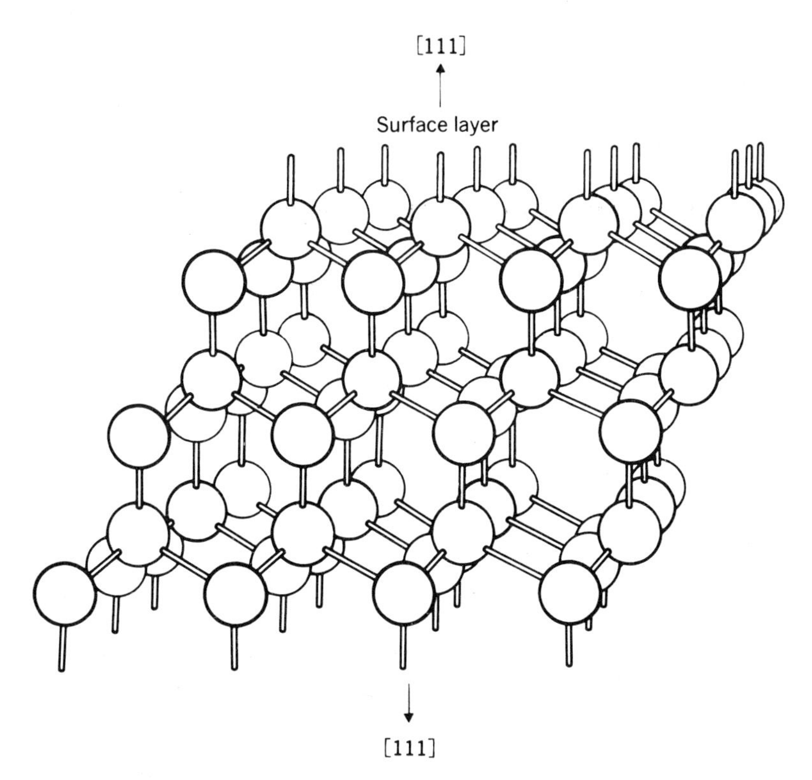

Figure 16-21 Representation of the (111) surface in a diamond crystal. The broken bonds are vertical to the surfaces.

if the lattice were unchanged. Apparently the energy of these bonds is too high, and they rearrange themselves so that the bond structure is saturated in the manner depicted in Fig. 16-22. This interpretation has been proposed to account for the extra features already described in Fig. 16-18, although the matter is still controversial at the time of writing.

With such large-scale rearrangement at the surface, it is not surprising that the theory of the surface states is still very tentative. In experiments on clean [111] surfaces of the semiconductors, the surface is strongly p-type, and various earlier theories of the electrical effects observed are couched in terms of dangling bonds. On the other hand, work on the observed structures seems to indicate a saturated structure without dangling bonds, so the current interpretation is clearly incomplete.

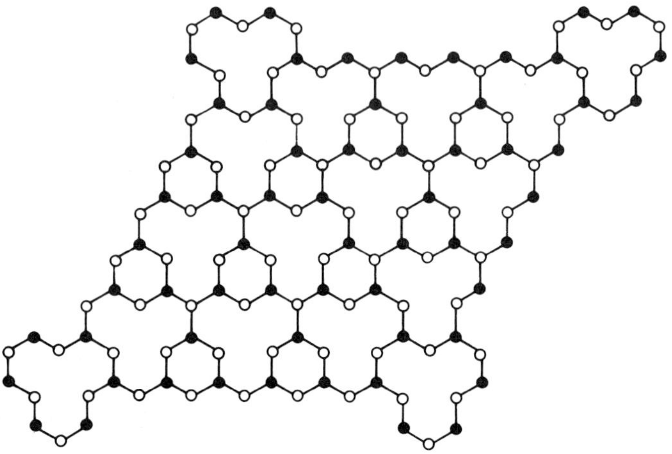

● Surface layer

○ Second layer

Figure 16-22 Proposed array of atoms on the Si (111) surface which explains the extra features of the diffraction results shown in Fig. 16-19.

Impurity films

Probably the greatest potential for the entire field of surface science lies in the study of various chemical reactions on the surface. Understanding the details of the atomic processes underlying corrosive reactions, oxide formation, and catalytic reactions cannot help but have tremendous economic benefits. Indeed, the hitherto puzzling complexity of surface reactions underlying such phenomena as catalytic effects is already showing its counterpart in the exceedingly rich structure apparent in studies of surfaces and the various restructurings observed for surfaces with impurity layers. As a general rule, surface restructuring (either ordered or amorphous) occurs starting with impurity concentrations amounting to much less than a monolayer.

Much work has been done with oxygen-surface reactions. The amounts and rates of oxygen absorption are found to depend strongly on the crystallographic orientation of the surface, because of the varying populations of jogs, steps, etc., on the different facets. A close-packed surface with no steps or jogs presents a relatively smooth aspect to incident atoms; diffusion is fast, and cohesion low. Along a step, however, the number of bonds to neighboring atoms increases, and at a corner of the step binding is at a maximum. On less closely packed surfaces, the atomic placements are more open, the surface is corrugated, and adsorption is stronger. Figure 16-23 shows a two-dimensional close-packed array of balls. The outer bounding edge turns through a smooth close-packed line of atoms and through a looser packing edge back again to the close-packed line. On the close-packed edge, the potential function is smooth, and the number of bonding atoms is small. On the open-packed portion, the surface is corrugated;

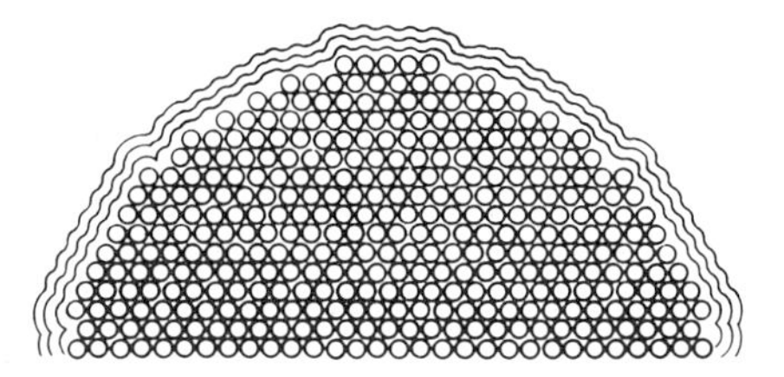

Figure 16-23 A two-dimensional hexagonal array of marbles bounded by edges of various crystallographic directions; and shows the difference between close packed and more open bounding edges.

the diffusion will be anisotropic at best (in three dimensions), and attachment will be easy. A three-dimensional sketch of marbles depicting a metal tip is shown in Fig. 16-24 and illustrates the obviously

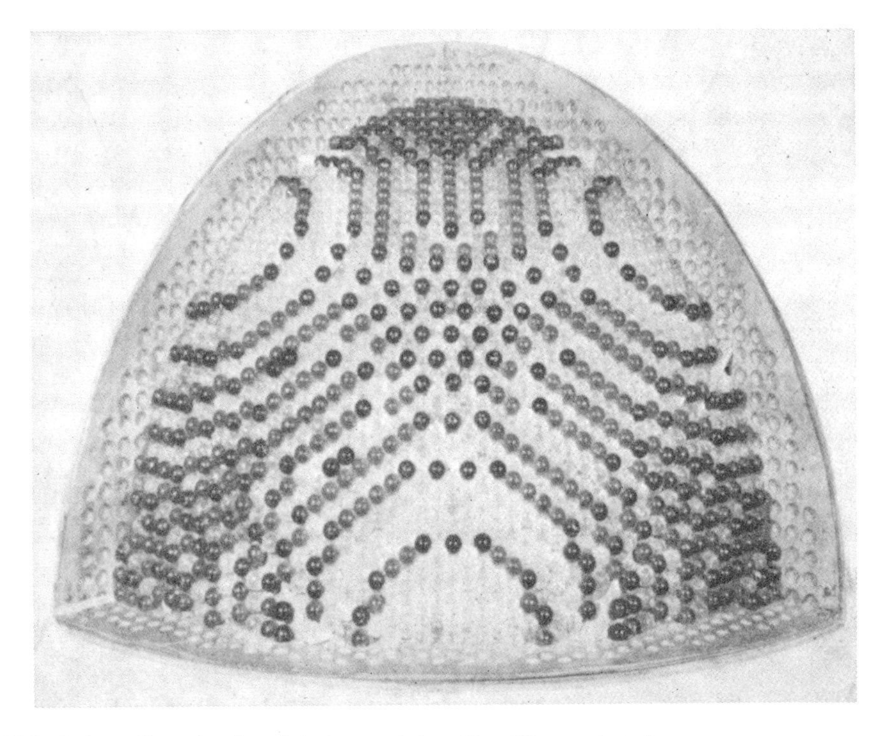

Figure 16-24 A three-dimensional model of a metal tip. The different shaded marbles have different numbers of nearest neighbors.

more complex structure to be expected in three dimensions. At the time of writing, the details of the results of surface diffusion are too incomplete, in view of the complex interpretation required, to be reviewed here.

As an example of some of the progress made to date in the study of oxidizing reactions on surfaces, clean surfaces of iron are found to form at least two distinct layers of oxide, with increasing oxygen content from the metal toward the outside. Likewise, the oxide layer on chromium is found to be amorphous, which probably accounts for the stability of chromium with respect to corrosion in the presence of an oxidizing environment.

16-5 Summary

Obviously this field is very incomplete. We have seen that the theory is inherently complex, and that little quantitative connection exists yet between the rich experimental findings and the still rudimentary theory. Likewise, in spite of the steady flow of new experimental results, the important field of surface reactivity still has far to go before an adequate atomic picture will be available. This entire field is thus obviously in a state of intense development, with the major problems yet to be solved and with the most important applications of the fundamental findings still in the future.

PROBLEMS

16-1 Calculate the current density emitted from a thoriated tungsten filament heated to 2000°C. An interesting effect here is that while the work function of pure tungsten is 4.5 ev and that of thorium is 3.35 ev, a monolayer of thorium on tungsten has a work function of only 2.63 ev. Apparently, the Th atoms on the surface are ionized, thereby causing an electric field which makes it easier to draw electrons from the metal.

16-2 Evaluate the integrals in Eq. (16-5) and obtain the result (16-6).

16-3 If the temperature is 1000°K, find the number of atoms per second which impinge on a square centimeter of a metal surface in a cavity containing a small pressure of a gas whose atomic mass is 10^{-22} gm. Perform an order of magnitude calculation.

16-4 In Prob. 16-3, if the atomic spacing on the surface is 3A, how long does it take to build up a complete monolayer, assuming that each atom colliding with the surface sticks to it.

16-5 A one-dimensional box is 1 cm long and has an infinite wall at the two ends. What are the values of k for the waves having 9 and 10 nodes, respectively, in the box?

16-6 If the box of Prob. 16-5 has a wall only 10 ev high at $x = 1$ cm, show that the wave with 10 nodes has a k value between the two values calculated in Prob. 16-5.

16-7 What is the distance which the electron "penetrates" outside the box of Prob. (16-6)?

16-8 Equations (16-26) were derived by simply rewriting Eqs. (16-17) and (16-18) in the proper way. For the wave function Eq. (16-21), go through the steps analogous to those leading from Eq. (16-15) to Eq. (16-18) and show directly that Eqs. (16-26) are correct.

16-9 For a square surface lattice of lattice spacing 4A, what is the smallest value of k for which one can get a diffraction spot in first order? What energy electrons does this correspond to?

17

Macroscopic electromagnetic behavior of solids

The point of departure of this chapter on the electric and magnetic properties of solids is Maxwell's electromagnetic equations, because through them the fundamental nature of the various fields quickly becomes apparent. Once Maxwell's equations are at hand, then the dielectric and magnetic properties of solids can be described in detail with consistency.

Maxwell's equations are written as

$$
\begin{aligned}
\nabla \cdot \mathbf{D} &= \rho \\
\nabla \times \mathbf{E} &= -\frac{\partial \mathbf{B}}{\partial t} \\
\nabla \cdot \mathbf{B} &= 0 \\
\nabla \times \mathbf{H} &= \frac{\partial \mathbf{D}}{\partial t} + \mathbf{J}
\end{aligned}
\tag{17-1}
$$

Since both **D** and **B** as well as **E** and **H** appear in these expressions, they are the equations of electromagnetism in the presence of matter. The charge density ρ and the current density **J** are both functions of position. **D** and **E** and **B** and **H** are related by the expressions

$$
\begin{aligned}
\mathbf{D} &= \epsilon \mathbf{E} \\
\mathbf{B} &= \mu \mathbf{H}
\end{aligned}
\tag{17-2}
$$

where ϵ and μ are the capacitivity† and permeability of the medium, respectively. Both ϵ and μ depend upon the medium.

† Commonly called the *dielectric constant* in the scientific literature.

The first Maxwell equation simply represents Coulomb's law. If a point charge q is at the origin, then integration of $\nabla \cdot \mathbf{D}$ over a sphere of radius r yields the field. Since

$$\int \nabla \cdot \mathbf{D}\, dv = \int \rho\, dv$$

then

$$\oint \mathbf{D} \cdot d\mathbf{S} = q$$

or

$$4\pi r^2 D = q$$

Finally

$$\mathbf{D} = \frac{q\mathbf{r}}{4\pi r^3} \tag{17-3}$$

The vector $\mathbf{r}$ points radially from the charge q.

The second of Maxwell's equations is Faraday's law of induction. If this equation is integrated over a surface in space,

$$\int \nabla \times \mathbf{E} \cdot d\mathbf{S} = -\int \frac{\partial \mathbf{B}}{\partial t} \cdot d\mathbf{S}$$

By using Stokes' theorem, this integral may be expressed in terms of a line integral along a curve bounding this surface.

$$\int \mathbf{E} \cdot d\mathbf{l} = -\frac{\partial \phi}{\partial t} \tag{17-4}$$

where ϕ is the total flux passing through the surface. The left-hand side of this equation is simply the emf.

The third of the Maxwell equations, $\nabla \cdot \mathbf{B} = 0$, states that magnetic lines of force must all close on themselves. Consequently there exists no entity in magnetism analogous to a free electric charge.

The last equation states that a magnetic field encloses any region containing a current. Thus, if a current passes through some surface in space, integration of this equation over the surface allows the calculation of this field. Thus

$$\int \nabla \times \mathbf{H} \cdot d\mathbf{S} = \int \frac{\partial \mathbf{D}}{\partial t} \cdot d\mathbf{S} + \int \mathbf{J} \cdot d\mathbf{S}$$

becomes

$$\int \mathbf{H} \cdot d\mathbf{l} = \int \frac{\partial \mathbf{D}}{\partial t} \cdot d\mathbf{S} + \mathbf{I} \tag{17-5}$$

Hence, the line integral of the magnetic field $\mathbf{H}$ about the boundary of the surface is given in terms of the real current $\mathbf{I}$ and any displacement current which might exist.

The reader may well have previously demonstrated for himself the foregoing illustrations and may be aware of their implications with regard to vacuum. However, we emphasize that these laws apply to fields in solid bodies as well. The specific electric and magnetic character of the medium can be brought out more clearly by writing Eq. (17-2) as

$$\mathbf{D} = \epsilon_v \mathbf{E} + \mathbf{P} \qquad (17\text{-}6)$$

$$\mathbf{B} = \mu_v \mathbf{H} + \mu_v \mathbf{M}$$

where ϵ_v = capacitivity of a vacuum
μ_v = permeability of a vacuum
$\mathbf{P}$ = electric dipole moment per unit volume
$\mathbf{M}$ = magnetic dipole moment per unit volume.

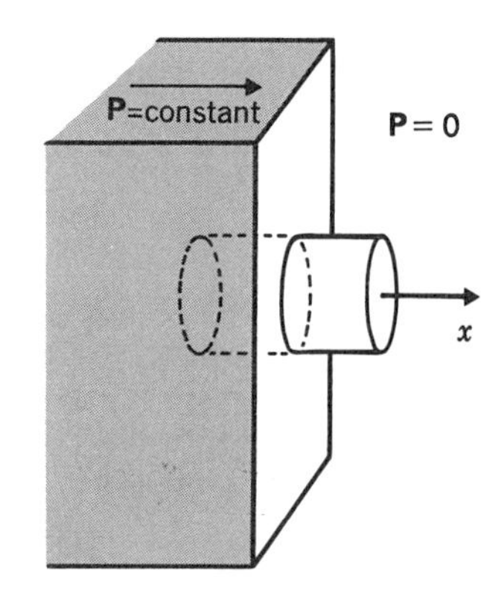

Figure 17-1 Interface between a solid with polarization **P** and a vacuum.

The equations of electrostatics are considered first, because they are somewhat simpler. If both ρ and $\partial \mathbf{B}/\partial t = 0$, the first two of Eq. (17-1) can be written [with the first of Eq. (17-6)].

$$\nabla \cdot \mathbf{E} = -\frac{1}{\epsilon_v} \nabla \cdot \mathbf{P} \qquad (17\text{-}7)$$

$$\nabla \times \mathbf{E} = 0$$

These equations are now similar to the first two Maxwell equations except that they contain only one unknown, **E**. In addition the right-hand side of the first of Eq. (17-7) represents a fictitious charge density which is related to the presence of matter and which is equal to $\nabla \cdot \mathbf{P}$. The precise meaning of **P** and the role of $\nabla \cdot \mathbf{P}$ in this equation are brought out in the following simple situation: Suppose all space is divided into two regions, one filled with matter and the other vacuum, the interface between these two regions being the $x = 0$ plane. Let **P** in the matter be a constant, and let it point in the x direction. In the vacuum, of course, $\mathbf{P} = 0$. Next construct a small pillbox in space, one face of the box being in the material, the other in the vacuum (Fig. 17-1). The flat faces of the pillbox are normal to the x axis and have area $A = 1\ \text{m}^2$. Now apply the first of Eq. (17-7) to the pillbox region and integrate over the volume of the box,

$$\int \nabla \cdot \mathbf{E}\, dV = -\frac{1}{\epsilon_v} \int \nabla \cdot \mathbf{P}\, dV \tag{17-8}$$

From Gauss' theorem, this equation becomes

$$\int \mathbf{E} \cdot d\mathbf{S} = -\frac{1}{\epsilon_v} \int \mathbf{P} \cdot d\mathbf{S} \tag{17-9}$$

where **S** is the surface of the pillbox.

Assume now that the height of the pillbox is decreased so that the integral over the cylindrical sides is small compared with that over

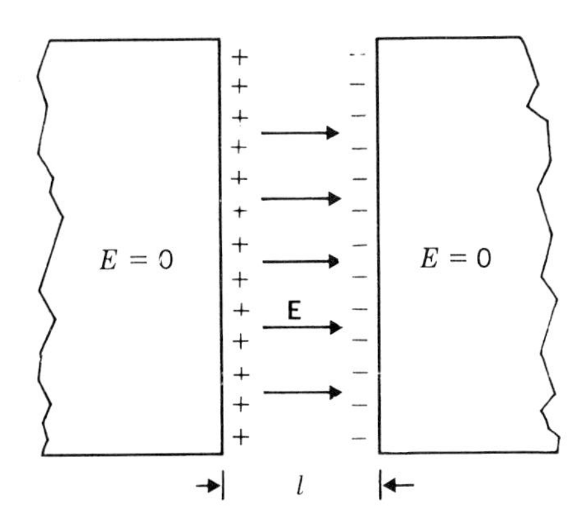

Figure 17-2 Field between charged plates of a parallel-plate capacitor.

the faces. Then the integration can be carried out to give

$$\Delta \mathbf{E} = \mathbf{E}_x - \mathbf{E}_x' = \frac{1}{\epsilon_v} \mathbf{P} \tag{17-10}$$

where $\mathbf{E}_x$ is the field in the vacuum region and $\mathbf{E}_x'$ that in the medium. Equation (17-10) is the desired result and states that the difference between **E** in the vacuum and **E** in the material is just equal to the value of $(1/\epsilon_v)\mathbf{P}$. Clearly, if the vacuum were replaced with a second material of polarization $\mathbf{P}'$, then the discontinuity of **E** at the interface would be equal to $(1/\epsilon_v)(\mathbf{P} - \mathbf{P}')$. (We have tacitly supposed for simplicity that **E** and **P** have the same direction, which is not true for all media.)

The interpretation of **P** is best seen by analogy with the situation in a parallel-plate capacitor. Let such a capacitor (Fig. 17-2) have a surface charge σ coulombs/m². From electrostatics we know that under such conditions there is a discontinuity in E which is given by σ/ϵ_v, that is,

$$\Delta E = \frac{\sigma}{\epsilon_v} \tag{17-11}$$

Thus by analogy we can see that a boundary on which the polarization **P** changes by a certain amount has the same effect on **E** as does the surface of a body on which there resides a real charge.

The next question is to ask what kind of charge distribution within the medium can give rise to a situation where there is uniform polarization but no net charge. Since matter is made up at every point in space by atoms which have an equal amount of positive and negative charge, the necessary charge can be induced on the surface

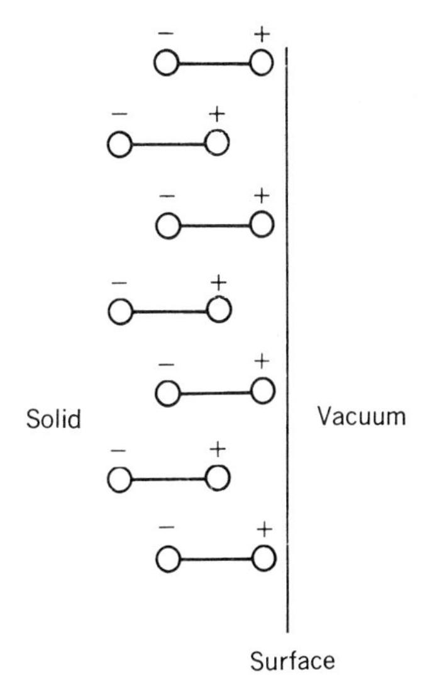

Figure 17-3 Generation of a surface charge by production of similarly oriented dipoles.

of the material simply by displacing the positive from the negative charge on each atom site. Thus a small dipole of moment **p** is produced at each atom site by displacement of charges $+q$ and $-q$ through a distance d,

$$\mathbf{p} = q\mathbf{d} \tag{17-12}$$

If at each point in space a similar dipole is formed, then no net charge is induced anywhere in the medium. On the other hand, at the surface of the material an excess charge is calculated by noting that all the dipoles within a distance d of the surface contribute a charge q to the surface. Figure 17-3 shows how this charge is generated for a particular geometry of charge distribution. From comparison of Eq. (17-10) with (17-11), we see that the magnitude of the surface charge produced in this way is just equal to the volume density of dipole moment.

This association of the gross polarization **P** with the sum of the atomic polarizations **p** is as far as simple notions can carry us. To go further requires that a detailed investigation be made of how the individual dipoles **p** are produced. The next chapter discusses several ways in which this can be done.

The magnetic-dipole moment is entirely analogous to the electric-dipole moment just discussed. The magnetostatic equations, using Eq. (17-6), are

$$\begin{aligned} \nabla \cdot \mathbf{B} &= 0 \\ \nabla \times \mathbf{B} &= \mu_v \nabla \times \mathbf{M} \end{aligned} \tag{17-13}$$

$\nabla \times \mathbf{H}$ is made to be zero by setting $\partial \mathbf{D}/\partial t$ and $\mathbf{J}$ to be zero.
These equations look similar to the last two Maxwell equations except that they contain only one field vector, $\mathbf{B}$, and that a fictitious current density $\mu_v \nabla \times \mathbf{M}$ exists. Again let these equations be applied to a space with the region to the left of the origin in Fig. 17-4 covered by a

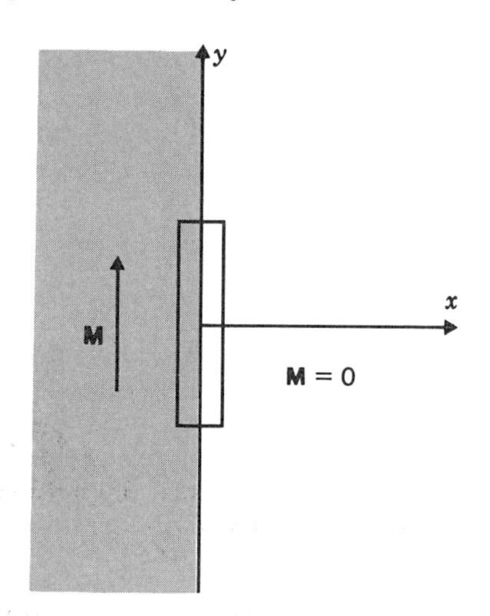

Figure 17-4 Interface between a vacuum and a solid with constant magnetization **M**.

medium with a constant magnetization $\mathbf{M}$. The region to the right of the origin is a vacuum with $\mathbf{M} = 0$. Assume that the magnetization vector $\mathbf{M}$ is in the y direction, parallel to the surface (Fig. 17-4). Integrate the second of Eq. (17-13) over a surface containing both the medium and the vacuum. We choose a rectangular area of integration with long sides parallel to y and short sides parallel to x.

The integration can be written formally as

$$\int \nabla \times \mathbf{B} \cdot d\mathbf{S} = \mu_v \int \nabla \times \mathbf{M} \cdot d\mathbf{S} \tag{17-14}$$

These surface integrals can be transformed to line integrals,

$$\int \mathbf{B} \cdot d\mathbf{l} = \mu_v \int \mathbf{M} \cdot d\mathbf{l} \tag{17-15}$$

If the rectangle is sufficiently long and narrow, the parts of the line integral taken in the x direction are unimportant. Hence,

$$\mathbf{B}_y(\text{medium}) - \mathbf{B}_y(\text{vacuum}) = \mu_v \mathbf{M} \tag{17-16}$$

Again a discontinuity exists in the field at the surface, as for the electrostatic case. Notice that, if the integral had been carried out over a surface lying entirely in either the medium or the vacuum, the right-hand side of Eq. (17-14) would have been zero, because the magnetization is constant everywhere in the medium or in the vacuum. The integral of Eq. (17-14) has a value different from zero only when the boundary of the integration surface encloses a portion of the surface of the medium. As has already been demonstrated in Eq. (17-5),

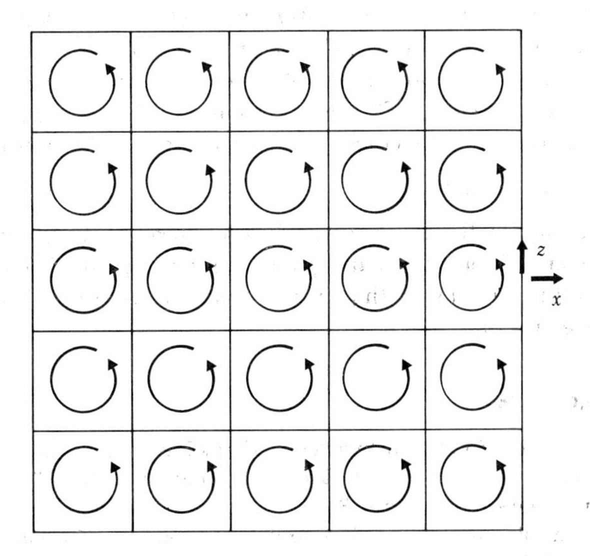

Figure 17-5 Generation of effective surface currents by circulating currents in cells in solids.

whenever such an integration circuit encloses a region with current passing through it, a net magnetic field circles the area. Thus Eq. (17-16) implies that there is a surface current in the z direction perpendicular to the page and to the integration parallelogram of Fig. 17-4.

Again, as in the electrostatic case, there is a physical basis for the current distribution which is found within the medium. The medium itself has no net current density, and yet a current density exists on the surface. In Fig. 17-5, a current distribution which satisfies these conditions is sketched; this is a section of the xz plane perpendicular to the section of Fig. 17-4. The region of the medium is divided into small cells; within each cell a small current circulates. If the current circulates in the same sense in each cell, the net current on the boundary of each interior cell is zero because of cancellation. However, on the surface of the medium this cancellation does not occur, and there is a net surface current in the z direction.

The previous discussion allows the interpretation of the magnetization vector $\mathbf{M}$. The existence of this vector implies that the medium contains many small regions (which are, of course, simply the atoms of the medium) in which charge circulates. This circulation of charge is the same in all parts of the medium. No net current flows within the interior of the medium because of current compensation, but on external surfaces of the medium where the magnetization changes from one value to another, an effective current density is set up. As before, the polarization vector is equal to the dipole moment per cubic meter, i.e.,

$$\mathbf{M} = \text{dipole moment/m}^3$$

The entire picture should, of course, be considered from an atomic point of view. An elementary magnetic-dipole moment can thus be pictured as current circulation in an atom. On a quantitative level, the expressions for the dipole moment of a circular current-carrying wire can be used. The dipole moment $\boldsymbol{\mu}_m$ is given as the product of the magnitude of the current, i, and the area of the circuit, A, that is

$$\boldsymbol{\mu}_m = \mathbf{n}iA$$

Recall that the unit vector $\mathbf{n}$ has the direction of the thumb when the fingers of the right hand point along the path of the current. In a succeeding chapter the origin of these elementary dipole moments is investigated.

PROBLEMS

17-1 From your knowledge that ρ has dimensions coulombs/m³, that $\mathbf{J}$ has dimensions coulombs/sec m², and that $\mathbf{E} = \nabla V$, determine the dimensions of $\mathbf{D}$, $\mathbf{B}$, and $\mathbf{H}$, using Eq. (17-1).

17-2 Using Eq. (17-2), find the dimensions of ϵ and μ.

17-3 Can you decide why the two Eqs. (17-6) are not symmetrical?

17-4 Using spherical polar coordinates, find expressions for the radial and angular components of $\mathbf{E}$ produced in a medium of capacitivity ϵ by a dipole of strength $Q\mathbf{d}$. Compare the magnitude and sign of the components of the field at the point $r = r_0$, $\theta = \theta_0$ with those at the point $r = r_0$, $\theta = \theta_0 + \pi$. (This observation is important in calculating the internal field in a solid in the next chapter.)

17-5 Compute the electric-field intensity at some point r due to a uniform spatial distribution of electric charge within a sphere of radius r_0 containing a total charge of Q coulombs. Especially, compare the fields for points for which $r > r_0$ and $r < r_0$. (Note that the results of these calculations were used as an approximation to screening in Chap. 7.)

17-6 Suppose that a voltage V exists across the plates of the capacitor in Fig. 17-2 and that the charge per unit area on the plates is $\pm q$. Let the area of each plate be A, and let the capacitance be C.

(*a*) If $\mathbf{D}$ and $\mathbf{E}$ are assumed to be constant within the capacitor and zero outside, find expressions for $\mathbf{D}$ and $\mathbf{E}$.

(*b*) If $\mathbf{D}$ and $\mathbf{E}$ are collinear vectors, show that $\epsilon = Cl/A$.

(*c*) Since the energy associated with the capacitor is $\frac{1}{2}qAV$, show that energy per unit volume, W_E, may be written

$$W_E = \tfrac{1}{2}DE$$

and that

$$W_E = \tfrac{1}{2}\epsilon_v \mathbf{E} \cdot \mathbf{E} + \tfrac{1}{2}\mathbf{P} \cdot \mathbf{E}$$

Note that the magnetic energy per unit volume in a static magnetic field, in which **B** and **H** are collinear, is given by

$$\begin{aligned} W_M &= \tfrac{1}{2}BH \\ &= \tfrac{1}{2}\mu_v \mathbf{H} \cdot \mathbf{H} + \tfrac{1}{2}\mu_v \mathbf{M} \cdot \mathbf{H} \end{aligned}$$

17-7 Look up or derive the following factors in cartesian, cylindrical, and spherical coordinates:

$$\nabla \cdot \mathbf{D}, \qquad \nabla \times \mathbf{E}, \qquad \nabla V$$

18

Static dielectric behavior

18-1 General Features

The identification of the macroscopic polarization **P** as being the net effect of many elementary dipoles in a piece of material shows the way to relating the behavior of atoms and molecules themselves to the macroscopic effects. The discussion is limited to a few of the simpler effects, to give a physical picture of the phenomena involved.

The basic starting points are the linear equations (17-2) and (17-6),

$$\mathbf{D} = \epsilon \mathbf{E} \tag{18-1a}$$

$$\mathbf{D} = \epsilon_v \mathbf{E} + \mathbf{P} \tag{18-1b}$$

The first of these equations can be rewritten as

$$\mathbf{D} = \epsilon_v \frac{\epsilon}{\epsilon_v} \mathbf{E}$$

from which $\epsilon_r = \epsilon/\epsilon_v$ is defined as the *relative capacitivity*, so that

$$\mathbf{D} = \epsilon_v \epsilon_r \mathbf{E} \tag{18-2}$$

The quantity ϵ_r is independent of E for many insulators in weak or moderate fields, at least up to 10^6 volts/m for most insulators. If ϵ_r is independent of E, Eqs. (18-1b) and (18-2) can be combined to give

$$\epsilon_v \epsilon_r \mathbf{E} = \epsilon_v \mathbf{E} + \mathbf{P}$$

from which

$$\mathbf{P} = \epsilon_v(\epsilon_r - 1)\mathbf{E} \tag{18-3}$$

The quantity $\epsilon_r - 1$ is termed the dielectric susceptibility χ_e. Then

$$\mathbf{P} = \epsilon_v \chi_e \mathbf{E} \tag{18-4}$$

The following discussion treats only those materials for which **D**, **P**, and **E** have the same direction (the isotropic case), so that χ_e is a pure number, as is ϵ_r. Furthermore **P** and **E** are assumed to have the same time dependence. The best way to ensure that they do is to make **E** and **P** completely independent of time. This chapter treats this static situation.

The polarization of the individual atoms or molecules in an electric field may arise from three distinct sources:

1. The electric field may cause a relative displacement of the positive and negative charges on an atom, *inducing a dipole moment* on the atom which originally possessed none.
2. Positively and negatively charged ions may undergo relative displacement by the action of the field, inducing an *ionic* polarization.
3. Permanent dipoles (i.e., dipoles existing in the absence of a field) may be rotated by the field from random directions into the field direction, causing polarization by *orientation of permanent dipoles*.

SUSCEPTIBILITY OF GASES

The susceptibility of dilute gases (for which $\chi_e \ll 1$) may be treated most easily, and so it is considered first. For such gases the behavior of each molecule is almost completely independent of the action of all the others, and the effective field at each atom or molecule is only the applied field **E**. Discussion of (1) and (3) above is given on the basis of relatively simple models; discussion of (2) is more difficult.

In the detailed discussion of these three factors the natural unit is the polarization per atom (or molecule). If the elementary dipole moment in the direction of **E** is called **p**, then

$$\mathbf{P} = N\mathbf{p} \tag{18-5}$$

where N is the number of particles per unit volume. Hence equating Eqs. (18-4) and (18-5) gives (after rearrangement)

$$\mathbf{p} = \epsilon_v \frac{\chi_e}{N} \mathbf{E}$$

from which

$$\mathbf{p} = \alpha \mathbf{E} \tag{18-6}$$

if α, the polarizability, is defined as

$$\alpha = \frac{\epsilon_v \chi_e}{N} \tag{18-7}$$

Through Eqs. (18-5) to (18-7) the detailed link is made between microscopic and macroscopic polarization phenomena.

18-2 Induced Polarization

The induced polarization of an atom can be described in the following way: An atom is pictured as a point nucleus of charge $+Z|e|$ and a smeared-out electron cloud of charge $-Z|e|$. The electron cloud is considered to be a spherical region of uniform charge of radius R. With no external field, the nucleus is at the center of this sphere of charge (Fig. 18-1*a*). Under the action of the field $\mathbf{E}$, the nucleus and the electron cloud shift with respect to one another by a distance d. (The electron cloud is assumed to remain spherical.) At the distance d the force on the electron cloud of the field $\mathbf{E}$ and the force on the electron cloud arising from the Coulombic attraction between the charges

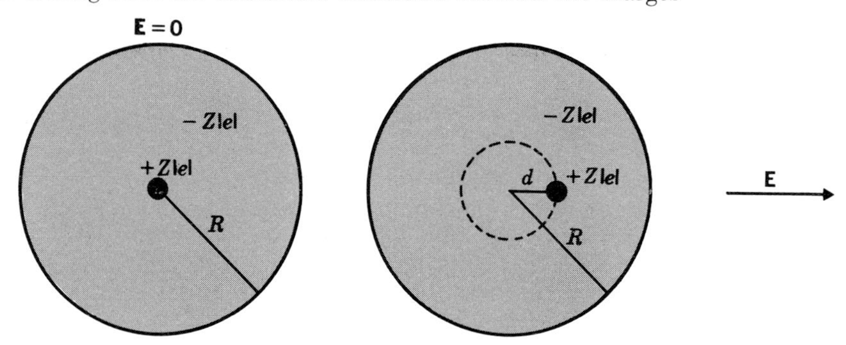

Figure 18-1 The displacement of the electron cloud by an applied field.

$+Z|e|$ and $-Z|e|$ balance, and stable equilibrium exists. The first of these forces, $\mathbf{F}_E$, is given by

$$\mathbf{F}_E = Z|e|\,\mathbf{E} \tag{18-8}$$

The second force, $\mathbf{F}_c$, is

$$\mathbf{F}_c = \frac{(Ze)^2}{4\pi\epsilon_v}\frac{\mathbf{d}}{R^3} \tag{18-9}$$

[The derivation of Eq. (18-9) is left as a problem.] At equilibrium the magnitudes of these forces must be equal, so that

$$Z|e|\,\mathbf{E} = \frac{(Ze)^2}{4\pi\epsilon_v}\frac{\mathbf{d}}{R^3}$$

from which

$$Z|e|\,\mathbf{d} = 4\pi\epsilon_v R^3\mathbf{E} \tag{18-10}$$

The dipole moment of the atom is just

$$\mathbf{p} = Z|e|\,\mathbf{d}$$

Hence

$$\mathbf{p} = 4\pi\epsilon_v R^3\mathbf{E} \tag{18-11}$$

Comparison of this expression with Eq. (18-6) shows that the polarizability in this instance (call this α_e, the *electronic* polarizability) can be written

$$\alpha_e = 4\pi\epsilon_v R^3 \tag{18-12}$$

This result states that the electronic polarizability depends, not on the atomic number Z, but only on the radius of the atom (or ion), R.

Comparison with experiment of calculations using Eq. (18-12) shows that this simple model represents the real case reasonably well. Considering argon, which has an atomic radius of 1.91 A, we obtain α_e to be about 7×10^{-40} farad m^2, using Eq. (18-12). The value of α_e

Table 18-1 The electronic polarizability of some of the inert gases†

Gas	farad m^2
He	0.23×10^{-40}
Ne	0.4
A	1.83
Kr	2.3
Xe	3.7

† The values for He and A are the most reliable; they are used as primary standards against which other gases are measured.

found from experiments is 1.83×10^{-40} farad m^2. The calculated value is somewhat too large, but the agreement is close enough, nevertheless, to indicate that the picture is generally correct.

Because the electronic structure of an atom is relatively temperature-independent, the variation of α_e with temperature is expected to be virtually zero on the basis of the theory. Experiments agree that this is true. Similarly the susceptibility χ_e of gases has been shown experimentally to be proportional to N, as Eq. (18-7) predicts (if α_e is independent of temperature). The measured values of α_e for a number of rare gas atoms are listed in Table 18-1. Observe the steady increase in α_e as the atomic number increases. This increase cannot be accounted for purely on the basis of increased R [as Eq. (18-12) predicts], since R^3 does not increase so rapidly as the atomic number increases.

18-3 Ionic Polarization

Ionic polarization exists only for those materials in which some of the ions in a molecule have a net positive charge and some have a net negative charge. For example, the HCl molecule has a permanent dipole moment $|e|\,x$ of order 1×10^{-29} coulomb m, where x is the distance of separation of the ions. In an electric field the resultant torque lines up the dipole parallel to the field at the absolute zero of temperature, as indicated in Fig. 18-2*a*. The field, however, produces forces on the two charges $\pm e$, as well as the torque on the dipole. The distance of separation x is increased by some small amount d_1 (Fig. 18-2*b*); then the dipole moment is $|e|\,(\mathbf{x} + \mathbf{d}_1)$. It is the added dipole moment $|e|\,\mathbf{d}_1$ which is the ionic moment sought.

The distance $\mathbf{d}_1$ has a value which is determined from the balance between the electrostatic force of magnitude eE and the interatomic force produced when the HCl bond is stretched. For calculation of the latter force, the force constant of the bond must be known. As a first estimate, it can be considered to be of the same order as the force constant between atoms in a solid, which was found to be about 25 newtons/m. This value is indeed of the right order for the HCl molecule, but a more accurate estimate from spectroscopic data gives about 75 newtons/m. Thus the restoring force which arises because of the stretching of the chemical bond has the value $75\mathbf{d}_1$ newtons. Hence

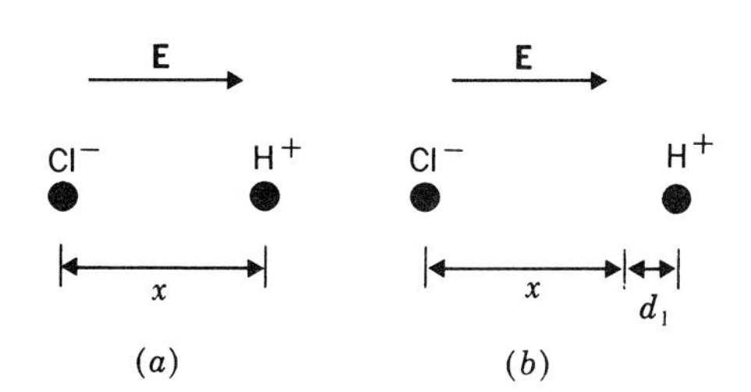

Figure 18-2 An external field causes the permanent dipole to rotate to a low-energy position and also causes the ionic separation of the ions to increase slightly.

$$75\mathbf{d}_1 = |e|\,\mathbf{E} \tag{18-13}$$

From this expression the ionic-dipole moment $\mathbf{p}_i$ is given by

$$\mathbf{p}_i = \frac{e^2}{75}\mathbf{E} \tag{18-14}$$

Finally the ionic polarizability a_i is given by

$$a_i = \frac{e^2}{75} \approx 3 \times 10^{-40} \text{ farad m}^2 \tag{18-15}$$

This is a little smaller than a_e calculated above for argon and of the same order as that for the heavier rare-gas atoms listed in Table 18-1. More refined estimates of a_i reduce its value considerably, as much as a factor of 10 or more, so that a_i is at least ten times smaller than a_e. These more refined estimates require an average over all possible orientations of the permanent dipole with respect to the field.

The simultaneous occurrence of both induced and ionic polarization $\mathbf{P}_e$ and $\mathbf{P}_i$, respectively, results in a volume polarization $\mathbf{P}$, which is the sum of the individual effects.

$$\mathbf{P} = \mathbf{P}_e + \mathbf{P}_i = N(\alpha_e + \alpha_i)\mathbf{E} \tag{18-16}$$

This expression is valid for any but the largest fields E, because the "deformation" of the electron cloud is very small; i.e., both d and d_1 are small compared with atomic dimensions. Note also that the distribution of electrons in both atoms and molecules is almost completely independent of temperature, so that both α_e and α_i are virtually independent of temperature, for moderate temperatures.

18-4 Orientation Polarization

The third type of polarization, *orientation of permanent dipoles*, can also be treated without difficulty for a dilute gas. The development

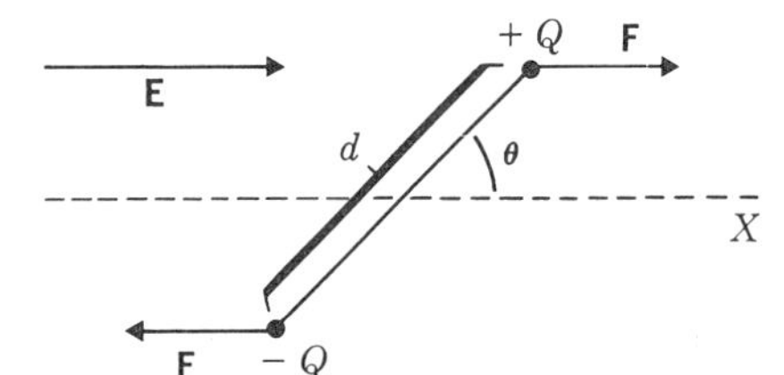

Figure 18-3 The turning couple produced on the dipole by the external field.

presented here was first accomplished in 1905 by Langevin for magnetic dipoles in a magnetic field. The results of his calculation are applicable for electric dipoles as well. They give a particular temperature dependence which has been verified experimentally many times. For moderate fields (up to at least 10^7 volts/m) the polarizability α_o can be written in terms of the magnitude p_p of a permanent electric dipole in the following way:

$$\alpha_o = \frac{p_p^{\,2}}{3kT} \tag{18-17}$$

Several factors are involved in the derivation of Eq. (18-17). The first of these is the calculation of the potential energy of a dipole in a uniform electric field; a calculation readily made from a consideration of the forces on the charges $\pm Q$ of the dipole (Fig. 18-3). The force on the charge $+Q$ is in the positive x direction and is of size QE. The force on the charge $-Q$ is in the negative x direction and is also of size QE. Thus, a couple of magnitude $QEd \sin \theta$ tends to turn the dipole into the direction of the field, i.e., to reduce θ. The energy of the dipole in the field is least when $\theta = 0$ and is maximum when $\theta = 180°$. If the energy is taken to be zero at $\theta = 90°$, the energy W as a function of angle is

$$W = -QEd \cos \theta = -p_p E \cos \theta \tag{18-18}$$

[Calculation of Eq. (18-18) is left to the student as a problem.]

In spite of the minimum-energy condition for $\theta = 0°$, all dipoles do not point in the direction of the field $\mathbf{E}$, for collisions of the gas molecules due to thermal motion constantly tend to misalign them into random directions. The calculation of the number of dipoles making various angles with the x axis can be carried out with the aid of Fig. 18-4. Here is shown one octant of a sphere of unit radius. The dipoles, all of moment p_p, are considered to be centered at the origin. Each dipole makes some angle with the x direction, the direction of $\mathbf{E}$. If the distribution of dipole moments were perfectly random, i.e., if $\mathbf{E}$ were reduced to zero, then the number of dipoles within a

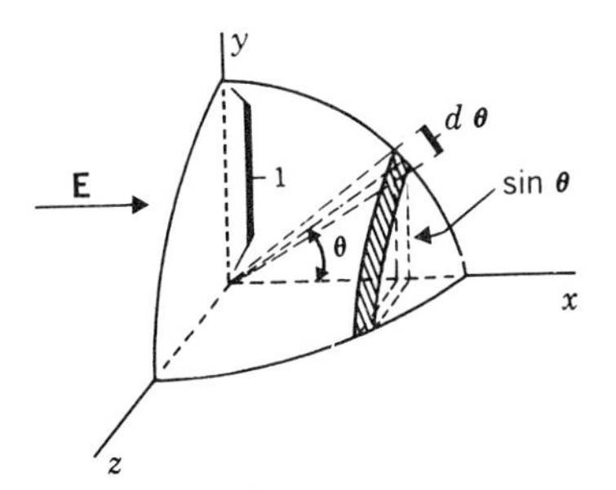

Figure 18-4 One octant of a sphere, showing how the number of dipoles in some range $d\theta$ may be calculated as a function of θ.

range $d\theta$ about an angle θ with the x axis [call this $N(\theta)\,d\theta$] would be proportional to the shaded area,

$$N(\theta)\,d\theta = C2\pi \sin\theta\,d\theta$$

where the constant C is related to the total number of dipoles, N. This equation must be modified if $\mathbf{E}$ is not zero. Then the number must be weighted by a factor containing the energy W necessary to turn a dipole into the angle θ. The appropriate factor is the Boltzmann factor,

$$e^{-W/kT} = e^{p_pE\cos\theta/kT}$$

Hence

$$N(\theta)\,d\theta = C2\pi \sin\theta\, e^{p_pE\cos\theta/kT}\,d\theta \tag{18-19}$$

The x component of each dipole which makes an angle θ with the x axis is $p_p \cos\theta$, and so the x component of all the dipoles in a range $d\theta$ about the angle θ is $p_p \cos\theta N(\theta)\,d\theta$. The net x component P_o due to all N dipoles is the sum of Eq. (18-19) over all angles θ,

$$P_o = \int_0^\pi 2\pi C p_p \cos\theta \sin\theta\, e^{p_pE\cos\theta/kT}\,d\theta$$

The constant C can be calculated from the knowledge that

$$N = \int_0^\pi N(\theta)\,d\theta$$

Hence

$$P_o = N \frac{\int_0^\pi p_p \cos\theta \sin\theta \, e^{p_p E \cos\theta/kT} \, d\theta}{\int_0^\pi \sin\theta \, e^{p_p E \cos\theta/kT} \, d\theta} \tag{18-20}$$

This formidable expression can be integrated without trouble by the substitutions $Z = p_p E \cos\theta/kT$ and $a = p_p E/kT$. (The details are left for the student as a problem.) The result is

$$P_o = N p_p \left(\coth a - \frac{1}{a} \right) \tag{18-21}$$

Commonly this expression has been written in briefer form by replacing

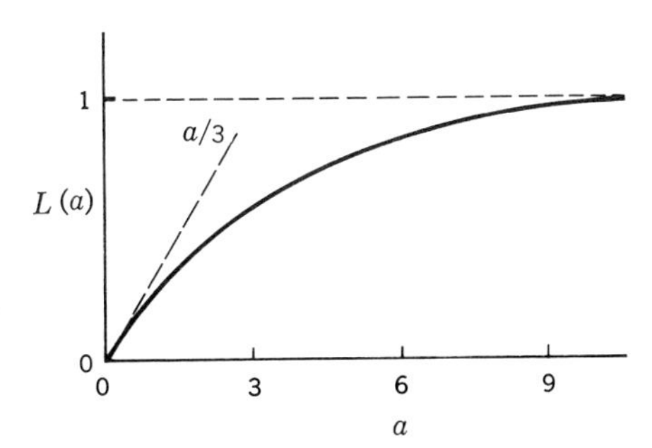

Figure 18-5 A plot of the Langevin function $L(a)$ as a function of a.

the factor in parentheses by the symbol $L(a)$ (the *Langevin function*). Then

$$P_o = N p_p L(a) \tag{18-22}$$

A sketch of the function $L(a)$ is shown in Fig. 18-5. For small values of a, $L(a)$ has a small value; for large values of a, it approaches 1. These limits correspond to weak polarization and saturation, respectively. In practice, conditions such that a is very large are never attained. For example, for a molecule such as HCl in a field of 10^6 volts/m, $a \approx 0.25/T$, so that at room temperature $a < 10^{-3}$. Thus $L(a)$ may be replaced without great loss of accuracy in Eq. (18-22) by the expression $L(a) = a/3$. Hence, for moderate fields and all but the very lowest temperatures, the orientational polarization $\mathbf{P}_o$ can be written

$$\mathbf{P}_o = \frac{N p_p^2 \mathbf{E}}{3kT} \tag{18-23}$$

From this expression the orientational polarizability α_o is defined as

$$\alpha_o = \frac{p_p^2}{3kT} \tag{18-24}$$

18-5 Combined Polarization

The *total polarization* of a dilute gas can now be written as the sum of the three components. In general

$$\mathbf{P} = \mathbf{P}_e + \mathbf{P}_i + \mathbf{P}_o = N\left(\alpha_e + \alpha_i + \frac{p_p{}^2}{3kT}\right)\mathbf{E} \tag{18-25}$$

Of course not all atoms or molecules possess each of the three types of polarizability. All apparently display induced polarization. Many molecules have ionic polarization to a greater or lesser extent, and a

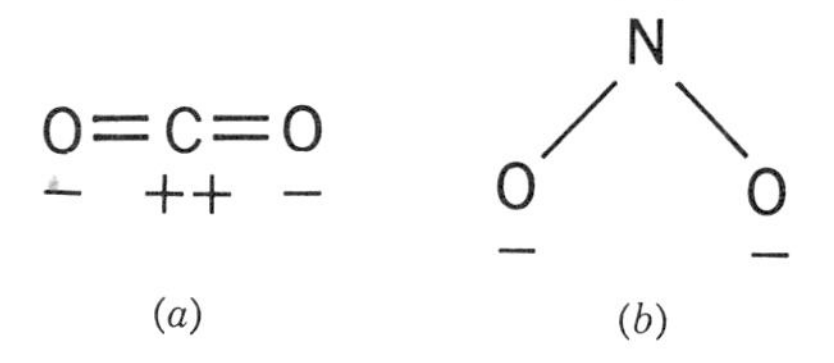

Figure 18-6 The geometry of a molecule determines in part the magnitude of its intrinsic dipole moment. (*a*) The CO_2 molecule has no dipole moment because of its symmetry; (*b*) the NO_2 molecule does have a dipole moment.

relatively large number have orientation polarization. The deciding factor for the existence of this last, of course, is simply whether or not the molecules have a permanent moment. The existence of a permanent moment is purely a matter of molecular geometry. Thus, CO_2 has no permanent moment at all, that is, $p_p = 0$, because its atoms are in line (see Fig. 18-6*a*). On the other hand, the different geometry of the NO_2 molecule (Fig. 18-6*b*) is such that p_p for this molecule has the value 1.3×10^{-30} coulomb m.

The use of Eq. (18-25) in determining α_e, α_i, and p_p separately does not look promising at first sight, since there is but one equation and three unknowns. However, appropriate experiments enable the parts to be separated. Because α_e and α_i are practically independent of temperature, whereas the third term is explicitly dependent on temperature, measurement of $\mathbf{P}$ as a function of temperature for a constant number of molecules per unit volume, N, permits separation of the first two terms from the last. As an example, data for the relative capacitivity of methyl amine, CH_5N, at several temperatures are plotted in Fig. 18-7. The intercept of the line for $1/T = 0$ is about 0.0008, and the slope is about 0.6 per degree Kelvin. Using Eqs. (18-3) and (18-25), one can write

$$\alpha_e + \alpha_i = \frac{(0.0008)\epsilon_v}{N} \tag{18-26}$$

and

$$p_p{}^2 = \frac{(0.6)3k\epsilon_v}{N}$$

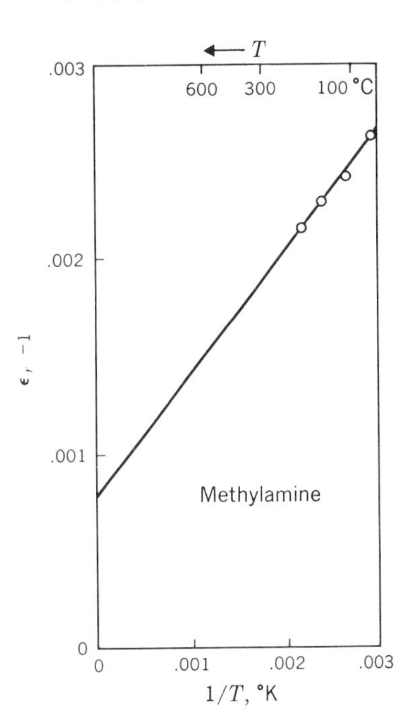

Figure 18-7 A plot of $\epsilon_r - 1$ as a function of $1/T$ for the molecule CH_5N. [*Data from R. Sänger, O. Steiger, and K. Gächter, Helv. Phys. Acta,* **5:** 200 (1932).]

From these expressions the following constants are obtained for this molecule:

$$\alpha_e + \alpha_i \approx 6 \times 10^{-40} \text{ farad m}^2$$

$$p_p \approx 4.2 \times 10^{-30} \text{ coulomb m}$$

Note that by this technique alone further separation of $\alpha_e + \alpha_i$ is not possible. In Table 18-2 representative values of p_p for several molecules are tabulated. The unit commonly used for the dipole moment is the *debye*, 3.33×10^{-30} coulomb m.

Table 18-2 Electric-dipole moments of some molecules

Molecule	Coulomb m	Debyes
HCl	3.5×10^{-30}	1.05
$CsCl$	35.0	10.5
H_2O	6.2	1.87
D_2O	6.0	1.80
NH_3	4.9	1.47
$HgCl_2$	0.0	0
CCl_4	0.0	0
CH_4O	5.7	1.71

SUSCEPTIBILITY OF LIQUIDS AND SOLIDS

Many phenomena which can be well understood for dilute gases become much more complicated when the atoms are condensed into the liquid or solid state, because interactions between the atoms cannot be neglected when distances of separation approach the atomic diameters. Furthermore, these interactions can generally be taken into account only approximately. The dielectric susceptibility is no exception to this rule. Consequently a study of the polarization of atoms and molecules in solids or liquids becomes much more complicated than the same study in gases. The new feature is not that the types of polarization change; in fact the atoms and molecules in a solid exhibit induced, ionic, and orientational polarizations as they do in the gaseous state. The difference is that the effective field on an atom, $\mathbf{E}'$, is not the same as the externally applied field $\mathbf{E}$. In general, $\mathbf{E}'$ is extremely difficult to calculate.

18-6 The Effective Field and Induced Polarization

Calculation of the effective field, often called *local*, or *internal*, *field*, cannot be done in a simple way except for the most symmetric crystal

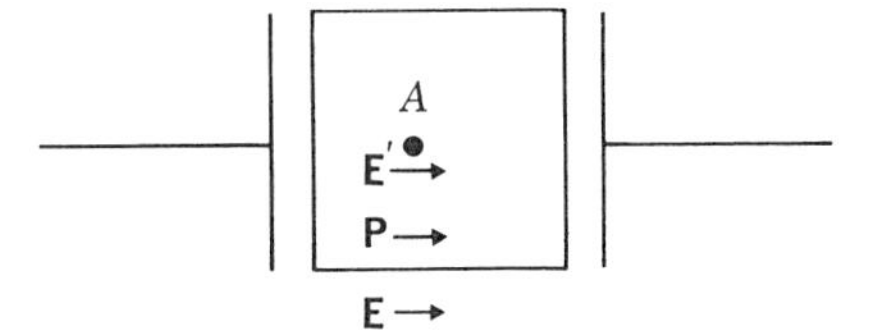

Figure 18-8 The field $\mathbf{E}'$ at point A inside a dielectric immersed in a uniform external field $\mathbf{E}$ has components arising both from the field $\mathbf{E}$ and from the polarization $\mathbf{P}$ of the dielectric produced by this external field.

lattices. The calculation is carried out here for the simplest case of all, the simple cubic crystal. Consider an atom at point A inside a piece of dielectric material which is between two plates of a parallel-plate capacitor (Fig. 18-8). The internal field at A comes from two sources. One is the external field $\mathbf{E}$ and the other the field of all the dipoles induced in the material by $\mathbf{E}$. At first sight calculation of the effect of the dipoles seems to be an endless task, because enormous numbers of them exist at various distances and angles from A. The calculation is possible, however, if it is divided into two parts.

The dielectric surrounding A is divided into two regions. One is a region close to A where the positions of the individual dipoles and their shapes must be considered as illustrated for the nearest-neighbor dipoles surrounding the point A in a simple cubic crystal (Fig. 18-9*a*). The effect of such an array of nearest-neighbor dipoles can be calculated, and the effect of further shells of second-nearest dipoles, third-nearest dipoles, etc., can be assumed by analogy. The second region about the point A is that for which the distances are so large that shape and position of the individual dipoles do not matter; these dipoles can be taken into account in a separate way. Let a sphere be cut in the dielectric about A, with a radius large compared with the interatomic spacing. Then a layer of surface charge appears at each point on the inside of this cavity, the charge of magnitude equal to the discontinuity in the normal component of $\mathbf{P}$ at that point (Fig. 18-9*b*). These bound charges produce a field at A. This field $\mathbf{E}_1$ and the net field of all the dipoles inside the sphere, $\mathbf{E}_2$, together with the applied field $\mathbf{E}$ make up the internal field $\mathbf{E}'$ at A.

$$\mathbf{E}' = \mathbf{E} + \mathbf{E}_1 + \mathbf{E}_2 \tag{18-27}$$

Calculations of $\mathbf{E}_1$ and $\mathbf{E}_2$ are not difficult for the simple geometry considered here, and the details are left for Probs. 18-5 and 18-6.

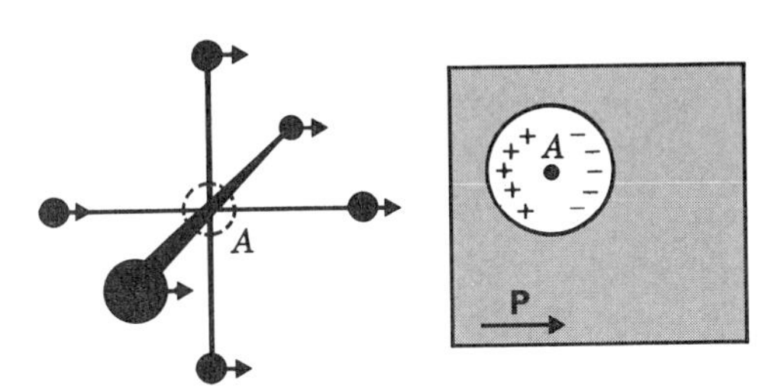

Figure 18-9 The dielectric about A can be divided into two parts: the close region, where the effect of individual dipoles must be considered, and the distant region, where the effect can be approximated by the surface charge on a spherical cavity centered at A.

The results of such calculations are that

$$\mathbf{E}_1 = \frac{\mathbf{P}}{3\epsilon_v} \tag{18-28}$$

and

$$\mathbf{E}_2 = 0$$

Hence the internal field *at a lattice position* can be written (for this case)

$$\mathbf{E}' = \mathbf{E} + \frac{\mathbf{P}}{3\epsilon_v} \tag{18-29}$$

The effective field on an atom at A is increased by the polarization of the medium.

Equation (18-29) actually has wider validity than this simple calculation implies. The term E_2 in Eq. (18-27) is found to be zero for any cubic arrangement of atoms about the point A.

The field $\mathbf{E}'$ calculated by Eq. (18-29) is called the *Lorentz field*. The induced-dipole moment $\mathbf{p}$ is defined as that moment produced by the actual field acting upon it, that is, $\mathbf{E}'$. Thus

$$\mathbf{p} = \alpha_e \mathbf{E}' \tag{18-30}$$

so that $\mathbf{P}$, which is equal to $N\mathbf{p}$, becomes

$$\mathbf{P} = N\alpha_e \mathbf{E}' \tag{18-31}$$

Combining Eqs. (18-29) and (18-31),

$$\frac{N\alpha_e}{3\epsilon_v} = \frac{\epsilon_r - 1}{\epsilon_r + 2} \tag{18-32}$$

This expression, called the *Clausius-Mossotti equation*, permits calculation of atomic polarizabilities α_e from measured dielectric constants. For dilute gases $\epsilon_r \approx 1$, so that $\epsilon_r + 2$ is about equal to 3. In this event

$$\alpha_e = \frac{\epsilon_v(\epsilon_r - 1)}{N} \tag{18-33}$$

in agreement with Eq. (18-7).

The polarizability of a simple element such as argon does not vary much between the element as a dilute gas and as a liquid, as seen from the data in Table 18-3. Such good agreement is probably not to be

Table 18-3 Polarizability of argon

Form	T, °K	Pressure, atm	ϵ_r	α_e, farad m²
Gas	293	1	1.000517	1.83×10^{-40}
Liquid	83	1	1.53	1.86

found for atoms and molecules which have more extensive changes in electronic structure when the gas is condensed to liquid or solid form.

18-7 Polarization of Ionic Crystals

Ionic polarization exists for solids, too. Such polarization depends again on the existence in the material of ions which are displaced by

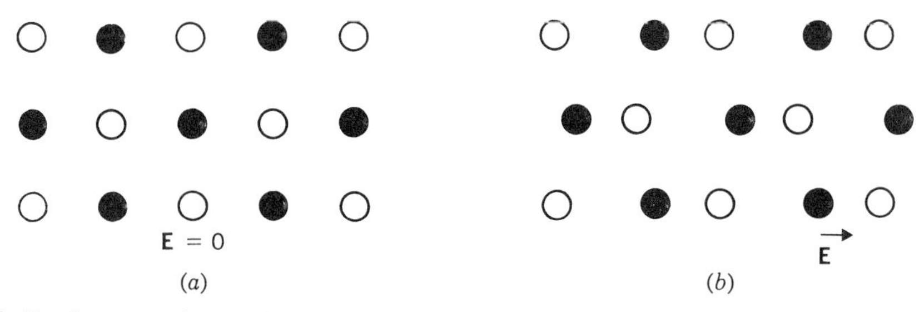

Figure 18-10 Displacement of ions of unlike sign in an electric field for a (100) plane of an NaCl crystal. The displacement is greatly exaggerated.

the field relative to one another. A simple case is illustrated in Fig. 18-10 for the (100) plane of a structure of the NaCl type. Such relative motion produces opposite surface charges on the two opposite faces of the crystal; hence a volume polarization exists. Unfortunately, calculation for these materials cannot be carried out simply.

An experimental method exists by which the polarizabilities α_e and α_i may be determined independently for a solid. This method is based on the fact that the dielectric constant of ionic crystals is a function of frequency of the applied field **E**. At low frequencies the static capacitivity depends on the polarization caused both by shifting of electron clouds (*induced* polarization) and by relative displacement of ions (*ionic* polarization). The capacitivity is a function of both α_e and α_i,

$$\epsilon_r(\text{static}) = f(\alpha_e,\alpha_i) \tag{18-34}$$

At high frequencies (say, for visible light) the heavy lattice ions cannot follow the applied field **E**, and so no ionic polarization exists. The electron clouds with their relatively low mass can do so; hence induced polarization remains. Then

$$\epsilon_r(\text{optical}) = F(\alpha_e) \tag{18-35}$$

Comparison of these two equations indicates that a difference between the static capacitivity and the optical capacitivity should exist. Such a difference is found experimentally, and the static capacitivity is found to be larger than the optical capacitivity. A few examples are listed in Table 18-4. The last column characterizes that part of the polarization from the shift of the electron clouds alone; the difference between the values in columns 2 and 3 is the part characteristic of the ionic polarization alone.

Table 18-4 Static and infinite-frequency capacitivity of some ionic crystals

Material	ϵ_r (static)	ϵ_r (optical)
LiF	9.27	1.90
LiC	11.05	2.68
NaCl	5.62	2.32
KCl	4.64	2.17
RbCl	5.10	2.18
NaI	6.60	2.96

Elementary calculations show why radiation of optical frequencies is necessary to suppress the ionic polarization. From Eq. (18-13) the static value of $\mathbf{d}_1$ for a typical molecule such as HCl is seen to be about $e\mathbf{E}/75$. Suppose that the ionic polarization can be neglected if $\mathbf{d}_1$ is less than 5 per cent of this value, that is, $e\mathbf{E}/1{,}500$. In the high-frequency field the displacement x during one half cycle as a function of time is given approximately by

$$\mathbf{x} \approx \frac{e\mathbf{E}t^2}{2m} \tag{18-36}$$

Then

$$\frac{e\mathbf{E}}{1{,}500} = \frac{e\mathbf{E}t^2}{2m}$$

so that

$$t = \frac{2m}{1{,}500}$$

For an average atom $m \approx 10^{-25}$ kg; hence

$$t \approx 10^{-14} \text{ sec}$$

This period corresponds to a wavelength of about 30,000 A, i.e., an infrared wavelength. For shorter wavelengths the ionic polarization is nearly fully suppressed. For wavelengths longer than infrared, those toward the radio spectrum, ionic polarization becomes more important, and the capacitivity approaches the static value.

18-8 Orientation of Dipoles

The static capacitivity of solids composed of dipolar molecules is similar in some ways to that of the corresponding gases. Rotation of the dipoles can take place in some circumstances, as in the gases, so that the effect of the permanent dipoles can be added to that of the induced and ionic polarization. In some cases, however, rotation of the molecules is hindered so that no contribution arises from the permanent dipole moments. This effect can be seen by comparison of the dielectric behavior of two substances, nitromethane, CH_3NO_2, and hydrogen chloride, HCl. The relative capacitivity for these substances is plotted as a function of temperature in Fig. 18-11.

The relative capacitivity of nitromethane in the liquid state is large because the moment of the molecule itself is large, about 11.1×10^{-30} coulomb m. In the liquid the molecules can rotate, and ϵ_r increases as the temperature of the liquid is lowered, as for this material in the gaseous state. Presumably all three types of polarization contribute to the dielectric behavior of the liquid. At the freezing point of the liquid, 244°K, the capacitivity drops abruptly, however. In solid form the molecules of nitromethane apparently are not free to rotate and hence cannot contribute to the polarization. The low value of ϵ_r is apparently a result only of induced and ionic moments, as indicated by the fact that ϵ_r for the solid does not change with temperature.

This type of behavior is not found for all solids, and one of the classic substances for which it is not true is HCl. In Fig. 18-11*b* the relative capacitivity is plotted for this substance as a function of temperature. In the liquid state, the capacitivity is relatively large and has a temperature dependence which indicates that rotation of the molecules does occur. At 165°K the liquid freezes, and at this point the capacitivity actually increases a little (because the density of the material increases), but the molecules in the solid are not prevented from orienting themselves along the field. This conclusion can be inferred by observing that the capacitivity still continues to increase

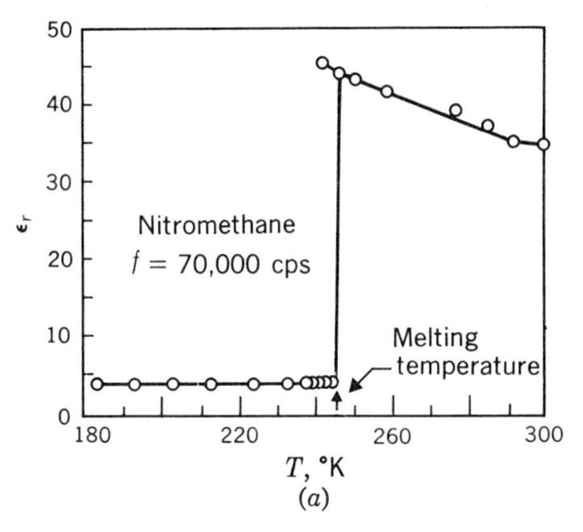

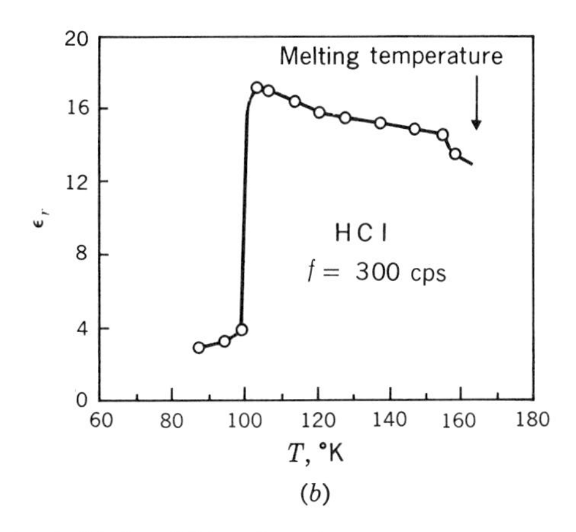

Figure 18-11 The behavior of ϵ_r versus T for completely hindered (a) and partly hindered (b) rotation of dipoles in the solid. (a) Nitromethane. [*Data of C. Smyth and W. Walls, J. Chem. Phys.*, **3:** 557 (1935).]; (b) Hydrogen chloride. [*Data of C. Smyth and C. Hitchcock, J. Am. Chem. Soc.*, **55:** 1830 (1933), *by permission of the American Chemical Society.*]

as the temperature is lowered. Finally, at about 100°K, the molecules become unable to rotate, and orientational polarization practically ceases. Below this temperature the polarization is partly induced, partly ionic, and slightly hindered-orientational in character.

18-9 Electrostriction and Piezoelectricity

The application of an electric field to a dielectric has been shown to produce distortion of the original electron configuration or rotation of permanent dipoles. These changes produce measurable dimensional changes in solids; the effect, though generally small, may be significant. In this section some of the features of this phenomenon are considered.

The dimensional changes of a solid resulting from the application of an electric field are defined as *electrostriction*. This is a small effect, and the dimensional change $\Delta l/l$ must be proportional to the square of $\mathbf{E}$, that is, $\Delta l/l = AE^2$, where A is a constant. This dependence of $\Delta l/l$ on E^2 results from the fact that $\Delta l/l$ must be independent of sign of $\mathbf{E}$. No inverse effect exists; i.e., the application of a stress does not produce an electric field in the material. Most dielectrics probably are electrostrictive, though the effect often may be extremely small. For glass, for example, the electrostrictive effect is so small that a field of 10^4 volts/m produces a length change of less than 10 A/m of glass.

A much more important electromechanical phenomenon is the *piezoelectric* effect. As the name implies, this phenomenon is a polarization of a solid on which forces are acting (Fig. 18-12). For reasonable forces the polarization is proportional to the magnitude of the applied force. If the external force is reversed in sign, the polarization changes sign. The relationship between **P** and **F** can be expressed (for particularly simple geometries) as

$$\mathbf{P} = \text{constant} \times \mathbf{F} \tag{18-37}$$

The piezoelectric effect has an inverse; i.e., a dimensional change (a strain) is produced in a crystal when it is placed in an electric field.

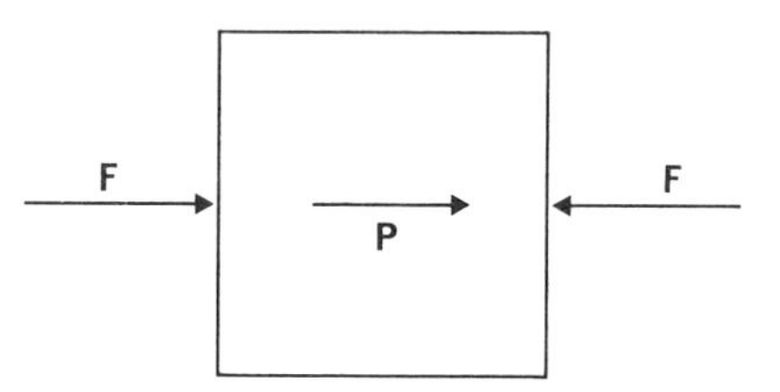

Figure 18-12 An external force **F** produces electric polarization in piezoelectric solids. Though this polarization is shown here collinear with **F**, in general **P** can make any arbitrary angle with **F**.

This effect is called the *inverse piezoelectric effect* and is somewhat analogous to the electrostrictive effect. The inverse piezoelectric effect, however, is a linear function of the first power of the applied field (for reasonable fields). Therefore the magnitude of the piezoelectric strain S_x in a specimen placed in a field **E** (again for particularly simple geometries) is given by the expression

$$S_x = \text{constant} \times E \tag{18-37a}$$

The geometry of this expression is given in Fig. 18-13. If the initial length (i.e., with no field) be l and the length in the field be $l + \Delta l$, then the strain $S_x = \Delta l/l$. The inverse effect senses the sign of the applied field. If the field is reversed, the strain is a contraction.

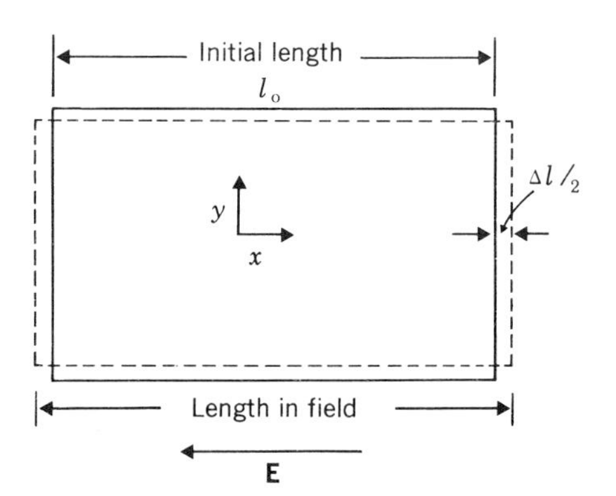

Figure 18-13 Dimensional changes of a piezoelectric solid in an external field.

To get an idea of the amount of strain involved in a typical piezoelectric crystal, consider a piece of crystal quartz. For a bar cut a particular way—an X cut—the constant connecting S_x and E is about 2.25×10^{-12} coulomb/newton. Consequently the longitudinal strain (Fig. 18-13) in such a bar in a moderate field (say, 10,000 volts/m) is

$$S_x = 2.25 \times 10^{-12} \times 10^4 = 2.25 \times 10^{-8} \text{ m/m}$$

That is, a bar initially 1 m long becomes 1.0000000225 m long. The piezoelectric effect is clearly small, but it is an important phenomenon (observe that it is much larger than the electrostrictive effect calculated above for glass). Figure 18-13 implies that a second strain in the y direction is also produced by the field in the x direction. This strain has the same magnitude as S_x (for this particular cut of crystal).

A number of crystalline materials are piezoelectric. A tabulation of the important ones is given in Table 18-5. The strength of the effect for the various materials is indicated in the third column. The chemical composition of most of these materials is not simple, and the crystal structures they assume in the solid state are not simple ones. As

Table 18-5 Representative piezoelectric crystals

Material	Chemical formula	Relative strength
α Quartz	SiO_2	Weak
Rochelle salt	$NaKC_4H_4O_6 \cdot 4H_2O$	Very strong
Ammonium dihydrogen phosphate	$NH_4H_2PO_4$	Strong
Potassium dihydrogen phosphate	KH_2PO_4	Moderate
Ethylene dihydrogen tartrate	$C_6H_{14}N_2O_6$	Moderate
Tourmaline	$(FeCrNaKLi)_4Mg_{12}B_6Al_{16}H_8Si_{12}O_{63}$	Weak

implied, crystals which have low symmetry are more likely to exhibit piezoelectricity than those which have high symmetry. From this criterion alone one can predict that certain groups of crystals *will not* be piezoelectric at all—one cannot, however, predict at the present time which *will* be piezoelectric. Since the crystals which are piezoelectric do have intricate crystal structures, their piezoelectric constants, elastic constants, and dielectric constants are all rather complicated functions of direction in the crystal.

18-10 Ferroelectricity

The discussion to this point has been limited (with one exception) to those materials which require an electric field to produce polarization. The single exception is the piezoelectric materials, in which polarization can be produced also by forces. Another important class of dielectrics are the *ferroelectrics*, which exhibit the phenomenon of self-polarization: adjacent dipoles in the solid tend to align themselves by mutual interaction. This aligning tendency transmits itself through the crystal from atom to atom so that large macroscopic regions may show a net polarization in a given direction. The polarization **P** of such crystals departs from a linear dependence on an external field **E**, and so the dielectric susceptibility of such a material cannot be defined in as simple a fashion as it can for the nonferroelectrics through Eq. (18-4).

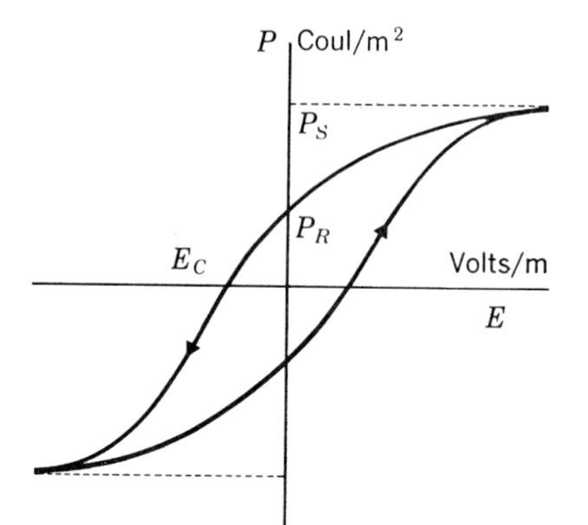

Figure 18-14 Hysteresis loop of a ferroelectric.

The most striking feature of the ferroelectrics is the way in which **P** does depend on **E**. The materials exhibit electric hysteresis, just as ferromagnetic materials exhibit magnetic hysteresis (hence the word ferroelectric). Such a hysteresis loop is sketched in Fig. 18-14. **P** is seen to be a multivalued function of **E**. The saturation value of P, called P_s, for a typical ferroelectric $BaTiO_3$ has the value 0.26 coulomb/m^2 at room temperature. The remanent polarization P_R is the polarization retained if the field E is reduced to zero after saturation. Furthermore, to reduce P to zero, a reverse field is required; this is the coercive field E_c. P_R and E_c depend not only on the general nature of the material but also on other factors such as impurities, crystallite size, and heat-treatment.

The general similarity of hysteresis loops of the ferroelectrics to those of the ferromagnetics led naturally to the search for *ferroelectric domains*, and they have been found in $BaTiO_3$ (see Fig. 18-15 for a sketch). Within a single domain the polarization has a common crystallographic direction. The net polarization of a massive piece of material consists, then, of the vector sum of polarization over all domains, the contribution of each domain being weighted according

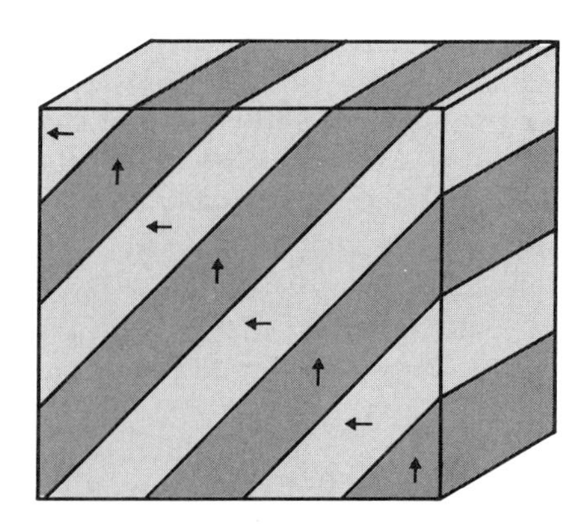

Figure 18-15 Ferroelectric domains.

to its volume. If an electric field is applied to such a crystal, several phenomena can occur: (1) the polarization may change in magnitude in a given domain, (2) the polarization of a domain may change direction, and (3) domains favorably oriented, i.e., those whose polarizations make least angles with **E**, may grow by movement of boundaries between domains. Any of these three changes modifies the total polarization of the entire solid.

The origin of the spontaneous polarization is only partially known. Indeed the details of the effect may be somewhat different for each of the common ferroelectrics. Further, understanding the effect is complicated by the fact that the various ferroelectric materials have

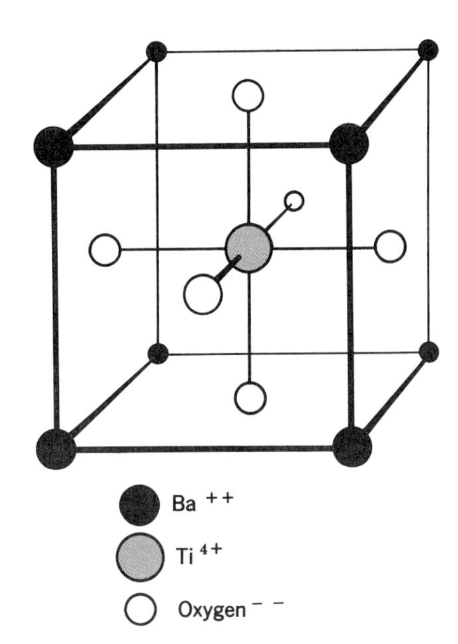

Figure 18-16 A unit cell for $BaTiO_3$.

complex crystal structures. However, for one substance, $BaTiO_3$, the structure is simple enough to show the atomic configuration which gives rise to the polarization. A unit cube of this material is sketched in Fig. 18-16. In this structure (the *perovskite* structure) the Ba ions are shown in the cube corners, the Ti ion in the cube center, and the O ions in the center of the cube faces. For this structure, the unit cube does not possess a dipole moment, because all charges are symmetrically disposed. If, however, the Ti and O ions are displaced relative to the Ba ions, a dipole develops for certain arrangements. Suppose that the Ti ion moves toward the oxygen at one of the cube faces; the unit cube then acquires a dipole moment in that direction. If this same movement of Ti ions occurs in all unit cubes in the solid, a *net* polarization of the entire solid occurs. From measured values of P_s for $BaTiO_3$, it can be calculated that a displacement of only about 0.1 A would suffice to cause the entire effect, and so this model is quite feasible. Recent studies have shown that displacements of both the Ti and O ions probably occur in opposite directions for the two ions.

The spontaneous polarization does not persist to the melting points of the ferroelectrics. For some materials, it exists at low temperatures up to a maximum temperature called the *ferroelectric Curie temperature*. Apparently the larger thermal motion of the atoms at temperatures higher than the Curie temperature T_c is sufficient to break down the effect of common displacement in adjacent unit cubes. For one material, rochelle salt, the effect exists only over a narrow region of temperature. The characteristics of a few of the common ferroelectrics are tabulated in Table 18-6.

Table 18-6 Properties of some ferroelectric materials

Material	Chemical formula	P_s, coulombs/m^2	T_c, °C
Rochelle salt	$KNaC_4H_4O_6 \cdot 4H_2O$	0.24×10^{-2}	Upper +24 lower −18
KDP	KH_2PO_4	4.95	−150
Potassium niobate	$KNbO_3$	3.78 (at 410°C)	+434
Barium titanate	$BaTiO_3$	26.0 (at 23°C)	+120

The relative capacitivity of ferroelectrics has less significance than that of the normal dielectrics, since it varies with E. However, if it is defined in this case, too, through Eq. (18-4), comparisons can be made with the normal dielectrics. Values of ϵ_r for the ferroelectrics are ordinarily much larger than for other solid dielectrics at the same field strength. For example, mica, an ordinary dielectric, has a relative capacitivity ϵ_r (independent of E) of about 7. $BaTiO_3$, a ferroelectric, has a relative capacitivity of several thousand (perhaps 3,000 to 5,000) at room temperature for a field of 10^6 volts/m. While not all ferroelectrics have values for ϵ_r as large as this, many of them do attain high values of P in relatively small fields.

Some interactions in ionic crystals lead to a type of spontaneous polarization termed *antiferroelectric*. Unlike the ferroelectrics, antiferroelectrics have no permanent moment, because of the geometry of the spontaneous polarization. As a possible model imagine lines of ions polarized in one direction, with adjacent lines of ions polarized oppositely. If the polarization in one direction is equal to that in the

Table 18-7 Properties of some antiferroelectric materials

Material	Composition	T_c, °C
ADP	$NH_4H_2PO_4$	−125
Sodium niobate	$NaNbO_3$	+638
Lead zirconate	$PbHO_3$	+233

other, the net polarization is zero. These solids have an antiferroelectric Curie temperature T_c above which the effect does not exist. They have a relative capacitivity somewhat larger than that of the normal dielectric. For the common antiferroelectrics ϵ_r is of order 50 to 150. Several of the antiferroelectric compounds are listed in Table 18-7.

PROBLEMS

18-1 Using the assumptions given in the first paragraph of Sec. 18-2, prove Eq. (18-9).

18-2 Prove Eq. (18-18) for the energy of a dipole in an electric field.

18-3 Show that Eq. (18-21) follows from the integration of Eq. (18-20).

18-4 Expand $L(a)$ in a series expansion. For a molecule with a dipole moment of 1×10^{-30} coulomb m, find the field E for which the second term of the expansion at 300°K makes a correction to the approximate expression (18-23) of 1 per cent.

18-5 Consider a dielectric polarized uniformly to a value $\mathbf{P}$. Let a spherical hole be made about a point A. Calculate the effective surface-charge density on the inner surface of the sphere. Show that the field at A produced by this surface charge is $\mathbf{E}_1 = \mathbf{P}/3\epsilon_v$. (See Fig. P18-5.)

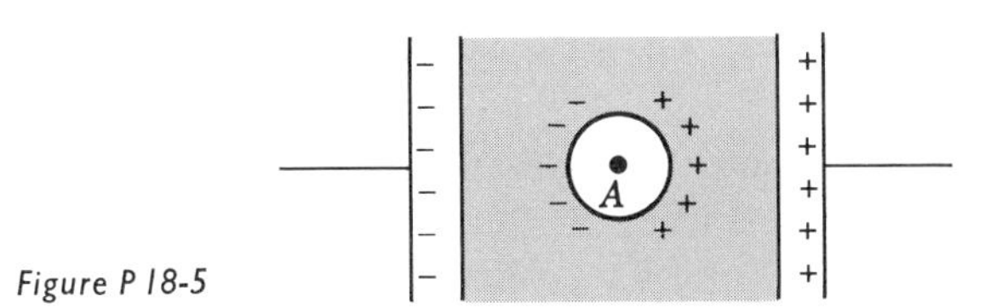

Figure P 18-5

18-6 Show that the field $\mathbf{E}_2$ at A caused by the symmetrical arrangement of dipoles of Fig. 18-9 is zero by cancellation. See Prob. 17-4.

18-7 Show that the torque $\mathbf{T}$ on a dipole $\mathbf{p}$ in a field $\mathbf{E}$ is given by

$$\mathbf{T} = \mathbf{p} \times \mathbf{E}$$

If N dipoles per unit volume, each of strength, are aligned, then show that the total torque $\mathbf{T}_N$ is given by

$$\mathbf{T}_N = N\mathbf{p} \times \mathbf{E} = \mathbf{P} \times \mathbf{E}$$

This expression permits us to describe $\mathbf{P}$ as the dipole moment per unit volume.

18-8 Suppose that an atom of induced polarizability α_e is placed in a field E. Show that the energy stored in the atom is $\alpha E^2/2$.

18-9 Let an atom of argon be placed 15 A from an electron. (*a*) Calculate the dipole moment induced in the argon atom. (*b*) Does a force exist between the atom and the electron? Of what magnitude? Of what sign?

18-10 Solid argon contains about 2.5×10^{28} atoms per cubic meter. Its polarizability is given in Table 18-3. Assuming that the Lorentz internal field is given by Eq. (18-29), calculate the ratio of the internal field to the applied field.

19 Dielectric loss

19-1 General

The dielectric behavior referred to in the previous chapter had two implicit attributes: The vectors **D**, **E**, and **P** were required to be collinear in space, and they were also required to be in the same time-phase with each other. Neither of these conditions need necessarily be true. To discard the first condition requires a more extensive treatment of the symmetry characteristics of the dielectric crystals than we wish to carry out, but we now examine dielectric behavior when the second condition is discarded. This problem has two important aspects, one concerned with the effects in materials containing permanent dipoles, the other with nonpolar crystals.

19-2 Dipole Relaxation

The phenomena occurring in materials containing permanent dipoles are quite distinct from those occurring in materials containing no dipoles, and so permanent dipoles are considered first. The principal effect in these solids is termed *dipole relaxation*. The physical picture of this relaxation is the following: Let a static field **E** be applied to a crystal at time $t = 0$ and held constant thereafter. The dipoles in the crystal do one of two things. They (1) align themselves rapidly with **E** (in so far as temperature fluctuations allow them to), or (2) gradually assume a final configuration. The first case is the static response we have already studied. In the second case **E** and **P** (and hence **E** and **D**) are not in time phase. The mathematical treatment of this problem is straightforward, though it is a long calculation.

If a field $\mathbf{E}$ is applied to matter containing permanent dipoles, the equation which gives $\mathbf{D}$ as a function of $\mathbf{E}$ must account for the dipole relaxation as a function of time. We take as the starting point the equation of electrostatics $\mathbf{D} = \epsilon\mathbf{E}$, where ϵ is a real, time-independent constant. Time dependence can be included by adding terms which contain time explicitly. We add terms in the first time derivatives of both D and E, so that

$$D + a\frac{dD}{dt} = bE + c\frac{dE}{dt} \tag{19-1}$$

where the constants a, b, and c have dimensions to make the corresponding terms dimensionally correct. Terms in second derivatives, third derivatives, etc., might be added; however, we assume that for moderate fields the higher-order terms can be neglected. The correctness of this choice comes, of course, when we see that solutions of this equation are consistent with experiments. The equation, however, cannot be derived from a more basic source.

The linear differential equation (19-1) has several features worth noting. We assume that $\mathbf{D}$ and $\mathbf{E}$ are collinear vectors. We suppose that a, b, and c are material constants which have physical significance. To see what these constants are, consider the following special cases:

1. Suppose that $\dot{D} = \dot{E} = 0$ and that $E =$ constant and $D =$ constant for all times. This is the static dielectric case. Equation (19-1) then becomes $D = bE$, so that b is the static capacitivity ϵ_s.† Equation (19-1) then becomes

$$D + a\dot{D} = \epsilon_s E + c\dot{E} \tag{19-2}$$

2. Suppose that a field of magnitude E is applied suddenly and held constant. D rises instantaneously to some characteristic value, then increases further with time as the dipoles align themselves. Since $\dot{E} = 0$, Eq. (19-2) becomes

$$D + a\dot{D} = \epsilon_s E \tag{19-3}$$

The solution of this equation is

$$\ln (D - \epsilon_s E) = -\frac{t}{a} + \text{constant}$$

Let $D = D(0)$ at $t = 0$. Then

$$D = \epsilon_s E - [-D(0) + \epsilon_s E]e^{-t/a} \tag{19-4}$$

† Considerable nonuniformity of language exists in the designation of the static capacitivity. Some authors call it ϵ_0, but others designate the capacitivity of a vacuum by ϵ_0. We choose to call the static capacitivity ϵ_s, the vacuum capacitivity ϵ_v, and the relative capacitivity ϵ_r.

Equation (19-4), when plotted, has the form of the curve in Fig. 19-1. After its initial rise to $D(0)$, D experiences an exponential rise to the asymptotic value $\epsilon_s E$. The quantity a is the time constant τ of this exponential. With this identification, Eq. (19-2) may be written as

$$D + \tau \dot{D} = \epsilon_s E + c\dot{E} \tag{19-5}$$

3. Suppose that a field of magnitude $E(0)$ is applied at a time $t = 0$. It produces a value of D in the material given by D_1. Let

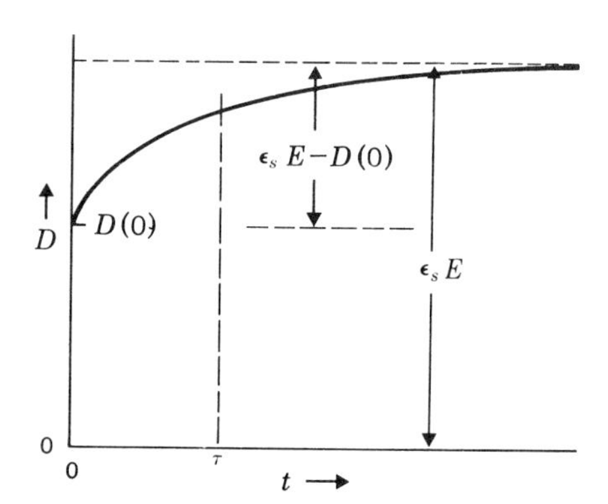

Figure 19-1 Variation of D with time at constant E.

now the field be adjusted with time so that D stays constant at the value D_1. Then Eq. (19-5) becomes

$$D_1 = \epsilon_s E + c\dot{E} \tag{19-6}$$

which may be integrated to give

$$\epsilon_s E = D_1 + [\epsilon_s E(0) - D_1]e^{-\epsilon_s t/c} \tag{19-7}$$

This expression has the form shown in Fig. 19-2. The curve is a decaying exponential with another time constant $\tau_1 = c/\epsilon_s$. Hence Eq. (19-5) can be written as

$$D + \tau\dot{D} = \epsilon_s E + \epsilon_s \tau_1 \dot{E} \tag{19-8}$$

This differential equation, then, expresses the relationship between D, E, and their first time derivatives in terms of constants which can be obtained from simple experiments.

Even more significance can be given to the constants by one simplification. Let a small increment in E, ΔE, be applied to the dielectric in a time Δt. A certain increment in D, call this ΔD, results. To see the effect of this change on Eq. (19-8), rewrite it in differential form, and integrate it term by term over the time interval Δt.

$$\int_0^{\Delta t} D\,dt + \tau \int_0^{\Delta D} dD = \epsilon_s \int_0^{\Delta t} E\,dt + \epsilon_s \tau_1 \int_0^{\Delta E} dE \tag{19-9}$$

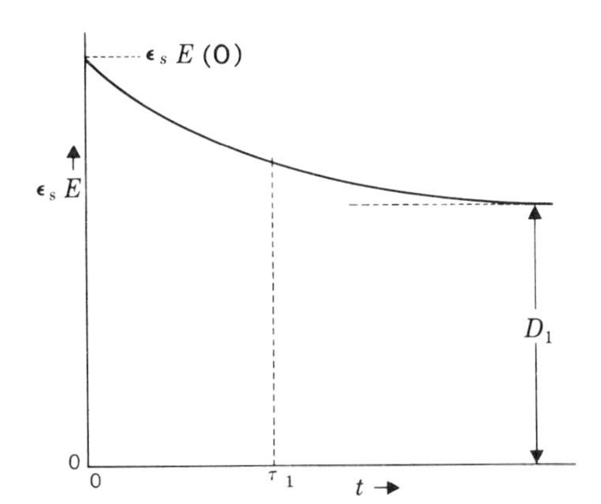

Figure 19-2 Variation of E with time at constant D.

As the time increment Δt is made shorter and shorter, the first term on each side of the equation approaches zero (since the limits approach each other). If ΔE and the corresponding ΔD do not change with this shortening, i.e., if the same increments are applied in the time intervals, these terms retain their constant values. Hence

$$\tau \, \Delta D = \epsilon_s \tau_1 \, \Delta E$$

or

$$\frac{\Delta D}{\Delta E} = \frac{\epsilon_s \tau_1}{\tau}$$

Define now a capacitivity ϵ_∞ through the expression

$$\Delta D = \epsilon_\infty \, \Delta E$$

The factor ϵ_∞ is clearly that capacitivity appropriate to an abrupt change in ΔE and ΔD before any movement of the permanent dipoles occurs. It corresponds to those frequencies which are infinite; hence the term ϵ_∞. With this definition

$$\tau \epsilon_\infty = \epsilon_s \tau_1 \tag{19-10}$$

Equation (19-8) can then be rewritten in its final form,

$$D + \tau \dot{D} = \epsilon_s E + \tau \epsilon_\infty \dot{E} \tag{19-11}$$

The most interesting case of dipole reorientation may now be studied through Eq. (19-11), a reorientation which occurs when a periodic field E of arbitrary frequency ω is applied to the dielectric. Let this periodic field be given by

$$E = E_1 e^{i\omega t} \tag{19-12}$$

The displacement D follows the field E, but not necessarily in phase with E. Consequently the capacitivity may have an in-phase component and an out-of-phase component. It may then be represented, in general, by a complex number. Let

$$D = \epsilon^* E \tag{19-13}$$

where ϵ^* may be complex; i.e.,

$$\epsilon^* = \epsilon_{\text{real}} - i\epsilon_{\text{imag}}$$

To find ϵ_{real} and ϵ_{imag}, substitute Eqs. (19-12) and (19-13) into Eq. (19-11). The following expression results (after complex fractions are cleared):

$$\epsilon^* = \left(\epsilon_\infty + \frac{\epsilon_s - \epsilon_\infty}{1 + \omega^2\tau^2}\right) + i\,\frac{(\epsilon_s - \epsilon_\infty)\omega\tau}{1 + \omega^2\tau^2} \tag{19-14}$$

The first term is a real number. It represents, when multiplied by E,

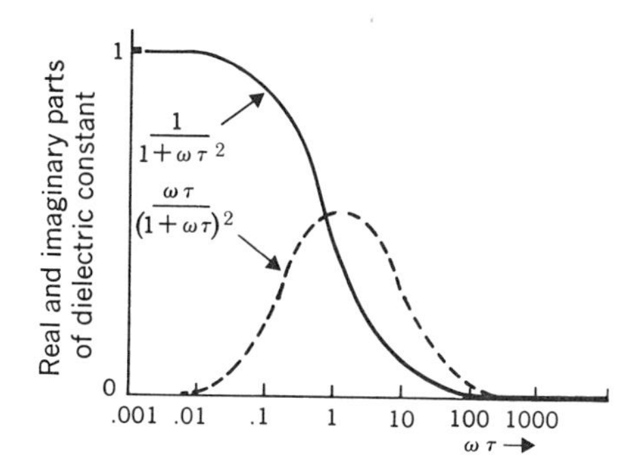

Figure 19-3 Frequency dependence of real and imaginary parts of the dielectric constant.

that part of D which is in phase with E. The second term is an imaginary number. It represents, when multiplied by E, that part of D which is 90° out of phase with E.

Plotting of the in-phase and out-of-phase parts of ϵ^* is instructive. For the in-phase term

$$\epsilon_{\text{real}} = \epsilon_\infty + \frac{\epsilon_s - \epsilon_\infty}{1 + \omega^2\tau^2}$$

which can be rearranged to be

$$\frac{\epsilon_{\text{real}} - \epsilon_\infty}{\epsilon_s - \epsilon_\infty} = \frac{1}{1 + (\omega\tau)^2} \tag{19-15}$$

Remember that τ is a property of the material, often a function of temperature, but not a function of ω. The right-hand side of (19-15) therefore goes through the range unity to zero as the frequency ranges from zero to ∞. This term is plotted as the solid line in Fig. 19-3 as a function of the variable log $\omega\tau$. At very low frequencies, i.e., those frequencies for which $\omega \ll 1/\tau$, ϵ_{real} is closely the same as ϵ_s, the static value, as expected. At high frequencies ($\omega \gg 1/\tau$), ϵ_{real} has the value

ϵ_∞, also expected. At intermediate frequencies, a smooth transition occurs over about three orders of magnitude in ω from the one limiting value to the other.

The out-of-phase component is also easy to plot. Write

$$\epsilon_{\text{imag}} = \frac{(\epsilon_s - \epsilon_\infty)\omega\tau}{1 + \omega^2\tau^2}$$

which can be rearranged to give

$$\frac{\epsilon_{\text{imag}}}{\epsilon_s - \epsilon_\infty} = \frac{\omega\tau}{1 + \omega^2\tau^2} \tag{19-16}$$

The right-hand side of this expression is also plotted in Fig. 19-3 as the dashed curve. This term goes through a broad maximum peaked near $\omega = 1/\tau$.

The existence of an out-of-phase term in D leads to energy dissipation in the dielectric. Recall that power dissipation $\mathscr{P}$ per unit volume is given by

$$\mathscr{P} = jE \tag{19-17}$$

where j is that part of the current in phase with E. The total currents (when there are no real currents) are given by

$$j = \frac{dD}{dt} = \frac{d}{dt}(\epsilon_{\text{real}}E - i\epsilon_{\text{imag}}E) = (i\omega\epsilon_{\text{real}} + \omega\epsilon_{\text{imag}})E \tag{19-18}$$

Thus the power loss is given by

$$\mathscr{P} = \omega\,|\epsilon_{\text{imag}}|\,E^2 \tag{19-19}$$

The power loss is in fact proportional to the out-of-phase part of D, and it follows rather closely the shape of ϵ_{imag} in Fig. 19-3 as a function of frequency. It is low at low frequencies and also at high frequencies, but at intermediate frequencies close to $\omega = 1/\tau$ it goes through a broad maximum. Commonly the magnitude of the energy loss is expressed through a loss angle δ, defined as

$$\tan\delta = \frac{\epsilon_{\text{imag}}}{\epsilon_{\text{real}}} = \frac{\epsilon_s - \epsilon_\infty}{\epsilon_s + \epsilon_\infty\omega^2\tau^2}\,\omega\tau \tag{19-20}$$

Then $\mathscr{P}$ may be written in terms of ϵ_{real} and $\tan\delta$.

$$\mathscr{P} = \omega E^2\epsilon_{\text{real}}\tan\delta \tag{19-21}$$

Many measurements have been made on the relaxation properties of dielectrics, mostly for liquids. First, many measurements have been made of relaxation times. Some typical data for several liquids are given in Table 19-1.

The critical frequencies for maximum energy loss in these liquids near room temperature are of order 1,000 Mc ($\lambda \approx 0.3$ m). The

Table 19-1†

Material	Temperature, °C	τ, sec
H_2O	19	1×10^{-11}
CH_3OH	19	6
C_2H_5OH	20	13

† From Böttcher, p. 383.

second type of data available concerns the variation of τ with temperature. Data for a few liquids at several temperatures are given in Table 19-2. Typically, relaxation times decrease with increasing temperature. Finally, data exist on the variation of the real part of the

Table 19-2†

Material	τ, $\times$ 10^{+11} sec		
	1°C	25°C	55°C
n-Propyl bromide	0.82	0.58	0.47
n-Amyl bromide	1.77	1.21	0.87
n-Octyl bromide	3.60	2.18	1.37

† From Böttcher, p. 386.

dielectric constant and the dissipation losses. Data for the real and imaginary parts of the capacitivity are shown for *i*-butyl bromide in Fig. 19-4. Observe that the shape of the curves is as predicted and

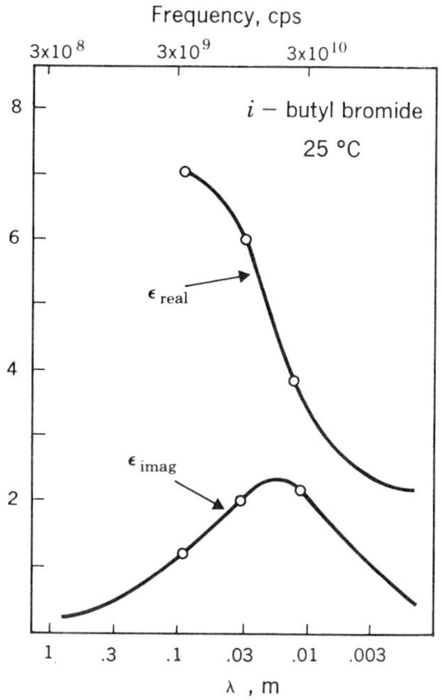

Figure 19-4 Variation of ϵ_{real} and ϵ_{imag} for *i*-butyl bromide as a function of frequency. [*After E. Hennelly, W. Heston, and C. Smyth, J. Am. Chem. Soc.*, **70:** 4102 (1948), *by permission of the American Chemical Society.*]

that the peak in ϵ_{imag} of about 2.5 units is about half the change in ϵ_{real} of about 5 units. The relaxation time τ for this material at 25°C is about 7×10^{-12} sec.

We would like to be able to show corresponding data for dielectric loss in solids, but such complete data for loss in solids are scant. In Prob. 19-4 data for loss in ThO_2 are presented.

19-3 Resonance Absorption

Another type of energy loss in dielectrics is characteristic of the induced and ionic parts of the polarization. This energy loss is similar in many respects to that of dipole relaxation. The capacitivity is a complex number, and the real and imaginary parts of the capacitivity vary in the vicinity of a critical frequency in somewhat the same way as for dipole relaxation. However, the energy loss itself has a different origin and requires a different analytical description.

The process leading to the resonance absorption may be described as follows: When an electron cloud is displaced from its normal configuration about its nucleus, a force exists which tends to restore the electron cloud to its normal position. The force is a complicated function of displacement, but suppose that for the small displacements present in both induced and ionic polarization the force is simply a linear function of displacement. With no applied field a charge distribution (or a pair of ions) oscillates if displaced according to the rules of simple harmonic motion. The differential equation governing the displacement x is

$$m\frac{d^2x}{dt^2} + Kx = 0 \tag{19-22}$$

A characteristic frequency ω_0 exists; it is related to the force constant K and the mass of the oscillator, m, by the expression

$$\omega_0{}^2 = \frac{K}{m}$$

In the presence of an alternating electric field, a driving-force term appears, due to the external field. Equation (19-22) becomes

$$m\frac{d^2x}{dt^2} + m\omega_0{}^2x = fE_1e^{-i\omega t} \tag{19-23}$$

Clearly the response of the oscillating system is a function of both ω and ω_0. In fact, oscillations build up without limit when $\omega = \omega_0$, though they are small at frequencies far from ω_0.

The damping effects can come from several sources. For example: (1) Charges being accelerated radiate electromagnetic energy. The charges in the simple harmonic oscillator described above are accelerated or decelerated constantly. Hence they radiate energy which must come from the energy of oscillation, and the oscillation is damped. (2) The atoms and ions of matter constantly make collisions with other atoms and ions. These collisions occur at random times during an oscillation cycle, and they are not elastic. Thus energy of "dielectric oscillation" can be transformed into heat by collisions. Both these losses can be approximately described by a frictional term proportional to dx/dt. Then Eq. (19-23) becomes

$$\frac{m\,d^2x}{dt^2} + mr\frac{dx}{dt} + m\omega_0{}^2x = fE_1e^{-i\omega t} \tag{19-24}$$

where r is a constant of the material called the *dissipation constant.* The solution of this equation for x gives the time dependence of the polarization of each individual atom; hence the polarizability per atom may be calculated. From the polarizability, the susceptibility and then the capacitivity may be calculated.

The solution for Eq. (19-24) can be most simply found by direct substitution of a trial solution,

$$x = ae^{-i\omega t}$$

This expression does satisfy the differential equation provided that

$$a = \frac{f}{m}E_1\left[\frac{\omega_0{}^2 - \omega^2}{(\omega_0{}^2 - \omega^2)^2 + r^2\omega^2} + \frac{ir\omega}{(\omega_0{}^2 - \omega^2)^2 + r^2\omega^2}\right] \tag{19-25}$$

The displacement x of the charge q has a real part in phase with E and an imaginary part out of phase with E. Therefore, the dipole moment of each atom has an in-phase and an out-of-phase component. The dipole moment of each unit dipole is $p = qx$, and the polarization of the medium (which consists of N dipoles per unit volume) is $P = Nqx$. The displacement D may now be written as

$$D = \left[\epsilon_v + \frac{Nfq}{m}\frac{\omega_0{}^2 - \omega^2}{(\omega_0{}^2 - \omega^2)^2 + r^2\omega^2} + i\frac{Nq^2}{m}\frac{r\omega}{(\omega_0{}^2 - \omega^2)^2 + r^2\omega^2}\right]E \tag{19-26}$$

The real part of the capacitivity is thus given by

$$\epsilon_{\text{real}} = \epsilon_v + \frac{Nfq}{m}\frac{\omega_0{}^2 - \omega^2}{(\omega_0{}^2 - \omega^2)^2 + r^2\omega^2} \tag{19-27}$$

and the imaginary part by

$$\epsilon_{\text{imag}} = \frac{Nfq}{m}\frac{r\omega}{(\omega_0{}^2 - \omega^2)^2 + r^2\omega^2} \tag{19-28}$$

These quantities when plotted have the forms shown in Fig. 19-5. For the static case, when $\omega = 0$, $\epsilon_{\text{real}} = \epsilon_v + Nfq/m\omega_0{}^2$ and, for $\omega \gg \omega_0$, $\epsilon_{\text{real}} = \epsilon_v$. ϵ_{imag} approaches zero for both $\omega \gg \omega_0$ and $\omega \ll \omega_0$; it goes through a maximum of height $(Nfq)(1/mr\omega)$ when $\omega = \omega_0$.

The out-of-phase component of ϵ again represents an energy loss. The magnitude of the loss may be calculated by the methods used in

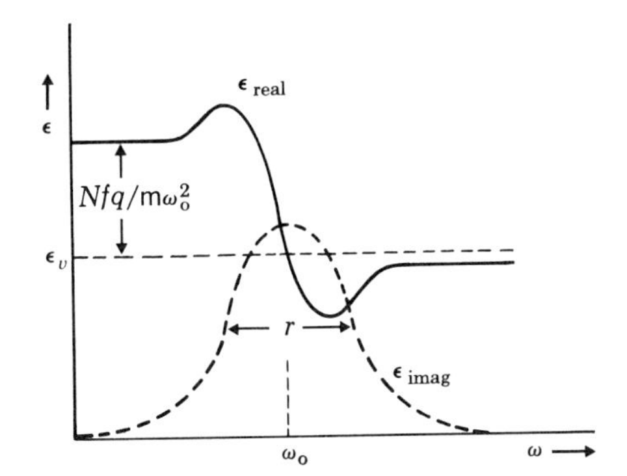

Figure 19-5 Plot of ϵ_{real} and ϵ_{imag} for loss from induced dipoles.

connection with Eq. (19-20). In the present case the power loss is again

$$\mathscr{P} = \omega\epsilon_{\text{imag}}E^2 \tag{19-29}$$

The similarity of the curves of Fig. 19-5 to those of Fig. 19-3 is noteworthy. *Keep in mind that the origin of the two effects is quite different,* however, and that the frequencies of the maxima in the several cases are likely to be quite different, as emphasized in the following section.

19-4 Summary

The energy losses cited in the two previous sections may all exist in a given material at the same time. To see how the capacitivity in such a case would depend on frequency, consider the frequency dependence of the various types of loss.

Dipolar materials

The time constant τ of typical dipolar liquids at room temperature is found to be about 10^{-11} sec, which gives characteristic values for ω at maximum energy loss of about 10^{11} radians/sec. This frequency of about 10^6 Mc/sec corresponds to a wavelength of about 1 cm.

Ionic molecules and solids

The characteristic vibrational frequencies of ionic molecules and ionic solids is about 10^{14} radians/sec, corresponding to wavelengths in the infrared ($\lambda \approx 10^{-3}$ cm).

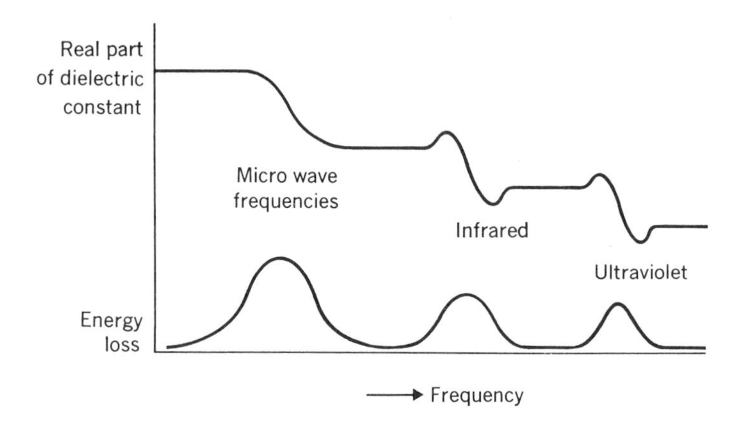

Figure 19-6 Characteristics of real dielectric materials as a function of frequency.

Induced polarization in all materials

The characteristic frequencies for the vibration of the electron clouds themselves are very high; they lie in the infrared and ultraviolet parts of the spectrum.

The dielectric constant of a solid which has all three modes of energy loss is a complicated function of frequency. Typical behavior is sketched in Fig. 19-6. Either or both of the lower-frequency loss peaks may be absent for particular materials.

REFERENCES

1. R. H. Cole, Theories of Dielectric Polarization and Relaxation, in J. Birks and J. Hart (eds.), *Progress in Dielectrics*, vol III, Heywood and Company, London, 1961.
2. C. J. F. Böttcher, *Theory of Electric Polarization*, Elsevier Publishing Company, Amsterdam, 1952.
3. A. R. von Hippel, *Dielectrics and Waves*, John Wiley & Sons, Inc., New York, 1954.
4. H. Fröhlich, *Theory of Dielectrics*, Oxford University Press, New York, 1949.

PROBLEMS

19-1 (*a*) Plot Eqs. (19·15) and (19·16) accurately for constant τ as a function of $\omega\tau$. (*b*) Determine the half-width of the absorption peak, i.e., the frequency range over which the term $\omega\tau/(1 + \omega\tau)^2$ falls to half its peak value. (*c*) Do the data of Fig. 19-4 have this ideally narrow range? (*d*) What does your answer to (*c*) tell you about the material?

19-2 Find the expression which describes the condition for the maximum of $\tan \delta$ [Eq. (19·20)] in terms of constants of the material and the variable angular frequency ω.

19-3 For some materials the factor τ depends exponentially on temperature according to the expression

$$\tau = \tau_0 e^{H/Rt}$$

(*a*) Insert this expression into Eqs. (19-15) and (19-16).

(*b*) Plot the left-hand side of the resultant equations as a function of τ and of $1/T$, assuming ω and τ_0 to be constants.

(*c*) Find the half-width in terms of $1/T$ of the curve of $\epsilon_{imag}/(\epsilon_s - \epsilon_\infty)$.

19-4 Below are data for dielectric loss in Ca-doped ThO_2 at two frequencies.

(*a*) Plot the data as a function of temperature to find the temperature of the point of maximum in tan δ for each frequency.

(*b*) The measured capacitivity at room temperature was ϵ_∞ (T = 25°C) $\approx 16.2\epsilon_v$. Show that these data are consistent with a value of $\epsilon_s - \epsilon_\infty \approx 3\epsilon_v$.

(*c*) Assuming that the event causing the loss obeys a temperature dependence $\tau = \tau_0 e^{H/RT}$, find H and τ_0.

(These data were taken from the Ph.D. thesis of J. Wachtman, University of Maryland, 1962.)

f = 695 cps		f = 6,950 cps	
T, °K	tan δ	T, °K	tan δ
555	0.023	631	0.026
543	0.042	621	0.036
532	0.070	612	0.043
524	0.086	604	0.055
516	0.092	590	0.073
509	0.086	581	0.086
503	0.081	568	0.086
494	0.063	543	0.055
485	0.042	518	0.025
475	0.029	498	0.010

19-5 Suppose that a steady electric field of magnitude E were applied to a specimen of Ca-doped ThO_2 at 295°K. Plot D as a function of time after the manner of Fig. 19-1.

20

Diamagnetism and paramagnetism

20-1 General

Many features of the magnetic characteristics of matter are formally similar to the dielectric characteristics. Induced magnetic polarization is analogous to induced dielectric polarization. Atoms and molecules with permanent magnetic dipoles exist, just as do those with permanent electric dipoles. Some materials possess spontaneous magnetization, just as some possess a spontaneous electric moment. These and numerous other similarities make natural a parallel study of the two phenomena.

One important difference must be pointed out between magnetic and electric properties of matter, however. Individual electric charges (monopoles) of one sign do exist; they may be either plus or minus charges. A corresponding magnetic monopole does not occur; all magnetic fields arise from current loops. To be sure, in elementary physics one commonly speaks of magnetic poles and considers that lines of **H** originate from them. However, this approach is just a convenience to eliminate the need of describing the magnetization **M**. Discussion of pole strength in describing the magnetic properties of matter is entirely unnecessary. Just as the main topics of the dielectric behavior of matter can be described in terms of **P**, so can the main topics of magnetization be described in terms of **M**.

The magnetic quantities in Maxwell's equations, **B** and **H**, have been related through Eq. (17·2),

$$\mathbf{B} = \mu \mathbf{H} \tag{20-1}$$

The factor μ, the permeability, is a constant for many materials. Since B has dimensions m/Qt and H, Q/lt, μ has dimensions ml/Q^2.

(For some materials the directions in space of the vectors **B** and **H** are not the same, and μ is then a more complicated quantity called a *tensor*. All materials are treated here as isotropic, however.) For a vacuum μ has a well-defined value μ_v, the magnitude of which is $4\pi \times 10^{-7}$ kg m/coulomb^2 (commonly termed henry/m).

The permeability is not the only quantity by which the magnetic features can be described. By defining the *relative permeability* μ_r as μ/μ_v, Eq. (20-1) can be written as

$$\mathbf{B} = \mu_v \mu_r \mathbf{H} \tag{20-2}$$

The relative permeability is clearly analogous in the dielectric case to the relative capacitivity ϵ_r. Furthermore, the magnetic state of a material body has been characterized in Eq. (17-6) by a vector called the magnetization **M** through the expression

$$\mathbf{B} = \mu_v \mathbf{H} + \mu_v \mathbf{M} \tag{20-3}$$

Finally, the magnetic susceptibility χ_m is defined through the expression

$$\mathbf{M} = \chi_m \mathbf{H} \tag{20-4}$$

Again **M** and **H** are considered to be collinear vectors so that χ_m is a pure number. Note that $\chi_m = \mu_r - 1$ in another analogy to the dielectric case.

Unfortunately, several systems of units defining these magnetic quantities are in common use. In addition to the rationalized mks units used above, another system, the cgs electromagnetic units (emu), is often used, especially in the scientific literature. To show how to go quickly from one system to another, we describe briefly the cgs emu system. As before, a factor termed the permeability μ' relates **B** and **H**.

$$\mathbf{B} = \mu' \mathbf{H}$$

In this system μ'_v is taken as unity; so $\mu' = \mu_r$. The magnetization is related to **B** and **H** through the expression

$$\mathbf{B} = \mathbf{H} + 4\pi \mathbf{I}$$

The magnetic susceptibility is defined as before, $\mathbf{I} = \chi'_m \mathbf{H}$. This means that the susceptibility is related to the permeability μ' through the expression

$$\chi'_m = \frac{1}{4\pi}(\mu' - 1)$$

The fields and quantities used to describe the magnetic effects in the two systems are tabulated in Table 20-1 for purposes of comparison.

Table 20-1 Magnetic units

Factor	Mks rationalized	Cgs emu	Conversion
B	Webers/m^2	Gauss	1 weber/m^2 $= 10^4$ gauss
H	Ampere-turns/m	Oersteds	1 amp-turn/m $= \frac{4\pi}{10^3}$ oersteds
Magnetization	M, amp-turns/m	I, maxwells/cm^2	1 amp-turn/m $= \frac{1}{10^3}$ maxwells/cm^2
Permeability of vacuum	μ_v, henrys/m	$\mu'_v = 1$	
Relative permeability	μ_r	μ'	$\mu_r = \mu'$
Susceptibility	χ_m	χ'_m	$\chi_m = 4\pi\chi'_m$

The important conversion factors are also listed. Fortunately the relative permeability in the two systems is identical. For consistency with the rest of the topics treated in this book, the rationalized mks system will be used in these chapters on magnetism.

20-2 Diamagnetism

The magnetic effect corresponding to the induced dielectric effect is termed *diamagnetism*. It is characterized by the fact that $\mathbf{M}$ is a linear function of $\mathbf{H}$ so that χ_m is indeed a constant, independent of field. The magnetization, however, is directed oppositely to $\mathbf{H}$. χ_m is therefore a negative quantity, and $\mathbf{B}$ in a diamagnetic material is reduced from its vacuum value. The effect is very weak, though, and μ_r is only slightly less than unity.

The physical origin of diamagnetism can be seen from the classical picture of an atom as electronic charges circulating around the nucleus in definite orbits. For simplicity, a circular orbit is usually assumed with a single electron having an angular velocity ω. This angular velocity is altered if a magnetic field is slowly applied, the radius of the orbit being unchanged. The change in angular velocity gives rise to a magnetic moment $\boldsymbol{\mu}_m$ of magnitude

$$\boldsymbol{\mu}_m = -\frac{e^2}{4m}\mu_v r^2\mathbf{H} \tag{20-5}$$

where e = charge of the electron
m = mass of the electron
r = radius of its orbit
$\mathbf{H}$ = field which is applied normally to the plane of the orbit

The atomic susceptibility is small: For orbits of radius of about 1 A, the factor $e^2\mu_v r^2/4m$ is about 10^{-34} m^3. Since an average solid has about 10^{28} to 10^{29} atoms per cubic meter, the volume susceptibility χ_m should be of order 10^{-5} to 10^{-6}, the order of magnitude actually measured for diamagnetic solids. As for the induced electronic polarizability, the diamagnetic susceptibility is independent of temperature.

The change in angular velocity of the electron which occurs when the field is applied is always such as to produce a negative susceptibility. The atom with its circulating electron acts like an inductance in which a counter emf is produced, i.e., in accordance with Lenz's law. Since no resistance is encountered by the electron in its motion about the nucleus, the increased velocity of the electron does not diminish with time.

Diamagnetism has no ordinary practical uses, and so the treatment here has not been thorough. The key to the quantitative understanding of the effect resides in the understanding of the physics behind the derivation of Eq. (20-5). Since this is presented elsewhere, the derivation here has not been carried out in detail.† The diamagnetic volume susceptibilities of several solids are listed in Table 20-2.

Table 20-2 Diamagnetic susceptibilities of some solids

Material	χ_m (volume)
Au	-3.7×10^{-5}
Cu	-0.95×10^{-5}
Hg	-2.9×10^{-5}
Ge	-0.8×10^{-5}
Si	-0.4×10^{-5}
He (NTP)	-0.5×10^{-9}
A (NTP)	-11.0×10^{-9}
Kr (NTP)	-16.0×10^{-9}
Xe (NTP)	-25.0×10^{-9}

Also listed are the volume susceptibilities of several of the rare gases at NTP.

One of the interesting diamagnets is the superconductor. This system has an infinite diamagnetic susceptibility, a property which has an important use in constructing superconducting magnets. However, the subject, being quite involved, is beyond the level of our discussion.

20-3 Paramagnetism

Some atoms and ions have permanent magnetic moments. In the absence of a magnetic field, these moments usually point in random directions, producing no macroscopic magnetization. However, in the

† See, for example, J. Bates, *Modern Magnetism*, 2d ed., Cambridge University Press, New York, 1948.

presence of a magnetic field, the moments tend to line up preferentially in the field direction and produce a net magnetization. When the atoms and ions are acted upon individually, with no mutual interaction between them, the effect is called *paramagnetism*. Since the moments line up in the direction of the field and enhance the external field, the susceptibility χ_m is greater than zero.

Two separate aspects characterize paramagnetism. One is the temperature dependence of the net magnetic moment of the entire specimen; the other is the origin of the moments themselves.

20-4 The Temperature Dependence of the Susceptibility

The behavior of the individual magnetic moments in a magnetic field is formally similar to that of permanent electric moments in an electric

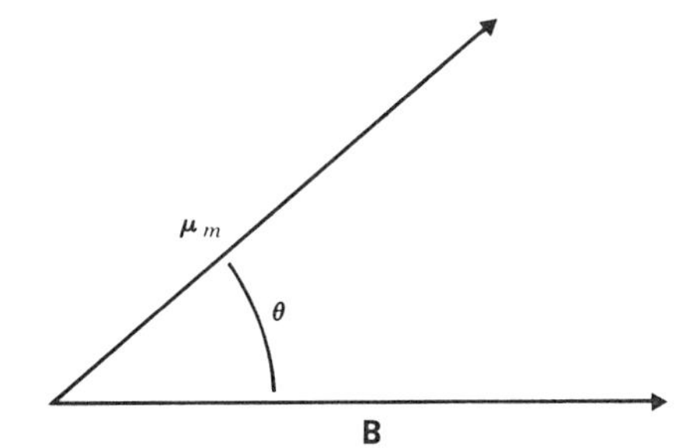

Figure 20-1 The energy of a magnetic dipole is least when the dipole points in the direction of **B**.

field. The energy W of the moments $\boldsymbol{\mu}_m$ in the field, **B**, depends on the angle θ between $\boldsymbol{\mu}_m$ and **B**.

$$W = -\mu_m B \cos\theta \qquad (20\text{-}6)$$

(see Fig. 20-1). When the susceptibility is small compared with one, this expression may be rewritten,

$$W = -\mu_m \mu_v H \cos\theta \qquad (20\text{-}7)$$

One might think that dipoles always point in the direction of H, which would be true if H could be made large enough (or if the temperature could be reduced to absolute zero). However, at any finite temperature the collisions of the atoms tend to randomize the directions of the moments, and the net moment of many dipoles is always something less than the possible maximum value. The number of dipoles in some range $d\theta$ making an angle θ with H is given by

$$N(\theta)\, d\theta = C 2\pi \sin\theta\, e^{(\mu_m \mu_v H \cos\theta)/kT} d\theta \qquad (20\text{-}8)$$

This equation is exactly analogous to Eq. (18-19) for dielectrics. $N(\theta)$ is a function such that $N(\theta)\, d\theta$ gives the density of dipoles in a range $d\theta$ about some angle θ. C is a constant which will be evaluated later. Again, the component of each dipole in the direction of H is given by

$\mu_m \cos\theta$, so that the net component of all dipoles in range $d\theta$ at angle θ is $\mu_m N(\theta)\cos\theta\, d\theta$. The net component of magnetization, M, due to N dipoles, is then

$$M = \int_0^\pi 2\pi C \mu_m \cos\theta \sin\theta\, e^{(\mu_m \mu_v H \cos\theta)/kT}\, d\theta \tag{20-9}$$

The constant C is again evaluated by noting that the total number of dipoles, N, is equal to the integral $\int_0^\pi N(\theta)\, d\theta$. Hence the final expression becomes

$$M = N \frac{\int_0^\pi \mu_m \cos\theta \sin\theta\, e^{(\mu_m \mu_v H \cos\theta)/kT}\, d\theta}{\int_0^\pi \sin\theta\, e^{(\mu_m \mu_v H \cos\theta)/kT}\, d\theta} \tag{20-10}$$

This expression corresponds exactly to Eq. (18-20) for dielectrics. It integrates in the same way to give

$$M = N\mu_m\left(\coth a - \frac{1}{a}\right) \tag{20-11}$$

where $a = \mu_m \mu_v H/kT$.

For small values of a Eq. (20-11) may be rewritten to be

$$\mathbf{M} = \frac{N\mu_v {\mu_m}^2 \mathbf{H}}{3kT} \tag{20-12}$$

This expression is written vectorially to show that **M** has the same direction as **H**.

The susceptibility may be seen from Eq. (20-12) to be

$$\chi_m = \frac{N\mu_v {\mu_m}^2}{3kT} \tag{20-13}$$

This statement, that the magnetic susceptibility of a paramagnetic arrangement of dipoles should vary as $1/T$, is known as the *Curie law*. It is well obeyed for many materials; in Fig. 20-2 data are shown for the susceptibility of $CuSO_4 \cdot K_2SO_4 \cdot 6H_2O$ as a function of $1/T$.

We mention in passing two complicating factors: (1) Paramagnetism does not exist alone in atoms or ions; diamagnetic behavior also exists at the same time. Since these two effects are in opposition, the net magnetic behavior is determined by the larger. For most materials χ_m(para) is very much larger than χ_m(dia) so that paramagnetism is usually dominant. (2) A deviation from the Curie law exists for many materials. Many materials obey a modification of Eq. (20-13) known as the *Curie-Weiss law*,

$$\chi_m = \frac{C'}{T - \theta} \tag{20-14}$$

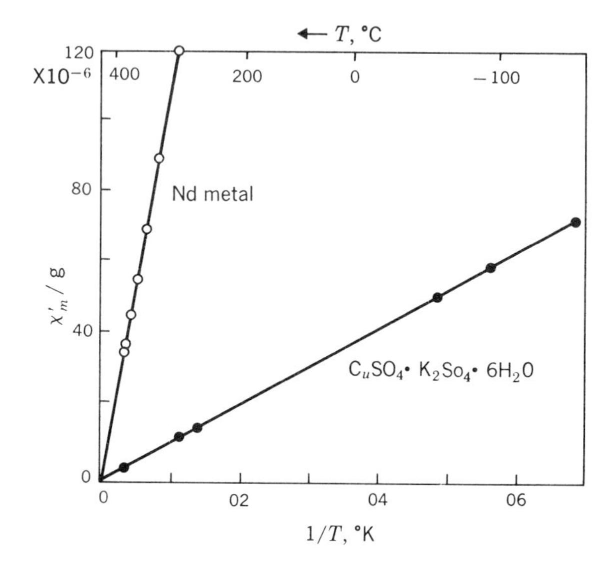

Figure 20-2 Susceptibility of two solids as a function of temperature. [*Data for* Nd *from J. Elliott, S. Levgold, and F. Spedding, Phys. Rev.*, **94:** 50 (1954); for $CuSO_4 \cdot K_2SO_4 \cdot 6H_2O$ from *J. C. Hupse, Physica*, **9:** 633 (1942).] Note that the ordinate is χ'_m per gram. To obtain χ_m of Eq. (20-13), multiply these values by 4π times the density of these materials.

where θ is some temperature, either positive or negative. The behavior of Ni at high temperatures fits this law. For Ni,

$$\theta = 678°\text{K}$$

This modification is related to the appearance of ferromagnetism at $T = \theta$, or to antiferromagnetism, which effects are discussed in later chapters. The linear plot of susceptibility of Nd as a function of temperature shows that it, too, obeys a Curie law. The value of N_B for Nd is about 3.6 Bohr magnetons, obtained by fitting Eq. (20-13) to the data of Fig. 20-2.

20-5 Origin of Permanent Moments

Since paramagnetism depends upon magnetic moments of atoms or ions, the origin of these magnetic moments is important. There are two chief origins for the moments, one from the orbital motion of the electrons, the other from the electron spin.

The *orbital moment* arises in principle because an electron in an orbit about an atom can be considered to be a small circulating current about the nucleus. The magnetic moment is related to the angular momentum by an important relation derived as follows: The primitive magnetic moment $\boldsymbol{\mu}_m$ of a circulating loop of charge is given by

$$\boldsymbol{\mu}_m = i\mathbf{A} \tag{20-15}$$

where i is the current in the loop and $\mathbf{A}$ is the area enclosed by the circulating current. Imagine, for simplicity, that the current is caused by a single electron in a circular orbit of radius r about the central nucleus. Then the current around the orbit is the amount of charge which passes any point in the orbit per unit time, i.e.,

$$i = \frac{|e|\, v}{2\pi r} \tag{20-16}$$

where v is the velocity of the electron in the orbit. The magnetic moment then has magnitude

$$\mu_m = \frac{|e|\, v}{2\pi r}\, \pi r^2 = \frac{|e|\, vr}{2} \tag{20-17}$$

The angular momentum of the electron in the circular orbit of radius r is equal to

$$\mathbf{L} = m\mathbf{v} \times \mathbf{r} \tag{20-18}$$

(Recall that m is the mass of the electron.) Hence $L = mvr$, so that

$$\boldsymbol{\mu}_m = \frac{e}{2m}\, \mathbf{L}$$

Notice that $\boldsymbol{\mu}_m$ and $\mathbf{L}$ are oppositely directed vectors, since the charge on the electron is negative. Thus the relation between the magnetic moment and the orbital moment is

$$\boldsymbol{\mu}_m = -\frac{|e|}{2m}\, \mathbf{L} \tag{20-19}$$

For this simple charge distribution the magnetic moment and the angular momentum are simply related through a constant $|e|/2m$ termed the *gyromagnetic ratio.*

In Chap. 8, the angular momentum of an electron in its orbit was found to exist in multiples of the unit $h/2\pi$. The magnetic moment of an electron with angular momentum $h/2\pi$ is called the *Bohr magneton* β, defined as

$$\beta = \frac{|e|}{2m}\,\frac{h}{2\pi}$$

$$= 9.27 \times 10^{-24} \text{ amp-m}^2 \tag{20-20}$$

The orbital moment in Eq. (20-19) can be related, of course, to the quantum numbers described in Sec. 8-1. Recall that the quantum numbers l and m_l refer to the angular momentum of the electron in its orbit. The number l refers to the total angular momentum by means of the equation

$$L = \sqrt{l(l+1)}\,\frac{h}{2\pi} \qquad l = 0, 1, 2, \ldots \tag{20-21}$$

The number m_l refers to the component of angular momentum in the z direction; it satisfies the following relation:

$$L_z = m_l \frac{h}{2\pi} \qquad m_l = 0, \pm 1, \pm 2, \ldots, \pm l \tag{20-22}$$

Combining Eqs. (20-19), (20-21), and (20-22), we can write

$$\mu_{m\ \text{orbit}} = \frac{|e|}{2m} \frac{h}{2\pi} \sqrt{l(l+1)} \tag{20-23}$$

and

$$\mu_{z\ \text{orbit}} = \frac{|e|\, h}{4\pi m} m_l \tag{20-24}$$

The *spin* of the electron also results in a contribution to the magnetic moment. Recall the hypothesis of Uhlenbeck and Goudsmit that the electron has an intrinsic angular momentum of magnitude $h/2\pi$. Thus the electron has an intrinsic magnetic moment due to this intrinsic angular momentum. The magnitude of this moment $\mu_{z\text{spin}}$ is found experimentally to be anomalous and is twice as large as Eq. (20-19) predicts. Thus

$$\mu_{z\,\text{spin}} = 2 \frac{|e|}{2m} (L_z)_{\text{spin}}$$

$$= \frac{|e|}{m} \frac{h}{2\pi} s \tag{20-25}$$

where s is the spin quantum number. Since s has the value $+\frac{1}{2}$ or $-\frac{1}{2}$, the magnitude of the magnetic moment due to the spin is always $|e|\, h/4\pi m$. Notice that the spin moment always has the magnitude of 1 Bohr magneton; the factor 2 appearing in the first of Eq. (20-25) just compensates for the fact that the spin quantum number is $\pm\frac{1}{2}$.

The magnetic moment of a multielectron atom is simply the sum of the magnetic moments of all the electrons, including both orbital and spin moments. Each electron contributes an independent vector quantity to the total magnetic moment of the atom. Since all filled shells have zero total angular momentum, they also have zero total magnetic moment. In particular, atoms or ions which possess only filled shells have no permanent moments, and hence they cannot be paramagnetic. No exceptions to this result have been observed. The inert gases He, A, Kr, etc., and ions such as Na^+ and Cl^- are all diamagnetic. Also many gases such as H_2, etc., are diamagnetic because the electrons are all completely paired. Other free atoms show paramagnetism if there are unpaired spins or a net orbital angular momentum.

The net spin angular momentum and spin magnetic moment to be expected of a partly filled shell are not obvious. However, *Hund's rule* states that spins of the electrons of a shell always add together in such a way as to contribute the maximum angular momentum and magnetic moment. Consider, for example, the electrons of a d shell. There are 10 states, 5 with spin up and 5 with spin down (Fig. 20-3). If a state contains 2 electrons, they have spins in the same direction, say, spin up. This gives the maximum spin for the 2 electrons, since if their spins were opposite, the total spin would be zero. If the state contains 5 electrons, they all have spin up (or down), which is the maximum spin angular momentum a d shell can have, since a sixth

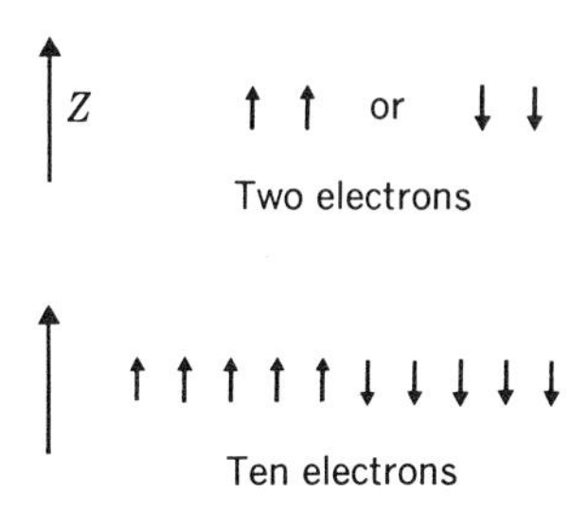

Figure 20-3 Spin alignment of a partly filled and filled *d* shell.

electron must have spin down (or up), canceling one of the first 5. Again, when 10 electrons are present, the total spin is zero, with 5 up and 5 down.

20-6 Diamagnetism and Paramagnetism of Solids

The calculation of paramagnetic susceptibility in Sec. 20-4 was based on the concept of matter as being a dilute gas, to avoid the complexities of interatomic interactions. Each atom or molecule was assumed to be independent of all others, and the effective field on each atom was considered to be the external field. However, when the susceptibilities of solids are considered, additional effects complicate the calculations. Recall that the same problem arose for calculation of dielectric susceptibilities.

Consider first the *salts*. The diamagnetic moment of the inner, filled shells is not affected by the nearness of other atoms. The contribution of these shells to the total moment is the same as for the individual atoms. The permanent moment of the unfilled shells of the ions is altered, however, so that the paramagnetic moment of a salt may not be calculated by summing over the moments of all the free atoms. In free atoms, the orbital moment contributes to the total magnetic moment, but not in solids. In the solid, the z component of the orbital momentum rotates back and forth from positive to negative values and averages to zero. L_z is then said to be *quenched*.

The spin moments are not affected in this way and do contribute to the permanent moment. The salts of the rare earths do not show this quenching effect, since their f shells are well shielded from the perturbations of neighboring ions by the outer valence shells. The number of Bohr magnetons per metallic ion in the rare-earth salts is generally higher than that of the salts of metals other than the rare earths.

The magnetic susceptibilities of the *metals* are affected by the fact that the outermost electrons are partially freed in the solid. Calculation of the susceptibilities of metals must therefore take account of the behavior of these electrons in the Fermi distribution. We discuss both the diamagnetic and paramagnetic contributions of these electrons but delay to a later chapter discussion of the ferromagnetic metals. The free electrons of a metal make both diamagnetic and paramagnetic contributions to the susceptibilities, but since the paramagnetic behavior is larger, the net contribution of the free electrons is paramagnetic.

The *diamagnetic* behavior of the valence electrons is a result of the effect of the external magnetic field on the motion of the electrons. The field produces a force on the moving electrons, causing them to move in spiral paths between collisions, rather than in straight lines. The effect is diamagnetic, but it is not as large as the paramagnetic effect described next.

The *paramagnetic* behavior of the valence electrons can be described in some detail from a relatively simple point of view; hence a more complete discussion of this effect is given. The paramagnetic susceptibility does not obey the Curie law, and the susceptibility is temperature-independent. Furthermore, the susceptibility decreases until it is only about the size of the underlying diamagnetism. The reason is that the wave functions of the valence electrons for the solid are not primarily atomic in character, and each distinguishable momentum state for the valence electron possesses two spin states. Recall that, whenever the density of states occurred in the formulas of Chaps. 9 and 11, a factor of 2 was introduced to account for spin degeneracy. In the presence of a magnetic field the individual spin states must be taken into account in a more detailed manner.

The manner in which the occupation of spin states is altered in the presence of an external field may best be seen by considering the sequence of events in Fig. 20-4. The density of states is plotted in Fig. 20-4a as a function of energy in accord with the free-electron model—$\rho(E) \sim E^{\frac{1}{2}}$. This curve may be split into two parts, one with spin up and one with spin down, as shown in Fig. 20-4b. Without an external field the distribution of the spin-up electrons is precisely the same as that of the spin-down electrons, because each momentum state is twofold degenerate in spin. However, in a magnetic field, the energy of each electron is shifted with respect to its former energy. The amount of shift, ΔE, is given by

$$\Delta E = -\boldsymbol{\mu}_m \cdot \mu_v \mathbf{H} = -|\mu_z|\,\mu_v H \tag{20-26}$$

The magnetic field is assumed to be in the $+z$ direction. The plot of density of states for the spin-up electrons ($s = +\frac{1}{2}$) is shifted to higher energy by the amount $|\mu_z|\,\mu_v H$, while the spin-down density of states curve is shifted down by the same amount. The situation is now as illustrated in Fig. 20-4*c*. Of course, after this shift in energy, the Fermi level for the two spin distributions is not the same, and electrons transfer from the distribution with spin up into the distribution with spin down until the two Fermi energies become equal (Fig. 20-4*d*).

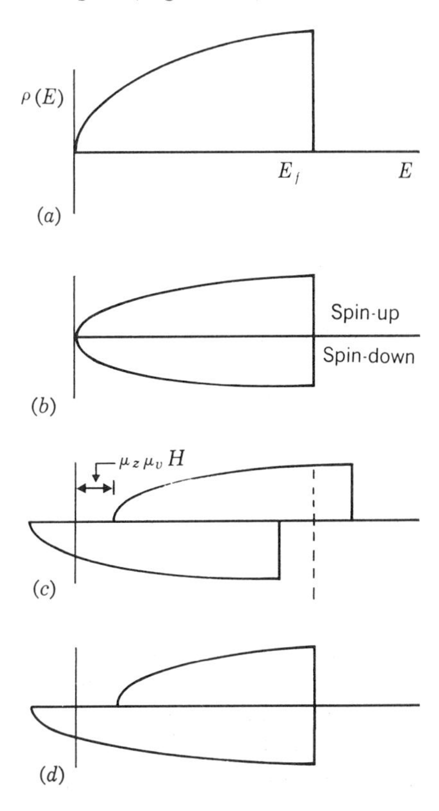

Figure 20-4 The effect of an external field on the number of electrons in the two half bands. (*a*) Density of states for full band; (*b*) density of states for each half band in no field; (*c*) shift of energy levels of each half band because of external field; (*d*) gain in population of one half band relative to the other because of the field.

Before application of the field, exactly as many electrons exist with spin up as with spin down; after the application of the field and the consequent electron transfer, more exist with spin down—hence more electrons exist with magnetic moments in the $+z$ direction.

The macroscopic magnetic moment **M** may be readily calculated. It is given as a product of the number of electrons which change their spin direction and twice the spin moment. This factor of 2 arises because each electron which transfers from one distribution to the other contributes twice its magnetic moment to the total magnetic moment of the system. Thus

$$M = (\text{number of transferred electrons}) \times 2\,|\mu_z|_{\text{spin}} \tag{20-27}$$

The total number of electrons which change their direction is equal to the density of states in one of the spin distributions times the change in energy,

$$\text{Transferred electrons} = \rho_+(E_f)\,\Delta E = \rho_+(E_f)\,|\mu_z|_{\text{spin}}\mu_v H \tag{20-28}$$

In this expression $\rho_+(E_f)$ designates the density of states in the spin-up distribution at the Fermi level, which is one-half the total density of states. The final expression for the total magnetic moment is

$$M = \rho(E_f)\,|\mu_z|^2_{\text{spin}}\mu_v H \tag{20-29}$$

The magnetic moment is thus proportional to the external magnetic field, and so the susceptibility is proportional to the density of states times the square of the spin moment for the electrons. Since nothing in this expression depends on temperature, the susceptibility of the free electrons in a metal is temperature-independent.

The magnitude of this effect is not very large. For a typical monovalent metal, the density of states is about 10^{47} per joule; hence χ has the value 10^{-5}. This value is in fact only slightly greater than the normal diamagnetism of the filled cores. Compare this with the susceptibility which would exist if all the valence electrons were able to respond to the external field. In such a case, the room-temperature susceptibility for the same monovalent metal, as given from Eq. (20-12), is about 10^{-3}. This difference of a factor of 100 to 1,000 is caused by the nature of the valence band in metals. Most of the electrons are not able to align themselves in an arbitrary direction in the external field, because the state of opposite spin is already occupied. Only a small fraction of the electrons near the top of the band are free to change their state of spin.

The expression derived here [Eq. (20-29)] has not been derived in a precisely rigorous way. To carry out such a derivation rigorously, one should use a density of occupied states at a finite temperature, not at absolute zero, since the magnetic excitation ΔE is small compared with kT at normal temperatures. For example, in an imposed field of 100 amp-turns/m, ΔE is only 10^{-5} as large as kT at room temperature. Thus at reasonable temperatures the susceptibility might actually depend on temperature. However, a rigorous derivation which takes account of thermal distributions of electrons at the top of the Fermi distribution gives the result that the Fermi distribution for each spin direction for a given temperature is shifted nearly rigidly, as in Fig. 20-4. Thus the zero-temperature calculation is nearly correct, and the temperature dependence is small at ordinary temperatures.

The overall susceptibility of a normal metal has been shown to be only slightly greater than zero. Possibly, then, some metals might be diamagnetic for special reasons. Although many of the metals are paramagnetic, several are weakly diamagnetic, as Fig. 20-5 shows. The diamagnetic contribution might exceed the paramagnetic contribution most simply if the free-electron paramagnetism were especially small, as when the density of states is low. The metal Be illustrates this phenomenon best. Recall that Be has two $2s$ electrons and that it would be an insulator if the $2p$ state did not overlap the $2s$ state. Overlap does occur, but not by much, and the density of states at the

Fermi level is indeed small. Thus χ (paramagnetic) is small, and the metal is diamagnetic.

The noble metals, like Cu, are diamagnetic for a different reason. Cu has one $4s$ electron. The density of states at the Fermi level is normal for a monovalent metal, which makes paramagnetic behavior of its valence electron quite normal. However, Cu comes at the end of the first long transition period, and it has a newly filled $3d$ shell. Since the radius of the d shell is large, the diamagnetic effect of each

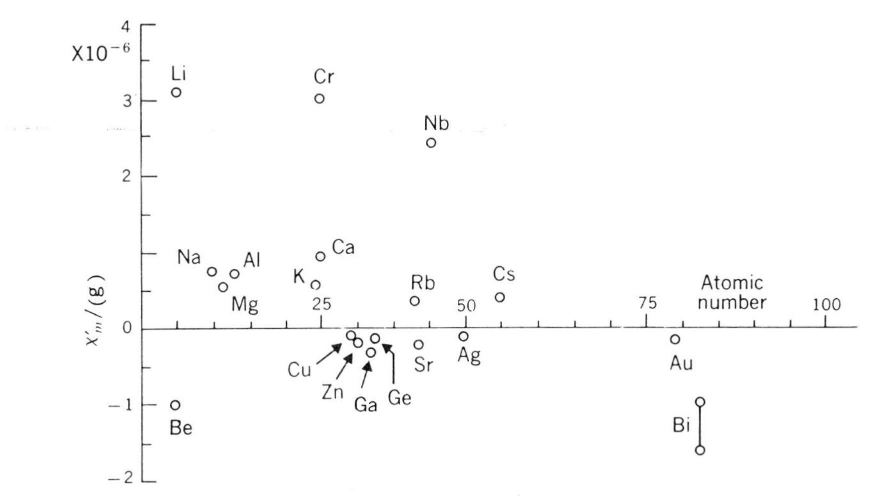

Figure 20-5 Room-temperature susceptibilities of some of the elements in the solid state.

of these $3d$ electrons is large [as Eq. (20-5) shows]. Furthermore, the $3d$ state contains 10 electrons, and so the total diamagnetic behavior is large. These two factors—large orbit of $3d$ electrons and large number of $3d$ electrons—make the diamagnetism of the closed shells of Cu larger than the paramagnetism of the free $4s$ electron. Cu is thus weakly diamagnetic. This effect persists to the metals following Cu, the metals Zn and Ga. Similar reasoning is valid for the other noble metals Ag and Au, which have single $5s$ and $6s$ electrons and large, closed d shells.

Of course, if the bands are either full or empty (as is true for insulators), the solid exhibits diamagnetic behavior. Thus, the semiconductors should be diamagnetic except for the small contribution due to the free electrons which are excited across the energy gap. Insulators like diamond are also diamagnetic for the same reason. Salts like sodium chloride have no free electrons and therefore have no paramagnetism if they are pure. Thus sodium chloride and other similar ionic salts with completely filled shells are diamagnetic.

According to Fig. 20-5, bismuth is more diamagnetic than any other element. The reason for this is unusual. Bi is a metal because empty states lie immediately adjacent to the Fermi level. However, the density of states is again small. In addition, the mass of the electrons at the Fermi level is small, and this small effective mass leads to the interesting effect. Since the effective mass is small, the distortion of the motion of the free electrons by the external magnetic field is large. The resultant diamagnetic susceptibility is larger than the paramagnetic susceptibility of the valence electrons, and so the solid, overall, is diamagnetic. This diamagnetism of the free electrons for Bi has an anomalous temperature effect as compared with the simple diamagnetism of closed shells. Not only is χ_m for Bi dependent on the temperature, but at very low temperatures it changes from large values to small values in an oscillatory fashion as the temperature is lowered, a phenomenon called the de Haas–van Alphen effect (Chap. 9).

PROBLEMS

20-1 The force $d\mathbf{F}$ on a length dl of a wire carrying a current $\mathbf{I}$ in a magnetic field $\mathbf{B}$ is given by

$$d\mathbf{F} = \mathbf{I} \times \mathbf{B}\, dl$$

Show that the torque $\mathbf{T}$ on a flat circular loop of radius r is given by

$$\mathbf{T} = \pi r^2 \mathbf{I} \times \mathbf{B}$$

By defining $\boldsymbol{\mu}_m = \pi r^2 I \mathbf{n}$, where n is the normal of the loop,

$$\mathbf{T} = \boldsymbol{\mu}_m \times \mathbf{B}$$

20-2 Show that the energy of the loop in the field is that given by Eq. (20-6),

$$W = -|\boldsymbol{\mu}_m|\, B \cos \theta$$

20-3 The susceptibility χ_m may be expressed in a variety of units. We use the symbol χ_m for the *volume* susceptibility, i.e., the quantity which gives the magnetic moment per unit volume when multiplied by H. The *mass* susceptibility $\chi_m(kg)$ gives the magnetic moment per kilogram when multiplied by H, and the *atomic* susceptibility $\chi_m(a)$ gives the magnetic moment per kilogram-mole when multiplied by H. The quantity $\chi_m = \chi_m(kg) \times$ density and $\chi_m(a) = \chi_m(kg) \times$ molecular weight in kilograms. The same units in the cgs emu system we designate χ'_m, $\chi'_m(g)$, and $\chi'_m(a)$, with corresponding relationships between them.

The quantity $\chi'_m(g)$ for Ni at 420°C is 177×10^{-6} per gram. Convert this number into the other five units.

20-4 Ge has a volume susceptibility of -0.8×10^{-5}. Calculate B for Ge in a field H of 10^5 amp-turns/m.

20-5 The O_2 molecule has a moment of 2.8 Bohr magnetons. (*a*) Calculate χ_m at room temperature for 1 m^3 of oxygen at 2,000 psi (the normal pressure of a tank of oxygen gas). (*b*) Calculate the magnetic moment per cubic meter of oxygen at this pressure in earth's magnetic field.

20-6 If a piece of Nb metal is placed in a field of 10^6 amp/m, find $\mathbf{M}$ and $\mathbf{B}$. (From Fig. 20-5 $\chi'_m(g) \approx 2.3 \times 10^{-6}$.) By what fraction is $\mathbf{B}$ changed by the presence of Nb over the vacuum value for $H = 10^6$ amp/m?

20-7 Make the same calculation for Ge for which

$$\chi'_m(g) \approx -0.1 \times 10^{-6}.$$

21

Ferromagnetism-physical basis

21-1 General Introduction

Some substances have a permanent magnetic moment even in the absence of an external magnetic field and are termed *ferromagnetic*. They have been recognized as interesting materials for centuries. The early knowledge of the lodestone led to the use of the magnetic compass as a navigational aid, the first use of magnetism in a practical way; it has been followed by an enormous expansion in usefulness. This growth in utility has occurred in spite of the fact that only a relatively small number of elements are ferromagnetic. Until recently, the only significant ferromagnetic materials in practical use were metals and alloys. Since 1940, however, use of nonmetallic ferromagnets has increased greatly for special purposes where their high electrical resistivity is important.

Two nearly separate lines of thought have developed in the understanding of ferromagnetism. One involves the attempt to understand what ferromagnetism is and why it exists: the *physical basis* of ferromagnetism. This basis, we shall see, is intimately connected with the electronic and band structure of the solids. The second aspect of knowledge of ferromagnets is related more directly to the practical use of magnetic materials and requires an understanding of the detailed way in which a ferromagnetic material behaves in an external field. This area of study has been termed *domain theory*. It is of great practical importance, since almost all use of ferromagnets concerns the behavior of magnetic domains in varying fields. These two topics are taken up in succession. (1) In this chapter attention is paid to the atomic nature of the physical basis of ferromagnetism. (2) In the next chapter the important features of domain theory are considered.

21-2 Properties of the Pure Ferromagnetic Materials

Only a few of the elements are ferromagnetic, Fe, Co, Ni, Gd, and several other of the rare earths. Since 1900 many experimental data concerning their magnetic behavior have become available to serve as a basis for understanding ferromagnetism. However, even with the development of band theory, the theory of ferromagnetism is not yet entirely satisfying. This difficulty is connected with the fact that the band structure of the transition elements is also, at this writing, not yet carried to the level of sophistication reached for the simpler metals. Ferromagnetism is due to the mutual self-alignment of groups of atoms carrying permanent magnetic moments in the same direction. These elementary permanent moments are those also responsible for paramagnetism (Sec. 20-5); hence the new aspect of a discussion of ferromagnetism is that of describing why the atomic moments should be self-aligning without the help of an outside field.

Table 21-1 Data for principal ferromagnetic elements

Element	Electronic configuration of free atom	Crystal type	Atomic radius, A	M (at 0°K), amp-turns/m	Curie temperature T_c, °K	Melting temperature T_m, °K
Fe	$3d^6 4s^2$	bcc	1.24	1.69×10^6	1043	1808
Co	$3d^7 4s^2$	hcp	1.25	1.36	1404	1753
Ni	$3d^8 4s^2$	fcc	1.25	0.47	631	1728
Gd	$4f^7 5d^1 6s^2$	hcp	1.78	5.66	289	1585

Before a study of the basis of the ferromagnetic interaction is begun, a review of some of the physical characteristics of the ferromagnetic elements is worthwhile. Some of these data are presented in Table 21-1. For each element the following types of information are listed:

1. Electronic structure of the free atom. Note that the inner rare-gas core is omitted for each.
2. The crystal type (at room temperature).
3. The atomic radius of the atom in the solid form.
4. The saturation magnetization at 0°K.
5. The Curie temperature, the maximum temperature for which ferromagnetism exists. Above the Curie temperature, the elements are paramagnetic.
6. The melting temperature.

An examination of these data shows that ferromagnetism is not exclusively characteristic of any particular crystal type, since all the simple metallic crystal types are represented. The elements have a wide variety of Curie temperatures and a considerable range in magnetic moment. The one common feature is the electronic structure. Three of these elements are found in the $3d$ transition group, the other in the $4f$ rare-earth transition group. The existence of partly filled d or f shells is essential in modern theories of ferromagnetism.

21-3 Origin of the Ferromagnetic State

Insight into the physical nature of ferromagnetism can be gained by considering in detail the mechanism by which the atomic moments in metals arise and how they couple in ferromagnets to produce a net

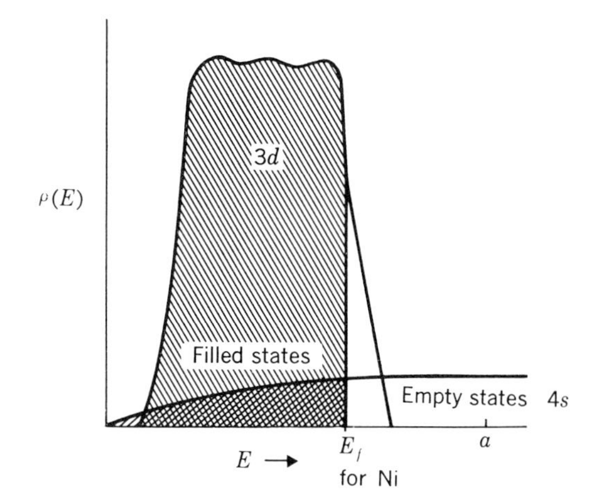

Figure 21-1 Schematic of the band structure of Ni. The $4s$ and $3d$ bands overlap completely. The ten $3d + 4s$ electrons of Ni fill the $4s$ and $3d$ states up to a level just below the top of the $3d$ band. For Cu, the states are filled to some level "a".

spontaneous magnetization. In particular, the three elements Fe, Co, and Ni are considered. These three elements have closed shells for the $1s$, $2s$, $2p$, $3s$, and $3p$ levels. Since a closed shell has no magnetic moment, clearly the 18 inner electrons do not participate in the magnetization. Attention must therefore be given to the $3d$ and $4s$ levels, which are split into bands by the overlap of the wave functions in the metal. Recall that the $4s$ splitting is considerable and the $3d$ somewhat less (see the density-of-states curve for a typical transition metal in Fig. 21-1). Since all states up to the Fermi level are full, the ten $3d + 4s$ electrons of Ni, for example, fill the states to the approximate position shown. Observe the fact that the density of states at the Fermi level for Ni is high; it is also high for Co and Fe, with nine and eight electrons, respectively, in these bands. For Cu, on the other hand, the eleven $3d + 4s$ electrons fill these levels to a point a in Fig. 21-1. To see how this configuration results in ferromagnetism for Fe, Co, and Ni, the energy of the electrons in these bands must be considered in more detail.

A summary of the paramagnetism of the valence electrons as discussed in Chap. 20 is helpful here. Recall that this paramagnetism has its origin in the interaction of the valence electrons with an external field. A lowering of the energy occurs for those electrons with spin moments parallel to the field compared with those with spin moments antiparallel to the field. This energy difference results in a transfer of some electrons from unfavorable to favorable states so that the Fermi energy of the two half bands is the same. Recall also that those electrons which do transfer actually gain kinetic energy, a gain offset by the lowering of the magnetic energy. The net magnetic moment of the resultant electron-spin distribution arises because the

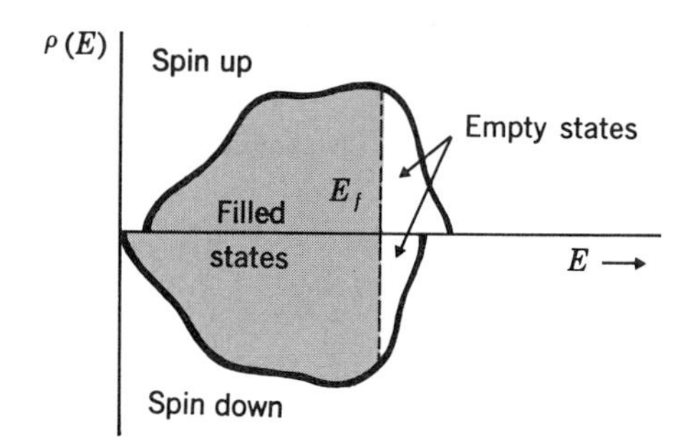

Figure 21-2 Density-of-states plot for a ferromagnetic solid for which the half bands are unequally filled because of a shift in energy between them.

spin moments of the two half bands no longer cancel, as they do in the absence of a field.

What has this to do with ferromagnetism? Simply that in a ferromagnetic material below the Curie temperature transfer of electrons must occur from one spin state to the other. That half band with the larger number of electrons has a spin moment greater than that of the other; hence the metal has a net magnetic moment. The electron transfer is sketched in Fig. 21-2. Electrons have transferred from the spin-up configuration to the spin-down configuration; hence a net magnetic moment exists in the up direction (recall that μ_m and $\mathbf{s}$ are antiparallel).

The important feature of ferromagnetism is that no external field is needed to effect the electron transfer sketched in Fig. 21-2: the transfer is *spontaneous*. Thus the magnetization itself is spontaneous. Understanding why ferromagnetism occurs is thus reduced to understanding why the electron transfer occurs. For this purpose consideration must be given to the energy change involved in producing an unbalance in population of the two half bands.

Several aspects of the energy balance of the ferromagnetic state need to be considered. One part of this energy is the kinetic energy of the electrons. Since the transfer is unlikely to change appreciably the density-of-states curve in either half band, clearly a transfer of electrons from the one half band to the other must involve a net increase in the kinetic energy of the $3d + 4s$ electrons. For the transfer process to occur spontaneously, then, a corresponding decrease of some other energy term must arise when transfer occurs. The source of this energy decrease is generally thought to be the exchange interaction between neighboring $3d$ shells, which lowers the energy when their spins are aligned. The balance between the populations of the two half bands, then, is fixed by the energy balance between these two terms; the electron transfer will cease either when one half band is filled or when the increase in kinetic energy of an electron is just equal to the decrease in alignment energy.

The first question concerns the nature of the $3d$ band in Fe, Ni, and Co. It is narrow compared with the s and p bands. Hence the translational kinetic energy of an electron in the $3d$ band is relatively small, since the states have more the character of the atomic states than do those of the s and p bands. The $3d$ electrons are thus more localized on the atoms than those of the s and p bands. One can thus speak with reasonable accuracy about interactions between the $3d$ electrons on adjacent atoms.

The relative narrowness of the $3d$ band has other important features: (1) the Fermi kinetic energy is small, and (2) the density of states at the Fermi level is relatively large. Thus a large number of electrons can be transferred from one half band to the other with a rather small overall increase in kinetic energy. Hence for a given magnitude of the exchange energy a large number of electrons can change their spin direction.

The direct interaction which is responsible for ferromagnetism is the exchange interaction between the electrons (Chap. 14). Complete calculations of this energy are difficult to make, for the reason explained above, that the bands of these transition solids have not yet been adequately worked out.

Some insight into the origin of the exchange energy can be gained by estimating the magnitude of the energy involved in the interaction, for example, by calculating the thermal energy necessary to overwhelm the magnetic alignment. At the Curie temperature of iron, for example, where the spontaneous alignment disappears, the thermal energy per atom kT_c is about 0.1 ev. The strength of the alignment interaction, the exchange energy per atom, must then be about this large.

This estimate of the strength of the magnetic interaction allows one to show easily that the interaction causing the magnetic alignment is not due to the magnetic interaction between dipoles and must be due to some stronger effect such as the exchange interaction. The energy E which aligns one dipole $\boldsymbol{\mu}_1$ in the field of another $\boldsymbol{\mu}_2$ is given by $E \approx \mu_v |\mu_1| |\mu_2|/r^3$, where r is the separation between the two. This energy for moments of about one Bohr magneton is of order 0.001 ev for appropriate nearest-neighbor distances in solids. Thus the direct dipole effect is at least two orders of magnitude too weak to account for the observed strength of the aligning interaction of about 0.1 ev. It must be concluded that the interaction energy arises from electronic interactions of a different sort.

The exchange interaction responsible for ferromagnetism cannot be described by classical models; it is purely a quantum-mechanical effect, as was described in Chap. 14. Detailed calculations of the interaction energy for an element as complicated as Ni ($Z = 27$) cannot be made except with great approximations and assumptions. Many years ago, Heisenberg suggested that ferromagnetism in solids is similar in character to two hydrogen atoms which bond to form a hydrogen molecule. In this simple case the two electrons cannot be assigned to a particular proton, but belong to the molecule as a whole. The bonding configuration is found to be one in which the spins of the two electrons are antiparallel (this must be so since the hydrogen molecule has no permanent magnetic moment). The exchange energy in this case is calculable for the antiparallel spins and is in fact of such a sign as to produce bonding of the H_2 molecule. The case for the multielectron ferromagnetic elements is thought to be similar, except that the exchange energy is lower in the case for *parallel* spins on adjacent (and hence on all) ions. An adequate theory of ferromagnetism must calculate the exchange interaction in the metal on the basis of a realistic d-band configuration and must also include other effects in the tendency to align or not to align, such as the simple Coulomb interaction between the electrons.

The preceding discussion is concerned entirely with electron spin and its contributions to the magnetic moment. The role of orbital angular momentum and its attendant *orbital magnetic moment* was not considered, because experiments on all ferromagnetic materials imply that the orbital contributions are not very important. The spin moments account for at least 90 per cent of the total magnetization in ferromagnetic materials.

21-4 The Temperature Dependence of the Magnetization

Thermal motion of the atoms affects ferromagnetic properties in several ways. One of the most important is the effect on the degree of magnetization. This effect is most pronounced in the vicinity of the Curie temperature, but it is also observable far below the Curie temperature.

The spontaneous magnetization of ferromagnets is caused, we have seen, by the interaction between neighboring atoms which tends to align their spins. When this effect is so strong that all adjacent spins are aligned, the magnetization of the material has its maximum value. This long-range alignment results from both the strong nearest-neighbor interactions and the continuity of the crystal. Thermal vibrations of the atoms, however, tend to misalign the spins. Hence the maximum magnetization (that which is characteristic of complete alignment of every spin) is observed only at the lowest possible temperature, i.e., absolute zero. At every higher temperature the magnetization has a lower value, until it becomes zero at the Curie temperature.

The magnetization as a function of temperature for iron and nickel is shown in Fig. 21-3. The magnetization decreases very slowly as the temperature is first raised above absolute zero. The curve drops more steeply at higher temperatures, until it finally falls precipitously to zero at the Curie temperature.

Above the Curie temperature the ferromagnetic solids exhibit paramagnetism. They have a large susceptibility just above T_c, but they do not all obey a Curie-Weiss law. Ni obeys a Curie-Weiss law

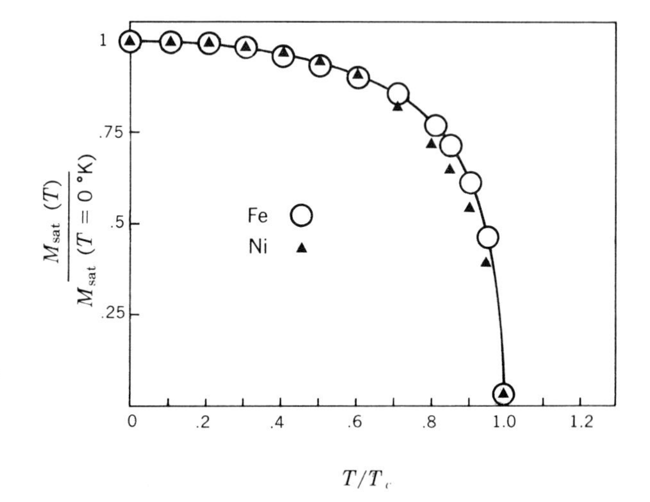

Figure 21-3 The relative magnetization $M_{sat}(T)/M_{sat}$ ($T = 0°K$) as a function of T/T_c. With these reduced coordinates, data for Fe, Ni, and Co plot on the same graph. Data are shown for Fe and Ni only.

best, but just above its Curie temperature its susceptibility deviates somewhat from exact agreement with this law (see Fig. 21-4). This behavior just above T_c has an interesting interpretation. It shows that, although thermal motion of the atoms has destroyed long-range order of the spin moments, some spin order of a much weaker character still persists. It is a kind of short-range spin order in which a given atom is surrounded by a small island in which the spins are more or less aligned. This phenomenon presumably exists for all ferromagnetic solids just above T_c.

The Curie-Weiss law was given in the previous chapter without justification. It may be thought of as a Curie law with the zero of temperature pegged at the Curie temperature. In the Langevin formula, the solid becomes completely aligned in the field at zero temperature, while, for the ferromagnet, this alignment occurs, of course, at T_c. Hence thermal fluctuations above T_c in a ferromagnet should play much the same role as they do in an ordinary paramagnet above $T = 0$.

As has been pointed out earlier, the energy of magnetization is about equal to the thermal energy required to destroy the spontaneous magnetization. Hence for Fe, Co, and Ni, for which T_c is of order 1,000°K, the energy of magnetization is of order 0.1 ev per atom or

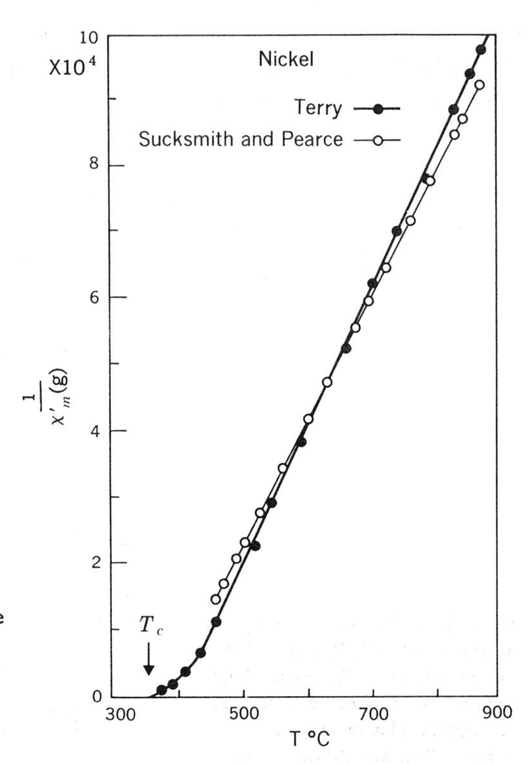

Figure 21-4 Susceptibility of Ni in the paramagnetic region above the Curie temperature obeys a Curie-Weiss law, with slight deviations just above T_c. Fe and Co do not obey this law as well. [*Measurements by E. Terry, Phys. Rev.,* **9:** 394 (1917), *and by W. Sucksmith and R. Pearce, Proc. Roy. Soc. (London),* **A167:** 180 (1938).]

about 2,000 cal/mole. Since this energy must be supplied to the lattice as the temperature of a ferromagnetic solid is raised from absolute zero, the heat capacity of these solids is higher than that of a nonferromagnetic solid. A heat-capacity curve for a typical ferromagnetic solid is sketched in Fig. 21-5. Although some of the energy required to misalign spins must be supplied at all temperatures from 0°K to T_c (actually slightly above), most of this energy must be supplied just below T_c, where the magnetization decreases most rapidly (see Fig. 21-3). Hence the heat capacity is peaked at T_c. Compare this curve with that for a nonferromagnetic metal (Fig. 11-2).

21-5 The Effect of Alloying

The average number of Bohr magnetons, N_B, per atom for the pure ferromagnets may be calculated from the experimentally determined values of saturation magnetization quoted in Table 21-1. Interestingly enough, none is an integer. From the point of view of band theory, no reason exists to assume that they should be; from the view of the electrons as being localized on atoms, the actual moment must fluctuate with time, being on the average the value quoted for N_B. For values of N_B, see Table 21-2.

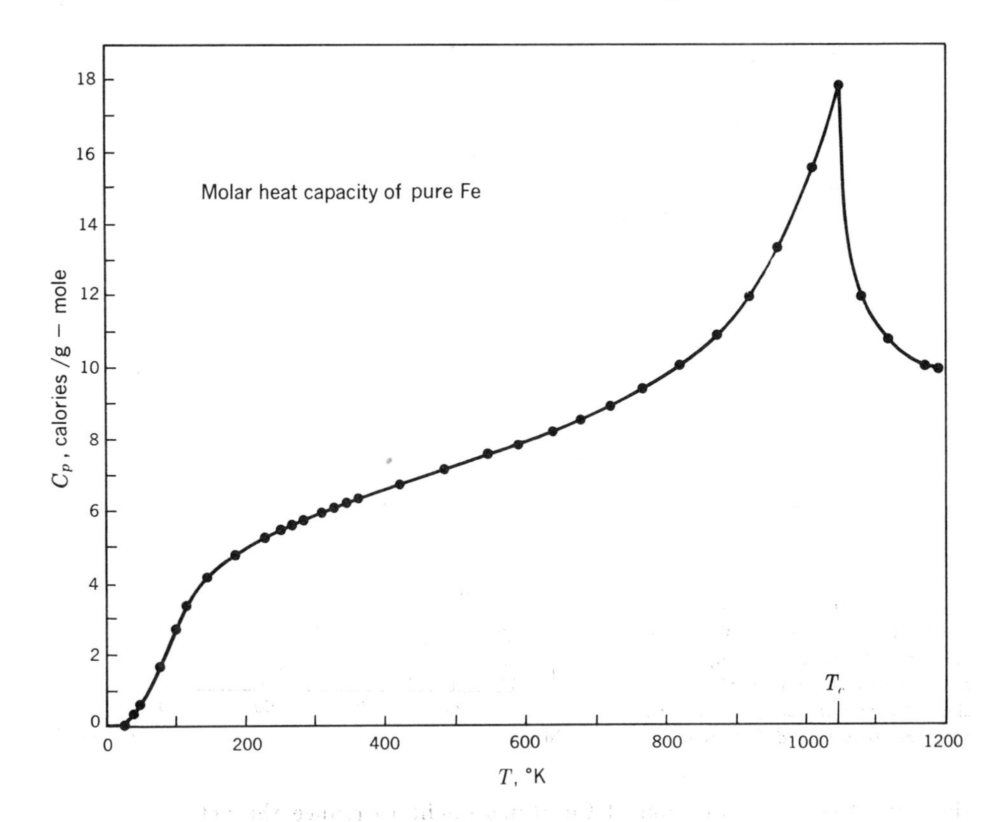

Figure 21-5 The heat capacity of a ferromagnetic element has a large peak at its Curie temperature. This peak is caused by the extra energy needed to misalign the magnetic moments responsible for ferromagnetism. The data shown here are for iron. [*Measurements from J. Austin, J. Ind. Eng. Chem.*, **24:** 1225 (1932).]

Alloying a ferromagnetic material with another element causes changes in the magnetic properties. One important change is that of the saturation magnetization; hence, with alloying, the number of Bohr magnetons averaged over the entire number of lattice atoms changes.

The way in which alloying affects the saturation magnetization depends in part on the type of structure formed upon alloying; perhaps the simplest structure is the solid solution. Of these, the Ni-Cu solution has some of the most easily interpreted features. Ni has

Table 21-2 N_B for the principal ferromagnetic elements

Metal	N_B
Fe	2.22
Co	1.72
Ni	0.60
Gd	7.12

ten $3d + 4s$ electrons which are thought to fill the overlapping d and s bands in such a way that one of the half bands of the $3d$ state is completely full and the other lacks 0.6 of an electron per atom of being filled. When an atom of Cu is substituted for an atom of Ni in the solid solution, eleven $3d + 4s$ electrons are substituted for the original ten; i.e., the solid has gained an extra electron by this substitution. This electron must go into the partially empty half band, filling it

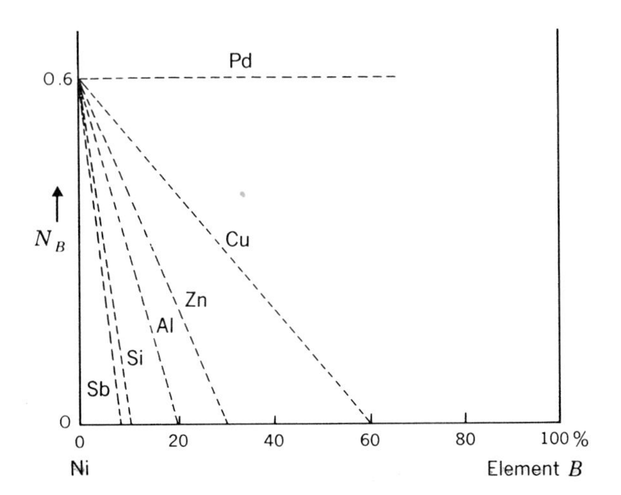

Figure 21-6 Alloying Ni with elements of different electronic structure changes N_B for the alloy according to the number of outer-shell electrons the alloying element contributes to the $3d + 4s$ band.

slightly. Hence the addition of Cu atoms ought to reduce the net magnetization. In fact, when Cu atoms have been substituted for 60 per cent of the Ni atoms, both half bands ought to be just full and the magnetization should have dropped to zero. These changes occur, as the curve marked Cu in Fig. 21-6 shows. Adding divalent Zn causes the decrease to occur more rapidly, trivalent Al even more rapidly, etc. Pd, which has an electronic structure in the $4d$ series analogous to that of Ni in the $3d$, causes no change in N_B.

Similar changes for other alloys of Co, Fe, and Ni with each other and with other metals of the $3d$ transition series have been measured. Some of these data are shown in Fig. 21-7, where N_B is plotted as a function of the number of electrons per atom. The data presented here fall rather nicely on a smooth curve for alloys of Ni-Cu and Ni-Zn on the high side to Fe-V and Fe-Cr on the low. If Ni is assumed to be the norm, values of N_B for the Ni-Fe and Ni-Co alloys rise linearly toward the value for pure Co. The value $N_B = 1.7$ for Co is consistent with the fact that it has one electron per atom fewer than Ni; hence its partially filled half band has one electron per atom fewer than Ni. This trend continues linearly toward the electron per atom ratio of Fe, which ought to have a value $N_B = 2.6$. But Fe does not have such a high value. In fact, at an electron-to-atom ratio corresponding to about 80 per cent Fe, 20 per cent Co, the trend reverses, and N_B *falls off* linearly as the electron-to-atom ratio is reduced further.

The existence of a maximum implies that the unbalance between the two half bands can be no greater than 2.4 electrons per atom for the 3*d* ferromagnetic alloys. This number represents the optimum balance which can be struck by the tendency to ferromagnetic alignment of the 3*d* electrons and is a number which cannot be predicted by present theories.

Considerable additional data of the type portrayed in Fig. 21-7 are available for other alloy systems. Some of these data fit the rules cited above, while other data deviate from them. For the alloy systems cited in Fig. 21-7, the number of $3d + 4s$ electrons for the constituent metals does not differ by more than one or two electrons.

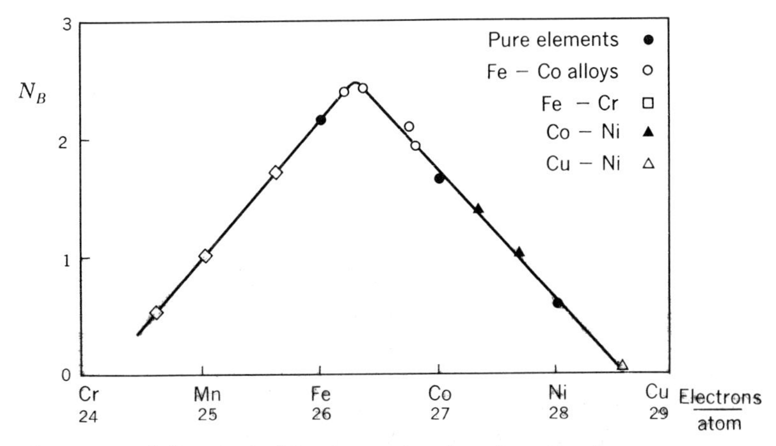

Figure 21-7 The atomic moment of alloys in the 3*d* series as a function of number of electrons per atom.

For those alloys where the difference is larger—say, three or four electrons—N_B departs from the curve. No completely satisfactory reasons can be given at present for this complication.

21-6 Antiferromagnetism

The electrostatic exchange interaction, which is the physical basis for ferromagnetism, was presumed to produce in ferromagnets a parallel alignment of adjacent spins. However, for some solids, the exchange interaction produces an antiparallel alignment of spins. One of the most striking examples is the compound MnO, in which the spins of the Mn ions are arranged in antiparallel or antiferromagnetic fashion. The magnetic structure of MnO is sketched in Fig. 21-8.

The crystal structure of this material is a simple one. Both the Mn ions and the oxygen ions occupy an fcc lattice. These two lattices are combined in such a way in MnO that the metal ions have only oxygen ions as nearest neighbors, and vice versa (this structure is the NaCl structure). The magnetic structure of the metal ions has the

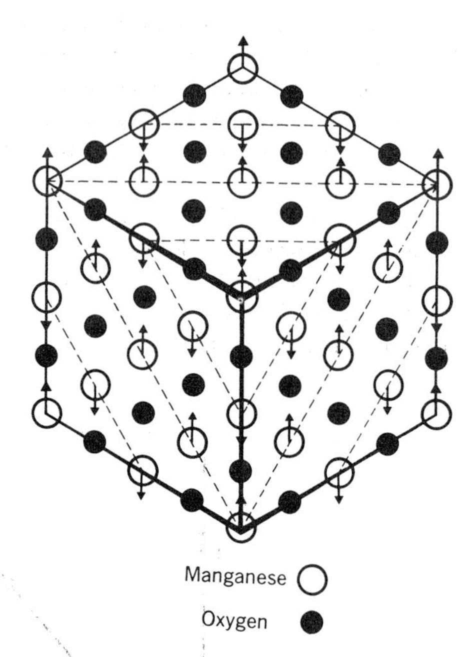

Figure 21-8 Crystal structure of MnO, showing magnetic moments on individual Mn ions.

attribute that ions in (111) planes (of the fcc Mn lattice) have parallel spins but that adjacent sheets have antiparallel spins. Clearly the moments of ions in the two directions cancel, and the solid as a whole has no net moment.

This magnetic structure, too, is affected by thermal vibration of the atoms. At low temperatures the thermal vibration is so small that the moments of the Mn ions are all antiferromagnetically aligned. As the temperature is increased, thermal motion tends to lessen the antiparallel ordering, until finally the ordered spin arrangement breaks down completely. The breakdown occurs abruptly, similarly to the breakdown of ferromagnetic order. The critical temperature for the antiferromagnetic crystals is called the *Néel temperature*.

Antiferromagnetic crystals display a positive magnetic susceptibility at all temperatures. Above the Néel temperature, they have a normal paramagnetic behavior and obey the Curie-Weiss law. Below the Néel temperature, the moments of the ions are not free to follow the field, and so the susceptibility does not obey the Curie-Weiss law (though it is positive). At absolute zero the susceptibility arising from the spins is zero.

21-7 Ferrites

One special group of compounds merit special consideration: a class of oxides called *ferrites*. They are ionic crystals of chemical composition $MeFe_2O_4$ (where the symbol Me means any one of a number of metal ions). The most common of this group is the ordinary Fe_3O_4, the lodestone of antiquity.

The chemistry of the compound is easy to deduce. Oxygen has the normal valence of 2; so the electron balance may be expressed by the formula $(Fe^{++}O^{=})\ (Fe_2^{3+}O_3^{=})$. Two kinds of iron ions are in the material, doubly charged ferrous ions and triply charged ferric ions. X-ray studies show that the unit cell contains altogether 56 atoms, 32 oxygen atoms and 24 iron atoms. Of the latter 8 are Fe^{++} and 16 are Fe^{3+}.

The permanent moment of this compound has its origin in the arrangement of the 24 iron ions in the unit cell. Additional X-ray evidence shows that the iron ions occupy two types of positions in the unit cell. In one type, the iron ion has 4 oxygen ions about it in tetrahedral coordination. The unit cell contains 8 positions of this type. In the other type, the iron ion has 6 oxygen atoms arranged about it in octahedral symmetry; the unit cell contains 16 of these positions. The 16 Fe^{3+} ions are in both types of position, 8 in the tetrahedral sites and 8 in half of the octahedral sites. The 8 Fe^{++} ions occupy the remaining octahedral positions.

The crucial point is the arrangement of net spin moments on these 24 iron ions. The spin moments of the 8 Fe^{3+} ions on the tetrahedral sites are directed oppositely to those of the 8 Fe^{3+} ions on the octahedral sites. These ions have their spin moments arranged antiferromagnetically. Hence these moments cancel, and the Fe^{3+} ions contribute no net moment to the solid as a whole. However, the 8 Fe^{++} ions have their net moments aligned in the same direction, and they give the net moment to the solid. Solids which have a net moment because of incomplete cancellation of antiferromagnetically arranged spin systems are called *ferrimagnets*.

The saturation magnetization of Fe_3O_4 is not difficult to calculate. Each Fe^{++} ion has a net moment of 4 Bohr magnetons. (The Fe^{++} ion has 24 electrons, 18 in the closed argon core and 6 in the 3*d* shell. These 6 spins are arranged $\uparrow\uparrow\uparrow\uparrow\uparrow\downarrow$ so that the ion as a whole has 4 uncompensated spins in the 3*d* shell, hence a net moment of 4 Bohr magnetons.) The unit-cube edge of Fe_3O_4, as determined by X-ray analysis, is 8.37 A. Hence the solid has $4 \times 8/(8.37 \times 10^{-10})^3$ net Bohr magnetons per cubic meter. Thus $M = 0.5 \times 10^6$ amp-turns/m. The measured value is 0.48×10^6 amp-turns/m.

Other metal ions may be substituted for some of the Fe ions in the compound. For example, the compound $NiFe_2O_4$ exists; here the Ni ions are Ni^{++} and occupy the same positions on the lattice as did the Fe^{++} ions. The moments of the Ni^{++} ions are all aligned. Since the Ni^{++} ion has eight electrons in the 3*d* shell, the net moment of this ferrite is only about 0.25×10^6 amp-turns/m. Many other ferrites of varying composition have been made, with such metals as Mn, Li, Zn, Cu, Mg, Co substituted for part of the iron ions.

The detailed crystallography of the ferrites has been passed over

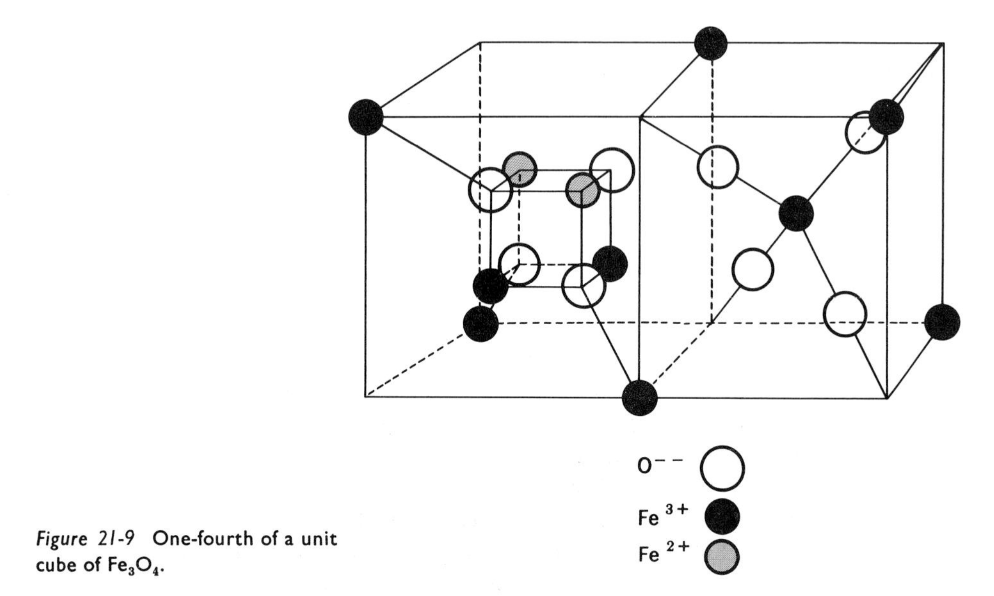

Figure 21-9 One-fourth of a unit cube of Fe_3O_4.

rather quickly so that the end result could be seen quickly. The crystal structure itself is called the inverted *spinel* structure (spinel is the compound $MgAl_2O_4$, a gem stone). The positions of the entire 56 ions in the unit cube of a typical ferrite with this structure can be shown only with difficulty, but the typical lattice positions in a fourth of a unit cube can be pictured. They are shown in Fig. 21-9 for Fe_3O_4. Note the spatial arrangement of the Fe^{3+} and Fe^{++} ions.

PROBLEMS

21-1 Over the temperature region shown in Fig. 21-4, the susceptibility of Ni fits a Curie-Weiss law except for the temperature region just above T_c. From the straight portion of the curve find the constant C' of Eq. (20-14) and, from it, the effective number of Bohr magnetons per atom of Ni.

21-2 Suppose that Ni were paramagnetic at all temperatures and that it obeyed a Curie-Weiss law with a critical temperature of 0°K. Further suppose that the approximation of Eq. (20-12) were valid to near saturation. Calculate the external magnetic field H necessary to produce at room temperature in this paramagnetic metal the same state of magnetization, about 0.44×10^6 amp-turns/m, that is spontaneously present at room temperature in ferromagnetic nickel. Is such a field attainable?

21-3 Values of $1/\chi_m'(g)$ for cobalt above the ferromagnetic Curie temperature are tabulated below. Show that the susceptibility obeys a Curie-Weiss law. Find the effective number of Bohr magnetons per atom. (Measurements of Sucksmith and Pearce.) $\chi_m'(g)$ is defined in Prob. 20-3.

Temperature, °C	$1/\chi'_m(g)$	Temperature, °C	$1/\chi'_m(g)$
1156	1.74×10^3	1247	5.49×10^3
1168	2.19	1288	7.46
1191	3.02	1370	11.50
1204	3.51	1397	12.75
1218	4.18	1440	14.75

21-4 Values of $1/\chi'_m(g)$ for bcc iron just above its ferromagnetic Curie temperature are tabulated below. Show that the susceptibility obeys a Curie-Weiss law over this temperature region. Find the value of θ and the effective number of Bohr magnetons per atom. (Measurements of Sucksmith and Pearce.)

Temperature, °C	$1/\chi'_m(g)$	Temperature, °C	$1/\chi'_m(g)$
824	1.125×10^3	884	3.390×10^3
838	1.650	894	3.680
846	2.430	904	3.970
872	2.860	909	4.260
875	3.030		

21-5 The Mn ion in $MnFe_2O_4$ replaces the Fe^{++} ion in the Fe_3O_4 crystal with almost no change in cube edge. Calculate the saturation magnetization of the manganese ferrite.

22

Ferromagnetic domains

22·1 General

The aspects of ferromagnetism covered in the last chapter are those which characterize a solid magnetized to saturation. However, most applications of ferromagnets have to do, not with saturation, but with the changes in net moment which occur when the external field is changed. Hence we must try to understand how the overall magnetization of a ferromagnetic solid can change without violating the principles discussed in Chap. 21.

A large difference exists between the magnetization in a given field of a paramagnetic solid and that of a good ferromagnet. For a paramagnetic salt, say, $CrCl_3$, the magnetic susceptibility is 1.5×10^{-3}. In a small field of 2 amp-turns/m, the magnetization $\mathbf{M}$ is 3×10^{-3} amp-turns/m. However, for a properly prepared ferromagnetic alloy of iron and nickel, the magnetization may change from zero in no field to a magnetization of order 10^6 amp-turns/m in the same field of 2 amp-turns/m. The change is a factor of order 10^8 over the paramagnetic solid. In all its quantitative details, an explanation of this behavior is very difficult, but fortunately it can be understood in a qualitative sense much more easily.

DOMAIN STRUCTURE

22·2 Principles of Domain Geometry of Ferromagnetic Solids

The basic problem is that of describing how the magnetization of a ferromagnetic solid can change from near zero in *no* external field to its saturation value in a *weak* external field (say, a field no larger than the earth's magnetic field). The complete answer to this problem clearly involves a description which would appear to be complicated by the strong interactions between atoms. The interaction forces are so strong, in fact, that they actually simplify the solution to the problem.

A ferromagnetic solid is subdivided into many regions called *domains*, each of which is magnetized to saturation; i.e., all the moments within each domain are aligned in some direction. The magnetization of the solid as a whole is the vector sum of the magnetizations of each of the domains, the contribution of each being weighted by its fraction of the total volume. The net magnetization of the solid thus

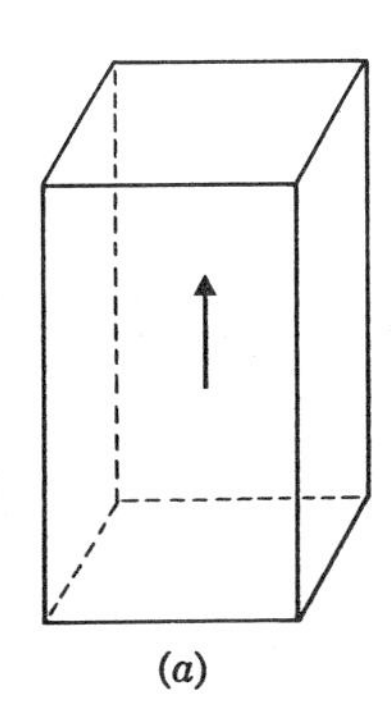

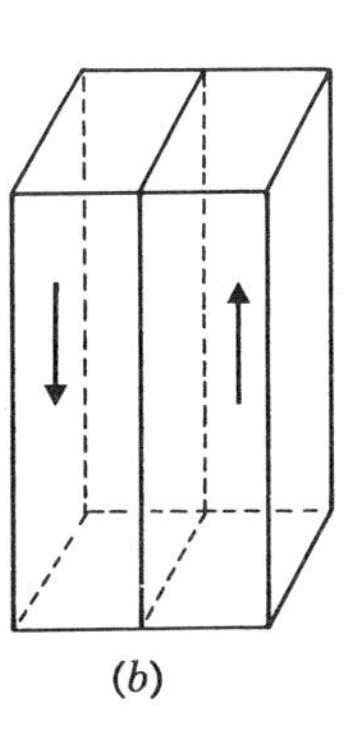

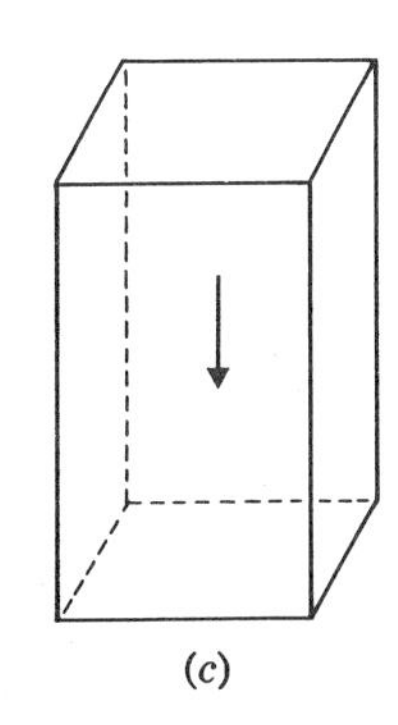

Figure 22-1 **Change in net magnetization of a solid, accompanying change in domain structure. (*a*) Net magnetization up; (*b*) net magnetization zero; (*c*) net magnetization down.**

ranges from zero, if the weighted vector sum is zero, to a maximum, if the solid is a single domain with all atomic moments aligned in a given direction.

Domain configurations which produce any desired state of magnetization in a solid up to the saturation value can easily be visualized. For example, three simple domain structures which produce quite different states of magnetization are sketched in Fig. 22-1. Intermediate stages of magnetization can be produced by causing either one or the other of the domains in (*b*) to be the larger.

Not all domain structures which might be visualized for a given degree of net magnetization are formed. The solid assumes structures which have the lowest energy, and considerable insight into the various contributions to the total energy is necessary to see which is the proper choice. For example, three configurations for which a bar (say, 1 cm on an edge) has zero magnetic moment are those of Fig. 22-2. Calculation shows that for zero net magnetization the total energy of (*a*) is greatest, that of (*b*) somewhat lower, and that of (*c*) lower yet. The pie-shaped domains, called *closure domains*, complete the flux path within the solid and cause poles not to be formed on the top and bottom surfaces. For most ferromagnetic materials of reasonable size (say, 1 cm in linear dimensions) the structure (*c*) is far from being optimum, and the division of the volume into even smaller domains reduces the energy further until an optimum size occurs, when some

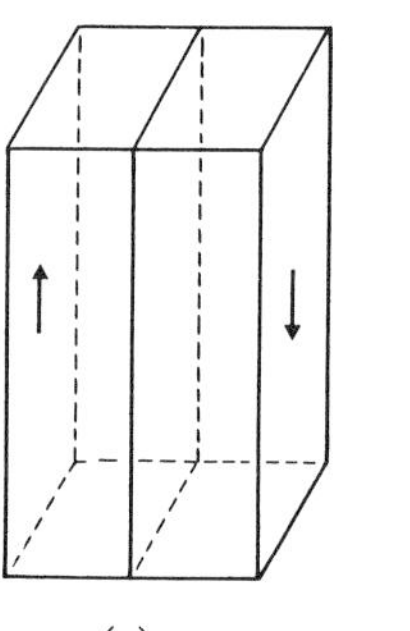

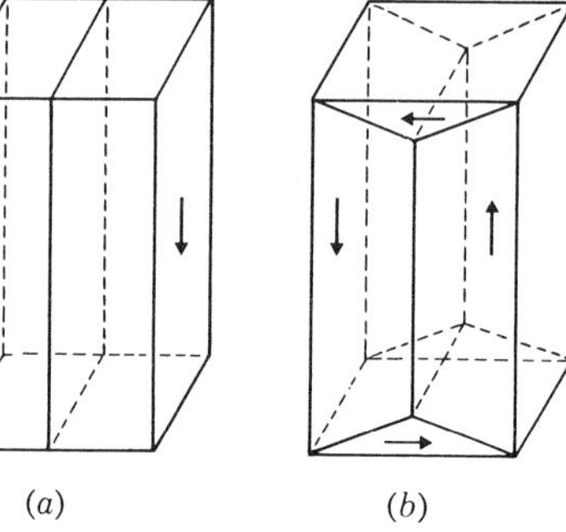

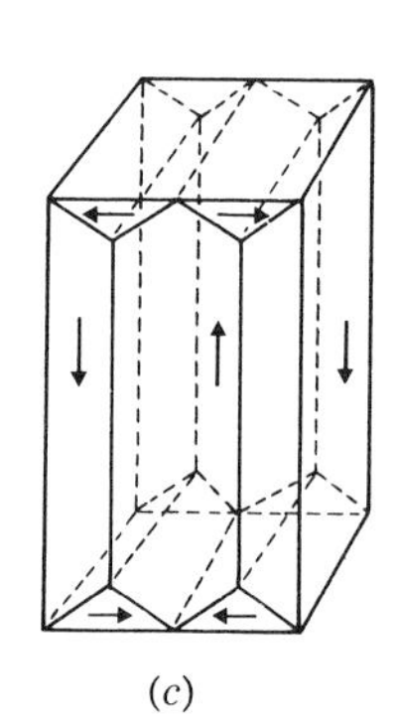

Figure 22-2 Several domain structures of a solid, each having zero net magnetization.

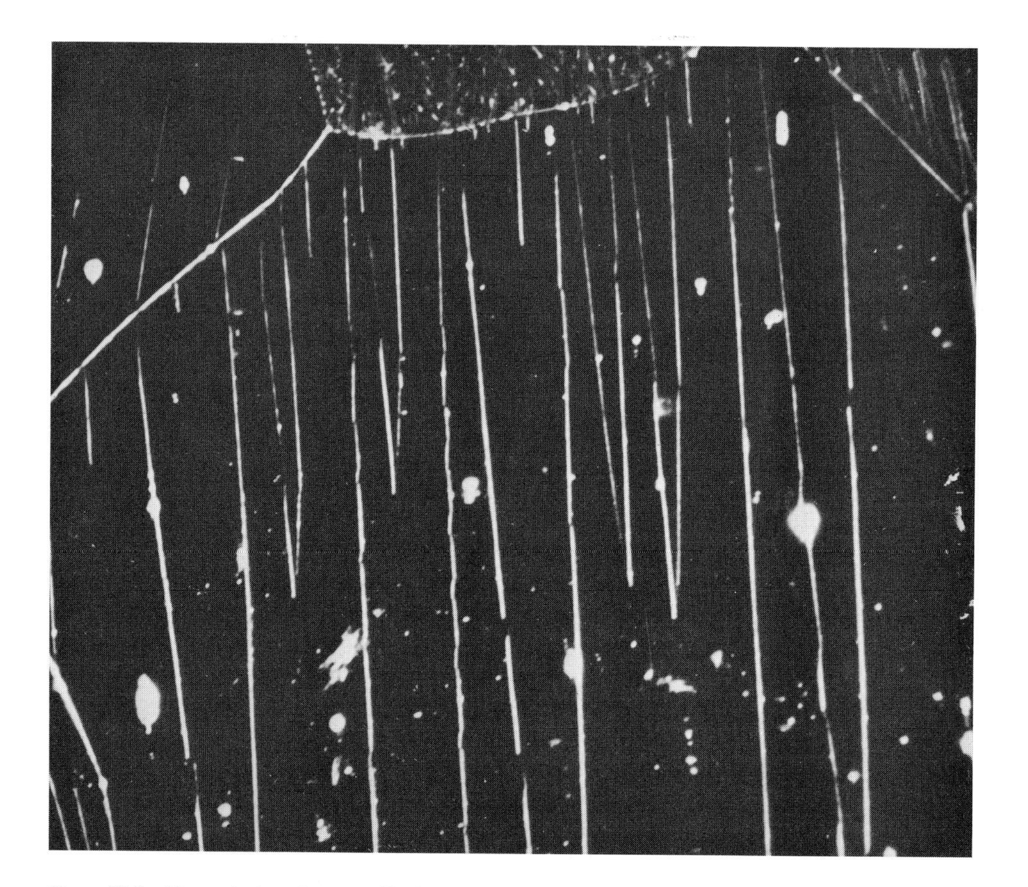

Figure 22-3 Magnetic domains in an Fe–3$\frac{1}{4}$ per cent Si alloy. The domains are dark bands separated by rather straight boundaries delineated in white. This is a coarse polycrystalline specimen, and several grain boundaries can also be seen. Since the magnetization is inclined about 2.5° from the specimen axis, some domains do not extend completely through the solid. *(Photograph courtesy of W. Paxton and T. Nilan, U.S. Steel.)*

hundreds or thousands of long domains exist in the solid. The detailed discussion of the actual structure is delayed to Sec. 22-8.

While the existence of domains was hypothesized at least as early as 1910 by Weiss, their existence was not proved till many years later by direct microscopic observation. Since then, further experiments combined with theoretical studies have shown many features of domains. A picture of domain structures in an iron-silicon alloy is shown in Fig. 22-3 at a magnification of about three hundred times. The dark bands which run vertically across this picture are domains. The magnetization of adjacent domains is oppositely directed. Some of the domains are V-shaped because the magnetization is not exactly parallel to the surface of the material.

The individual moments on atoms within a single domain must all be in the same direction. Regions must then exist between adjacent domains where the moments change direction rather abruptly. These regions, termed *domain boundaries*, are the white lines in Fig. 22-3. In later sections, some of the fundamental attributes of the boundary will be discussed.

22-3 The Hysteresis Loop

The foregoing discussion of the magnetic domains can be extended in several directions. One of the most important is the examination of the state of magnetization of a solid as a function of the strength and

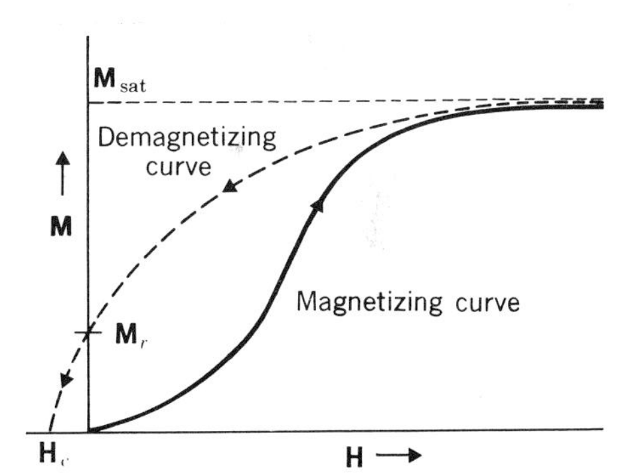

Figure 22-4 For a piece of virgin material, the processes of magnetization and demagnetization are not reversible.

direction of a magnetizing field. Two particular cases of interest are (1) the magnetization of a solid which initially has zero net magnetization and (2) the alternate magnetization and demagnetization of a solid in an alternating field.

The first of these, the magnetization of a virgin solid, is the simpler of the two. The behavior seen for a typical ferromagnetic solid is sketched in Fig. 22-4. The net magnetization $\mathbf{M}$ is initially zero in zero external field. It increases as $\mathbf{H}$ increases, slowly at first, then more rapidly. Finally it levels off, reaching a constant value, called $\mathbf{M}_{sat}$, above which it does not rise with increasing $\mathbf{H}$.

A picture of what must occur inside the solid during the magnetizing process can easily be constructed. In zero field as much domain volume exists with magnetization in one direction as exists in the other, and the net magnetization is thus zero. When a field **H** is applied, however, this canceling effect is reduced and some of the domain volume antiparallel to **H** becomes parallel to **H** because the magnetic energy is higher for moments in the antiparallel direction. If **H** is increased further, more domain volume reverses itself, increasing **M**

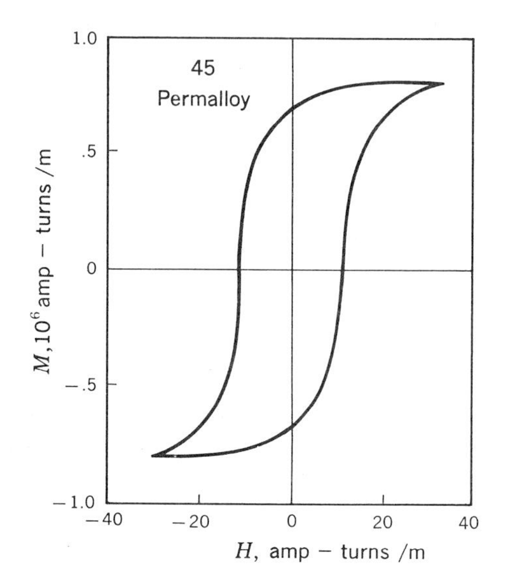

Figure 22-5 The hysteresis loop of a ferromagnetic material. The loop is the path retraced on successive alternations of H. (*After Bozorth.*)

still further. Finally, no antiparallel domains are left, and all atoms have their magnetization in the direction of **H**. Saturation has been reached.

As **H** is reduced from its high value, **M** does not retrace the same curve (Fig. 22-4). When **H** is reduced to zero, **M** still has a net positive value (the remanent magnetization, or remanence). In fact, a reverse magnetizing field, the coercive field, is required to reduce **M** to zero. If **H** be increased to a high negative value and again to a high positive value, **M** follows a path called the *hysteresis loop*, shown for an Fe–45 per cent Ni alloy in Fig. 22-5. This solid curve is retraced upon successive alternations of **H**.

The characteristic feature of this curve is that the demagnetization process lags behind the decreasing magnetizing field; hence the curve is known as the hysteresis loop. The lag shows that the energy introduced into the solid during magnetization is not completely recovered during demagnetization, so that some energy is lost. The magnitude of the magnetic energy loss in a complete cycle can be demonstrated to be proportional to the area enclosed by the hysteresis loop. Consider a piece of material being magnetized from the state $H = 0$, $M = 0$. The magnetic energy stored in the material as **H** increases from zero

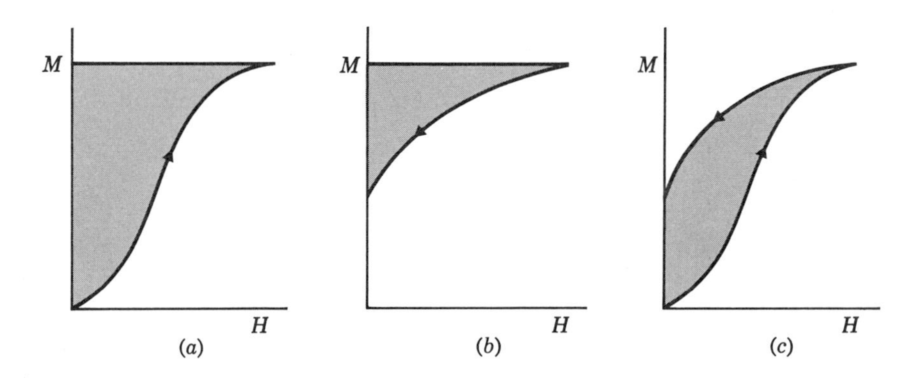

Figure 22-6 (*a*) The energy of magnetization; (*b*) that recovered during demagnetization; (*c*) a plot of the unrecovered energy.

to $\mathbf{H}'$ is given by

$$\text{Energy} = \int_0^{H'} \mu_v \mathbf{H} \cdot d\mathbf{M} \tag{22-1}$$

(This expression is analogous to the expression for storage of potential energy in a mechanical system when the force F changes from zero to F' with an accompanying displacement in coordinate x.) The total energy stored as H' is increased to saturation is proportional to the shaded area of Fig. 22-6*a*. If $\mathbf{H}$ is reduced to zero, the magnetization traces out the line shown in Fig. 22-6*b* and the potential energy recovered during the demagnetization is proportional to the shaded area in this plot. The net energy still remaining in the solid is proportional to the difference of these two areas, hence to the shaded area in Fig. 22-6c. In an analogous manner, this calculation can be continued to the other parts of the hysteresis loop to show that the entire energy lost in a complete cycle is proportional to the area of the entire loop.

From an energy point of view, then, knowledge of the $\mathbf{M}$ versus $\mathbf{H}$ curve for any material is important. Also important is knowledge of the physical mechanism by which the energy is dissipated, to be considered in more detail in succeeding sections.

Magnetization curves are not always plotted with variables $\mathbf{M}$ and $\mathbf{H}$. Characteristic curves of transformer materials are sometimes plotted with coordinates $\mathbf{B}$ and $\mu_v\mathbf{H}$ so that the values of the relative permeability may be read directly off the graph (see Fig. 22-7*a*). The magnetization curves of permanent magnets are sometimes plotted with coordinates $\mathbf{B}$ and $\mathbf{H}$, so that a quantity called the *energy product* may be read directly from the graph (Fig. 22-7*b*).

22-4 The Exchange Energy

The exchange energy depends upon the spins of the interacting atoms and their relative orientation. We shall thus write this energy w_E as

$$w_E = -2\mathbf{S}_1 \cdot \mathbf{S}_2 J \tag{22-2}$$

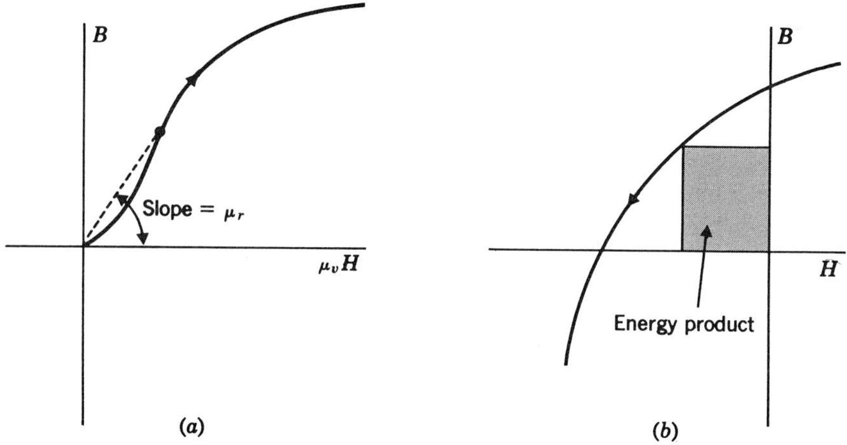

Figure 22-7 Alternative ways of plotting magnetization curves for special purposes.

where $\mathbf{S}_1$ and $\mathbf{S}_2$ are the spin vectors on the two atoms and J is the constant of proportionality called the exchange integral (p. 311). The calculation of J obviously depends upon the details of the theory, which, as we have mentioned, have not yet been adequately worked out. This energy can be written in a form in which the relative angle between the spin directions is made explicit.

$$w_E = -2JS^2 \cos\theta \tag{22-3}$$

where the spins on the atoms are assumed equal in magnitude.

For small angles, Eq. (22-3) can be simplified. Cos θ can be expanded in a power series,

$$w_E(\theta) = -2JS^2\left(1 - \frac{\theta^2}{2} + \frac{\theta^4}{24} + \cdots\right) \tag{22-4}$$

For small θ, $w_E(\theta) - w_E(0)$, defined as Δw_E, can be written as

$$\Delta w_E \approx JS^2\theta^2 \tag{22-5}$$

This expression gives the increase in potential energy, Δw_E, when two moments of spin S are rotated from exact parallelism to make some small angle θ with each other.

The energy increase Δw_E, given by Eq. (22-5), has significance in determining the nature of the domain boundary, since the spins must necessarily change as $\mathbf{M}$ changes from that of one domain to the adjacent one. Hence a certain number of spins must be misaligned in the boundary. The misalignment has associated with it a definite energy, the amount of which is given per spin pair by Eq. (22-5). This energy is a part of the energy of the domain boundary.

22-5 The Crystalline Energy

In addition to the exchange interaction between the spins of two neighboring atoms of a ferromagnet, which depends only upon the degree

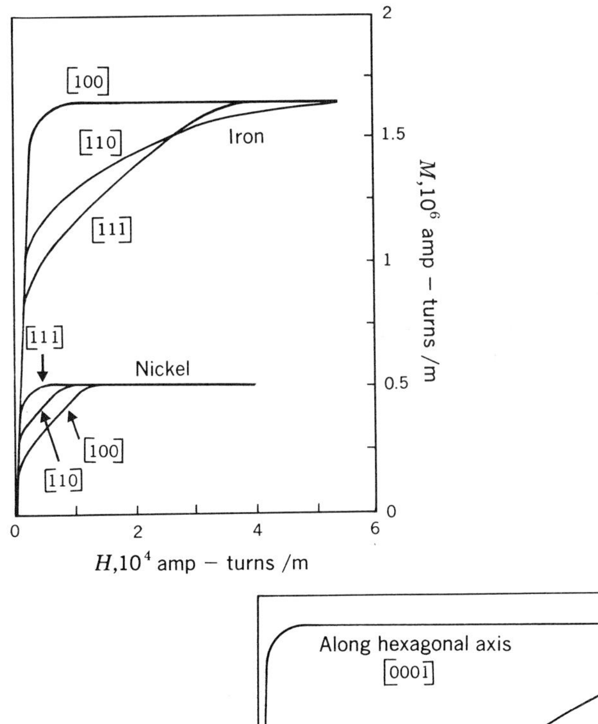

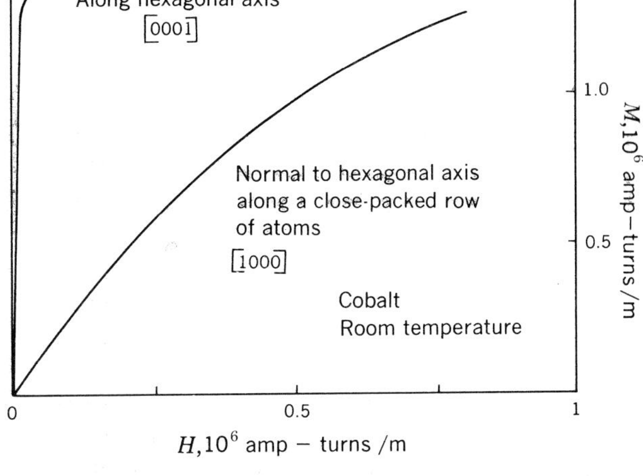

Figure 22-8 The energy of magnetization $\int \mu_v H\, dM$ is a specific function of crystal orientation. The lines in these graphs show the relative ease of magnetizing Fe, Ni, and Co in important crystal directions.

of relative alignment of the spins, the spins also interact with the lattice atoms. This interaction is demonstrated if it is assumed that all the spins are completely aligned, but the direction of alignment is rotated relative to the crystal axes. Preferred directions of the magnetic moment are then observed. For bcc iron these preferred directions are along the cube edge, the $\langle 100 \rangle$ directions (called *easy directions*). For fcc nickel the cube diagonals, the $\langle 111 \rangle$ directions, and for hcp cobalt the c axis, the $\langle 0001 \rangle$ directions, are the preferred directions. Magnetization curves are shown in Fig. 22-8 for Fe, Ni, and Co. For each metal the excess energy required to magnetize it in the most difficult direction may be calculated by subtracting $\int u_v \mathbf{H} \cdot d\mathbf{M}$ for that direction from $\int u_v \mathbf{H} \cdot d\mathbf{M}$ for the easy direction. For Fe, for example, the excess work done in magnetizing to saturation a specimen in the [111] direction (compared with the easy [100] direction) is about 1.4×10^4 joules/m^3. This energy is called the *crystal anisotropy energy* and is important in determining the character of a domain boundary.

22-6 The Domain Wall

The electron spins on the atoms in the boundary region between two adjacent domains are in a special type of position. While the geometry of the spin moments cannot be described in a perfectly general way for all crystal structures, details can be worked out for individual crystal types. The following discussion describes the most common type of domain wall of iron.

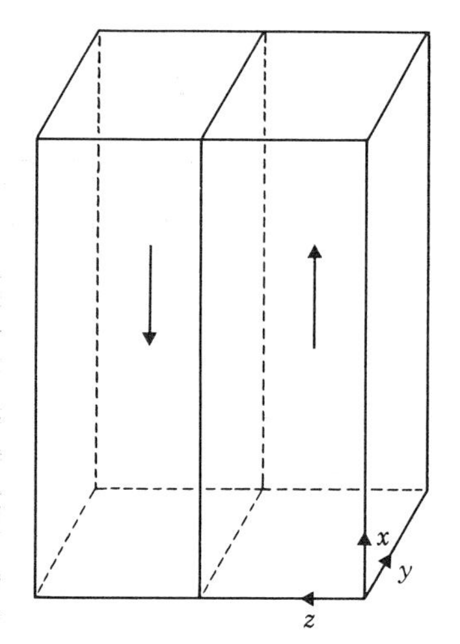

Figure 22-9 Domain pattern of a single crystal of iron with no net magnetization. One domain is magnetized in the [100] direction, the other in the [$\bar{1}$00] direction. The domain wall lies in the (001) plane.

The crystalline energy of iron is least when the magnetization of the individual domain of iron lies in one of the $\langle 100 \rangle$ directions. The wall for two adjacent domains, one magnetized along [100], the other along [$\bar{1}$00], is sketched in Fig. 22-9. The boundary between the two domains is one across which the spins must change direction by 180°; so this is called the 180° wall. This change might take place in one atom spacing, or it might be spread out over a number of atom planes.

To decide whether the domain wall is as thin as one atom spacing, consider the energy of the various spin configurations. The thinner the domain wall, the larger the angle between adjacent spins, and hence the larger the interaction energy as given by Eq. (22-5). This term in the energy forces the domain wall to be as thick as possible; the number of pairs involved increases linearly as the wall thickness increases, but the energy per pair decreases as the square of the thickness because of the factor θ^2. Overall, the energy from this source decreases with increasing thickness.

The anisotropic crystal energy has an opposite effect. Clearly, if the spin direction goes around from one direction to the other in a number of steps, some of the individual magnetic moments must be in hard directions; i.e., not all the moments can be in a [100] direction. These moments have higher energy than they would in the easy directions. Since the crystalline energy increases as the thickness of the domain wall increases, this tends to make the domain wall as thin as possible.

The domain wall assumes a thickness which balances these two energies to give the lowest total energy. The minimum total energy

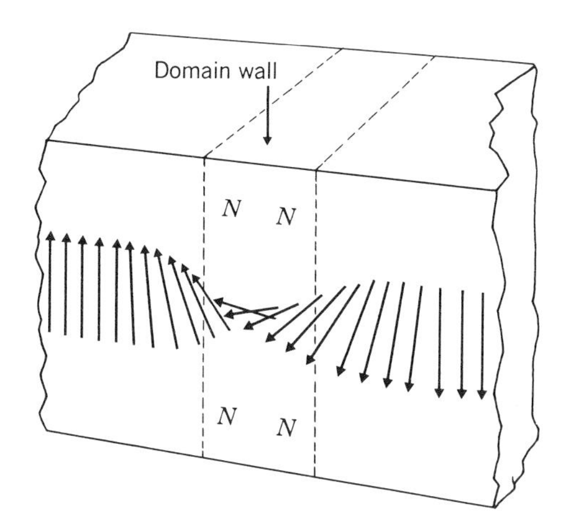

Figure 22-10 Change in atomic-dipole orientation through a domain wall. All moments lie in the plane of the wall. The *N*'s represent poles which are formed on the surface of the material.

occurs for the 180° wall in iron for a thickness of about 1,000 A, and the energy of the wall is about 10^{-3} joule/m^2. The angle between adjacent ion moments is not constant across the domain wall, since the crystalline energy is a strong function of the angle between the atomic moment and a $\langle 100 \rangle$ direction, but an average angle between adjacent spins is about $\frac{1}{4}°$.

The domain wall itself is very nearly a plane. The arrangement of the individual moments in this plane is represented by the sketch of Fig. 22-10. Notice that the moments always lie in the plane of the domain wall; hence no internal poles exist in the material. Poles are formed only at the surface of the material. These poles are regions which are delineated by the bright lines in Fig. 22-3.

22-7 Magnetostriction

Just as dielectrics change their length when they are polarized, so do most magnetic materials change their length when they are magnetized. Nickel, for example, contracts in the direction of magnetization and expands in the transverse direction by about 40 parts per million (ppm) at saturation magnetization. In spite of the small size of the

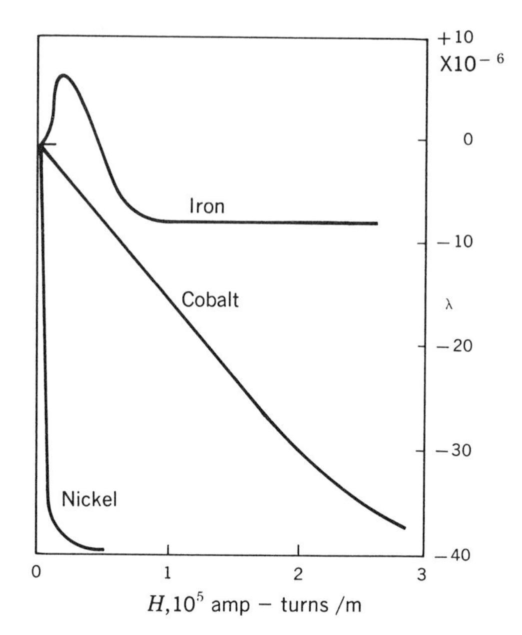

Figure 22-11 Magnetostrictive constant at room temperature as a function of applied field for polycrystalline ferromagnetic metals.

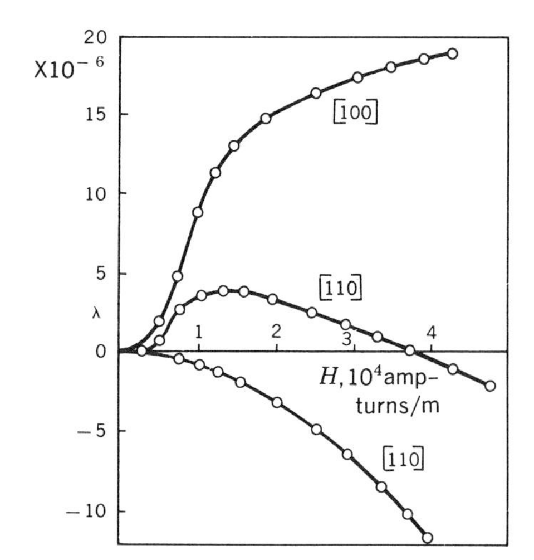

Figure 22-12 Magnetostrictive constant for several crystal directions in Fe as a function of *H*. [*After W. Webster, Proc. Roy. Soc. (London),* **A109:** 570 (1925).]

effect, the change in length, called *longitudinal magnetostriction*, is important, and it must be considered, because it plays an important role in domain geometry and in the practical use of transformer materials.

The dimensional changes accompanying magnetization are not always of the same sign. In low fields, Fe expands in the direction of magnetization, whereas Ni and Co both contract. In high fields, all three contract. This behavior is illustrated in Fig. 22-11, where λ_s, the coefficient of magnetostriction, defined as $\lambda_s = \Delta l / l_0$, is plotted as a function of H. Not only is λ_s a function of field strength, but it is also a function of direction in the unit cube. For example, for single crystals of Fe magnetized in three particular crystal directions, the magnetostrictive coefficient even has opposite signs, as Fig. 22-12 shows.

The volume of a ferromagnetic solid usually does not remain fixed as a solid is magnetized. Iron, for example, undergoes a small expansion in all fields, an expansion of about 1 part in 10^6 for the relatively strong field of 250,000 amp-turns/m. Up to this field strength the volume change, called the *volume magnetostriction*, is almost a linear function of field strength.

One of the important features of magnetostriction is that it has an inverse: since the dimensions of a magnetic material are changed when it becomes magnetized, its magnetic properties are changed when its dimensions are changed, i.e., when it is strained elastically. The solid does not necessarily acquire an overall magnetic moment when it is strained, for magnetostriction, like electrostriction, is insensitive to a change in direction of magnetization by 180°. However, through this inverse effect, internal strains in the material influence the direction in which a given domain will become magnetized. Internal strains of random magnitude and direction, in fact, greatly affect the shape of the hysteresis loop, often broadening it considerably.

The importance of magnetostrictive effects in domain configuration may be seen by considering the metal iron. Since domains in iron are magnetized to saturation along a $\langle 100 \rangle$ direction, domains in Fe are slightly expanded in the direction of magnetization, say, [100], and slightly contracted in a transverse, [010] or [001] direction. Because λ_s is independent of reversal of field, magnetization in [100] producing the same strain as magnetization in $[\bar{1}00]$, two domains such as those of Fig. 22-13*a* have no elastic-strain energy. Consider, however, these same domains intersecting the surface with a domain of flux closure (Fig. 22-13*b*). The closure domain (the pie-shaped region) is magnetized in the [010] direction at right angles to the original two domains; hence it tends to expand slightly in the [010] direction. This was the direction of contraction of the two slablike domains. Since the three domains are constrained to be continuous across their boundary, the domain of closure must be the seat of some elastic-strain energy. Detailed calculation shows that this energy is (for iron) of order of magnitude 50 joules/m³. Since this energy is roughly proportional to the volume of the domains of closure, it can be reduced if the closure domain can be reduced in total volume. For the configuration of Fig. 22-13*c*, the closure domain volume has been decreased by the introduction of more slablike domains. The configuration (*c*) thus has lower magnetoelastic energy than (*b*) but more domain wall energy (since the total area of domain wall has increased). If no factors other than these two are important, a massive piece of single-crystal iron would be subdivided into domains in such a way that a minimum in the sum of these two energies is attained.

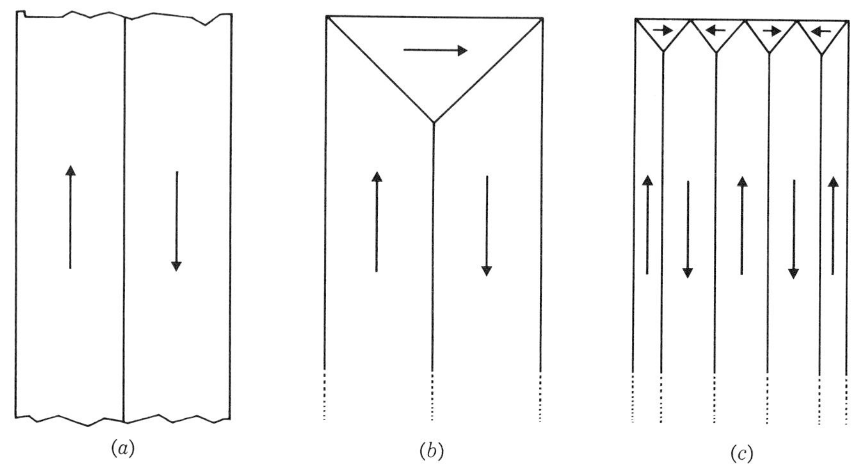

Figure 22-13 The domain structure of (*a*) has no elastic energy as a result of magnetostriction. (*b*) The introduction of a domain of closure results in the introduction of considerable elastic energy. (*c*) Elastic energy can be reduced in total amount by reducing the volume of closure domain, but with an increase in domain-wall energy.

22-8 The Magnetostatic Energy

The energy of a ferromagnetic solid is made up in part of *magnetostatic energy*, which is not related to the domain walls themselves, but which has a great deal to do with the configuration of domains. The origin of this energy can be demonstrated in a qualitative way, but detailed quantitative calculations are impossible except for special geometries.

The quantity of importance is the energy of domains which results from their being in a magnetic field. The field may be an externally applied field, the field of adjacent domains or the field of the domain under observation. To begin with, consider the energy of a solid uniformly magnetized to polarization $\mathbf{M}$ and placed in a uniform external magnetic field $\mathbf{H}_{\text{ext}}$. The energy per unit volume of the magnet in this field is

$$W_{\text{mag}} = -\mu_v \mathbf{M} \cdot \mathbf{H}_{\text{ext}} \tag{22-6}$$

In this expression, the zero of energy is that for $\mathbf{M}$ and $\mathbf{H}$ at right angles to each other.

Even if the material is not in an external field, it is still in its own field, $\mathbf{H}_{\text{self}}$. Thus a domain has a *self-energy* which is given by

$$W_{\text{mag}}(\text{self}) = -\tfrac{1}{2}\mu_v \mathbf{M} \cdot \mathbf{H}_{\text{self}} \tag{22-7}$$

(The $\frac{1}{2}$ in this equation is characteristic of all self-energy computations.) While this equation does give the formal solution to the problem, its usefulness is limited. In general, for a permanent magnet, the field $\mathbf{H}_{\text{self}}$ reacts on the magnetization $\mathbf{M}$ in such a way as to reduce $\mathbf{M}$ in some regions of the solid. Hence $\mathbf{M}$ is not uniform in the material, and the relationship between **M** and **H** can be determined only in special cases, one of which is an ellipsoid magnetized along one cf its principal axes. Then the self-energy is given by

$$W_{\text{mag}}(\text{self}) = \tfrac{1}{2}\mu_v N\mathbf{M} \cdot \mathbf{M} \tag{22.8}$$

where N is the number called the *demagnetization factor*. It is large for oblate ellipsoids magnetized along a short axis and small for prolate ellipsoids magnetized along a long axis. As an example of calculation of the self-energy of a permanent magnet, consider a long iron bar magnet, of square cross section, magnetized in an external field, which is then returned to zero. Suppose that this is a magnet of convenient laboratory size, say, 8 in. long and 0.5 in. on an edge. Treating this as though it were an ellipsoid of revolution, one finds the demagnetization factor to be about 0.1. The self-energy for this magnet is about 15 joules.

The demagnetizing effect has a number of important consequences. One is the fact that permanent magnets retain their magnetic moment longer if they are stored with a soft ferromagnetic link between pole faces. This "keeper" completes the flux linkage around the loop and reduces the reverse field through the magnet, which tends to demagnetize it.

Reduction of the self-energy is also a driving force for the partial demagnetization of a ferromagnetic solid which occurs when the external magnetizing field is removed. The magnetostatic energy is reduced if domains magnetized in antiparallel directions form throughout the solid. The magnetostatic energy is further reduced (in fact nearly to zero) if domains of closure are formed. Closure domains act as keepers, closing the flux lines completely in the material. The domains multiply in number until the energy of the resulting domain walls (some 10^{-3} joule/m^2) is balanced by other energies, mainly magnetostatic and magnetoelastic.

22-9 Summary of Energy Considerations

The equilibrium domain structure is that which gives the lowest total energy. The contributions to this energy are several in number, and although they have been discussed in detail, their important features are now summarized.

An important energy is that of the magnet as a permanent dipole in an external field, $\mathbf{H}_{\text{ext}}$. If the permanent moment is $\mathbf{M}_p$, then the magnetic energy is

$$W_{\text{mag}}(\text{perm}) = -\mu_v \mathbf{M}_p \cdot \mathbf{H}_{\text{ext}}$$

For a typical ferromagnet, say, Fe, magnetized to saturation, this energy is of order $2.2\mathbf{H}_{\text{ext}}$ joules/m^3. For typical fields, say, 10^5 amp-turns/m, this energy is large.

The self-energy of a dipole in its own field may also be of significant size. This energy may be written formally as a product of **M** and **H**, but the detailed calculations are difficult to carry out for a solid of arbitrary shape. For special geometries the calculation is possible; it may be done in these cases because **H** can be related to **M**. The simplest case of this type is for an ellipsoid of revolution magnetized along one of the principal axes; then **H** is just a constant times **M**. The energy may be written

$$W_{\text{mag}}(\text{self}) = \tfrac{1}{2}\mu_v N M^2 \tag{22-9}$$

where N is a number which depends on the relative sizes of the major and minor axes of the ellipse. For an oblate ellipsoid of revolution magnetized along the short axis, N is near unity, and the energy is high. For a prolate ellipsoid of revolution magnetized along the long axis, N is much less than 1, and the energy is correspondingly smaller. The self-energy of the magnet, nevertheless, is a large energy for any except extremely long, thin magnets. The existence of this energy even in the absence of an external field is cause for the material to break up into domains magnetized in geometries which reduce this magnetostatic energy.

The presence of domains requires the existence of transition regions in which the magnetization goes smoothly from one direction to the other. These transition regions, analogous in crystal geometry to grain boundaries, have well-defined properties. They are generally straight, since in this way the formation of internal poles, which increase the magnetostatic energy, is avoided. The energy of the domain walls comes chiefly from two sources, exchange energy and crystalline energy; each of these contributes about half to the total energy of about 10^{-3} joule/m^2 for Fe.

Domains in a demagnetized material are usually arranged in such a way that the magnetostatic energy is small, by ensuring that paths of flux closure exist within the solid. The resulting domains must be magnetized in a variety of directions. Consequently, the magnetostrictive energy can be high if the closure domains have large volume. The domains thus tend to be narrow to reduce the volume of the closure domains.

SOFT MAGNETIC MATERIALS

22-10 Principles

Most use is made of ferromagnetic materials either in transformer and motor cores or in permanent magnets. The magnetic qualities necessary for these two purposes are quite different. The remaining sections of this chapter have the function of describing the principles involved in these uses and of showing the state of development of material for either use.

Transformer cores present problems typical of the soft magnetic materials. Here the primary concern is large power-handling capacity with low-energy loss. Great strides have been made in reducing loss, and indications are that still more can be done. The losses themselves come prinçipally from two sources, hysteresis loss and eddy-current loss. While they are not entirely independent of each other, they are different enough so that they can be studied separately.

A more detailed understanding of the mechanism of *hysteresis loss* than was presented in Sec. 22-3 requires further understanding of the processes of magnetization and demagnetization. With the picture of domains and domain boundaries presented in previous sections, a fairly accurate picture can be made of these processes. Consider, as before, that a solid has a great variety of domains of different sizes and shapes. If the material has no net moment, i.e., if it is completely demagnetized, the vector sum of the moments of the individual domains is zero. The energy of the solid is then the sum of the self-energy (if there are internal poles), the magnetostatic energy, and the domain-wall energy. In an external field another term must be added, the energy of each domain in the field, that is, $-\mu_v \mathbf{M} \cdot \mathbf{H}$. Clearly those domains making a small angle with the field are in a lower energy state than are those making a large angle. To reduce the total energy to a minimum requires a redistribution of domain volume to put the net moment more in the direction of **H**. This redistribution occurs if favorably oriented domains, i.e., those for which $|\mathbf{M} \cdot \mathbf{H}|$ is largest, increase their volume at the expense of those less favorably oriented by the sidewise motion of domain walls.

The curves of **M** versus **H** are examined in more detail. A single crystal is again considered, for simplicity, but the same principles apply to polycrystals. In the demagnetized state, the domains are magnetized in general along all possible easy directions in such a way that the net moment is zero. If a steadily increasing external field be applied at some angle θ (not zero) with one of the easy directions, the net magnetization changes steadily. At low fields the domains with magnetization along the particular easy direction having lowest magnetic energy in the field grow in size at the expense of other domains. Growth continues as the field is increased until the solid is one single domain magnetized along this particular easy direction. At this time **M** in the direction of **H** has the value $\mathbf{M}_{\text{sat}} \cos \theta$, corresponding to the point a in Fig. 22-14. Further increases in **H** rotate $\mathbf{M}_{\text{sat}}$ out of the easy direction into the direction of **H**. At sufficiently strong fields the angle between $\mathbf{M}_{\text{sat}}$ and **H** becomes very small, and **M** attains the value $\mathbf{M}_{\text{sat}}$.

Demagnetization after saturation is in part the reverse of this process. The large single domain first rotates back as the field is reduced. After the magnetization is in the closest easy direction, the material breaks up into the individual domains. The reason why the demagnetization curve does not retrace the magnetization curve is not immediately apparent. Numerous experimental and theoretical studies show that the reason for hysteresis is found in the details of the growth of domains. Since they grow by the displacement of domain walls, the solid must present resistance to the passage of these walls. Impurities and lattice strains are commonly believed to impede this motion, and extra energy must be supplied to overcome the restraining

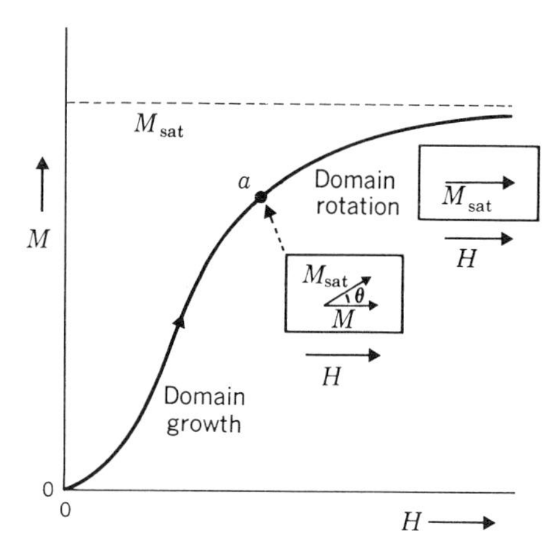

Figure 22-14 The magnetization curve is composed of two parts. One is a region of domain growth, where favorably oriented domains grow and unfavorably oriented domains shrink. When the solid is a single domain magnetized to saturation, rotation of this single domain out of its easy direction into the direction of *H* begins.

action. This energy is not recovered and is the magnetic loss discussed in Sec. 22-3.

The second kind of loss, from electrical *eddy currents* in the material, is also undesirable. It can be reduced by one of several methods. The loss comes, of course, from real currents which are induced in the core itself when the magnetic flux ϕ in the core changes. Since the power loss is proportional to V^2/R, where V, the locally induced voltage, is proportional to $d\phi/dt$, increasing the resistance of the core material reduces this loss. The macroscopic eddy currents flowing completely around the core are reduced by lamination of the core. When done properly, lamination increases but little the magnetic reluctance of the core but increases the electrical resistance enormously. Hence the loss due to macroscopic eddy currents can be made nearly zero. Microscopic eddy currents, i.e., currents within a single lamination, are more difficult to reduce. Commonly they are reduced by increasing the inherent resistivity of the core material itself by adding impurities. Addition of multivalent substitutional impurities such as Al and Si to Fe increases the resistivity without adversely affecting the hysteresis loss. Examples of this are given in the next section.

22-11 Transformer Materials

The engineering problems which must be solved in attaining high-quality transformer material are those of obtaining high permeability and low energy loss. Cost is, of course, a real production problem as well. The development of soft magnetic materials has been accomplished mainly with three types of materials: (1) Fe-Si alloys for low-frequency high-power applications; (2) Fe-Ni alloys for high-quality low-power uses (e.g., for audio transformers); and (3) ferrites for high-frequency (megacycle) uses.

The first requirement, that of high permeability, is important in designing the driving coil of the transformer. High permeability permits saturation to be attained with a small applied field; hence the primary turns and current may be kept small. Second, a high permeability commonly implies a narrow hysteresis loop; hence the direct magnetic losses are low. The low energy is then obtained by making the eddy-current losses low by alloying.

Table 22·1 Properties of typical soft magnetic materials

Material	Chemical composition	M_{sat} 10^5 amp-turns/m	μ_r (max)	ρ, ohm-meters	Hysteresis loss, joules/Kg/cycle	Typical use
Fe	Nominally pure	17.5	5,000	10×10^{-8}	0.03	
Fe-2%Si	Fe–2%Si	16.7	7,500	35×10^{-8}	0.02	Motors
G.O. Si-Fe	Fe+4%Si	15.5	30,000	55×10^{-8}	0.005	High-quality power transformers
Hipernik	Fe+50%Ni	12.7	60,000	45×10^{-8}	0.003	Audio transformers
78-Permalloy	Fe+78%Ni	8.7	100,000	16×10^{-8}	0.0005	Audio transformers
Supermalloy	Fe+5%Mo+79%Ni	7	1,000,000	60×10^{-8}	0.0001	Mainly laboratory use
Mn-Zn ferrite	$Mn_{\frac{1}{2}}Zn_{\frac{1}{2}}Fe_2O_4$	3	2,500	20×10^{-2}	0.001	Cathode-ray deflection pulse transformers
Nickel ferrite	$NiFe_2O_4$	~0.8	~100		~2	

Characteristics typical of several common ferromagnetic materials are listed in Table 22-1. Pure iron, which is listed for comparison purposes, is seen to have rather ordinary properties compared with those of the alloys listed. Silicon additions to pure iron improve its magnetic quality in two ways—both by increasing the resistivity so that eddy current losses are lower and by decreasing the magnetic hysteresis loss. Two grades of Si-Fe alloys are listed—one of the common Fe–2 per cent Si used for ordinary-quality induction motors, and the other of higher quality, the grain-oriented Fe–4 per cent Si alloy. The latter material is specially heat-treated so that the polycrystalline grains in the sheet are closely enough aligned for the material to behave nearly like a single crystal; hence the domains are larger and more regular in shape. These differences reduce the hysteresis loss greatly. The larger addition of Si also makes the eddy-current losses low. Both these effects could be accentuated by using even more Si, but unfortunately the alloy is so hard above 6 per cent Si that it becomes almost unworkable.

The alloys of Fe and Ni, the permalloys, are very useful. Two compositions are commonly used, one around 50 per cent Ni, which is typified by the commercial material Hipernik, and the other around 78 per cent Ni, usually called 78-permalloy. Both have high permeability and low hysteresis loss. One of the reasons for the excellent quality of the 78-permalloy is found in the fact that the magnetostrictive coefficient is nearly zero for this composition. Hence, the magnetostrictive energy is also near zero, and the domain configuration is favorable for high permeability. The material Supermalloy is basically this same material, with 5 per cent of the iron replaced by Mo. This material has both low magnetic loss and eddy-current loss, the latter because the resistivity is greatly increased by the Mo addition.

A typical ferrite, the Mn-Zn ferrite, is listed mainly for one reason. This ferrite, like all the rest of the ferrites, is a good insulator so that ρ is relatively high. Thus the eddy-current loss is vanishingly small, even in unlaminated sections. The ferrites are extremely useful at ultrahigh frequencies, where metallic solids become virtually useless. Thus ferrite cores can be used in high-Q antennas (the ferri-loop-stick), in pulse transformers operating in the megacycle range, in magnetic deflection yokes for cathode-ray tubes, and the like.

These examples are only a few of the many magnetic materials commonly used. For more information the interested reader is referred to the books listed at the end of the chapter.

PERMANENT MAGNETS

22-12 Desired Characteristics

Some of the qualities of magnetic solids desired in transformer materials are also useful in permanent magnets, e.g., high Curie temperature and high saturation magnetization. Permanent-magnet materials, however, must have some additional special features.

Most important in most permanent-magnet applications is the strength of the field in an air gap. Such uses as d-c meters, sound transducers, some electron tubes, and the like, all require an air gap in which currents can circulate. Since the field in the gap must usually be a strong one, materials with a high saturation moment must be used. At the same time the magnet design must be such as to put most of the flux in the air gap, since this reduces the demagnetizing field, which tends to decrease the magnetic moment.

The magnetic qualities desired in such a magnet may be most easily seen by reference to Fig. 22-15. Here is sketched part of the hysteresis loop for the material Alnico V, one of the best permanent-magnet alloys. The remanent induction B_r is high, more than 1 w/m^2, and the coercive field is also high, about 60,000 amp-turns/m. While the remanence of some transformer materials may be this high, the coercive fields are always much less than this (78 permalloy has a coercive field of about 10 amp-turns/m). The high coercive field for

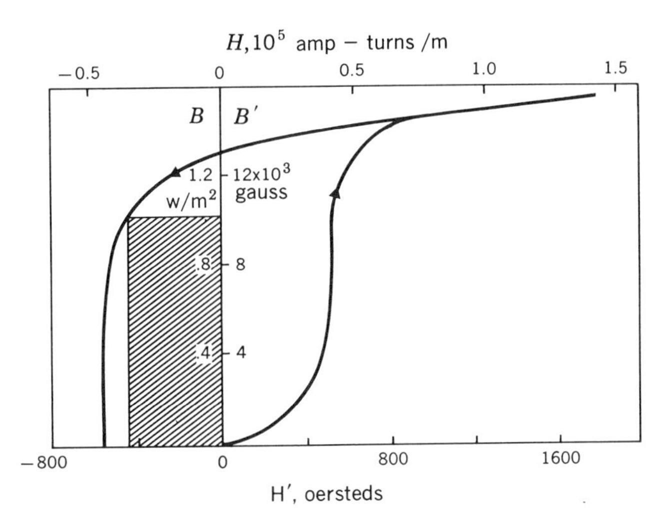

Figure 22-15 Part of the hysteresis loop of one of the good permanent-magnet materials, Alnico V, for which $B_r \approx 1.2$ weber/m^2 and $H_c \approx$ 500,000 amp-turns/m. The shaded area is the energy product. *(After Bozorth.)*

the permanent magnets is desirable for two reasons: (1) so that the demagnetizing field does not readily demagnetize the magnet and (2) so that stray external fields do not reduce the moment. Obtaining the combination of high remanence and high coercive field is the principal metallurgical problem in designing magnetic alloys.

A single quantity, the *energy product*, is commonly used to describe the quality of a permanent-magnet alloy. It is defined in terms of the behavior of the material in the second quadrant. Let rectangles be formed under the B–H trace in the manner sketched in Fig. 22-15. These rectangles have different areas depending on which point on the trace is used for the corner of the rectangle. One of these rectangles has the maximum area $(BH)_{\max}$. This product, which has the dimensions of energy, is called the energy product.† High-quality permanent-magnet alloys have an energy product of several tens of thousands joules per cubic meter. Observe that, in general, high values of $(BH)_{\max}$ require high values of both remanence and coercive field.

The characteristics of typical permanent-magnet alloys are listed in Table 22-2, along with those values which are typical of pure iron.

† In cgs units the energy product $(B'H')_{\max}$ is $40\pi(BH)_{\max}$.

Table 22-2 Approximate magnetic characteristics of typical permanent-magnet materials†

Name	Year introduced	Composition	H_c, amp-turns/m	B_r, webers/m²	$(BH)_{max}$, joules/m³
Fe			75	1.5	100
Tungsten steel	1885	6%W, 1%C, 93%Fe	5,000	1.0	3,000
K. S. steel	1917	35%Co, 7%W, 2%Cr, 0.6%C, bal. Fe	10,000	1.0	7,000
Remalloy	1931	12%Co, 17%Mo, 71%Fe	20,000	1.0	12,000
Alnico II	1934	12%Co, 17%Ni, 10%Al, 6%Cu, 55%Fe	50,000	0.7	17,000
Alnico V	1940	24%Co, 14%Ni, 8%Al, 3%Cu, 51%Fe	50,000	1.2	45,000
Indox I	1954	$BaFe_{12}O_{19}$	140,000	0.22	8,000
Indox V	1956	$BaFe_{12}O_{19}$ (oriented)	160,000	0.38	28,000

† Some of the data on the recent materials were kindly supplied by Indiana General Corporation.

A steady improvement in quality has occurred over the years until the high quality of Alnico V has been reached, one of the most commonly used commercial materials at this time. Alloys used in the beginning years of this century were hard steels such as those used for cutting tools even today. These alloys secured their desired properties in two ways: (1) large quantities of Fe and Co served to keep B_r high, and (2) carbide-forming elements such as W and Cr put large quantities of precipitates in the structure. Because precipitates impede domain-wall motion and thereby increase the coercive field, $(BH)_{max}$ is increased for these steels over that of a pure material.

The later improvements in magnet alloys typified by the two Alnico alloys tabulated here have an additional desirable feature. They secure high B_r from the same source, i.e., from large fractions of Fe and Co. They also have precipitates which affect domain-wall motion. However, they have one new feature, which takes advantage of the fact that bar magnets resist being magnetized in their short dimensions, i.e., crosswise. The microstructure of the Alnico alloys consists of rows of tiny rods of ferromagnetic alloy aligned in the same direction. These rods are embedded in a nonmagnetic matrix (see Fig. 22-16). The rods are so small (perhaps 100 A in lateral dimensions) that domain walls cannot exist in them (since domain walls themselves have thickness of some hundreds of angstroms). Hence each little rod must be magnetized along its length, either one way or the other. A very high-energy state exists for them when they are magnetized across their short dimension, since this configuration has enormous magnetostatic energy. If all are magnetized initially in one direction, say, up, then an enormous external field is required to reverse the direction of magnetization. The resultant high coercive force contributes to a high energy product.

Some of the ferrites can be used for permanent magnets. While their properties are not commonly as good as the best of the Alnico

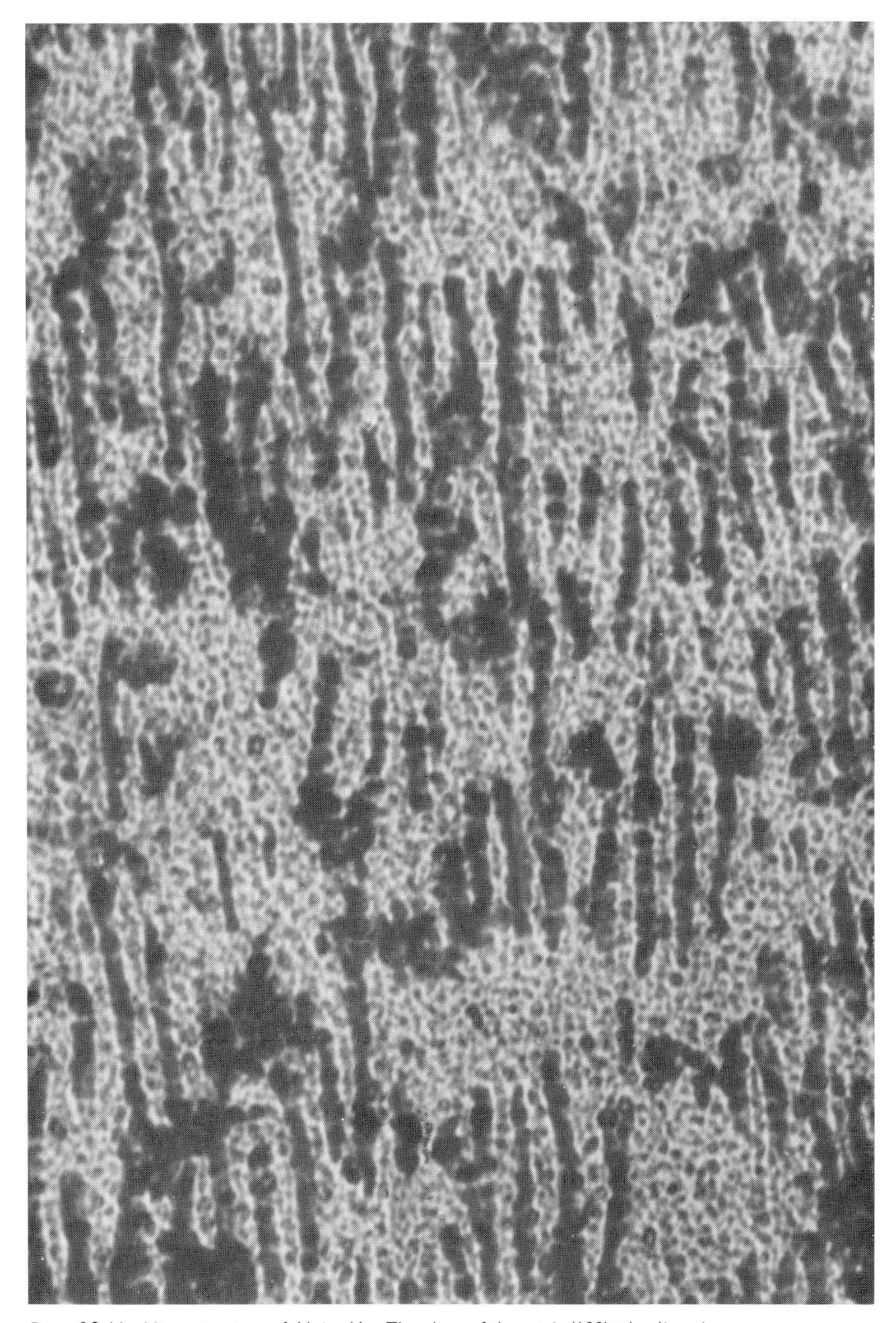

Figure 22-16 Microstructure of Alnico V. The plane of the cut is (100); the direction of the rodlike precipitate is [010]. ×22,000. (*Courtesy of R. Heidenreich.*)

alloys, they are lighter and cheaper. For descriptions of these and other permanent-magnet alloys, the reader may consult the References.

REFERENCES

1. R. Bozorth, *Ferromagnetism*, D. Van Nostrand Company, Inc., Princeton, N.J., 1951.
2. L. Bates, *Modern Magnetism*, Cambridge University Press, New York, 1948.
3. F. Brailsford, *Magnetic Materials*, John Wiley & Sons, Inc., New York, 1960.

PROBLEMS

22-1 (*a*) Estimate the magnetic energy loss per cycle for 45-permalloy under the conditions of Fig. 22-5. Express this loss both in joules per cubic meter per cycle and in joules per kilogram per cycle.

(*b*) If this hysteresis loss was all the loss in a transformer core made of 45 permalloy and all the energy loss was retained in the transformer as heat, how rapidly would the transformer heat up at 60 cps under the magnetizing conditions of Fig. 22-5?

22-2 Estimate from the curves of Fig. 22-8 the energy required to saturate Fe in the [100], [111], and [110] directions. Using these values, evaluate the constants K_0, K_1, and K_2 in the following equation, which has been found to give the energy of magnetization accurately in an arbitrary direction [*hkl*] which makes angles with the crystallographic cube edges having direction cosines α_1, α_2, and α_3.

$$E_{hkl} = K_0 + K_1({\alpha_1}^2{\alpha_2}^2 + {\alpha_1}^2{\alpha_3}^2 + {\alpha_2}^2{\alpha_3}^2) + K_2{\alpha_1}^2{\alpha_2}^2{\alpha_3}^2$$

(This method of determining the anisotropy energy is fraught with uncertainty, as your calculations will show. More reliable experimental methods yield, at room temperature, $K_1 \approx 40{,}000$ joules/m^3 and $K_2 \approx 15{,}000$ joules/m^3.)

22-3 Make the calculations of Prob. 22-2 for Ni at room temperature to show that $K_1 \approx -5{,}000$ joules/m^3 and $K_2 \approx +5{,}000$ joules/m^3. (The last number is not highly reliable.)

22-4 For Co, the magnetization energy obeys a simple relation; in a direction making an angle θ with the hexagonal *c* axis,

$$E_\theta = K_0 + K_1 \sin^2\theta + K_2 \sin^4\theta$$

From the data of Fig. 22-8 show that $K_1 + K_2 \approx 6 \times 10^5$ joules/m^2. Why can K_1 and K_2 not be separated with the data of Fig. 22-8?

22-5 The domain-wall energy of iron is about 10^{-3} joule/m^2 (i.e., about 1 erg/cm^2). In a typical specimen of demagnetized polycrystalline iron the domains might be parallelepipeds 0.1 by 0.01 by 0.01 cm. Calculate the total domain-wall area of a small piece of iron of, say, 500 g. Also calculate the total energy of all this domain wall.

22-6 (*a*) About half the domain-wall energy in iron is spin-spin interaction energy. Using the approximation that the average angle between adjacent spins is about 0.25°, estimate for iron the quantity JS^2 in Eq. (22-5).

(*b*) How does this number compare with that which you can calculate from the energy of the magnetized state, 2,000 cal/mole for iron?

22-7 In Sec. 22-7 the statement is made that the magnetoelastic energy in closure domains for iron is of order of magnitude 50 joules/m^3. Verify that this is so.

22-8 Suppose that long, thin, demagnetized rods of soft magnetic materials were placed with the long axis tangential to the earth's magnetic field. Estimate the magnetization which would be produced in the soft magnetic materials listed in Table 22-1.

23
Resonance

23-1 General

Reference has been made all through the previous chapters to experimental verification of many of the theories which describe the electronic structure of solids. Many of the experimental methods are of long standing; among these are methods of measuring specific heats, conductivities, and dielectric susceptibility. Others are more recent; among these are methods of measuring effective masses, the geometry of defects, and magnetic domain-wall properties. The latter methods demand, as a group, more refined experimental technique; at the same time they yield information on details which the simple methods cannot detect. One of the most recently developed experimental techniques for the analysis of certain details about materials is that of magnetic resonance.

The term *resonance* refers to the fact that at certain critical frequencies the materials under study absorb energy from an incident oscillating electromagnetic field. Resonance is employed in several different experimental techniques. The most prominent of these are *nuclear magnetic resonance*, in which the nuclei absorb energy, and *electron spin resonance*, in which electrons absorb energy. The first of these is the simpler of the two and is considered in detail first.

23-2 Nuclear Magnetic Resonance (NMR)

This phenomenon, searched for in solids in the 1930s (and not seen then), has been under intense investigation since the late 1940s when it was discovered. Many studies using NMR have been made since, but even brief mention of most of them is out of place in a general book. The following discussion is therefore limited largely to its application to the study of metals. The topics discussed are the following: (1) the nature of the phenomenon, (2) a typical experimental technique, (3) parameters used to characterize the resonance, (4) the Knight shift in metals, (5) application to the study of diffusion in solids, (6) applications to the study of local electric fields in metals and alloys, and (7) application to the study of deformed metals.

The most extensive use of nuclear magnetic resonance is not for solids at all, but for structure determination of organic molecules. The use of the proton resonance (and to a lesser extent the resonance of other nuclei) is extremely successful in determining the locations of protons (and hence of entire radicals) in such molecules. This topic is treated in detail in some of the references.

The nature of the phenomenon

Two principal methods exist for studying the magnetic properties of nuclei in solids. First is the nuclear induction method. The success of this technique depends on the fact that material which is otherwise diamagnetic exhibits a weak nuclear paramagnetism if the nuclei have a magnetic dipole moment. If the net magnetic moment of the nuclei, initially magnetized in some direction by a strong steady field, is changed abruptly, a weak but measurable emf is produced in a coil appropriately wound around the specimen. Changes in the moment can also be produced if an alternating magnetic field of proper frequency is imposed upon the solid, in addition to the steady field. The second method, and the one which is discussed in detail here, involves measurement of the absorption of energy from an electrical circuit when nuclear moments are induced to go from a low to a high energy level.

The magnetic moment of a nucleus is defined in exactly the same way as that of an atom: it is given in terms of the energy necessary to cause it to turn in a field [see Eq. (23-2)]. Also, as is true for atoms, an elementary moment is defined. The *nuclear magneton* is used to fix the units of nuclear moments. The nuclear magneton μ_0 is given by the expression

$$\mu_0 = \frac{|e|\,h}{4\pi m_p} \tag{23-1}$$

where m_p is the mass of the proton. In mks units, μ_0 has the value 5×10^{-27} joule/weber m^2. The magnetic moments μ of nuclei are generally not integral multiples of μ_0. Even the proton does not have the moment of one nuclear magneton. The nucleus has a spin quantum number I (analogous to s) and a magnetic quantum number m. While the electron-spin quantum number s may take only the values $\pm\frac{1}{2}$, I may have any integral or half-integral value greater than (or equal to) zero. Values larger than $I = 6$ are known, while for many nuclei $I = 0$. Spin numbers of 1 or greater introduce complexity into the description of nuclear magnetic resonance; nuclei with zero spin do not exhibit the effect, since they have no magnetic-dipole moment. Therefore we select for detailed discussion nuclei for which $I = \frac{1}{2}$. Since the proton has spin $\frac{1}{2}$, we use it frequently in the following discussion.

The magnetic-moment vector of the proton may point in a magnetic field so that its projection in the direction of the field is either $+\mu$ or $-\mu$. The difference in energy, ΔE, between these two directions is

$$\Delta E = 2\mu B_0 \tag{23-2}$$

where B_0 is the magnetic induction at the nucleus. If the induced magnetization M is small compared with the external field H_0, an

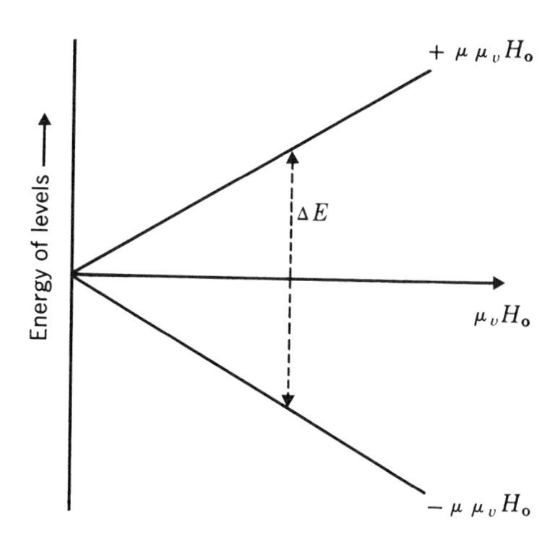

Figure 23-1 Splitting of levels by an external field.

infinitesimal error is made if B_0 is replaced by $\mu_v H_0$. Therefore Eq. (23-2) can be written as

$$\Delta E = 2\mu\mu_v H_0 \tag{23-3}$$

(This change of nomenclature is especially important in correlating this discussion with published work on NMR, all of which is written in cgs emu.)

The nuclear moment μ can be written (for $I = \frac{1}{2}$) in terms of the splitting factor g and the nuclear magneton by the equation $\mu = (g/2)\mu_0$; so Eq. (23-3) can be expressed as

$$\Delta E = g\mu_0\mu_v H_0 \tag{23-4}$$

The g factor for electron spin was introduced in Chap. 20; it equals 2. For the proton g is 5.58. The energy difference between states is plotted as a function of H_0 in Fig. 23-1. For a reasonable value of $\mu_v H_0$, say, 1 weber/m^2 (10,000 gauss), ΔE for a proton has the value 27.9×10^{-27} joule or 18×10^{-8} ev. This energy difference is relatively small, as atomic energies go, but it has one extremely important attribute. If a nucleus is induced to make the transition from one level to the other by electromagnetic radiation, the frequency of the required radiation has the convenient value given by the Einstein relation,

$$h\nu = g\mu_0\mu_v H_0 \tag{23-5}$$

For the proton in a field $\mu_v H_0 = 1$ weber/m^2,

$$\nu = 42.6 \text{ Mc/sec} \tag{23-6}$$

which is a convenient working frequency.

Other nuclei have moments of the same order and for all the critical frequencies fall in the megacycle range for suitable choice of external field, $\mu_v H_0$. *Nuclear magnetic resonance is the technique of determining* ν *(and hence* μ*) for a given nucleus in a given field.* From this information and from other, more subtle factors, details about the structure of molecules and solids are deduced.

The experimental method

Many types of apparatus are used to detect the resonance. The particular apparatus used depends chiefly on which of the two above-mentioned manifestations of resonance is to be used. For solid materials, the energy-absorption method is most widely used.

A block diagram of a typical experimental arrangement is shown in Fig. 23-2. The specimen S, about 1 cc of the material being investigated, is placed between the poles of an electromagnet. This magnet produces the field H_0. An r-f coil surrounding the specimen is carefully positioned to produce a second field perpendicular to H_0. An r-f generator not only serves to drive the coil but also supplies a signal to auxiliary circuitry which measures the r-f power absorbed by the specimen. To trace out the absorption line, an auxiliary low-frequency oscillator supplies power to secondary coils wound on the main magnet core. These coils permit the main field H_0 to be varied by a small amount on both sides of the resonant field.

The circuitry involved in making the measurement is complex, but the experiment is simple in principle. The field H_0 and the r-f frequency are fixed at values which are close to those which satisfy the resonant condition [Eq. (23-4)]. The low-frequency modulator is then used to vary the main magnetic field, causing it to sweep through the resonant condition twice each cycle. Associated electronic circuitry display on an oscilloscope a signal proportional to the energy absorbed as a function of field. Such a signal is sketched in Fig. 23-3*a*. This experimental technique is quite satisfactory for many absorption lines. Other techniques have been developed for experimental convenience. The most important of these measures the derivative of the absorption curve; by its use a curve such as that shown in Fig. 23-3*b* is obtained. From such a curve the parameters which characterize the resonance are easily deduced.

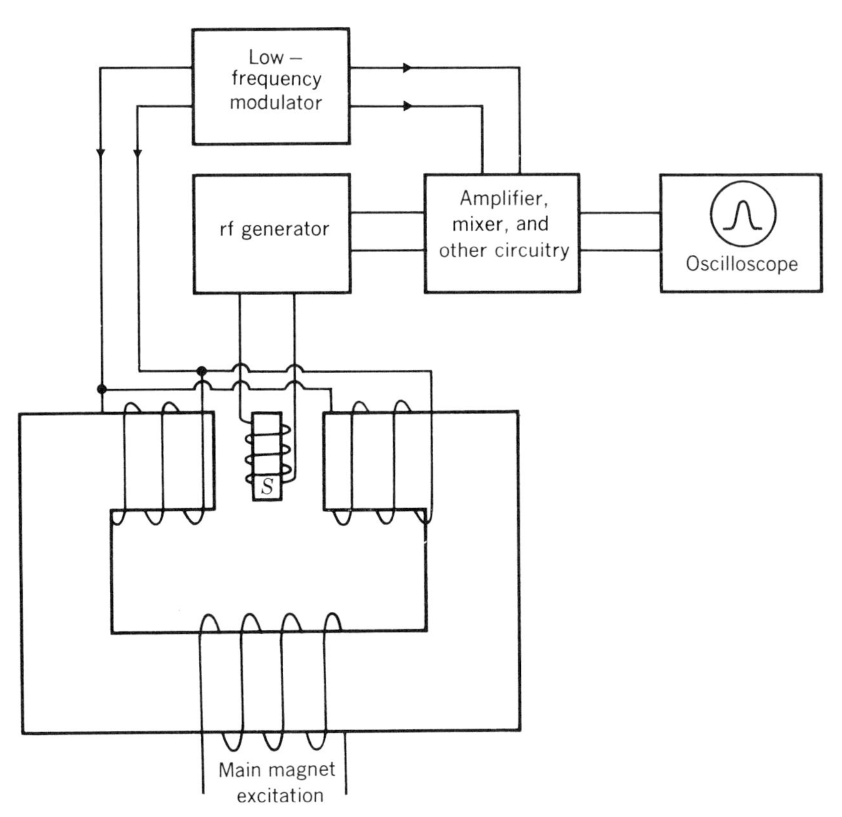

Figure **23-2** NMR spectrometer.

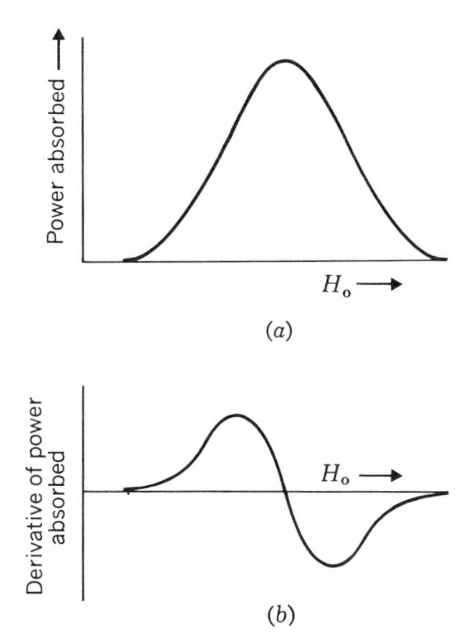

Figure **23-3** (*a*) Power absorbed as a function of field; (*b*) derivative of the absorption curve.

Details of the condition of the specimen vary from material to material. Specimens are usually about 1 cc in volume. Specimens much larger than this pose a difficulty in producing uniform fields; smaller specimens do not provide enough signal for coils of convenient size.

Liquids can be encapsulated in glass. Solid specimens of insulating materials may be single chunks of material. Solid specimens of good conductors, including all metals and alloys, may not have dimensions greater than the depth of penetration of the r-f signal. This requirement dictates the use of colloids, thin foils, or filings for these materials.

Characteristics of an absorption line

Perhaps the most significant feature of a line is its position in frequency for a given field. From this frequency the value of g can be determined, hence the value of μ, if I is known. Such information, of only mild interest in the study of materials, is of more importance to nuclear physicists.

The second feature of the absorption line is its magnitude. This is a complicated function of both the population difference of the two states and the induced and spontaneous transition probabilities between the two states. The population difference itself is relatively small. Since the energy difference between the two states, ΔE, is only of order 10^{-7} ev, the population difference Δn between the two states is small. In fact, by expanding Eq. (3-18) for $\Delta E \ll kT$, $\Delta n/n_1$ is found to be

$$\frac{\Delta n}{n_1} \approx -\frac{\Delta E}{kT}$$

where n_1 is the population of the lower state. For room temperature, then, $\Delta n/n_1 \approx -10^{-5}$. The population difference is thus less than 1 in 10^5 moments. Since the r-f signal induces transitions down as well as up, the net change in population which can be induced by the signal is small.

An important factor in considering the population of states concerns the rate at which the population of the system of nuclear spins adjusts to a temperature change of the lattice. This rate is described in terms of a quantity called T_1, the *spin-lattice relaxation time.* T_1 can be defined explicitly in the following way: Let the equilibrium population difference of the lower state be Δn_0. By some means (say, by an r-f field) let this excess be reduced to some number $\Delta n'$. If the r-f field is turned off, the spin system gradually reverts to equilibrium, and the excess population at any time t is given by the expression

$$\Delta n_0 - \Delta n = (\Delta n_0 - \Delta n')e^{-t/T_1} \tag{23-7}$$

The time T_1 varies widely for different materials. For pure compounds it may be of order seconds or even hours; for metals it is of order 1 to 10 msec.

This time has an important experimental significance. If a large r-f field of the critical frequency is left on indefinitely, induced transitions may cause the population difference to dwindle rapidly, since spontaneous emission cannot maintain the initial population difference. To avoid this occurrence, experimenters commonly vary the magnetic field constantly, sweeping it through the critical field rapidly and giving the material a chance to relax between sweeps.

The shape and width of the absorption lines are important parameters. For the most part we consider only symmetrical absorption lines, i.e., lines such as those in Fig. 23-3. The width is considered in more detail. A measure of the width commonly used is the difference in frequency between maximum and minimum slope on the absorption curve. Line widths commonly fall in the range $\Delta\nu \approx 1$ to 10 Kc/sec for nuclear spins in solids. Line widths may alternatively be expressed in terms of $\mu_v \, \Delta H_0$ at constant frequency; they are usually of order 10^{-4} weber/m^2 (this is of order 1 gauss). Another definition of line width commonly used is related to the reciprocal of $\Delta\nu$. The quantity T_2, often approximated by $1/2\pi \, \Delta\nu$, lies in the range 0.1 to 1 msec.

Absorption lines generally are broad for two reasons. First, the field at the nucleus is not exactly $\mu_v H_0$, since it is changed slightly by local perturbations caused by the nuclear-dipole moments of neighboring nuclei. Since these perturbations may add or subtract small variable increments to or from the field at all nuclei, the critical frequency of resonance is not sharp but has a small spread. Second, the lifetime in a state is not infinitely long, since nuclei are caused to make transitions from one state to another by adjacent nuclei. According to the uncertainty relation, the two energy levels themselves are not perfectly sharp; hence some width must exist in the absorption line itself.

The Knight shift in metals

The conditions for resonance of a given nucleus depend somewhat on the surroundings of the nucleus, since atomic or molecular orbital electrons produce small fields at the nucleus under study which add to or subtract from the external field. Thus the critical field for resonance (at a given frequency) varies from compound to compound. This shift is called the *chemical shift*, and from this effect NMR methods are used to determine the structure of compounds. Chemical shifts for the proton resonance are small—relative to the resonance in water the resonance of the proton occurs at a magnetic field about 6.7×10^{-4} per cent lower for the radical —SO_3H and about 2.5×10^{-4} per cent higher for the unit —CH_2—. Chemical shifts for other nuclei are often much larger. For oxygen and nitrogen in a number of compounds they are about one hundred times larger, of order 5×10^{-2} per cent.

A shift much larger than the chemical shift occurs in metals and alloys. This shift, termed the *Knight shift*, is toward higher r-f frequencies in the fixed field H_0, is proportional to the field, and is such that $\Delta\nu/\nu_0$ is of order 0.1 per cent to 1 per cent of the frequency at resonance. These and other features of this shift have led to the conclusion that it is caused by the paramagnetism of the conduction electrons in the metal.

An approximate calculation of the line shift can be made as follows: In addition to the external field H_0, an additional increment of field, ΔH, is produced at the nucleus by the conduction electrons. For a nucleus with magnetic moment μ, the energy of the interaction is $\mu\mu_v\,\Delta H$; the extra energy of separation caused by ΔH is therefore $2\mu\mu_v\,\Delta H$. Hence, the shift could be calculated if ΔH were known. As a first guess, we might suppose that ΔH is just the value calculated by using Eq. (20-29); for a typical metal, use of this equation yields a value for $\Delta\nu/\nu_0$ of order 10^{-6}. This estimate is almost 10,000 times smaller than is observed, and so the field at the nucleus, effective on the spins, must not be simply the spatial average of the paramagnetism of the conduction electrons. The answer to this dilemma is found in Sec. 9-2, where the point was made that, though the state functions of the conduction electrons spread over all the crystal, they still are peaked at the nuclei. In fact by supposing that the probability density of the valence electrons at the nuclei in the metal is the same as that for the free atoms, a factor of 10^2 to 10^4 is gained (varying somewhat from metal to metal). The theory does therefore agree fairly well with the experimentally observed facts. The Knight shift is thus assumed to be caused by the incremental field at the nucleus produced by the unpaired electrons in the conduction band.

The study of diffusion in solids

The line widths of all materials are not the same, being generally much less for liquids than for solids. For some liquids the line widths can be as narrow as 0.5 cps, but for most solid metals the line width at room temperature is 5,000 cps or more. This difference suggests that some phenomenon occurs in liquids which does not occur in these metals at room temperature. This phenomenon is assumed to be rapid atomic diffusion.

Recall that line broadening has been related to the influence on a given nucleus of its neighbors. The two effects described were the local change in magnetic field caused by neighboring dipoles and the transitions induced by neighboring dipoles. For these effects to be important, detailed analysis shows that a nucleus must remain in the same surroundings for a period of time about equal to the reciprocal of the line width itself. If the nuclei make diffusion jumps at a faster rate, the effects of the neighboring moments, random as they are, average to zero and the line then is narrow.

Consider first liquids of low viscosity. The diffusion coefficient D is of order 10^{-9} m^2/sec for such liquids near their melting points (see Table 4-2). In a period of time of 1 sec (about the reciprocal of the line width) a given atom makes about 10^{10} atom jumps of length about 1 A. The influence of the neighbors must surely average to zero over 1 sec for such a liquid, and these two effects on line broadening must not be important.

For solids, however, the atom jumping is much slower. For Al, for example, D at room temperature is about 10^{-30} m^2/sec; hence about 10^{10} sec is required for each place change of an Al atom on the

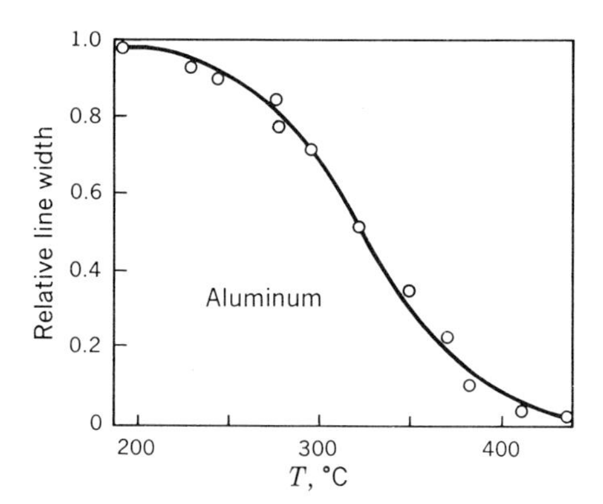

Figure 23-4 Motional narrowing of absorption line in Al.

lattice. The Al atoms are therefore static, and both contributions to line broadening occur at this temperature.

If the above-mentioned examples are valid, then the width of a resonance line for nuclei in metals should become narrow if the temperature can be increased enough to make D about 10^{-16} m^2/sec. For Al a temperature of about 360°C is required [using Eq. (4-14)]. Evidence for this narrowing is given for Al in Fig. 23-4: the narrowing actually occurs between 300 and 350°C. Below this temperature the line breadth of some 10^4 cps is contributed largely by perturbations caused by the effects cited above; at higher temperatures they are unimportant.

Diffusion in some metals is rapid enough so that even at room temperature the absorption line is narrow. The narrowing occurs in Na at about 190°K, a temperature which agrees well with that calculated from known diffusion data obtained by radioactive tracer methods.

In conclusion:

1. Motional narrowing is useful in determining approximate diffusion coefficients. The relation between reciprocal line widths before narrowing to mean time between atom jumps is not precise, however. Diffusion coefficients cannot therefore be determined as precisely by NMR as by other means.

2. By using NMR, diffusion coefficients can be measured for solids for which D is not easily measured by other means, i.e., for some organic crystals and for metals for which usable radioactive isotopes do not exist.

Quadrupole effects; the influence of internal electric fields

While the detailed discussion of energy levels of nuclei in a magnetic field in Sec. 23-2 was limited to nuclei for which $I = \frac{1}{2}$, many nuclei have nuclear spins which are larger. Examples for a few of the elements are listed in Table 23-1, along with values of μ for these elements. (Another factor, the nuclear quadrupole moment, is listed in the last column. This property of nuclei is described later.)

The possible orientations of a nuclear magnetic moment for nuclei with $I > \frac{1}{2}$ are more in number than just parallel and antiparallel to the external field. In fact $2I + 1$ orientations are possible. Since

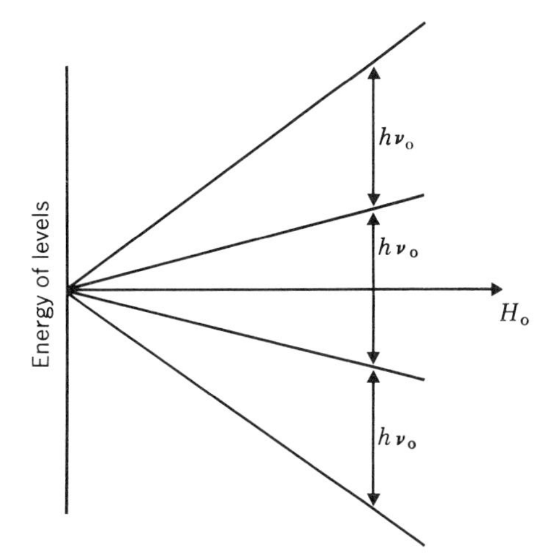

Figure 23-5 Splitting of $2I + 1$ levels in a magnetic field for a nucleus for which $I = \frac{3}{2}$.

the energy of interaction of the nuclear moment with the magnetic field is different for each orientation the single level in zero field splits into $2I + 1$ levels with the field. (See Fig. 23-5 for splitting of nuclei for which $I = \frac{3}{2}$.) Selection rules permit transitions only between adjacent levels, so that all transitions have the same critical frequency ν_0 and all absorption peaks superpose. This superposition holds true for all nuclei, regardless of I, if all the splitting is caused by magnetic fields. For many crystals, the only splitting is caused by magnetic fields, but, for some, an additional splitting is caused by interaction of the nuclei with electric fields. This interaction exists for nuclei which have an electric quadrupole moment.

A nuclear electric quadrupole moment arises from a particular nonspherical arrangement of charge on the nucleus. It is characterized by a particular angular and radial dependence of the electric field produced by its arrangement of charge. A few pertinent facts are the following:

1. It exists only for nuclei for which $I > \frac{1}{2}$.
2. Nuclei may have either positive or negative quadrupole moment (Table 23-1). Note that the unit used here to designate the

Table 23-1 Nuclear moments and magnetic moment of several of the elements†

Nucleus	I	μ	Q
H^1	$\frac{1}{2}$	2.79	0
Li^6	1	0.82	4.6×10^{-4}
Be^9	$\frac{3}{2}$	−1.18	2×10^{-2}
C^{13}	$\frac{1}{2}$	0.70	0
Na^{23}	$\frac{3}{2}$	2.22	0.1
Al^{27}	$\frac{5}{2}$	3.64	0.15
Si^{29}	$\frac{1}{2}$	−0.55	0
Cu^{63}	$\frac{3}{2}$	2.22	−0.16
Ge^{73}	$\frac{9}{2}$	−0.88	−0.2
As^{75}	$\frac{3}{2}$	1.43	0.3
Se^{77}	$\frac{1}{2}$	0.53	0
Ag^{109}	$\frac{1}{2}$	−0.13	0

† More complete listings are given in the References appended to this chapter. In this table, the spin number I is given in units of $h/2\pi$, the magnetic moment μ in units of the nuclear magneton and the quadrupole moment in units of 10^{-28} m^2.

quadrupole moment is 10^{-28} m^2, roughly the cross-sectional area of a nucleus.

3. Gradients in electric fields arising outside the nuclei interact with nuclear quadrupole moments to produce a splitting of energy levels. The magnitude of the splitting is a function of the product of the quadrupole moment Q and the gradient of the electric field at the nucleus. This splitting is independent of and in addition to that caused by the magnetic field. Furthermore the effect is not the same for all intervals; so the absorption lines for the several transitions do not superpose.

The effects of the quadrupole interaction for a nucleus for which $I = \frac{3}{2}$ are shown in Fig. 23-6 for an axial electric field. The effect of the electric field is to alter the energy levels so that three lines appear

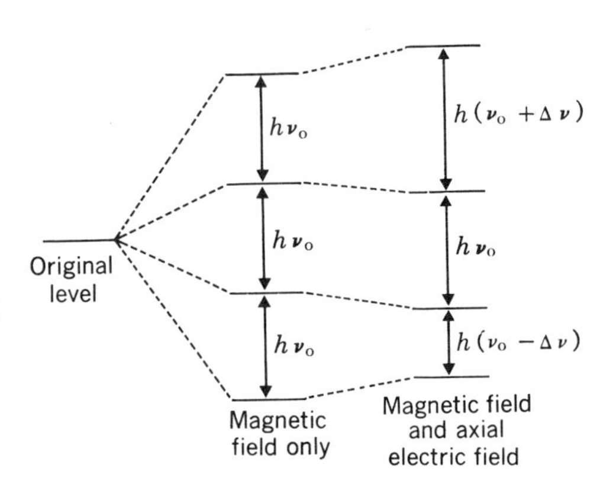

Figure 23-6 Magnetic and quadrupole splitting of the levels for a nucleus for which $I = \frac{3}{2}$ in an axial electric field. The frequency of the central line is unchanged (to a first approximation); the frequencies of the satellite lines are displaced from the central line by $\pm\Delta\nu$.

where one existed before. One of these lines is at the original frequency ν_0 and the other two (often termed *satellites*) appear $\Delta\nu$ above and below ν_0. For single crystals of the material, the positions of the satellite lines relative to the central line depend on the orientation of the crystalline axes relative to the direction of the field. For metals, randomly oriented polycrystals must be used for actual measurement, since the specimens must be powders or filings. The satellites are

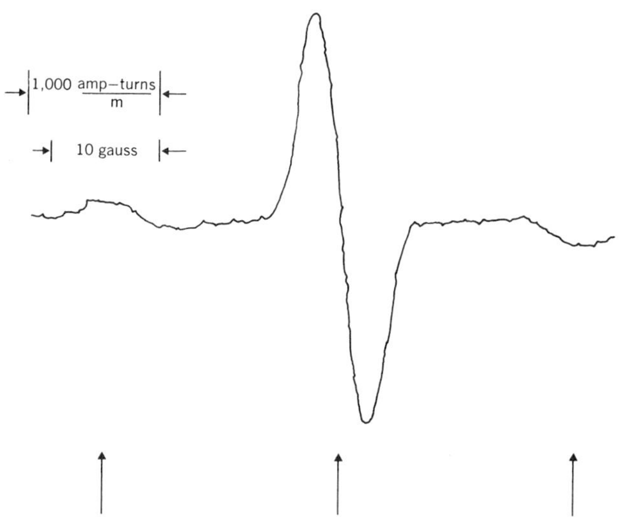

Figure 23-7 Quadrupole splitting in Be^9 in an axial electric field. The strong central line is accompanied by satellite lines equally displaced from it. The "wiggles" on the trace are circuit noise. [*T. Rowland, Progr. in Materials Sci.*, **9** (1) (1961).]

thereby smeared out somewhat. Actual data for Be^9 are shown in Fig. 23-7. Arrows point to the strong central line and the weaker satellite lines in this derivative curve. Quadrupole effects are of considerable importance, for they permit the probing of electric fields in solids. Unfortunately, the results of experiments can be determined only in terms of the product of Q and the electric-field gradient, two quantities about which little is known.

Electric fields in solids are produced by inhomogeneous charge distribution within the solid itself. Impurities, especially those of different valence from the lattice atoms, provide a point-charge inhomogeneity which causes an electric-field gradient approximately radially divergent from the impurity. Hence impurities can cause quadrupole splitting of transitions of the lattice atoms. Because the electric-field gradients are nonuniform over the sample, the splitting varies widely from atom to atom, being largest for atoms adjacent to the impurity and near zero for atoms sufficiently far removed from the impurity. The satellite lines are thereby so smeared out as to be unobservable. However, information can be derived from analysis of the central line.

The only complete study of quadrupole effects in a wide series of alloys has been made by Rowland for alloys of Cu. He measured the properties of the central line when Cu (which is monovalent) was alloyed with: (1) monovalent impurities (Ag and Au), (2) divalent impurities (Cd and Zn), (3) trivalent impurities (Ga and In), (4) tetravalent impurities (Ge and Si), and (5) pentavalent impurities (Sb and As). His extensive work and its interpretation can be summarized as follows:

1. The central line is not appreciably broadened by impurity additions, but its amplitude is decreased.
2. The amplitude of the central line decreases rapidly with increasing impurity additions. For a typical alloy, Ge in Cu, the amplitude drops to 10 per cent of its initial (pure Cu) value for a Ge addition of less than 4 atomic per cent.
3. The larger the valence difference between Cu and the impurity, the larger the effect per impurity atom.
4. Even Au and Ag, which have no valency difference with Cu, produce an effect. This amplitude change is thought to be caused by the local distortion of the crystal lattice when these large ions are inserted in the lattice.

The conclusions which Rowland draws from his observations concern the nature of the conduction electrons themselves. The electrons in the Fermi distribution tend to "bunch up" around impurities. This bunching up can be caused by the attraction of conduction electrons either to the positive impurity ion cores (screening) or to regions of lattice distortion. The electric-field gradients extend far out from the point of the impurity: a Ge impurity in Cu sensibly affects about 100 Cu nuclei (that is, 25 unit cells).

Alloying affects the central line in still another way: it modifies the Knight shift for the pure metal. That this possibility exists may be seen from the following argument: The Knight shift is caused by the effect on the nuclei of the paramagnetism of the conduction electrons (Sec. 23-2). Since alloying with atoms of a different relative valency can change the density of states at the Fermi level, it can change the magnitude of the paramagnetic moment [see Eq. (20-29)]; hence alloying should alter the Knight shift. The effect of alloying on the Knight shift has been observed for a number of alloys, but line broadening complicates interpretation of the results. The most extensive study was carried out on Ag alloyed with a number of elements (Ag^{109} was selected for a base metal because it has $I = \frac{1}{2}$ and consequently no quadrupole effects to trouble the measurement). Variations in Knight shift on alloying were observed. They could not, however, be interpreted in the simple way suggested above; apparently the screening effect produces local variations in electron concentration which are much more important. The conclusion is that Knight-shift measurements apparently do not yield density-of-states information about alloys, at least not in a simple way.

Application to the study of deformed metals

Quadrupole splitting cannot occur in pure, perfect crystals which are cubic. It can occur, however, in cubic crystals if some regions of the crystal have noncubic symmetry. Such regions are found in the vicinity of dislocations. Hence mechanical deformation, which produces many dislocations, does affect the resonance absorption. The

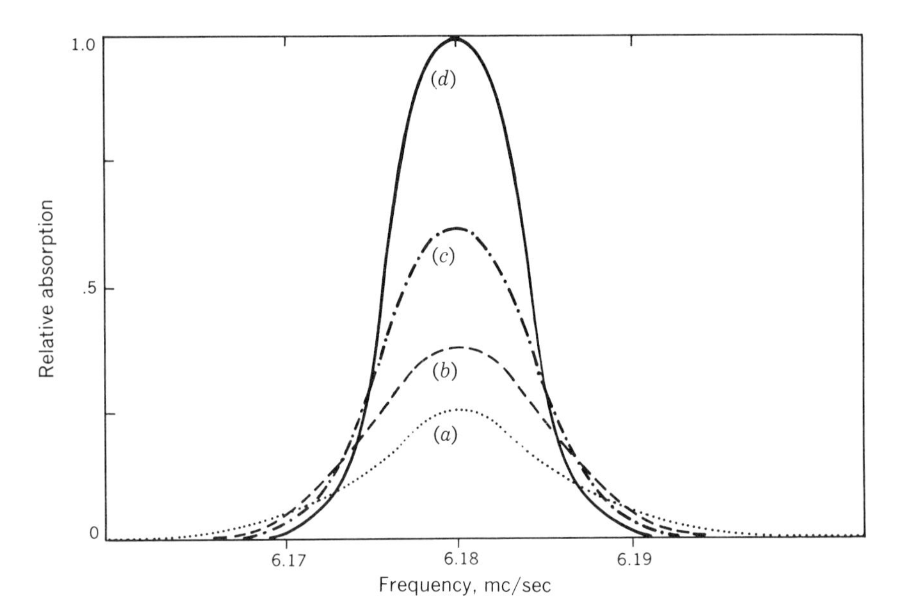

Figure 23-8 Experimental result of annealing of cold-worked specimen of Al + 0.64 atomic per cent Mg. The absorption curves were obtained after the following treatments: (*a*) As filed; (*b*) after a 250°C 87-hr anneal; (*c*) after a 480°C 2-hr anneal; (*d*) pure, well-annealed Al, included for comparison. [*After T. Rowland, Acta Met.,* **3**: 74 (1955).]

strain field about the dislocations produces electric-field gradients which cause the quadrupole splitting. As for the impurity additions, the electric-field gradient is not the same at each lattice site, and so well defined satellite lines are not produced.

The effect of mechanical deformation on the resonance absorption of a metal is illustrated in Fig. 23-8. Here are shown data for an alloy of Al-Mg as mechanically deformed by filing (curve a). The curve is broad and low, indicating considerable quadrupole splitting. After an intermediate anneal at 250°C, the curve is sharper and higher (curve b). The annealing presumably removed some of the dislocations, thereby reducing the fraction of Al atoms in local field gradients. A higher temperature anneal increases the central line absorption of this alloy even more (curve c), but not so much as is true for annealed filings of pure Al (curve d).

Though not many studies of this kind exist, they generally show the same features as are described above.

Summary of applications of NMR to metals

The nuclei of some atoms have magnetic moments. The moment of a given atom may assume positions in a magnetic field which do not have the same energy. Transitions between adjacent positions can be induced by radio-frequency fields; for magnetic fields of convenient size, the frequency ν is between 1 and 100 Mc. This phenomenon is termed resonance. The condition for resonance is that $h\nu$ is equal to the energy differences between levels. The nucleus is selective, and frequencies only a little removed from this critical frequency do not excite the transition; i.e., the breadth of the resonance is small.

From knowledge of the critical frequency and other details of the frequency dependence of the transition, information can be obtained about several features of solids:

1. The nuclear magnetic moment can be accurately determined.
2. The effect of the paramagnetism of conduction electrons on the resonance can be deduced.
3. The breadth of the resonance depends mainly on perturbations of the resonance conditions by neighboring nuclei. From a study of these perturbations information about diffusion can be obtained.
4. In alloys and cold-worked materials, internal, local electric fields cause perturbation of the resonance conditions. Measurements of the resonance in such materials therefore can be used to study these fields.

23-3 Electron Spin Resonance (ESR)

The electron, too, has a magnetic moment; therefore all matter might be expected to exhibit resonance similar to NMR, since all matter contains electrons. However, only very particular materials exhibit a resonance caused by absorption of energy by electrons. Nevertheless for the materials in which it is found, ESR can be immensely important.

The following discussion parallels somewhat that of the preceding section. Topics to be discussed are (1) the nature of the effect, (2) experimental methods, (3) an application to metals, (4) applications to ionic crystals, (5) applications to semiconductors, and (6) characteristics of resonance in ferromagnetic materials. Unfortunately the uses to be cited in (3) to (5) are only specific examples; the uses of electron spin resonance are sufficiently different in detail so that generalities cannot be made with the same ease as for NMR. At the same time, the number of research papers in the literature utilizing ESR is larger than those utilizing NMR. The selection of examples for detailed discussion in this section was therefore difficult, and those selected were chosen on the basis of their applicability to problems discussed in earlier chapters.

The nature of the effect

The theory of energy absorption from an r-f field in a magnetic field is strikingly similar to that for a nucleus in a magnetic field (Sec. 23-2). The single electron, which has only two quantized states in the field, can be induced to make transitions if the frequency ν of the r-f field satisfies the equation

$$h\nu = 2\mu_e \mu_v H_0 \tag{23-8}$$

where μ_e is the magnetic moment of the electron. Again the factor g is introduced so that

$$h\nu = g\beta\mu_v H_0 \tag{23-9}$$

For free electrons the factor g is equal to 2; however, in metals it can vary somewhat, as discussed below. For normal fields H_0, say, such that $\mu_v H_0$ is about 1 weber/m^2, critical frequencies are of order 30,000 Mc/sec ($\lambda \approx 1$ cm). β is the Bohr magneton.

The condition on resonance given by Eq. (23-9) is based on an ideal configuration for a single electron. Two things can serve to upset applicability of this expression: (1) a set of N coupled electrons may be involved in the resonance and (2) the field H_0 may not be the only influence which establishes the energy levels for the electron between which transitions can take place.

The first of these contributions might be investigated by inquiring how a partly filled shell of electrons in an ion behaves in a magnetic field. Consider as an example the electrons in the $3d$ shell of a Mn^{++} ion. Mn has outer-electron configuration $4s^2 3d^5$; so Mn^{++} has the configuration $3d^5$. The net spin of this state is $\frac{5}{2}$, and the net spin vector and hence the net magnetic moment may take six quantized positions in the field H_0. Five distinct transitions between these states are allowed (since again transitions are allowed only between adjacent levels). However, all these transitions have equal energy and each satisfies the same relation [Eq. (23-9)]. So, for the ideal situation, all effects in multielectron states superpose (as they do for NMR).

The ideal case is upset if other interactions affect the energy levels between which transitions occur. One of the most important of these interactions is the influence of the nucleus itself. The orientation of the magnetic moment of the nucleus in the field relative to that of the moment of the electron influences the energy levels of the electrons. Since the nuclear moment may take $2I + 1$ quantized orientations in the magnetic field, each of which alters the electron level by a different amount, each original level is split up into $2I + 1$ sub-levels (for $I < S$). Thus each "ideal" spectral line is split into $2I + 1$ lines by the nuclear-electron interaction, termed *hyperfine splitting*. The hyperfine splitting is in the range of 1 to 10 per cent in $\Delta\nu/\nu$ (or $\Delta H_0/H_0$) and hence is readily observable. A schematic of the effect of hyperfine splitting is drawn in Fig. 23-9 for a hypothetical ion with $I = \frac{5}{2}$.

Other effects also alter the position of the observed lines, such as the internal electric fields arising from inhomogeneities in the material, crystalline defects, and impurities. Through interpretations of these effects, in fact, comes the greatest use of ESR for solids.

One final requirement for the existence of electron spin resonance which is implicit in the above discussion concerns the necessary electron configuration. Electrons which are paired two by two in atomic or metallic states produce no resonance, since the electron pair has no net spin. Consequently, filled shells of atoms and paired electrons in valence bands produce no effect. One large group of atoms or solids which can show the effect are those which are paramagnetic. Hence ESR in these materials is often called *electron paramagnetic resonance* (EPR). Another group with unpaired spins is the group of ferromagnetic metals and alloys. ESR in these materials is called *ferromagnetic resonance.* The corresponding resonance in ferrites is called *ferrimagnetic resonance.* Finally the antiferromagnetic materials, which are paramagnetic, also display spin resonance. (These latter topics are discussed in several of the References.)

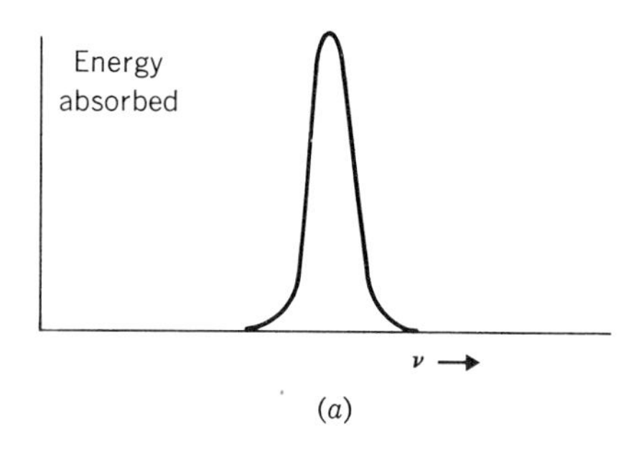

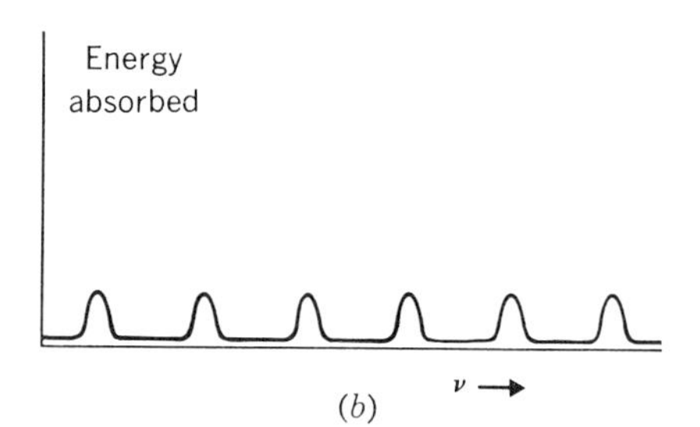

Figure 23-9 Effect of hyperfine structure splitting on an ion for which $I = \frac{5}{2}$. (*a*) Ideal case, one line; (*b*) effect of hyperfine structure, six lines.

Experimental methods

Since the frequencies at which resonance occurs for reasonable fields are in the kilomegacycle-per-second region, measurements are commonly (but not always) made with the specimen in a waveguide or a resonant cavity. A typical schematic diagram is that shown in Fig. 23-10. It consists of a variable but highly stable r-f generator in the

10,000 Mc/sec range which feeds a waveguide containing a resonant cavity into which the sample is placed. A detector measures the amount of r-f energy reaching it; its output is amplified and passed through other circuitry, finally being recorded in some fashion. An external magnet supplies the main field H_0, which closely satisfies the resonance condition. A modulating field serves to sweep the main field through the resonance.

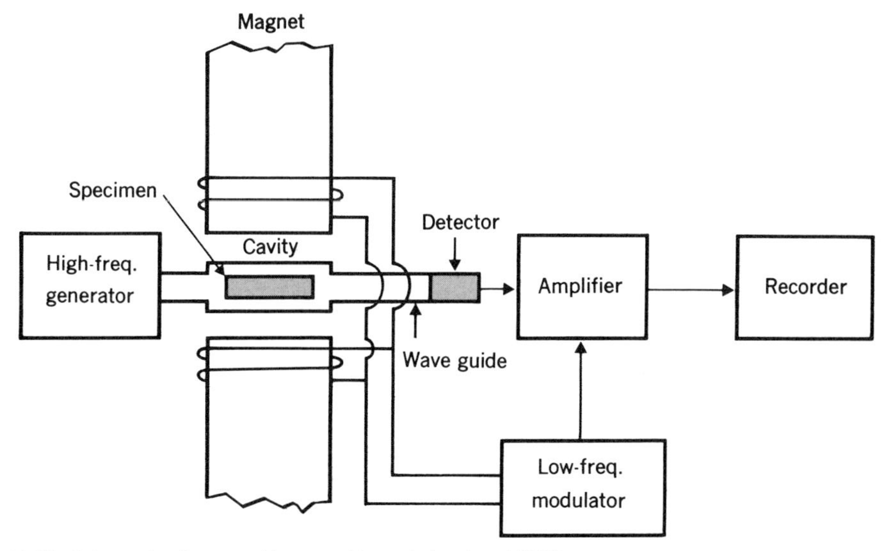

Figure 23-10 Schematic of waveguide assembly and circuitry of ESR apparatus.

This schematic must be modified somewhat to get proper impedance matching, attenuation of signal to the proper incident level, stability of operation, and the like. These details vary from experiment to experiment and are not discussed here.

ESR in metals

Relatively few electron-resonance studies on metals have been carried out. One reason is that the skin depth is strikingly small in the 10^4 Mc/sec region; therefore the metal must be divided into small particles. A second reason is that the valence electrons are paired and no resonance effects occur. However, in a magnetic field some electrons are unpaired, and paramagnetism of the valence electrons themselves is very important.

The nature of the paramagnetism of valence electrons (the Pauli paramagnetism) is described by Eq. (20-29). The induced moment is proportional to the applied field, and the susceptibility is about 10^{-5} for many metals. Experimental verification of this value is difficult to achieve, since any conventional measurement yields the total susceptibility χ_T, which is the sum of χ_p and the diamagnetic susceptibility χ_d (which latter may have a significant contribution from the diamagnetism of the valence electrons).

$$\chi_T = \chi_p + \chi_d \tag{23-10}$$

A method which would measure either χ_p or χ_d separately would be of interest in unraveling the entire magnetic character of the nonferromagnetic metals.

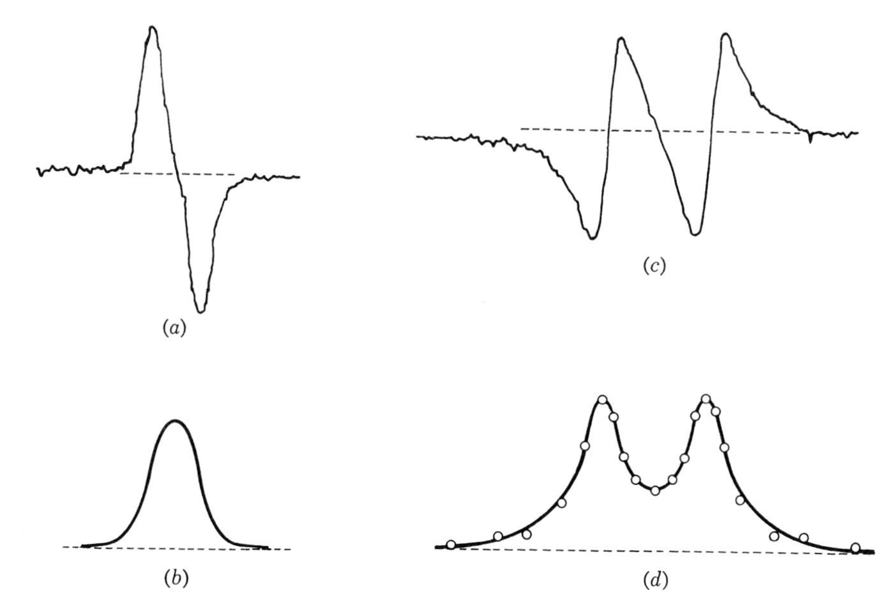

Figure 23-11 Resonances in Na^{23} metal. (*a*) Nuclear magnetic resonance in Na at 11.5 Mc/sec. The range of magnetic field is about -2×10^{-3} weber/m^2 to $+2 \times 10^{-3}$ weber/m^2. (*b*) Absorption peak calculated from curve *a*. (*c*) Electron spin resonance in Na at 11.5 Mc/sec. The range of magnetic field is about 2×10^{-3} weber/m^2. (*d*) Absorption peaks calculated from the above experimental derivative curves. The two peaks come at about -4 and $+4 \times 10^{-4}$ weber/m^2. *(After Schumacher and Slichter.)*

Two classic examples of the measurement of the resonance of the paramagnetism of valence electrons are those made for Na and Li (Ref. 6). The study was an absolute measurement of χ_p alone for each of these two metals and was carried out along the following lines: Finely divided particles of metal were suspended in oil contained in a small capsule. The capsule was placed between the poles of a large electromagnet, and the electron spin resonance of the valence electrons was measured. The measuring frequency was rather low, about 11.5 Mc/sec, and so the critical field for resonance was low [the value calculated from Eq. (23-9) is $\mu_v H_0 \approx 4 \times 10^{-4}$ weber/m^2]. Figure 23-11*c* shows the resonance absorption (the derivative curve) as the field was swept from the -2×10^{-3} weber/m^2 to $+2 \times 10^{-3}$ weber/m^2. Two resonances were observed, one centered at about -4×10^{-4} weber/m^2, the other at $+4 \times 10^{-4}$ weber/m^2; the two correspond to magnetization of the sample in the $-$ direction and in the $+$ direction. The resonance energy-absorption peaks calculated from the derivative curves are shown in Fig. 23-11*d*. Recall that this absorption is energy absorbed from the 11.5 Mc/sec signal as spins of valence electrons flip from one direction to the other in the magnetic field.

The volume susceptibility can be calculated by measuring the areas under the energy-absorption peaks if the number of atoms per unit volume in the suspension is known. To measure this number, a special technique was devised. The nuclear magnetic resonance of the undisturbed sample was measured (also at 11.5 Mc/sec) by increasing $\mu_v H_0$ to its appropriate value of about 1 weber/m^2. The derivative curve obtained by varying the critical field about the resonance condition is shown in Fig. 23-11*a*. The corresponding energy-absorption curve is plotted in Fig. 23-11*b*. The area under this absorption peak is proportional to the nuclear paramagnetic susceptibility, which is calculable. Thus a comparison of the area under peak *b* compared

Table 23-2 Data for the susceptibilities of Na and Li metals† (Mks units)

Metal	χ_p (measured)	χ_p (calculated)	χ_T (measured)	χ_d
Na	1.19×10^{-5}	0.99×10^{-5}	0.88×10^{-5}	-0.31×10^{-5}
Li	2.61	1.19	2.38	-0.23

† Probable errors and further analysis of the data can be found in Ref. 6.

with that under peak *d* permits calculation of χ_p for the valence band of Na.

The value calculated for χ_p for Na^{23} from the data of Fig. 23-11 is tabulated in Table 23-2 along with the data on χ_T for Na measured by standard methods. Also listed are χ_d calculated from these two values [by using Eq. (23-10)] and the theoretical value of χ_p calculated by using Eq. (20-29). Similar data for Li^7 are also tabulated.

Examples of applications to ionic crystals

No electron spin resonance occurs for pure, perfect ionic crystals, since all electron shells are filled. Imperfections (including impurities) in such crystals, however, often have uncompensated spins, which produce an absorption line or lines. Analysis of these absorptions permits both an electrical and a geometrical description of the imperfections. The present section shows two examples of this use of resonance.

1. A V center in KCl

When a pure, perfect crystal of KCl is bombarded with X rays at low temperature (i.e., in liquid nitrogen), electrons are ejected from the filled $3p$ shell of some of the Cl atoms. Many of the ejected electrons immediately recombine with the holes in the $3p$ shell, but some do not. Those which do not are presumed to be trapped in other parts of the crystal, leaving some Cl atoms with a permanent electron deficiency. These Cl atoms are said to be *V_K centers*.† Such Cl atoms exhibit spin resonance.

The resonance of this V_K center is shown in Fig. 23-12. It consists of seven main lines (indicated by the arrows in the figure) and a

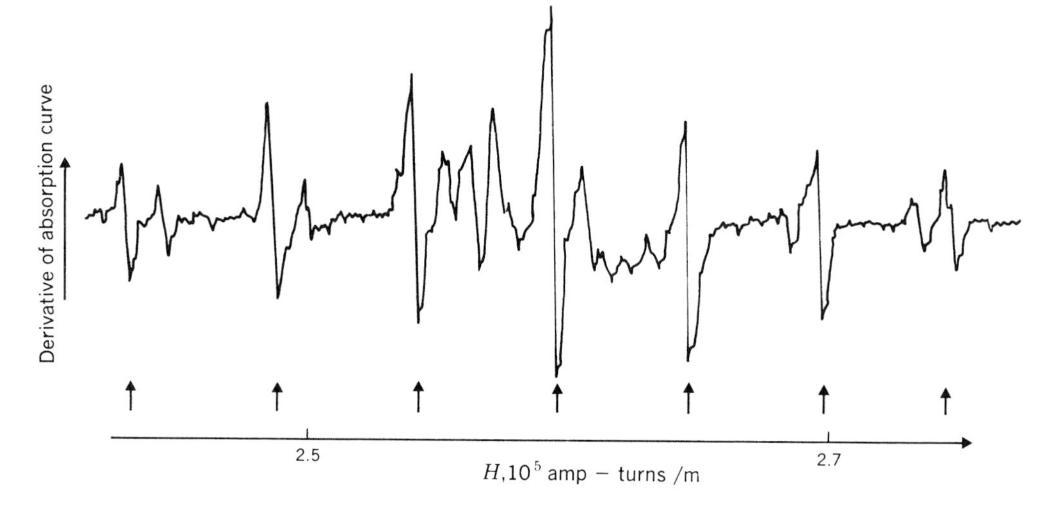

Figure 23-12 The resonance spectrum of the V_K centre.

number of other lines. From analysis of the position and intensity of these lines and of the way in which the splitting changes as a function of angle of the magnetic field in the crystal, a complete description of the geometry of the V_K center has been furnished (Ref. 7).

The statement made above that a V_K center consists of an electron-deficient Cl atom is not strictly true. Rather, the deficiency is shared by two adjacent Cl atoms, the two forming a Cl_2^- molecular bond. Furthermore they draw closer to each other than normal, so that the lattice is locally strained elastically. Such a pair is sketched in Fig. 23-13 in a (100) plane.

The spectrum of the V_K center displayed in Fig. 23-12 is easily explained in terms of this model. The lines indicated by the arrows are the hyperfine lines of the electron as split by the coupled Cl nuclei of the molecule. The nuclei for the most abundant Cl isotope have atomic weight 35; I for this isotope is $\frac{3}{2}$. Since the Z components of the angular momentum combine algebraically, not only are seven

† This V_K center is a special case of the V centers mentioned in Chap. 14.

hyperfine lines in the spectrum explained, but also the relative intensities of the lines are explained (for the details of this explanation, see Ref. 7). Thus, the existence of the defect (Cl^{35}—Cl^{35}) seems certain.

Other lines in the spectrum are a result of other combinations. Some are caused by the ion $(Cl^{35}—Cl^{37})^-$ and others by the $(Cl^{37}—Cl^{37})^-$; these have different splittings and different intensities and account for some of the other lines seen in the spectrum.

2. The M_n^{++} *ion in doped* NaCl

A second way in which resonance can be observed in salts is

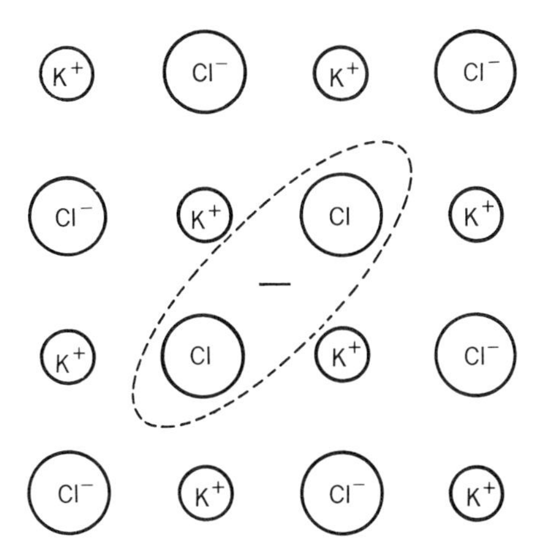

Figure 23-13 The geometry of the V_K centre.

through the replacement of the monovalent metal ion by a divalent or trivalent metal ion. Such an addition is $MnCl_2$ to NaCl. Manganese goes into the lattice as an Mn^{++} ion. With its incomplete 3*d* shell, the Mn^{++} ion is paramagnetic, and hence it can be detected by spin resonance. Furthermore, the crystal, to produce local charge neutrality around the divalent ion, has a vacant metal-ion site near some of the dissolved Mn ions.

The nature of the ESR resonance for such a material can be quite complex. Data for an NaCl single crystal containing a small amount of dissolved Mn are shown in Fig. 23-14. (These data were obtained by Watkins, Ref. 3.) Though this spectrum at first sight looks hopelessly complicated, using these data and others obtained for several orientations of the magnetic field, Watkins was able to show that the Mn ions exist in the crystal in several configurations.

The most prominent feature of the spectrum is the six evenly spaced lines of equal magnitude, indicated by the arrows in the figure. These lines are the hyperfine lines of Mn^{++} ions not associated with a vacancy (Fig. 23-15*a*). The six lines are produced by coupling of the net electron spin to the nuclear moment of $I = \frac{5}{2}$ for the Mn^{55} ions. Since these lines are rather strong (compared with the others), many of the Mn^{++} ions are not associated with vacancies.

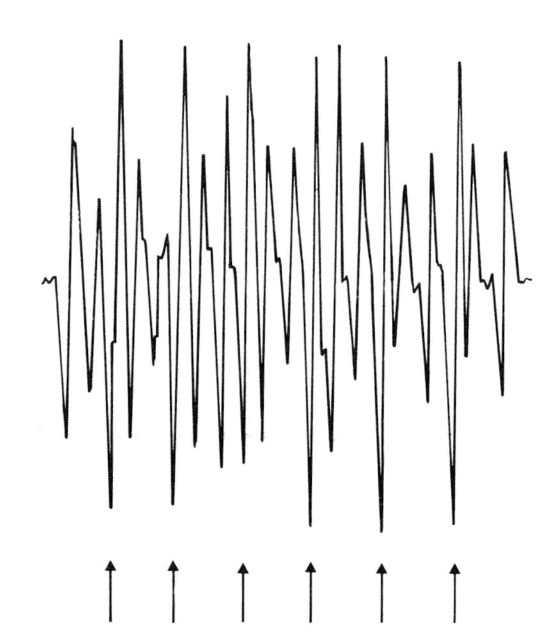

Figure 23-14 Resonance spectrum of NaCl containing Mn^{++} ions. Specimen quenched from high temperature. Measurement made at 20,000 Mc/sec, with magnetic field along a [100] direction. (*After Watkins.*)

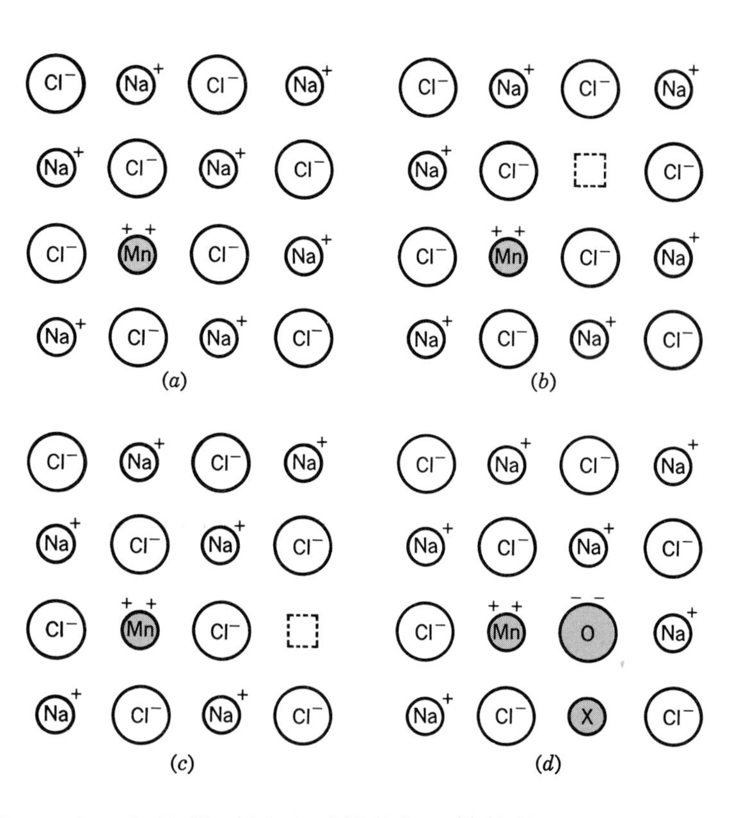

Figure 23-15 Nature of Mn complexes in NaCl. (*a*) Isolated Mn^{++} ions; (*b*) Mn^{++}-vacancy pair; (*c*) second type of Mn^{++}-vacancy pair; (*d*) $Mn^{++}O^{=}X$ defect.

Other lines in the spectrum have been associated with Mn-vacancy complexes, at least two of which have been identified. One is thought to be an Mn^{++} ion with a vacancy in an adjacent metal-ion site (Fig. 23-15*b*). The stability of this pair is fairly large; the energy necessary to separate the two is about 0.39 ev. A second complex is thought to be the Mn^{++} with a vacancy in a next nearest-neighbor metal-ion site (Fig. 23-15*c*). The energy necessary to separate this pair is a little less, about 0.34 ev.

A third complex has been postulated, too, though it is not on such firm ground as are the other two. This is an Mn^{++} associated with the $O^{=}$ or $S^{=}$ which replaces a nearest-neighbor Cl^{-} ion (Fig. 23-15). A second monovalent impurity might also be present in a nearest-neighbor metal-ion site. This is a more stable defect than either of the other two.

Several other factors of the spectrum have been used to describe other details of the defect structures, but they are not discussed here.

Example of application to semiconductors

Intrinsic semiconductors have completely paired electrons and show no spin resonance. If impurities or crystalline defects are present,

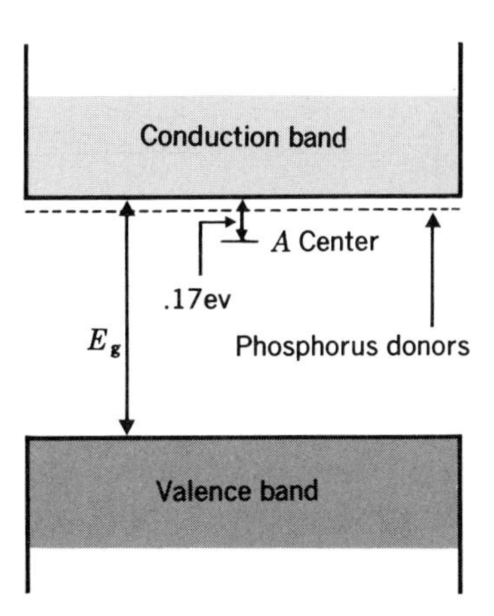

Figure 23-16 Energy level of the Si *A* center.

resonance can be seen. One example of this occurs in the *A* center in Si. The results described in the following were taken from a paper by Watkins and Corbett (Ref. 9).

The Si *A* center can be produced in *n*-type Si (in the study cited, phosphorus donors were present to the extent of about 10^{19} to 10^{20} per cubic meter), after it is irradiated with high-energy electrons. Oxygen also must be present in the crystal. The *A* center is a deep-lying electron trap about 0.17 ev below the bottom of the conduction band (Fig. 23-16). It does produce a spin-resonance spectrum, which has been analyzed to give a detailed geometrical model of the defect.

The model of the defect proposed by Watkins and Corbett can best be described by first outlining how it is produced. Irradiation of the Si with electrons at ordinary temperatures knocks some Si atoms out of their regular lattice sites, producing both Si interstitials and vacancies; the A center is formed out of the latter. The energy of the vacancies can be reduced somewhat if bonds can be made to some atom which can drop into the vacancy. Oxygen, which is normally present in Si, does this in part, since it has four $2p$ electrons. Therefore it occupies

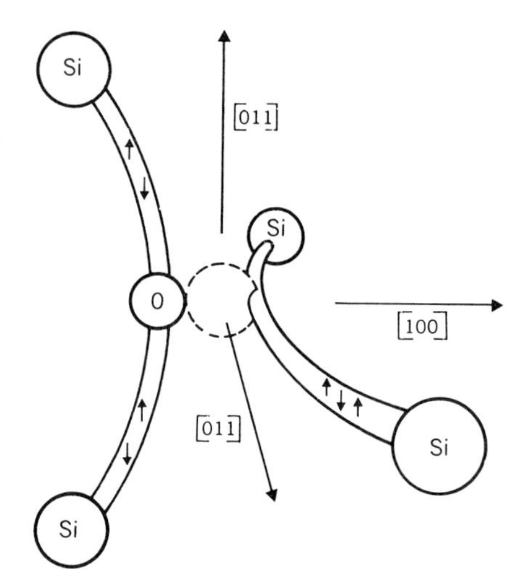

Figure 23-17 Geometry of the Si *A* center. (*After Watkins and Corbett.*)

the vacant site and forms a bond with two of the Si atoms, leaving only two unbonded. These two form a molecular bond, occasionally with an additional electron which is trapped from the P donors. Neither state of the defect, that with only two electrons shared or that with three electrons shared, is a true covalent bond. The two-electron bond does not show resonance, but the three-electron bond has an uncompensated electron. It makes the defect visible to ESR and permits detailed analysis of the geometry.

The defect so formed is shown in Fig. 23-17 in one of its six possible configurations in the lattice. All six are energetically equivalent in an unstrained crystal, and so thermal fluctuations may switch the oxygen from one pair of silicon atoms to another. Exchange does not occur rapidly at ordinary temperatures, however, since the potential-energy barrier between the two states is of height 0.4 ev, much higher than kT at ordinary temperatures.

Ferromagnetic resonance

Some mention must be made of resonance in materials in which the uncompensated electrons are coupled tightly from ion to ion. The ferromagnetic metals and alloys show strong resonance absorption, as do the ferrimagnetic and antiferromagnetic solids.

The resonance energy absorption in these solids is similar to that in the paramagnetic solids. Resonance occurs at frequencies of some 20,000 to 30,000 Mc/sec for values of $\mu_v H_0$ of order 0.5 weber/m^2. The resonance lines, however, are broad; their width of 0.01 weber/m^2 and more makes them immensely broader than absorption lines of NMR or ESR. A resonance line of the ferrite $NiOFe_2O_3$ is shown in Fig. 23-18.

Experimental difficulties and problems in interpreting experimental results have plagued this field from the beginning. Because of

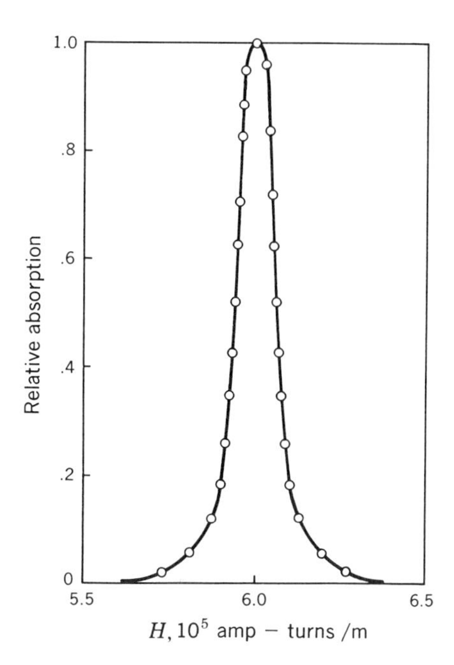

Figure 23-18 Resonance absorption curve of single crystal of $NiFe_2O_4$ at 24 KMc/sec. *(After W. Yager, J. Galt, F. Merritt, and E. Wood, Ref. 12.)*

the extraordinarily small skin depth at such frequencies, experiments on metals and alloys are of necessity carried out on thin films or fine powders. This problem fortunately does not exist with the low-conductivity and antiferromagnetic materials. The high external fields almost completely saturate the magnetization of the sample, and so the effects of demagnetization must be considered. The role of the strong coupling between spins on adjacent ions has required clarification. The spin-lattice relaxation time (analogous to T_1 in NMR) and spin-spin relaxation time (approximately analogous to T_2 in NMR) have demanded extensive study. All these and many more problems have required a great deal of effort; nevertheless some interesting results have emerged.

Accounts of experiments on ferromagnetic resonance (and the companion ferri- and antiferromagnetic resonance) are myriad. To pick out typical experiments and to state typical results is not easy. Two areas in this field have particular bearing on our treatment of ferromagnetism. (1) By these techniques, values of g [in Eq. (23-9)] can be determined. They fall consistently above 2 (the value for free electrons). For example, the g value determined from the data of Fig. 23-19 for the ferrite $NiOFe_2O_3$ is about 2.2 (after a small correction to $\mu_v H_0$ is made to take account of demagnetizing effects). (2) Constants which describe the anisotropy of magnetization can be obtained if single crystals of magnetic materials are used in resonance measurements. For example, data reported in Ref. 12 show that the energy of magnetization of $NiOFe_3O_4$ is least in the [111] and largest in the [100] direction—the excess energy required for magnetization (at room temperature) in [100] over that in [111] is about 2.2×10^3 joules/m^3. [For comparison, the other ferrite which was mentioned earlier, Fe_3O_4, has $g \approx 2.1$, and the excess energy of magnetization in [100] is about 3.7×10^3 joules/m^3 (both at room temperature). These data are taken from Ref. 13.]

Summary of applications of ESR to solids

Electrons can be induced to change their spin states in a magnetic field by absorbing energy from an r-f wave of critical frequency. The phenomenon, termed electron spin resonance, provides a useful technique for measuring a number of properties of solids. The effect is limited to materials having uncompensated electron spins, and many problems are not amenable to study by use of this method. However, this very restriction has made the method a powerful tool in studying defects in some materials, since the great majority of lattice atoms (with their compensated spins) do not respond and only the special sites, defects, and impurities interact with the field. Electron spin resonance has had its widest use in the study of specific properties of insulators and semiconductors. Generalizations about applications are hard to draw, since each application presents its own problems of interpretation. Ferromagnetic and ferrimagnetic resonance are used extensively to study these magnetic materials at microwave frequencies in high magnetic fields. Again the techniques and interpretations cover a wide range.

REFERENCES

1. T. J. Rowland, Nuclear Magnetic Resonance in Metals and Alloys, *Prog. in Materials Sci.*, vol. **9** (1) (1961), General editor, Bruce Chalmers, Pergamon Press, New York.
2. E. R. Andrew, *Nuclear Magnetic Resonance*, Cambridge University Press, New York, 1955.
3. K. K. Darrow, Magnetic Resonance, *Bell System Tech. J.*, **32:** 74–99, 384–405 (1953).
4. J. D. Roberts, *Nuclear Magnetic Resonance*, McGraw-Hill Book Company, Inc., New York, 1959.
5. G. E. Pake, Fundamentals of Nuclear Resonance Absorption, *Am. J. Phys.*, **18:** 438–452, 473–486 (1950).

6. R. Schumacher and C. Slichter, Electron Spin Paramagnetism of Li and Na, *Phys. Rev.*, **101**: 58–65 (1956).
7. T. C. Castner and W. Kanzig, The Electronic Structure of V-centers, *Phys. Chem. Solids*, **3**: 178–195 (1957).
8. G. Watkins, Electron Spin Resonance of Mn^{++} in Alkali Chlorides: Association with Vacancies and Impurities, *Phys. Rev.*, **113**: 79–90 (1959).
9. G. Watkins and J. Corbett, Defects in Irradiated Silicon: I. Electron Spin Resonance of the Si-A Center, *Phys. Rev.*, **121**, pp. 1001–1014, 1961.
10. E. Abrahams, Relaxation Processes in Ferromagnetism, *Advances in Electronics and Electron Phys.*, **7**: 47 (1955).
11. G. T. Rado, Ferromagnetic Phenomena at Microwave Frequencies, *Advances in Electronics*, **2**: 251 (1950).
12. W. Yager, J. Galt, F. Merritt, and E. Wood, *Phys. Rev.*, **80**: 744–748 (1950).
13. L. J. Bickford, Jr., Ferromagnetic Resonance Absorption of Magnetic Single Crystals, *Phys. Rev.*, **78**: 449–457 (1950).
14. C. P. Slichter, *Principles of Nuclear Magnetic Resonance*, Harper & Row, Publishers, Inc., New York, 1963.

PROBLEMS

23-1 Calculate the magnetic field in which a proton would have a resonance frequency of 1 Mc/sec; of 100 Mc/sec. Are fields of this size readily attainable?

23-2 From the data of Table 4-2, calculate the temperature at which motional narrowing of a resonance line would occur for Ge; for Li.

23-3 Plot the energy of splitting of Al^{27} in a uniform applied field H_0. Suppose that no quadruple splitting occurs. Find the resonance frequencies of the several allowed transitions in a field H_0 of 10^6 amp-turns/m.

23-4 Find the field at which electron resonance would occur for a frequency of 10,000 Mc/sec (assuming that $g = 2.00$).

23-5 (*a*) Go through the calculations of Schumacher and Slichter (Ref. 6) to show that the absorption lines *b* and *d* are obtained from the experimental data *a* and *c* (Fig. 23-11).

(*b*) Calculate χ_p (in Table 23-2) for Li and Na, using Eq. (20-29).

(*c*) Are the values of χ_d in column 5 of Table 23-2 reasonable for these elements?

Appendix: units and dimensions

The units used throughout this book are the rationalized mks units. This set of units has most of the advantages of the cgs system and has in addition the advantage that the electrical units of resistance and power are the practical ones. It has the minor disadvantage that the units of volume (the cubic meter) and of density (the kilogram per cubic meter) are inconveniently large. The average engineer and scientist often finds himself dropping back to the cubic centimeter and grams per cubic centimeter in describing these quantities. No weakness of character should be felt in doing this; in fact, each engineering and scientific discipline ultimately devises and uses a convenient body of units whose sizes are about those commonly met in its activity. Thus, the astronomer measures distance in light-years, the crystallographer in angstroms; the power engineer measures energy in kilowatthours, the atomic physicist in electron volts.

The complete set of units must contain five basic units. They are chosen to be the meter (m), kilogram (kg), second (sec), degree centigrade (°C), and coulomb. All secondary quantities are described in terms of these units through the use of simple physical equations. This set of five units is not the only group which could be selected, but it is chosen as the most convenient group for this subject.

The dimension of length is designated l. The unit of volume is a cube 1 m on an edge; volume has the dimension l^3. Mass has dimension m. From these two quantities the density of matter is described by the equation: density = mass/volume. Density therefore has dimensions m/l^3.

The dimension of time is denoted by t. Several other secondary units may now be described. Velocity of translation is defined through the equation

Velocity = distance/time

Velocity thus has dimensions l/t. At once the units of acceleration may be defined through the expression

$$\text{Acceleration} = \text{velocity change/time}$$

Acceleration has dimensions l/t^2. With this definition dimensions of force are found by using Newton's law,

$$F = ma$$

The dimensions are obviously ml/t^2. Unit force, i.e., that force which corresponds to setting both m and a equal to unity, is termed the *newton*. Energy (or work), defined as the product of a force and the distance through which it acts, has dimensions ml^2/t^2. The unit of work is the *joule*. Other mechanical quantities may be defined, momentum, torque, and the like.

A unit of temperature is needed to describe thermodynamic quantities. The most used temperature scale is the centigrade scale, the unit of which is 1 per cent of the temperature difference (as measured by an ideal gas thermometer) between the freezing and boiling points of water. The zero of temperature on this scale is taken to be the freezing point of water.

The calorie, an arbitrarily defined unit with the dimension of energy, or work, was initially designated as the energy necessary to raise the temperature of one gram of water from 15.5 to 16.5°C. By this definition the relationship between the calorie and the joule is dependent on the refinements of experiment. The calorie is approximately 4.18 joules. The formal relationship between heat energy E and mechanical work is contained in the mathematical statement of the first law of thermodynamics, and so no differentiation of units is made between heat and work.

Most factors based on statistics of particles require use of the absolute scale of temperature, where the zero point is set at -273.16°C. This absolute scale, designated degrees Kelvin (°K), has the same size unit as the centigrade scale. The proportionality constant in the ideal-gas law which has dimensions energy/degree has the value of about 2 cal/°K mole, or the value 1.4×10^{-23} joule/°K atom.

The units of entropy, fixed by the second law of thermodynamics, are joules (or calories) per degree. Hence entropy has dimensions ml^2/t^2T, where the dimension of temperature is designated T.

A thermal quantity which is worth noting is the thermal conductivity K, defined through the expression

$$\text{Heat flux (joules/m}^2\text{ sec)} = \frac{K\,dT}{dx}$$

By inserting proper dimensions in this expression, one finds that K has dimensions ml/t^3T.

The mks system is superior to other systems of units in describing electrical phenomena. One additional unit is required, chosen here to be a charge of size one coulomb. If charge be given dimension Q, then current I has dimension Q/t and current density J dimension Q/tl^2.

Two physical laws are used to define the field quantities. The potential in volts at a point is defined as the work in joules necessary to bring a small positive charge from some point chosen to have zero potential to the point in question, divided by the charge in coulombs. Hence the volt has dimensions ml^2/t^2Q. The electric intensity $\mathbf{E}$ is defined as the gradient of the potential V, and so it has dimensions ml/t^2Q. The electric displacement $\mathbf{D}$ is defined through one of Maxwell's equations,

$$\nabla \cdot \mathbf{D} = \rho$$

where ρ is the charge per unit volume. D then has dimensions Q/l^2.

With these definitions, the dimensions of a number of secondary quantities can be written. The resistance R is defined through Ohm's law, its dimensions are V/I or ml^2/tQ^2. Capacitivity ϵ may be defined through the expression $\mathbf{D} = \epsilon\mathbf{E}$; its dimensions are Q^2t^2/ml^3. The capacitance C of a capacitor is defined as the charge stored on the capacitor with one volt across it; here C has dimensions t^2Q^2/ml^2. The unit of capacitance is the farad; a capacitor of 1 farad stores 1 coulomb under a potential of 1 volt.

The quantities describing the dielectric properties of matter are also defined through the simple physical laws given in Chap. 15. Polarization P has the dimensions D, that is, Q/l^2. Defining a dipole moment as the product of the displaced charge and the distance of displacement, we see that the dipole moment has dimensions Ql. The elementary dipole moment p, on an atom or molecule, defined through the expression $p = P/N$ (where N is the number of dipoles per unit volume) also has dimensions Ql. Dielectric susceptibility χ_e, which relates P to E through the equation

$$P = \epsilon_v \chi_e E$$

is a dimensionless quantity. Finally the polarizability α, which is the magnitude of the dipole moment produced on an atom in a field of 1 volt/m, has dimensions Q^2t^2/m.

The dimensions of magnetic quantities may be found easily by use of Maxwell's equations. Since

$$\frac{\partial \mathbf{B}}{\partial t} = -\nabla \times \mathbf{E}$$

the dimensions of $\mathbf{B}$ must be m/tQ. $\mathbf{B}$ can also be defined in the following way: Let an element of conductor dl carrying a current $\mathbf{I}$ be located in a magnetic field whose induction is $\mathbf{B}$. A force $d\mathbf{F}$ exists which is given by the expression

$$d\mathbf{F} = \mathbf{I} \times \mathbf{B}\, dl$$

This expression also gives the dimension for $\mathbf{B}$ of m/tQ. Since the unit of magnetic flux, the weber, has dimensions ml^2/tQ, the magnitude of $\mathbf{B}$ can be expressed as webers per square meter. The magnetic intensity $\mathbf{H}$ is expressed as

$$\nabla \times \mathbf{H} = \frac{\partial \mathbf{D}}{\partial t} + \mathbf{J}$$

$\mathbf{H}$ then has dimensions Q/tl. $\mathbf{H}$ may also be expressed through an equation which gives the field of a current,

$$\int \mathbf{H} \cdot \mathbf{dl} = \mathbf{I}$$

This likewise gives **H** the dimensions Q/tl. Since many magnetic fields are produced by solenoids, the field **H** is often expressed in ampere-turns per meter.

With these definitions of **B** and **H**, the permeability μ has dimensions ml/Q^2. The magnetic polarization **M** has the same dimensions as **H**, that is, Q/tl; hence the susceptibility χ_m is a dimensionless quantity again. Note that the parallelism between the dielectric and magnetic quantities breaks down here, since **P** and **D** have the same dimensions. The magnetic moment $\boldsymbol{\mu}_m$ may be defined as the moment produced by a current loop through the expression

$$\boldsymbol{\mu}_m = \mathbf{n}IA$$

where **n** is the normal to the area of the (flat) loop. From this expression $\boldsymbol{\mu}_m$ has dimension Ql^2/t, so that **M** is the magnetic moment per unit volume.

A summary of principal units and dimensions is given in Table A-1 along with some important conversion factors. To go from a quantity

Table A-1 Summary of principal units and dimensions, with some of the important conversion factors

Quantity	Symbol or abbreviation	Dimensions	Mks unit	To obtain the number of cgs emu, multiply the number of mks units by	To obtain the number of cgs esu, multiply the number of mks units by
Force	F	mlt^{-2}	Newton	10^5	10^5
Energy	E	ml^2t^{-2}	Joule	10^7	10^7
Power		ml^2t^{-3}	Watt		
Charge	q	Q	Coulomb	10^{-1}	3×10^9
Current	I	$t^{-1}Q$	Ampere	10^{-1}	3×10^9
Charge density	ρ	$l^{-3}Q$	Coulomb/m³	10^{-7}	3×10^3
Current density	J	$l^{-2}t^{-1}Q$	Ampere/m²	10^{-5}	3×10^5
Resistance	R	$ml^2t^{-1}Q^{-2}$	Ohm	10^9	$\frac{1}{9} \times 10^{-11}$
Conductivity	σ	$m^{-1}l^{-3}tQ^2$	Mho/m	10^{-11}	9×10^9
Electric potential	V	$ml^2t^{-2}Q^{-1}$	Volt	10^8	$\frac{1}{3} \times 10^{-2}$
Electric-field intensity	E	$mlt^{-2}Q^{-1}$	Volt/m	10^6	$\frac{1}{3} \times 10^{-4}$
Capacitance	C	$m^{-1}l^{-2}t^2Q^2$	Farad	10^{-9}	9×10^{11}
Dielectric displacement	D	$l^{-2}Q$	Coulomb/m²		$1/(3\epsilon_v \times 10^4)$
Electric polarization	P	$l^{-2}Q$	Coulomb/m²		3×10^5
Capacitivity	ϵ	$m^{-1}l^{-3}t^2Q^2$	Farad/m		
Magnetic flux	ϕ	$ml^2t^{-1}Q^{-1}$	Weber	10^8	
Flux density	B	$mt^{-1}Q^{-1}$	Weber/m²	10^4	
Magnetomotive force	mmf	$t^{-1}Q$	Ampere-turn		
Magnetic-field intensity	H	$l^{-1}t^{-1}Q$	Ampere-turn/m	$4\pi \times 10^{-3}$	
Inductance	L	ml^2Q^{-2}	Henry	10^9	
Permeability	μ	mlQ^{-2}	Henry/m		

in the rationalized mks units to cgs emu or cgs esu system, multiply by the number in the appropriate space in columns 5 or 6. Following are examples:

1. 1 coulomb $= 10^{-1}$ abcoulomb $= 3 \times 10^{9}$ statcoulombs
2. 1 ohm $= 10^{9}$ abohms $= \frac{1}{9} \times 10^{-11}$ statohm
3. 1 weber/m^2 $= 10^{4}$ gauss

and, of course

1′. 2 coulombs $= 2 \times 10^{-1}$ abcoulombs, etc.

Index

Index

Page numbers in **boldface** refer to extensive discussions of a topic.

COLLEGE OF MARIN LIBRARY
3 2555 00020223 9